CRC SERIES IN EXPERIMENTAL ORAL BIOLOGY

Editor-in-Chief

Barnet M. Levy, D.D.S., M.S.
Professor and Director
University of Texas Dental Science Institute
Professor of Pathology
University of Texas Dental Branch and
Graduate School of Biomedical Sciences
Houston, Texas

HANDBOOK OF EXPERIMENTAL STOMATOLOGY

Editor
Samuel Dreizen, D.D.S., M.D.
Professor and Chairman
Department of Dental Oncology
Professor
Department of Pathology
University of Texas
Dental Branch
Professor
University of Texas
Dental Science Institute
Houston, Texas
and
Barnet M. Levy, D.D.S., M.S.

HANDBOOK OF EXPERIMENTAL ORAL PHARMACOLOGY

Editor
Frederick A. Curro, D.M.D., Ph.D.
Associate Professor and Chairman
Department of Pharmacology
Fairleigh Dickenson University
School of Dentistry
Hackensack, New Jersey

HANDBOOK OF EXPERIMENTAL ASPECTS OF ORAL BIOCHEMISTRY

Editor
Eugene P. Lazzari, Ph.D.
Professor of Biochemistry
University of Texas
Dental Branch
Professor
University of Texas
Graduate School of Biomedical Sciences
Houston, Texas

HANDBOOK OF EXPERIMENTAL DENTAL CARIES

Editor
Leon M. Silverstone, Ph.D., D.D.Sc.
Associate Dean for Research
University of Colorado
School of Dentistry
Denver, Colorado

CRC Handbook of Experimental Aspects of Oral Biochemistry

Editor
Eugene P. Lazzari, Ph.D.
Professor of Biochemistry
University of Texas Dental Branch
Professor
University of Texas Graduate School
of Biomedical Sciences
Houston, Texas

CRC Series in Experimental Oral Biology

Editor-in-Chief
Barnet M. Levy, D.D.S., M.S.
Professor and Director
University of Texas Dental Science Institute
Professor of Pathology
University of Texas Dental Branch and
Graduate School of Biomedical Sciences
Houston, Texas

CRC Press, Inc.
Boca Raton, Florida

Library of Congress Cataloging in Publication Data

Main entry under title:

Handbook of experimental aspects of oral biochemistry.

(CRC series in experimental oral biology)
Bibliography: p.
Includes index.
1. Mouth. 2. Biological chemistry. I. Lazzari,
Eugene P. II. Series.
QP146.H36 1983 612'.31 82-17737
ISBN 0-8493-3162-5

This book represents information obtained from authentic and highly regarded sources. Reprinted material is quoted with permission, and sources are indicated. A wide variety of references are listed. Every reasonable effort has been made to give reliable data and information, but the author and the publisher cannot assume responsibility for the validity of all materials or for the consequences of their use.

Direct all inquiries to CRC Press, Inc., 2000 Corporate Blvd., N.W., Boca Raton, Florida, 33431.

International Standard Book Number 0-8493-3162-5
Library of Congress Card Number 82-17737
Printed in the United States

PREFACE

The moving finger writes; and, having writ,
Moves on; nor all your Piety nor Wit
Shall have it back to cancel half a line,
Nor all your Tears wash out a Word of it.

Rubiyat of Omar Khayyám, Stanza

When Dr. Barnet M. Levy, the Editor-in-Chief of the *CRC Series in Experimental Oral Biology,* asked me to be responsible for the publication of a *Handbook of Experimental Aspects of Oral Biochemistry,* my response was twofold. First, flattery that such an eminent scholar would select me for the project and second, dumbfoundedness by the intent and meaning of the title. Was this to be an actual "How to" book with exact experimental detail or one which outlined the experimental techniques, gave copious references, and critically reviewed the procedures used? The latter question consumed the greatest amount of thought resulting in the decision that the book would have to be multiauthored in order to be authoritative and written for the researcher who wished to gain knowledge in areas outside his own scope of expertise. The extent of our success must be left to the reader.

An attempt was made to obtain the premier experts in the world of Dental Biochemistry to write each section of a major subject; each subject to be presented in four sections, a short history, composition, metabolism, and disease. The success of this endeavor was limited either due to the lack of clear understanding between us or my inability to obtain a sufficient number of qualified authors before the prescribed deadline. The latter reason also explains why there are no chapters on maxilla and mandible, cementum, or taste present herein.

The contributors deserve more praise and gratitude for their genuine professionalism, generous giving of time, and genial and gracious acceptance of suggestions than a bullying editor has any reason to expect. I wish to thank Dr. Levy for his constancy in bolstering my sometimes faltering ego during this project and to Mrs. Lillie Jean for her arduous and devoted secretarial assistance.

To all those who have toiled mightily in the vineyards of the scientific garden, our deepest gratitude and if I may paraphrase Sir Isaac Newton's comment in a letter to Robert Hooke on February 5, 1675/6, if we "have seen further... it is by standing upon the shoulder's of Giants."

EDITOR-IN-CHIEF

Dr. Barnet M. Levy is Professor of Pathology, University of Texas Dental Branch at Houston; Adjunct Professor of Veterinary Anatomy, College of Veterinary Medicine, Texas A&M University at College Station; Ignacio Chavez Visiting Professor, Universidad Nacional Autonoma de Mexico (University of Mexico); and Adjunct Foundation Scientist, Southwest Foundation for Research and Education at San Antonio, Texas. He is the Director of the University of Texas Dental Science Institute at Houston. He has A.B. and D.D.S. degrees from the University of Pennsylvania and an M.S. from the Medical College of Virginia. He has been President of the International Association for Denal Research, the American Academy of Oral Pathology, and the American Board of Oral Pathology. He is Editor of the *Journal of Dental Research.* Dr. Levy has been the recipient of the I.A.D.R. Isaac Schour Memorial Award for Research, the Fred Birnberg Research Medal of Columbia University, and the Alumni Award of Merit from Washington University and the University of Pennsylvania. He is or has been a consultant to many government agencies and research hospitals. He has published more than 150 research papers in the scientific literature.

THE EDITOR

Dr. Eugene P. Lazzari is Professor of Biochemistry, University of Texas Dental Branch at Houston and Professor, University of Texas Graduate School of Biomedical Sciences at Houston. Born in and an early resident of Archbald, Pa., Dr. Lazzari attended the University of Scranton, Pa. (B.S. Chem.) on a football scholarship, Williams College, Williamstown, Mass. (M.A. Org. Chem.), and Iowa State University, Ames (Ph.D.). He received post-doctoral training with Dr. C. H. W. Hirs at the Brookhaven National Laboratory and Dr. Milton Winitz at the City of Hope Medical Center. Dr. Lazzari has been at the University of Texas Dental School since 1963 and he serves on the Editorial Boards of the *Journal of Dental Research* and *Journal of Herpetology*; has been a reviewer for two other journals and a consultant for a dictionary of dentistry. He has been a Supervising Professor for six Master Degrees at the Dental School; served on advisory/supervisor committees for six students at the Graduate School, and Supervisory Professor for one Ph.D. student. Dr. Lazzari has published more than 30 research and scholarly reports in the scientific and dental literature and has edited 2 textbooks on Dental Biochemistry.

CONTRIBUTORS

Steven I. Altchuler, Ph.D.
Nutritional Physiologist
Biomedical Laboratories
NASA/Johnson Space Center
Adjunct Assistant Professor of Nutrition
University of Houston
Houston, Texas

Ronald C. Auvenshine, D.D.S., Ph.D.
Clinical Assistant Professor
Department of Orthodontics
University of Texas Health Science Center
Dental Branch
Houston, Texas

Alain B. Belcourt
Chargé de Recherches INSERM
Faculté de Chirurgie Dentaire
Université Louis Pasteur
Strasbourg, France

William T. Butler, Ph.D.
Professor
Department of Biochemistry
Senior Scientist
Institute of Dental Research
University of Alabama in Birmingham
Birmingham, Alabama

M. E. J. Curzon, B.D.S., M.S., Ph.D.
Chairman
Department of Caries Research
Eastman Dental Center
Rochester, New York

Michael T. DiMuzio, Ph.D.
Research Assistant Professor
Department of Biochemistry
Investigator
Institute of Dental Research
University of Alabama in Birmingham
Birmingham, Alabama

Frank Dowd, D.D.S., Ph.D.
Associate Professor and Chairman
Department of Pharmacology
Creighton University
School of Medicine
Omaha, Nebraska

Harold H. Draper
Professor and Chairman
Department of Nutrition
University of Guelph
Guelph, Ontario, Canada

J. D. B. Featherstone, M.Sc., Ph.D., F.N.Z.I.C.
Senior Research Associate
Department of Caries Research
Eastman Dental Center
Rochester, New York

Alan G. Fincham, Ph.D.
Director
Medical Learning Resources Unit
University of the West Indies
Mona, Kingston, Jamaica

Donald I. Hay, Ph.D.
Senior Member of Staff
Forsyth Dental Center
Associate Professor
Harvard School of Dental Medicine
Boston, Massachusetts

Taro Hayakawa, M.D., Ph.D.
Head and Professor
Department of Biochemistry
School of Dentistry
Aichi-Gakuin University
Nagoya, Japan

Max O. Hutchins, Ph.D.
Professor and Chairman
Department of Physiology
University of Texas Dental Branch
Houston, Texas

G. Neil Jenkins, MSC, Ph.D., FSCRCS
Emeritus Professor
University of Newcastle Upon Tyne
Newcastle Upon Tyne, England

Racquel Z. LeGeros, Ph.D.
Associate Professor
Department of Dental Materials Science
New York University
New York, New York

Mary L. Mailman, Ph.D.
Assistant Professor
Biochemistry Department
University of Texas Health Science Center
Dental Branch
Houston, Texas

Walter A. Orlowski, D.D.S., Ph.D.
Professor of Pathology
Fairleigh Dickenson University
School of Dentistry
Hackensack, New Jersey

David H. Pashley, D.M.D., Ph.D.
Professor of Oral Biology
Medical College of Georgia
School of Dentistry
Augusta, Georgia

Paul B. Robertson, D.D.S., M.S.
Professor and Chairman
Department of Periodontology
School of Dental Medicine
University of Connecticut
Farmington, Connecticut

Gunnar Rølla, Ph.D., D.D.S.
Professor and Chairman
Department of Pedodontics and Caries Prophylaxis
University of Oslo
Geitmyrsun, Oslo, Norway

Ira L. Shannon, D.M.D., M.S.D.
Director
Oral Disease Research Laboratory
Professor
Biochemistry Department
University of Texas Dental Branch
Houston, Texas

John W. Simpson, Ph.D.
Associate Professor
University of Texas Health Science Center
Dental Branch
Houston, Texas

G. M. Siu, D.D.S., M.Sc.
Department of Nutrition
University of Guelph
Guelph, Ontario, Canada

Harold C. Slavkin
Professor of Biochemistry and Nutrition
School of Dentistry
University of Southern California
Los Angeles, California

Jaro Sodek, Ph.D.
Associate Professor of Biochemistry
MRC Group in Periodontal Physiology
Faculty of Dentistry
University of Toronto
Toronto, Ontario, Canada

Richard P. Suddick, D.D.S., Ph.D.
Professor of Periodontics
Professor of Physiology
University of Texas Health Science Center
San Antonio, Texas

John D. Termine, Ph.D.
Chief, Skeletal Matrix Biochemistry Section
Laboratory of Biological Structure
National Institute of Dental Research
National Institutes of Health
Bethesda, Maryland

Arthur Veis, Ph.D.
Professor of Biochemistry and Molecular Biology
Chairman
Department of Oral Biology
Northwestern University Dental School
Chicago, Illinois

Eileen L. Watson, Ph.D.
Research Assistant Professor
Departments of Oral Biology and Pharmacology
University of Washington
Schools of Dentistry and Medicine
Seattle, Washington

Ira Wolinsky, Ph.D.
Associate Professor of Nutrition
Department of Human Development and
Consumer Sciences
University of Houston
Houston, Texas

Margarita Zeichner-David
Research Associate Professor
Department of Basic Sciences
School of Dentistry
University of Southern California
Los Angeles, California

TABLE OF CONTENTS

Calcium and Phosphorus Metabolism

CALCIUM

Ira Wolinsky and Steven I. Altchuler

CALCIUM INTAKE AND OUTPUT

Calcium (Ca) metabolism can be viewed as a series of interactive compartments, all communicating with each other and providing feedback information to each other. Using this concept, it is possible to present models of Ca metabolism and isolate the particular sections.[1] One such model of Ca metabolism is shown in Figure 1.[2] In this model, five general sections are included: gastrointestinal (GI) intake and output, renal excretion, hormone regulation, bone storage, and a circulatory compartment which integrates the information from all of the remaining compartments. This model can be used as a simplified synopsis of the actual metabolism of Ca within the body.

Gastrointestinal Intake and Output

A summary of the GI Ca fluxes is shown in Figure 2. Ca is taken in exogenously. This Ca can then be absorbed either actively, in a saturable process, or passively. Endogenous Ca is secreted into the lumen of the gut. The sum of exogenous Ca plus endogenous Ca minus Ca absorbed actively and passively, yields the total amount of fecal Ca.

The maximal amount of Ca absorption in humans appears to occur in the proximal small intestine.[3] The quantity of Ca absorbed has been related to the quantity ingested in a curvilinear fashion.[4] Such a curvilinear relationship indicates that actual Ca absorption is composed of two components: one is a saturable component, the active transport mechanism, and the other is a nonsaturable component, the passive diffusion mechanism. This conclusion is supported by work done earlier with perfusion studies also suggesting two mechanisms.[5]

The amount of Ca absorbed by the active process appears to decrease across the length of the small intestine.[6-8] The absorption of Ca from the intestine is significant even when intraluminal Ca concentrations are lower than extracellular Ca concentrations, indicating that the active, saturable process of Ca absorption is important in the proximal small intestine. However, as the intraluminal Ca concentration is raised, absorption also increases, indicating that the passive absorption mechanism is present in tandem with the active mechanism.

The majority of Ca excreted into the intestine occurs proximal to Ca absorptive sites and hence is absorbed along with the exogenous Ca. This assumes that the endogenous Ca is handled similarly to the exogenous Ca. However, it appears experimentally that approximately 15% of the Ca so excreted will not be absorbed.[9] The ratio of absorption to excretion does not remain constant along the gut. In the ileum, endogenous secretion may actually be larger than absorption.[10] The average human total intestinal Ca secretion is 0.194 $\pm$ 0.073 g/day, and this quantity is independent of dietary Ca intake, caloric intake, urine Ca, or plasma Ca. The Ca which is secreted consists of two fractions: a dissolved fraction, consisting of about two thirds of the secreted Ca, and a precipitable, acid-soluble fraction.[11]

Absorption of Ca appears to be dependent upon the Ca status of the body. After Ca deprivation, the active transport mechanism for Ca is facilitated, resulting in more efficient Ca transport.[7] This adaptive phenomenon may help to explain the wide range of Ca intakes on which humans can maintain a zero or positive Ca balance. The phenomenon itself appears to be related to vitamin D metabolites.

The actual flux of Ca from the small intestine to the serum is in the range of 2 to 7 μmol/min/50 cm of gut,[12] when the intraluminal concentration is in the physiologic range of 0.5 to 3.5 μmol/mℓ. Absorption is delayed for about 15 to 20 min after consumption of a Ca-containing meal on an empty stomach. Maximal Ca absorption occurs 40 to 60 min after

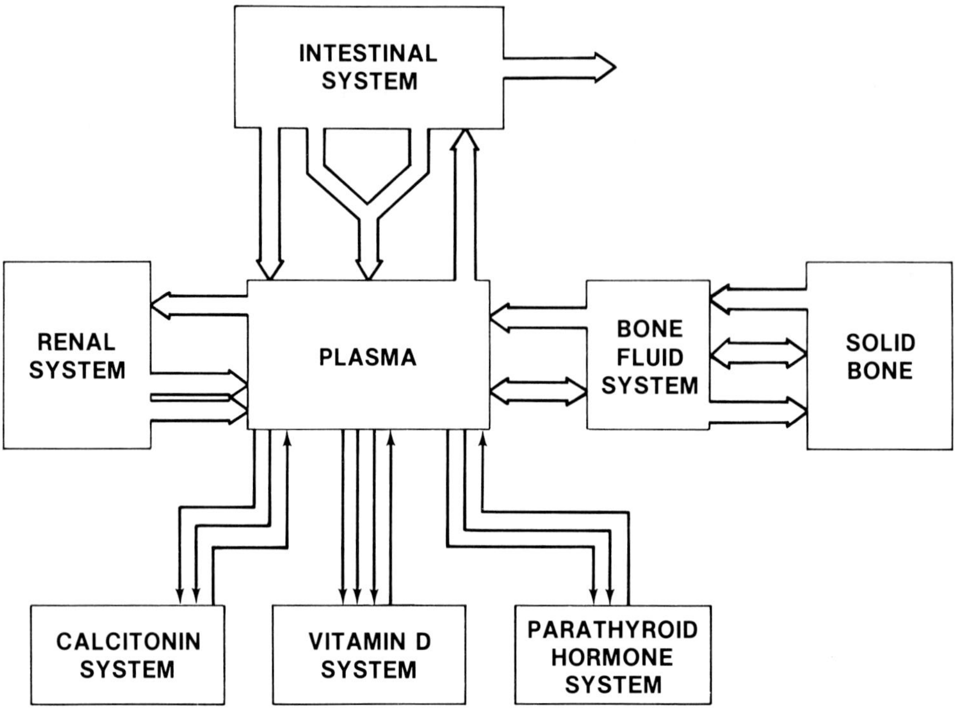

FIGURE 1. Ca metabolism can be modeled by a system consisting of several different blocks: Intake and Output — intestinal system, and renal system; Storage — bone fluid system, and solid bone; Regulatory System — parathyroid hormone, vitamin D, and calcitonin; Distribution System — plasma. Figure courtesy of NASA/Lyndon B. Johnson Space Center, Houston.

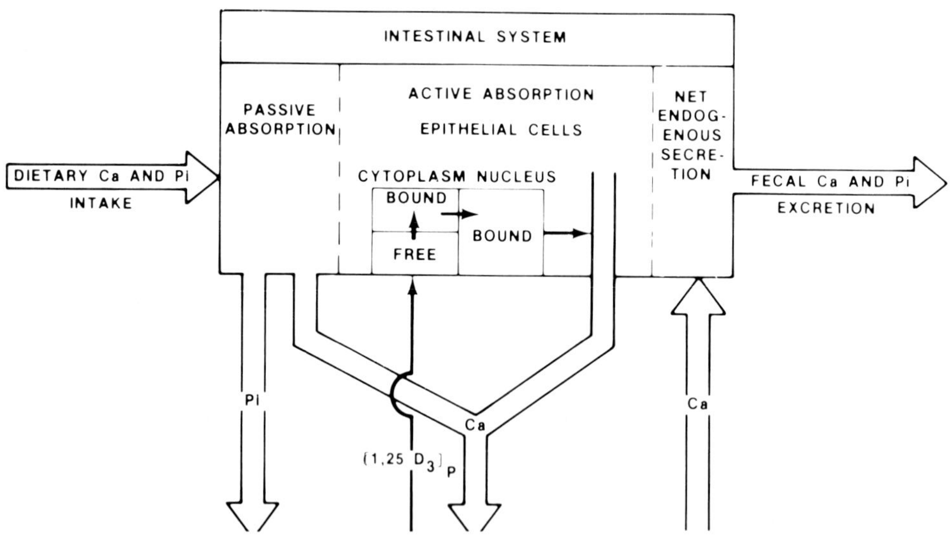

FIGURE 2. The intestinal system takes in dietary Ca. This Ca is then absorbed through both an active, saturable process, which is modulated by 1,25-dihydroxyvitamin D in the plasma, and through a passive, concentration-dependent process. Endogenous Ca is secreted into the gut, where the majority of it is also reabsorbed through the above two processes. The remaining, nonabsorbed Ca is excreted in the feces. (Pi = phosphorus). Figure courtesy of NASA/Lyndon B. Johnson Space Center, Houston.

ingestion, and 95% of the Ca which will be absorbed is actually absorbed by 2.5 hr after ingestion.[13] When Ca intake is below 800 mg/day, the majority of this absorption occurs through the saturable, active process; but when intake reaches 2 to 3 g/day, the linear, passive absorption process is the more important, and the active process is of negligible import.[14]

Of practical note is the fact that daily Ca absorption appears to be maximized when dietary Ca intake is spread over the course of a day, instead of being given as a single bolus.[15] This may occur as the active mechanism may be saturated when the Ca is given as a bolus, whereas saturation is not as likely in numerous smaller doses.

Renal Output

Ca and phosphorus (P) are filtered through the glomerulus; reabsorption of both occurs in the tubules. That portion which is filtered but not reabsorbed is excreted in the urine.

Elevation of plasma Ca levels has been reported to reduce the glomerular filtration rate.[16,17] This reduction appears to be a result of a decrease in the glomerular capillary ultrafiltration coefficient, which is the product of the total glomerular capillary surface area and the effective capillary permeability to water.[18] It appears that it is a reduction in the capillary water permeability which results in the decrease of the glomerular capillary ultrafiltration coefficient.

Ca concentration has been determined by micropuncture studies to increase from the glomerulus through the proximal tubule to the loop of Henle, where it reaches a maximum.[19] Some Ca appears to be reabsorbed in the proximal convoluted tubule, but this reabsorption appears to be passive and occurs secondarily to the absorption of water and sodium (Na).[20] In studies of isolated section of the loop of Henle, it was determined that the entire loop is relatively impermeable to Ca and that reabsorption which occurs takes place in the thick, ascending limb.[21] Further studies have determined that the transport of the Ca cation across the thick ascending limb is an active process, across a chemical gradient, which requires aerobic metabolism. It also appears to be generally associated with chloride transport, but can be dissociated, such as after administration of chlorthiazide.[22,23] The majority of Ca reabsorption hence occurs in the distal nephron. The reabsorption is sensitive to metabolic acidosis. Studies using infusion of bicarbonate showed that bicarbonate infusion reduced urinary Ca excretion where the glomerular filtration rate was either not altered or actually increased. On the other hand, infusions of L-lysine monohydrochloride, which induces proximal renal tubular acidosis, produced no alterations in Ca excretion.

The reabsorption of Ca appears to be sensitive to several hormones, particularly parathyroid hormone (PTH) and vitamin D metabolites. PTH results in increased renal tubular reabsorption of Ca.[24] Micropuncture studies show that this is probably due to a selective increase in Ca reabsorption in and after the distal tubule.[25,26] In studies of monkey kidney cell cultures, the effect of PTH was observed not to be inhibited by the metabolic inhibitor, dinitrophenol. This indicates that PTH activates a passive process of Ca translocation.[27] Effects similar to those obtained by PTH can be obtained with infusion of cyclic AMP and dibutyrl-cyclic AMP, raising the possibility that these nucleotides are intermediaries in the reabsorption process controlled by PTH.[28]

There is evidence to indicate that the glomeruli themselves should be sensitive to PTH. Isolated rat glomeruli showed an increase in adenylate cyclase activity after incubation with purified PTH.[29] Studies with labeled PTH or with labeled antibodies to PTH also demonstrated the presence of such receptors.[30] Studies of glomerular kinetics showed an increase in the glomerular capillary ultrafiltration coefficient in thyroparathyroidectomized (TPTX) rats. Infusions of PTH to the TPTX animals resulted in a decrease in the single nephron glomerular filtration rate and a return of the glomerular capillary ultrafiltration coefficient to baseline levels.[31]

Infusion of vitamin D (cholecalciferol) or of 25-hydroxyvitamin D produces a decrease in Ca excretion.[32] Measurements of the effects of 1,25-dihydroxyvitamin D showed a decrease in fractional Ca excretion, but alterations in neither renal hemodynamics nor in serum ultrafilterable Ca, indicating that the vitamin D metabolite's actions were most likely on the tubules.[33] Studies with other metabolites of vitamin D suggest that renal effects are only present when the 1-hydroxy group is present and that vitamin D metabolites lacking that moiety are ineffective.[34] Other studies of the vitamin D effect show that in TPTX rats, the vitamin D effect is not manifest unless both PTH and vasopressin are present. The suggestion is made that both PTH and vasopressin act as permissive substances for vitamin D's actions.[35]

Calcitonin (CT) has also been demonstrated to increase renal clearance of Ca.[36] In thyroidectomized rats, urinary Ca values were below the levels in control groups until the experimental animals were injected with CT, which brought their urinary Ca levels back to normal.[37] The suggestion has been made that CT causes a decrease in the tubular reabsorption of Ca.[38] It is of note however, that in TPTX rats, CT infusion was observed to lead to a reduced renal Ca excretion, which was attributed to increased Ca reabsorption within the loop of Henle and the distant convoluted tubule.[39]

Dermal Output

Dermal losses of Ca have been measured in a variety of circumstances. Although some investigators have found no loss of Ca through the skin, others have found losses of over 200 mg/day.[40-42] As may be expected, when subjects are in either a warmer or a more humid environment, sweating increases, resulting in an increased dermal loss of Ca.[42] Although some studies have indicated an increased dermal loss of Ca following exercise, the studies were not well controlled.[43]

BONE STORAGE OF CALCIUM

Over 99% of the body's Ca is stored within the bones, the rest being found in the soft tissues and blood. Bone is composed of a collagen matrix which has become calcified with Ca-phosphate salts, principally hydroxyapatite, $Ca_{10}(PO_4)_6(OH)_2$. Within the bone are three principal types of cells: osteocytes, osteoclasts, and osteoblasts. In general, osteoblasts secrete an amorphous material which eventually becomes fibrous osteoid. Precipitation of the apatite salts occurs on the osteoid, resulting in bone. Osteoclasts are large, multinucleated cells, which are thought to be responsible for bone resorption, such as occurs during Haversian remodeling.[44]

Isolated bone cells release Ca when bathed in a medium of low Ca concentration.[45] Conversely, when bathed in a medium of high Ca concentration, there is a net uptake of Ca by the bone. Further studies with metabolic inhibitors have shown that the uptake appears to be a passive process, dependent upon the ambient Ca concentration. The release of Ca from bone also appears to be principally a passive process, but as there is some reduction in the release when metabolic inhibitors are applied, some active metabolic component appears to be involved.[46] Some level of control appears to be present in the cellular layers, since removing the periosteum and endosteum from a chick calvarium culture results in increased influx of Ca.[47]

Ca uptake within isolated bone cells has been shown to be stimulated by both PTH and CT.[48] The effects of the PTH and CT are not additive, however, suggesting that they both may operate through a single mechanism. PTH has also been shown to stimulate specifically the efflux of Ca from bone. This effect can be saturated in vitro by raising the concentration of Ca in the medium.[49]

Bone is sensitive to the overall acid-base balance of the body. In studies with guinea pigs exposed to high ambient carbon dioxide levels, bone bicarbonate was found to rise immediately after introduction to the atmosphere, decline after 2 weeks, and then to rise again after 6 to 8 weeks. Bone carbonate decreased during the first 2 weeks, then rose, eventually remaining stable at an elevated level. Bone Ca content fell in association with the increasing bone bicarbonate and rose in association with increasing bone carbonate.[50] In studies with devitalized bone culture, loss of Ca was linearly related to $[H^+]$. In live bone cultures, bone resorption was independent of $[H^+]$, $[CO_2]$, and $[HCO_3^-]$. This suggests that any effect of acidosis on bone is due to an effect on devitalized bone and not on bone cells themselves.[51]

Bone activity is correlated with regional blood flow within the bone. In rabbits, blood flow in epiphyses with high levels of osteoblastic activity is significantly higher than in those with moderate or low levels of activity. Diaphyseal blood flow is higher in young, rapidly growing animals than in mature animals.[52] In rats made hypocalcemic from a low Ca diet, the red cell space of the femur was found to be markedly increased.[53] One study found that there was no increase in blood flow in the area around a healing diaphyseal tibia fracture in the rat.[54] However, when defect were made in the distal tibias of dogs, there was a highly significant correlation between new endosteal bone formation and between the blood flow.[55]

REGULATORY COMPOUNDS

Ca metabolism is regulated by numerous compounds. Principal among these are the two hormones PTH and vitamin D and their metabolites. In addition, there are other compounds which play a role in Ca metabolism, such as CT, calcium-binding proteins, and the γ-carboxyglutamic acid-containing proteins.

Parathyroid Hormone

PTH secretion is regulated principally by the plasma concentration of Ca.[56,57] In cows, there is an inverse linear relationship between serum PTH and Ca levels, when plasma Ca is in the range of 4 to 12 mg/100 mℓ. Hypercalcemia rapidly suppressed PTH secretion.[56] A basal level of PTH secretion is maintained, on the order of 0.3 ng/kg/min in calves, however, even when plasma Ca exceeds 10.5 mg/100mℓ.[58] The response to hypocalcemia is marked: when serum Ca levels are in the critical range for turning on PTH secretion, a drop of plasma Ca of 0.1 mg/100 mℓ can result in a significant increase in immunoreactive plasma PTH of 0.2 ng/mℓ.[59]

Vitamin D metabolites appear to have an inhibitory effect on PTH secretion. Unhydroxylated vitamin D_3, when administered to dogs, had no effect on PTH secretion. However, 25-hydroxyvitamin D and 24,25-dihydroxyvitamin D both suppressed PTH secretion when given in physiologic doses. 1,25-Dihydroxyvitamin D stimulated PTH secretion, but only when given in pharmacologic doses. Equal doses of 1,25-dihydroxyvitamin D and 24,25-dihydroxyvitamin D resulted in strong suppression of PTH secretion. In addition, when PTH secretion was stimulated by hypocalcemia, it could be suppressed by simultaneous administration of 24,25-dihydroxyvitamin D.[60] In a rat study, however, physiologic doses of 1,25-dihydroxyvitamin D, administered intraperitoneally, resulted in a 43% decrease in serum immunoreactive parathyroid hormone and also inhibited the PTH response to hypocalcemia.[61] In another study, TPTX rats were either vitamin D-depleted or -repleted. Studying various parameters, it was determined that an infusion of 0.81 to 0.84 USP units per hour of PTH was necessary to maintain homeostasis in the repleted animals, whereas 5.1 to 8.6 USP units per hour were necessary in the depleted animals.[62]

PTH secretion appears to occur inversely to magnesium (Mg) levels. In vitro studies show an inverse linear response between PTH secreted by cultured parathyroid glands and the concentration of Mg in the medium.[63, 64] In these in vitro studies, it appears that both Ca and Mg are equally effective in inhibiting PTH secretion and that the effects of the two divalent cations are additive.

In in vivo studies in calves, however, differences between Ca and Mg were noted. When plasma Mg alone was elevated by infusion, PTH secretion that had been induced by hypocalcemia was decreased. When plasma Mg was increased equimolarly to the decrease in plasma Ca, though, the PTH secretion continued in response to the decreased plasma Ca. Careful measurements of the magnitudes of the change indicate that in this model, the Mg was about 0.3 to 0.5 times as effective as Ca in inducing changes in PTH secretion.[65]

PTH secretion is sensitive to several other substances. PTH secretion has been induced by infusions of CT, but the secretion appears to be in response to the resultant hypocalcemia, not to the CT itself.[66] Catecholamines also appear to increase PTH secretion. Infusions of both epinephrine and isoproterenol rapidly raise levels of PTH. The response could be saturated, but recovers after a wait of more than 100 min between infusions. In addition, hypocalcemic animals show a more dramatic response than do normocalcemic animals.[67] Importantly, it appears that there is a basal PTH secretion which cannot be suppressed by hypercalcemia. In calves receiving an infusion of $CaCl_2$, a basal secretion was maintained even after more than 35 hr of infusion, yet the parathyroid glands were fully responsive to hypocalcemia at the termination of the infusion.[68]

The parathyroid gland stores and secretes a polypeptide hormone, 84 residues long, with a molecular weight of 9500 daltons. The parathyroid glands actually synthesize a precursor hormone, preproparathyroid hormone, which is subsequently shortened to proparathyroid hormone, and then to PTH, prior to secretion. This polypeptide has a half-life of less than 10 min.[69-71] The 9500-dalton protein is broken down into two shorter fragments, an amino-terminal fragment consisting of residues 1 to 34 and weighing about 2500 daltons and a carboxy-terminal fragment consisting of the remainder of the molecule, with a molecular weight of 7000 daltons. The amino-terminal fragment is present in only low concentration in the plasma and appears to be the immunologically active form of PTH. The carboxy-terminal fragment is the major circulating form of PTH, but appears to be biologically inert.

PTH has effects on numerous systems involved with Ca metabolism. When rats undergo a parathyroidectomy, their bone turnover decreases.[72] As discussed above, PTH alters Ca reabsorption in the renal tubules. In terms of overall shifts, pair-fed intact and parathyroidectomized rats were compared. The only significant difference in Ca metabolism in the parathyroidectomized rats occurred during low Ca intake, in which case the parathyroidectomized rats absorbed less Ca, excreted more urinary Ca, and had decreased levels of serum Ca.[73] These data are similar to those found in humans with primary hyperparathyroidism, in which the afflicted subjects had increased Ca retention over normal subjects.[74] In another study, subjects with primary hyperparathyroidism who were treated with parathyroidectomy had increased plasma Ca levels.[75]

Vitamin D

No more exciting area in the field of nutritional biochemistry is to be found than the elucidation of the intermediary metabolism of vitamin D and its metabolites and the application of this knowledge to practical clinical use. Authoritative reviews[76,77] as well as a discussion of the vitamin from antiquity to modern times have appeared recently.[78]

Endogenous synthesis of vitamin D occurs in the skin, primarily the epidermis, although somewhat in the dermis, where 7-dehydrocholesterol is converted, in the presence of UV B radiation, to previtamin D_3. The previtamin undergoes a temperature-dependent isomerization over the course of the next several days to convert it to vitamin D_3. The vitamin D_3 is then mobilized from the skin into the circulation by transport on a specific vitamin D-binding protein.[79] The quantity of vitamin D_3 ultimately produced is regulated by several criteria. As the concentration of previtamin D_3 increases, some of it undergoes a photochemical conversion to lumisterol or tachysterol. This appears to be a more important regulator than the limitations imposed either by skin pigmentation or by the latitude at which solar exposure occurs.[80]

The vitamin D produced is then hydroxylated in the liver by an endoplasmic reticular enzyme to 25-hydroxyvitamin D.[81,82] The resulting 25-hydroxyvitamin D is the most prevalent circulating vitamin D metabolite. The 25-hydroxyvitamin D undergoes a further hydroxylation in the kidneys, on either the 1- or the 24-carbon, activated by a mitochondrial enzyme which is cytochrome P_{450}-dependent.[83,84] 1-Hydroxylation predominates when serum Ca levels are low, but as they rise, 24-hydroxylation becomes more prevalent.[85,86] Much of the control implied by serum Ca concentrations may actually occur through parathyroid hormone.[87] Other metabolites of vitamin D also occur, including 25,26-dihydroxyvitamin D and 1,24,25-trihydroxyvitamin D.[88,89]

Vitamin D elevates plasma Ca by ultimately increasing GI Ca absorption and by mobilizing bone Ca. Vitamin D stimulates an active transport mechanism for gut Ca absorption, through an uncertain route.[90] Administration of 1,25-dihydroxyvitamin D to rats results in increased absorption of Ca in all sections of their gut, with no alterations in Ca secretion[91] and in increased release of Ca from already-formed bone mineral, elevating serum Ca levels.[92]

Calcitonin

CT is a polypeptide, with a molecular weight of about 3500 daltons, secreted by the parafollicular cells of the thyroid. Although manipulations with CT alter Ca status, absence of CT appears to have no impact on any measured parameter of Ca status.[93,94] Newer hypotheses suggest that CT may play an important role in routing dietary Ca into the bone fluid where it is used to maintain serum Ca levels between meals.[95] In clinical studies of humans with hypercalcemia resulting from malignancy, CT treatment reduced the serum Ca levels, but they returned to pretreatment levels within a matter of days unless glucocorticoids were given simultaneously.[96] Similar results were obtained in an animal model where hypercalcemia was induced by treatment with 1,25-dihydroxyvitamin D.[97] In other studies, infusions of Ca into humans caused an elevation in serum CT levels.[98] There are data which suggest that CT secretion is not modulated as strongly by serum Ca, but rather by catecholamines.[99,100]

Calcium-Binding Proteins

Numerous different carrier proteins for Ca have been recognized. At least two have been identified in rat intestine, one in the proximal small intestine and one in the distal small intestine.[101] The proximal protein has a molecular weight of about 12,500 daltons and has a high affinity for Ca, whereas the distal protein is over twice as large and has a lower affinity for Ca. Both are vitamin D dependent. The rat Ca-binding protein is similar to that found in humans and pigs.[102] The protein has been localized intracellularly in absorptive cells and is absent from the secretory goblet cells.[103] The protein is induced by the addition of vitamin D to chick embryo intestine cultures. As the protein is induced, Ca uptake also increases.[104] A Ca-binding protein has also been isolated from rat renal cortex.[105] This protein is vitamin D dependent and is twice as heavy as that found in the rat proximal intestine. In chicks fed a vitamin D-free diet, the levels of the renal Ca-binding protein were found to decrease far slower than those in the intestine and were found to be correlated with serum Ca levels.[106]

γ-Carboxyglutamic Acid

In the mid-1970s, a new amino acid, γ-carboxyglutamic acid (Gla) was demonstrated.[107] Proteins containing this amino acid were dependent upon vitamin K for their synthesis.[108] Some of these proteins bind Ca. An important example of such a protein is osteocalcin, isolated in 1975.[109] Osteocalcin is rich in Gla and accounts for up to one fifth of noncollagenous proteins in bone. It is formed in bone tissue itself, at the same time and place that mineralization occurs.[110] As bone density increases, so does the osteocalcin content.[111] Evidence now indicates that there are several different Gla-containing proteins in bone;

several may be precursors of osteocalcin.[110] The precise role that the Gla-containing proteins play in Ca metabolism is continuing to be investigated.

NUTRITIONAL FACTORS ALTERING CALCIUM METABOLISM

Calcium to Phosphorus Ratio

Absorption of Calcium

In past years a relationship between dietary P and Ca absorption was claimed.[112] However, it is generally accepted today that, in man, if the diet is sufficient in Ca and vitamin D, changing the Ca to P ratio over a wide range will not greatly affect the intestinal absorption of Ca.[112-117] Spencer et al.[115] demonstrated that varying the Ca to P ratio from 2:1 through 1:8 by the addition of glycerophosphate to both a low-intake (22 mg/day) and a high-intake (1500 mg/day) Ca diet had no consistent effects on Ca balance in human subjects. Intestinal Ca absorption was, however, monitored indirectly in this study. In a confirming follow-up study in 19 male patients,[117] the effect of varying P and Ca intakes on Ca absorption, as determined by using oral tracer doses of ^{47}Ca, was monitored. The intestinal absorption of ^{47}Ca did not vary significantly when the P intake was 800 mg/day or 2000 mg/day with daily Ca intake levels of about 200 through 2700 mg (Ca to P ratios = 2.9:1 through 1:10). Several laboratories, employing different animal species and sundry analytical techniques in addition to the balance method, support the findings of Spencer's laboratory[112] while others do not.[118-120]

Bone

Not only the absolute amount of dietary Ca and P is important for normal skeletal growth,[121-125] but so are their proper proportions, particularly in rapidly growing animals. Steenbock and Hertig[126] conducted a series of experiments on rats measuring growth and bone ash when dietary Ca was fixed and P varied or with fixed P and varying Ca. When the diet contained 0.45% Ca, optimal growth was achieved with 0.32% P (Ca to P = 1.4) and highest bone ash concentration was achieved with 0.2 to 0.6% P (Ca to P = 0.8 to 2.2). With a P level of 0.315%, best growth and highest bone ash were achieved when the diet had 0.43 to 0.83% Ca (Ca to P = 1.4 to 2.7). In another study using rats,[127] highest femur weights and Ca contents of bone were obtained with a diet containing 1.6% Ca and 1.6% P (Ca to P = 1.0); lowest values were obtained when the diet contained 0.2% Ca and 0.1% P (Ca to P = 0.5). In young rats[128] breaking strength and limit and modulus of elasticity were less affected by feeding a diet limited in P (0.06%) than by a Ca-deficient (0.11%) diet when the corresponding mineral was present in adequate concentration (0.56 and 0.34% Ca and P, respectively) Deficiency of both Ca and P (Ca to P = 1.73) had less of a weakening effect on bone strength than deficiency of Ca with a Ca to P ratio of 0.34. The levels of Ca and P in the rations of young pigs were apparently of less importance than their mutual proportions.[129] They could vary widely with little affect on growth as long as the Ca to P ratio remained about 1.3. Shah and associates[122] clarified a relationship between dietary P intake expressed either in absolute terms or in terms of a Ca to P ratio and rate of bone loss in senescence. When the ratio was high (2.0), increases in dietary Ca were associated in increases in ash content, cortical thickness, and femur breaking load. At a lower ratio (1.0), these measures were depressed and not responsive to dietary Ca increases. The highest values for bone mineral were obtained by feeding a diet containing 1.2% Ca and 0.6% P; the lowest by a diet with 1.2% Ca and 1.2% P. It has been reported that excess dietary P or a low Ca to P ratio causes a marked bone loss or porosity in adult animals,[114,118,119,123,124,130,131] likely via a PTH-mediated mechanism. In a recent study using mature mice,[124] a diet high in P (1.2%) and adequate in Ca (0.6%, Ca to P ratio = 0.5) resulted in lower long bone weight, bone mineral content, and femoral cortical thickness than mice fed a 0.6% Ca, 0.3% P (Ca to P = 2.0) diet. These detrimental changes could be overcome, at least in part, by increases in dietary Ca while keeping the Ca to P ratio at 2.0; forced exercise tended to mitigate these changes, too.

Researchers have speculated[132-134] that excess dietary P or a low Ca to P ratio may also cause bone loss in humans, although this has not been substantiated under controlled conditions. In a long-term study of up to 7 years with young, growing and mature wild-caught ringtail monkeys, Anderson et al.[135] observed that feeding diets with Ca to P ratios of 1.4 and 1:2.1 resulted in only minor microscopic changes suggestive of bone loss. This research team suggests that a nonhuman primate, in view of their taxonomic proximity to humans, may be an appropriate animal model for the investigation of nutritional bone loss in man where bone resorption seems to be a slowly progressive process. They further suggest, warranted in our opinion, that studies employing lower animal species should be reevaluated with respect to the effect of dietary imbalance of Ca and P on bone loss in man. The Ca to P ratio in human foods may be estimated to be about 1:1.5 to 1.6,[136] without taking into account the ingestion of P from food additive.[137] Large intakes of P from soft drinks[138,139] by teenagers accompanied by low Ca (milk) intakes have raised serious concern about the possible effect of low Ca to P ratios on bone loss.

Dietary Protein

Protein-induced hypercalciuria is a well-documented nutritional response of undefined mechanism. It may occur where meat is the primary protein source;[140-149] or other protein sources including egg albumin, soya, casein, lactalbumin, cottage cheese, wheat gluten, and gelatin.[150-152] The relationship between dietary protein intake and the urinary Ca appears to be approximately linear. For example, an increase in dietary protein from 47 to 95 to 142 g resulted in an increase in urinary Ca from 168 to 240 to 301 mg.[146] Similar results have been found when purified amino acids were used.[150]

Measurements have been made of the changes in fecal Ca with dietary protein level, but they are not consistent. Linkswiler's group reported in 1972[144] that increasing the dietary protein via mixed food sources resulted in decreased fecal Ca, but they later reported in 1974 no effect.[146] Serum Ca levels have appeared to be independent of the dietary protein levels.[145,152] Similarly, dermal Ca losses have appeared to be independent of dietary protein levels.[140,153]

Studies of intestinal Ca absorption as affected by dietary protein levels have demonstrated usually that, in man, increased dietary protein results in increased intestinal Ca absorption.[143,151,154] However, some longer studies, i.e., over 60 days, have demonstrated no change in the Ca absorption.[149]

Studies of the overall change in Ca balance as affected by dietary protein have generally shown that increasing the dietary protein level results in an increasingly negative Ca balance.[146,147,149,155] These studies have been performed with both mixed foods serving as the protein source, and with parentally administered amino acid mixtures. Cummings et al.,[141] however, found no alteration in overall Ca balance, although there was hypercalciuria in their study using meat as the protein source. Chu et al.[140] found that subjects receiving 12 g of N per day had a less negative Ca balance than subjects receiving 24 g of N per day, but they also had a less negative Ca balance than subjects receiving 0.9 g of N per day. Linkswiler et al.[143] have found no effect of dietary Ca content on the hypercalciuric action of dietary protein.

Spencer et al.[156] have found no or slight hypercalciuric effect of meat as the dietary protein. That group speculates that this may occur since using meat as a protein source, the increased P content of this protein source may have a protective effect. Hegsted et al.[147] did find a hypercalciuric effect of dietary protein, but they also demonstrated that increasing the P level of the diet did indeed reduce the magnitude of the effect.

Most laboratories have reported that increasing dietary protein content alters the renal handling of Ca. Several have demonstrated that increasing the dietary protein content results in an increased glomerular filtration rate,[145,148,149,157] although at least one group has not found this change.[152] All of these researchers, however, along with others, have demonstrated a decreased renal tubular fractional reabsorption of Ca with an increased dietary protein content.[145, 148, 149, 152, 157, 158] Schuette et al.[157] have demonstrated an increased renal acid and ammonium excretion along with these renal changes. It has been proposed that these changes may be due to the role of the sulfur-containing amino acids. Zemel et al.[148] fed subjects a diet with either 50 g of protein, 150 g of protein, or 50 g of protein plus additional sulfur-containing amino acids to equal the sulfur-containing amino acid content of the 150-g protein diet. They demonstrated that on the 150-g protein diet, there was a doubling of the urinary Ca level, an increase in the glomerular filtration rate, and a decrease in renal tubular Ca resorption. In subjects on the diet containing 50 g of protein plus sulfur-containing amino acids, they demonstrated changes in the same direction, but of only about half the magnitude as the subjects on the 150-g protein diet. This raises the possibility that, although the sulfur-containing amino acids may play a role in the hypercalciuria, they may not be totally responsible for the increase, and other components of a high-protein diet may play some role. Opposing these data, though, are those of Block et al.[158] who fed subjects diets consisting of 15 g of protein, 45 g of protein, or 15 g of protein plus sulfur-containing amino acids. They found no difference in urinary Ca levels between those subjects fed the diet with 15 g of protein and those fed the diet with 15 g of protein plus sulfur-containing amino acids.

Other mechanisms to explain a dietary protein-induced hypercalciuria have been tested. Increased dietary protein does not appear to affect levels of circulating PTH[145, 157, 159] or 1,25-dihydroxyvitamin D.[157] Recent work indicates that insulin may, however, be affected.[159]

Studies using animal models have shown similarities and differences to observations from human studies. Increasing dietary protein levels does produce hypercalciuria in the rat.[161-165] In one study using pregnant rats, high dietary protein caused both hypercalciuria and a hyperphosphaturia.[164] There tends to be a marked drop in fecal Ca excretion[160, 165] coupled with an increased intestinal Ca absorption,[160, 163] even showing in one study an increasingly positive Ca balance in rats with increased dietary protein.[165] In those studies which examined bone status, animals fed increased levels of dietary protein did have widening of their epiphyseal plates,[162] but other studies showed no change in the bone wet weight, ash weight, Ca content, specific gravity, or length.[161, 163, 164]

Weiss et al.[166] reported that in young rats, a high-protein (80% casein), normal Ca (0.61% Ca; 0.40% P) diet reduced osteogenesis, as measured by ^{45}Ca incorporation by 78%. The researchers hypothesize that this protein-induced "osteoporosis" was due to a failure of osteogenesis at the stage of ossification, possibly a result of restricted availability of Ca at the site of mineralization.

Other studies on the mechanism of the hypercalciuric effect in rats have shown, by using different protein sources, that the increase in urinary Ca is proportional to the sulfur-containing amino acid levels in the diet.[167] In the same studies, supplementing the casein with the sulfur-containing amino acids methionine and cystine also resulted in an increase in urinary Ca, proportional to the sulfur-containing amino acid content and proportional to the sulfate excretion.

One study on the mechanism of the increased intestinal Ca absorption in rats fed a high-protein diet demonstrated that in the high-protein animals, the mucosal dry weight per centimeter of length in the duodenum and the ileum was increased in animals fed a 39% casein diet, compared to those fed a 0 or 89% diet.[168] However, the Ca absorption per gram of mucosa was the same in all three groups. This study suggests that the mechanism of the increase in GI Ca absorption observed is due to an increase in the mucosal surface area.

Lactose

It has been known for some decades that when lactose is incorporated into diets of animals[123, 169-171] and man[170, 172-174] it enhances the intestinal absorption of Ca. In humans, improved absorption of Ca would appear to depend on the relative activity of intestinal lactase.[172-174] In a study on healthy infants, Kobayashi et al.[173] found that the intestinal absorption of Ca and Mg was enhanced by about 24% when fed a proprietary milk product and by about 35.4% when given a lactase preparation, compared to a group of infants fed a lactose-free milk. Condon et al.[172] determined the effect of dietary lactose on Ca metabolism in four patients with normal lactose intolerance. Beneficial effects of lactose on Ca (and P) balance were observed in these lactose-tolerant individuals but not in the lactose-intolerant volunteer. Kocian and associates[174] reported similar findings. Greenwald and colleagues,[175] also using balance techniques, did not detect improvement in intestinal Ca absorption in three patients, although radioisotope studies in one of them suggested a 10% improvement.

The promoting effect of lactose on Ca in the rat could occur in vitamin D deficiency and was most marked in the ileum, where Ca absorption is largely passive.[6-8, 176, 177] Lactose had to be present in the same intestinal segment as the Ca in order to exert its effect.[171, 178-180] Measuring the uptake of ^{45}Ca by rat ileum everted gut sacs, Ambrecht and Wasserman[171] found no evidence for the cotransport of lactose and Ca into the tissue. These experiments and others[181] indicate that lactose does not interact with Ca in solution but, rather, with the absorptive cells of the intestine to increase their permeability to the ion. Wasserman[181] postulated that the experimental evidence indicates that lactose removes a hypothetical metabolically dependent inhibition to Ca absorption across the absorption cells. However, it would appear that lactose promotes Ca absorption by a mechanism other than inhibiting aerobic metabolism.[171] Dupuis et al.[182] claims that lactose, among several other carbohydrates, may interact directly with the brush border membrane of intestinal cells to increase their permeability. The precise way in which lactose promotes Ca absorption is as yet not known; evidence supporting different hypotheses on the mechanism of action have been presented in informed reviews.[170, 171, 181] Most of our understanding of the dietary lactose-Ca interrelationship has come from work using animal models; long-duration studies on the possible benefits of lactose ingestion on bone health and the relationship of lactose intolerance and onset of osteoporosis[172, 183, 184] would be valuable.

Oxalic Acid

From time to time concern has been raised that ingestion of plant foods containing large amounts of oxalic acid or oxalates (e.g., spinach, rhubarb, Swiss chard, tea) may interfere with Ca absorption and metabolism by forming insoluble Ca oxalate which passes through the intestine without being absorbed.[185-189]

Pingle and Ramasastri[187, 188] recently investigated the availability of Ca in *Amaranthus* leaves, a commonly consumed food in India, one rich in both Ca and oxalic acid. The absorption of Ca was low from this vegetable; further, the leaves also caused a decrease of Ca absorption from milk ingested with the vegetable. In view of these results, this research team advised that recommending green leafy vegetables as a Ca source to sections of the Indian population which are Ca deficient should be considered with caution. In experiments on rats when potassium oxalate was included in the diet to a level of 2.5% (equivalent to 13% dried spinach!), Mackenzie and McCollum[190] observed a slight reduction of bone ash without any change in growth rate; using high dietary levels of oxalate, but lower than that employed by Mackenzie and McCollum,[190] Tarjan et al.[191, 192] noted no effect on the development of the skeleton in the rat. Some ruminants and rodents are able to tolerate a considerable amount of soluble oxalates and to absorb Ca from insoluble Ca oxalate, apparently by degrading oxalate in the intestine by microbial action.[193]

The deleterious effect of oxalate on Ca metabolism is a real one under certain conditions.

However, this effect is limited because it depends on the quantity of oxalic acid which is ingested with foods. No human foods eaten in ordinary amounts provide enough oxalates to produce serious effects to any significant degree. Also, at least in the West, the amounts of dietary Ca and vitamin D are such that Ca losses through insoluble oxalate formation are not serious.[194]

Fiber

In the last decade or so there has been a terrific escalation of interest in the role and significance of dietary fiber in human nutrition.[195-198] Pertinent to the subject of this chapter is the ability of fiber to bind, among other cations, Ca.[199-204] Since many diets are rich in both fiber and phytate (see below), both of which can affect Ca metabolism, the distinction between the effects of fiber and phytate on mineral absorption has been difficult,[197,203] and in evaluating experimental records, one must discern between fiber effects, phytate effects, or additive effects. McCance and Widdowson[205] found Ca absorption was lower when dephytinized brown flour (high fiber) was fed as bread than when white flour (low fiber) was fed to human subjects. James et al.[199] reported the binding of ^{45}Ca, in vitro, to extracted fiber from 29 different food plants low in phytate. There was a range in binding depending on the plant source of the fiber residue. In a controlled experiment with weanling rats, Harmuth-Hoene and Schelenz[200] reported, using balance studies, that agar-agar and carrageenan in the diet at the 10% level reduced Ca absorption from about 60% of dietary intake in control rats to 50 and 6%, respectively, over a period of 8 days. Other dietary fiber preparations (gum guar, carob bean gum, sodium alginate) were without significant effect. Ismail-Beigi et al.[201] describe experiments in which fiber intakes in human subjects were increased without an increase in the intake of P (presumably, then, without an increase in phytate). Addition of 10 g of filter paper cellulose in apple sauce to a daily diet of low- or fiber content of healthy adult Iranian men over a 2- to 3-week period caused the fecal excretion of Ca to increase. Ca balance became negative and plasma concentration decreased. The quantity of cellulose employed in these experiments approximated that consumed in the form of bread by male villagers in Iran. In healthy adolescent girls the inclusion of 21 g of pure cellulose for 21 days to a low-fiber, adequate Ca diet resulted in increased fecal excretion of Ca and decreased serum Ca from 10.6 mg% in control subjects (low fiber diets) to 9.6 mg%.[206] In a well-executed study using fruits and vegetables as a virtually phytate-free source of fiber, Kelsay's research group[202] studied their effect on Ca balance in 12 men consuming controlled diets containing a high- (59.2 g/day fiber from fruits and vegetables) or low- (9.2 g/day fiber from fruit and vegetable juices) fiber content for 26 days. Balances were significantly lower on the high fiber diet than on the low fiber diet (+72 mg vs. −122 mg, respectively). There is a need for studies to determine if these short-term negative Ca balances[201,202] would eventually resume a positive value by some physiological adaptation such as reduced urinary loss of the mineral.[200]

The studies cited above suggest that dietary fiber may influence adversely the absorption of Ca, at least over the short term. The effect is likely to be related to binding of Ca by some component(s) of dietary fiber. The magnitude of these effects are related to not only the type,[200,207] but the quantity of ingested fiber. As such, these effects may not be of significance in the average American diet,[197] but may assume more importance in that sector of our health-conscious population ingesting increasing amounts of dietary fiber preparations and fiber enriched foods and in certain Middle East and North India populations ingesting relatively high-fiber diets.[208]

Phytate

One of the components of cereal fiber may be phytic acid, inositol hexaphosphoric acid. Anywhere from 60 to 90% of the P content of mature cereal grains and seeds is present as phytates, mainly in the bran and germ portion of the seed.[209] Interest in phytates was first aroused by the anticalcifying and rachitogenic factor present in cereals as phytic acid. It was subsequently shown that this compound could inhibit absorption of Ca, among other minerals,[203,204,211-213] from the diet, probably by forming insoluble Ca salts, thus rendering it unavailable.[214] Hydrolytic products of phytic acid have been shown to inhibit certain biological calcifications.[215,216] As in the case of fiber or oxalate (see above), the possible harmful effects depend primarily on the phytate intake; increased Ca ingestion can partially or wholly ameliorate. Walker et al.[217] suggested that adults, in time, can accustom themselves to increased phytate intakes. In long-term balance experiments, humans ingesting bread made from almost unmilled (phytate rich) flour displayed an initial negative Ca balance which changed to a positive one in about 6 weeks.

By 1955, it was generally conceded that when the intake of unrefined cereals is relatively low and that of Ca is relatively high, nutritional problems due to phytate digestion are negligible.[211,218] Thus, the controversy subsided only to reemerge with vigor in the 1970s.

Phytates are a potential nutritional hazard in those areas of the globe where dietary calcium intake is marginal and whole-grain cereals are consumed. This is particularly true where unleavened whole-meal preparations are used, as leavening reduces the phytate content of the meal.[219] Indeed, whole-meal bread may account for the main energy intake in many parts of the world. In the U.S., the diet has been one of refined cereals and adequate Ca consumption; however, in some quarters there has been a trend toward increased consumption of "natural" and unprocessed cereal food products. Here phytate and its effect on Ca metabolism may become a greater issue. High phytate intake in the form of unleavened breads has been suspect as one contributing factor in rickets and osteomalacia in Pakistani and Indian groups in England[220,221] and in osteomalacia in Israeli Bedouins.[222] On the other hand, Walker[223] points out that the Bantu regularly consume high-fiber, high-phytate, and low Ca diets, but osteomalacia is virtually unknown. Endogenous synthesis of vitamin D is probably adequate due to this population's exposure to the sun.

Reinhold's group[203,204,208,213,219,224] has done considerable work in Iran on the effect of purified phytate and phytate-rich bread on Ca metabolism measured by the balance method. Administration of 2.5 g *purified* sodium phytate to three young men for 28 days caused a negative Ca balance in one, significantly less positive in another, and unchanged in the third; urinary Ca excretion decreased significantly in all. When tanok, the phytate-rich unleavened whole-meal flat wheat bread that is the staple food of many Iranian villages, replaced purified phytate for 32 days, there were strong and stronger negative Ca balances in all three subjects. Losses of Ca averaged 200 to 300 mg/day. When the subjects resumed a control diet in which phytate-free white bread replaced tanok, there was a return to positive Ca balances. The authors concluded that high-phytate intakes can cause disturbances of Ca metabolism in Iranian villagers and point out that since tanok resembles in composition the chapattis of India and Pakistan, the results of their studies may be applicable to other populations eating large amounts of unleavened whole bread. The absence of adaptation in the subjects under study was claimed by the researchers to be related to the large amounts of phytate they consumed or, we feel, it could be that the duration of the study wasn't long enough to detect adaptations under their experimental conditions. In another recent study by Reinhold's group[208] utilizing these male Iranian volunteers it was demonstrated that the switch over from tanok to white bread resulted in improved Ca balances despite the fact that the intake of Ca was greater with the tanok diet than with white bread. Although disturbances of Ca metabolism, at intakes that might otherwise be considered adequate, may be prevalent in parts of the Middle East and India, they have not attracted the attention that the effects of deficiencies of certain nutrients have perhaps because it is not life threatening, ordinarily. It may, though, be an underestimated source of poor health, discomfort, and disability.

Fluoride

The benefits of fluoride (F) use in preventing dental caries certainly cannot be denied.[225-227] Also, it has been used, albeit guardedly, in the treatment of osteoporosis.[228,229] Bone and tooth F uptake depends, among other factors,[230] on the Ca content of the diet.[128,231-234] In one study,[231] young rats and mice were fed low- and adequate-Ca diets (0.02 and 0.5%, respectively) with or without 50 ppm F in the drinking water, for a period of 4 weeks. The F content of rat femur ash reached the very high value of 13,166 ppm when Ca was limiting, as compared to 3916 ppm when Ca was sufficient. For mice the F contents were 5777 and 2833 ppm, respectively. These data clearly demonstrate the occurrence of a secondary skeletal fluorosis when dietary Ca is low. It is data such as these that fuel the arguments of antifluoridationists who claim that artificially fluoridated drinking water is dangerous whereas naturally fluoridated water (even at levels higher than that used in artificially fluoridated water) is not because the latter contains naturally higher amounts of Ca. This argument ignores the fact that the Ca and F intakes in the study such as cited above[231] are considerably lower and higher, respectively, than ever would be consumed in a typical American diet. However, the effect of dietary Ca on F uptake by bone cannot be ignored entirely and should be taken into account when establishing F dosage regimens for prophylaxis of dental caries.[235] Ca may reduce the uptake of F by bone by an inhibitory action on F absorption and retention, possibly mediated through the formation of insoluble CaF_2.[233,234] In a short-term metabolic study on mature humans,[232] increasing Ca intake from about 200 to 1500 mg/day during low or high F intake (4 or 14 mg/day, respectively) had little effect on fecal F excretion; skeletal uptake of F was not measured, however. The different results found in animal studies[128,231,233,234] as opposed to the human study[232] may be due to the fact that young, growing animal were used in the former and mature adults (42 to 65 years old) in the latter.

The effect of F administration on Ca metabolism was studied both in the rat[236] and in man.[237] Chang et al.[236] kept groups of young rats on two levels of dietary Ca (0.2 or 1.0%) without or with F in the diet up to a concentration of 200 ppm for 4 weeks. No beneficial effect of F supplementation on Ca absorption, measured by intestinal transport of ^{45}Ca, was observed, but excessive intake of dietary F (200 ppm, a toxic dose) resulted in increased urinary Ca excretion regardless of dietary Ca level. The ingested F was virtually all deposited in bone. However, this research group failed to observe the well-documented effect[128,231-234] of a low Ca diet on promoting F uptake by bone as described above. Spencer and associates[237] studied Ca balance and absorption in nine healthy males receiving diets containing either 206 mg Ca or 1448 mg Ca daily with or without a F supplementation (19.6 to 21.3 mg of NaF per day) for up to 42 days. Ca balance was not improved with F supplementation but, in most of the subjects, there were indications of a decreased Ca absorption (lowered ^{47}Ca levels in serum after oral tracer doses of the isotope). These two reports stand in contrast to others studying the effect of F supplementation on Ca absorption and retention in patients suffering from skeletal disorders[238,239] or chronic renal failure,[240] where evidence is presented to indicate increased Ca absorption and retention.

REQUIREMENTS

A number of informative reviews on Ca requirements have been written.[241-249] It is clear from them that we cannot state with certainty what is the optimum Ca intake for man. Almost all point out that it is extremely difficult to determine Ca requirements. This is because other dietary components interact with Ca, affecting its utilization; humans can adapt to Ca deprivation over time and restore normal balance, and because there is a lack of an indisputable Ca deficiency disease, which excludes the assessment of how much Ca is needed to prevent or cure the disease.[7,112,127,243,244,247]

In 1962 the FAO/WHO Expert Group on Calcium Requirements proposed that high intakes of Ca were not necessary and that requirements could be met safely with less than half the customary intakes in most European countries and in North America.[243] They based this claim on the well-documented ability of man to be in Ca balance on low dietary intakes and the absence of nutritional disability in populations on Ca intakes of at, or even below, 300 mg/day. This was done at a time when dietary Ca deficiency was felt to be widespread and policy favoring Ca fortification of cereals and other foods was advocated. The FAO/WHO suggested a practical allowance, defined as one likely to meet the needs of the great majority of persons ingesting an adequate amount of vitamin D. For adults the allowance suggested was 400 to 500 mg/day; for children the allowance ranges from 400 to 700 mg/day depending on age; during pregnancy and lactation an allowance between 1000 and 1200 mg/day is suggested. The recommendations appeared realistic to many and a number of nations have adopted them with minor modifications including, among others, the industrialized countries of the U.K., Canada, and Australia.[248]

There are those, however, who consider that higher intakes of Ca are desirable.[245] In 1941 the Food and Nutrition Board of the U.S. National Research Council recommended 800 mg of Ca per day as a suitable daily allowance for an adult, with larger amounts for adolescents and during pregnancy and lactation. It has remained at that high level since then. The recommendation has been based in part on balance studies conducted with groups of individuals accustomed to ample intakes of foods high in Ca. It was, however, recognized by the Food and Nutrition Board in 1980[245] that children can grow healthy bones, adults remain in Ca balance with intakes as low as 200 to 400 mg/day, and that a higher proportion of Ca is utilized when intake is low then when it is generous. However, they were reluctant to lower the recommendation feeling that "no advantage accures from such low intakes, and for a variety of reasons it seems unwise to recommend such low calcium intake." These reasons include the possible effects of high dietary intakes of protein and P on Ca (see above) and the possibility of reduced Ca absorption during aging.[250] A number of nations have found the arguments of the U.S. Food and Nutrition Board convincing and have, with some modifications, adopted their recommendations, including, among others, Sweden and the Netherlands.[248]

Clearly more experimentation and exposition are required before the controversy between the Ca "minimalists"[243] and "maximalists"[245] can be resolved. Inappropriate concern about meeting artificially inflated Ca requirements may divert attention and resources from perhaps more important nutritional problems. Perhaps, if the accumulated data of the many studies available to us were reevaluated so that Ca requirements were expressed in terms of intake of another critical nutrient (e.g., protein or P), some of the controversy would be dispelled. Or, it may be that evidence garnered from population groups habitually consuming liberal Ca intakes and with large Ca stores may have little relevance or applicability to the majority of this planet's population and, hence, the two extreme views are irreconcilable.

REFERENCES

1. **Neer, R., Berman, M., Fisher, L., and Rosenberg, L. E.,** Multicompartmental analysis of calcium kinetics in normal adult males, *J. Clin. Invest.*, 46, 1364, 1967.
2. **Altchuler, S. I., Brand, S. N., and White, R. J.,** A mathematical model of calcium metabolism, *Fed. Proc.*, 40, 921, 1981.
3. **Szymendera, J., Heaney, R. P., and Saville, P. D.,** Intestinal calcium absorption: concurrent use of oral and intravenous tracers and calculation by the inverse convolution method, *J. Lab. Clin. Med.*, 70, 570, 1972.
4. **Heaney, R. P., Saville, P. D., and Recker, R. R.,** Calcium absorption as a function of calcium intake, *J. Lab. Clin. Med.*, 85, 881, 1975.

5. **Ireland, P. and Fordtran, J. S.,** Effect of dietary calcium and age on jejunal calcium absorption in humans studied by intestinal perfusion, *J. Clin. Invest.*, 52, 2672, 1973.
6. **Wensel, R. H., Rich, C., Brown, C., and Volwiler, W.,** Absorption of calcium measured by intubation and perfusion of the intact human small intestine, *J. Clin. Invest.*, 48, 1768, 1969.
7. **Kimberg, D. V., Schachter, D., and Schenker, H.,** Active transport of calcium by intestine: effects of dietary calcium, *Am. J. Physiol.*, 200, 1256, 1961.
8. **Bronner, F.,** Calcium homeostasis, in *Disorders of Mineral Metabolism*, Vol. 1, Bronner, F. and Coburn, J., Eds., Academic Press, New York, in press.
9. **Heaney, R. P. and Skilman, T. G.,** Secretion and excretion of calcium by the human gastrointestinal tract, *J. Lab. Clin. Med.*, 64, 29, 1964.
10. **Walling, M. W. and Kimberg, D. V.,** Active secretion of calcium by adult rat ileum and jejunum in vitro, *Am. J. Physiol.*, 225, 415, 1973.
11. **Schedl, H. P., Osbaldiston, G. W., and Mills, I. H.,** Absorption, secretion, and precipitation of calcium in the small intestine of the dog, *Am. J. Physiol.*, 214, 814, 1968.
12. **Wensel, R. H., Rich, C., Brown, A. C., and Volwiler, W.,** Absorption of calcium measured by intubation and perfusion of the intact human small intestine, *J. Clin. Invest.*, 48, 1768, 1969.
13. **Birg, S. J., Peck, W. A., Berman, M., and Whedon, G. D.,** Study of calcium absorption in man: a kinetic analysis and physiologic model, *J. Clin. Invest.*, 48, 1705, 1969.
14. **Heany, R. P., Saville, P. D., and Recker, R. R.,** Calcium absorption as a function of calcium intake, *J. Lab. Clin. Med.*, 85, 881, 1975.
15. **Kales, A. N. and Phang, J. M.,** Effect of divided calcium intake on calcium metabolism, *J. Clin. Endocrinol.*, 32, 83, 1971.
16. **Edwards, B. R., Sutton, R. A. L., and Dirks, J. H.,** Effect of calcium infusions on renal tubular reabsorption in the dog, *Am. J. Physiol.*, 227, 13, 1974.
17. **Vanherweghem, J. L., Ducobu, J., D'Hollander, A., and Toussaint, C.,** Effects of hypercalcemia on water and sodium excretion by the isolated dog kidney, *Pflügers Arch.*, 363, 75, 1976.
18. **Humes, H. D., Ichikawa, I., Troy, J. L., Anderson, A. J., and Brenner, B. M.,** Influence of calcium on the determinants of glomerular ultrafiltration, *Trans. Assoc. Am. Physicians*, 90, 228, 1977.
19. **Lassiter, W. E., Gottschalk, C. W., and Mylle, M.,** Micropuncture study of renal tubular reabsorption of calcium in normal rodents, *Am. J. Physiol.*, 204, 771, 1963.
20. **Suki, W. N.,** Calcium transport in the neprhon, *Am. J. Physiol.*, 237, F1, 1979.
21. **Rocha, A. S., Magaldi, J. B., and Kokko, J. P.,** Calcium and phosphate transport in isolated segments of rabbits' Henle's loop, *J. Clin. Invest.*, 59, 975, 1977.
22. **Imai, M.,** Calcium transport across the rabbit thick ascending limb of Henle's loop perfused in vitro, *Pflügers Arch.*, 374, 255, 1978.
23. **Edward, B. R., Baer, P. G., Sutton, R. A. L., and Dirks, J. H.,** Micropuncture study of diuretic effect on sodium and calcium reabsorption in the dog nephron, *J. Clin. Invest.*, 52, 2418, 1973.
24. **Talmage, R. V. and Kraintz, F. W.,** Progressive changes in renal phosphate and calcium excretion in rats following parathyroidectomy or parathyroid administration, *Proc. Soc. Exp. Biol. Med.*, 87, 263, 1954.
25. **Sutton, R. A. L., Wong, N. L. M., and Dirks, J. H.,** Effects of parathyroid hormone on sodium and calcium transport in the dog nephron, *Clin. Sci. Mol. Med.*, 51, 345, 1976.
26. **Agus, Z. S., Chiu, P. J. S., and Goldberg, M.,** Regulation of urinary calcium excretion in the rat, *Am. J. Physiol.*, 232, F545, 1977.
27. **Borle, A. B.,** Effects of purified parathyroid hormone on the calcium metabolism of monkey kidney cells, *Endocrinology*, 83, 1316, 1968.
28. **Burnatowska, M. A., Harris, C. A., Sutton, R. A. L., and Dirks, J. H.,** Effects of PTH and cAMP on renal handling of calcium, magnesium, and phosphate in the hamster, *Am. J. Physiol.*, 233, F514, 1977.
29. **Sraer, J., Ardaillou, R., Loreau, N., and Sraer, J. D.,**Evidence for parathyroid hormone sensitive adenylate cyclase in rat glomeruli, *Mol. Cell Endocrinol.*, 1, 285, 1974.
30. **Sraer, J., Sraer, J. D., Chansel, D., Jueppner, H., Hesch, R. D., and Ardaillou, R.,** Evidence for glomerular receptors for parathyroid hormone, *Am. J. Physiol.*, 235, F96, 1978.
31. **Ichikawa, I., Humes, H. D., Dousa, T. P., and Brenner, B. M.,** Influence of parathyroid hormone on glomerular ultrafiltration in the rat, *Am. J. Physiol.*, 234, F393, 1978.
32. **Puschett, J. B., Moranz, J., and Kurnick, W. S.,** Evidence for a direct action of cholecalciferol and 25-hydroxycholecalciferol on the renal transport of phosphate, sodium, and calcium, *J. Clin. Invest.*, 51, 373, 1972.
33. **Puschett, J. B., Fernandez, P. C., Boyle, I. T., Graw, R. W., Omdahl, J. L., and Deluca, H. F.,** The acute renal tubular effects of 1,25-dihydroxycholecalciferol, *Proc. Soc. Exp. Biol. Med.*, 141, 379, 1972.
34. **Winaver, J. and Puschett, J. B.,** Structural determinants of the renal tubular activity of vitamin D_3 derivatives: studies with 1-α-hydroxy, 24R,25-dihydroxy, and 1-α,24R,25-trihydroxy vitamin D_3, *Proc. Soc. Exp. Biol. Med.*, 159, 204, 1978.

35. **Puschett, J. B. and Kuhrman, M. S.,** Renal tubular effects of 1,25-dihydroxy vitamin D_3: interactions with vasopressin and parathyroid hormone in the vitamin D-depleted rat, *J. Lab. Clin. Med.*, 92, 895, 1978.
36. **Haas, H. G., Dambacher, M. A., Guncaga, J., and Lauffenburger, T.,** Renal effects of calcitonin and parathyroid extract in man, *J. Clin. Invest.*, 50, 2689, 1971.
37. **Talmage, R. V., Vanderwiel, C. J., Decker, S. A., and Grubb, S. A.,** Changes produced in postprandial urinary calcium excretion by thyroidectomy and calcitonin administration in rats on different calcium regimes, *Endocrinology*, 105, 459, 1979.
38. **Cochran, M., Peacock, M., Sachs, G., and Nordin, B. E. C.,** Renal effects of calcitonin, *Br. Med. J.*, 1, 135, 1970.
39. **Quamme, G. A.,** Effect of calcitonin on calcium and magnesium transport in rat nephron, *Am. J. Physiol.*, 238, F573, 1980.
40. **Consolazio, C. F., Matoush, L. O., Nelson, R. A., Hackler, L. R., and Preston, E. E.,** Relationship between calcium in sweat, calcium balance, and calcium requirement, *J. Nutr.*, 78, 78, 1962.
41. **Isaksson, B., Lindholm, B., and Sjorgren, B.,** A critical evaluation of the calcium balance technic. II. Dermal calcium losses, *Metabolism*, 16, 303, 1967.
42. **Mitchell, H. H. and Hamilton, T. S.,** The dermal excretion under controlled environmental conditions of nitrogen and minerals in human subjects, with particular reference to calcium and iron, *J. Biol. Chem.*, 178, 345, 1949.
43. **Consolazio, C. F., Shapiro, R., Masterson, J. E., and McKinzie, P. S. L.,** Energy requirements of men in extreme heat, *J. Nutr.*, 73, 126, 1961.
44. **Pritchard, J. J.,** General histology of bone, in *The Biochemistry and Physiology of Bone*, 2nd ed., Bourne, G. H., Ed., Academic Press, New York, 1972, 1.
45. **Messer, H. H., Yuen, S. Y., and Copp, D. H.,** Net uptake and release of calcium and phosphate by bone in vitro: effects of medium calcium and phosphate concentrations, *Calcif. Tissue Res.*, 19, 1, 1975.
46. **Neuman, W. F., Neuman, M. W., and Myers, C. R.,** Blood:bone disequilibrium. III. Linkage between cell energetics and Ca fluxes, *Am. J. Physiol.*, 236, C244, 1979.
47. **Scarpace, P. J. and Neuman, W. F.,** The blood:bone disequilibrium. II. Evidence against the active accumulation of calcium or phosphate into the bone extracellular fluid, *Calcif. Tissue Res.*, 20, 151, 1976.
48. **Nijweide, P. J. and Van Der Plas, A.,** Regulation of calcium transport in isolated periosteal cells, effects of hormones and metabolic inhibitors, *Calcif. Tissue Int.*, 29, 155, 1979.
49. **Ramp, W. K. and McNeil, R. W.,** Selective stimulation of net calcium efflux from chick embryo tibiae by parathyroid hormone in vitro, *Calcif. Tissue Res.*, 25, 227, 1978.
50. **Schaefer, K. E., Pasquale, S., Messier, A. A., and Shea, M.,** Phasic changes in bone CO_2 fractions, calcium and phosphorus during chronic hypercapnia, *J. Appl. Physiol. Respir. Environ. Exercise Physiol.*, 48, 802, 1980.
51. **Dominguez, J. H. and Raisz, L. H.,** Effects of changing hydrogen ion, carbonic acid, and bicarbonate concentrations on bone resorption in vitro, *Calcif. Tissue Int.*, 29, 7, 1979.
52. **Whiteside, L. A., Simmons, D. J., and Lesker, P. A.,** Comparisons of regional bone blood flow in areas with differing osteoblastic activity in the rabbit tibia, *Clin. Orthop. Relat. Res.*, 124, 267, 1977.
53. **Lien, J. and Kaye, M.,** Changes in red cell space of bone in hypocalcemic states: studies in rat femur, *Am. J. Physiol.*, 234, E6, 1978.
54. **Schumichen, C., Mundriziewski, L., Tischler, E., and Hoffmann, G.,** Relationship between blood flow and radiostrontium uptake in the healing bone fracture, *Eur. J. Nucl. Med.*, 4, 413, 1979.
55. **McInnis, J. C., Robb, R. A., and Kelly, P. J.,** The relationship of bone blood flow, bone tracer deposition, and endosteal new bone formation, *J. Lab. Clin. Med.*, 96, 511, 1980.
56. **Sherwood, L. M., Mayer, G. P., Ramberg, C. F., Kronfeld, D. S., Aurbach, G. D., and Potts, J. T.,** Regulation of parathyroid hormone secretion: proportional control by calcium lack of effect of phosphate, *Endocrinology*, 83, 1043, 1968.
57. **Reitz, R. E., Mayer, G. P., Deftos, J. L., and Potts, J. T.,** Endogenous parathyroid hormone response to thyrocalcitonin-induced hypocalcemia in the cow, *Endocrinology*, 89, 932, 1971.
58. **Mayer, G. P. and Hurst, J. G.,** Sigmoidal relationship between parathyroid hormone secretion rate and plasma calcium concentration in calves, *Endocrinology*, 102, 1036, 1978.
59. **Blum, J. W., Fischer, J. A., Schwoerer, D., Hunziker, W., and Binswanger, U.,** Acute parathyroid hormone response: sensitivity, relationship to hypocalcemia, and rapidity, *Endocrinology*, 95, 753, 1974.
60. **Canterbury, J. M., Lerman, S., Claflin, A. J., Henry, H., Norman, A., and Reiss, H.,** Inhibition of parathyroid hormone secretion by 25-hydroxycholecalciferol and 24,25-dihydroxycholecalciferol in the dog, *J. Clin. Invest.*, 61, 1375, 1978.
61. **Chertow, B. S., Baylink, D. J., Wergedal, J. E., Su, M. H. H., and Norman, A. W.,** Decrease in serum immunoreactive parathyroid hormone in rats and in parathyroid hormone secretion in vitro by 1,25-dihydroxycholecalciferol, *J. Clin. Invest.*, 56, 668, 1975.
62. **Takahashi, H., Shimazawa, E., Horiuchi, N., Suda, T., Yamashita, K., and Ogata, E.,** An estimation of the parathyroid hormone secretion rate in vitamin D deficient rats, *Horm. Metab. Res.*, 10, 161, 1978.

63. **Targovnik, J. H., Rodman, J. S., and Sherwood, L. M.,** Regulation of parathyroid hormone secretion in vitro: quantitative aspects of calcium and magnesium ion control, *Endocrinology*, 88, 1477, 1971.
64. **Morrissey, J. J. and Cohn, D. V.,** The effects of calcium and magnesium on the secretion of parathormone and parathyroid secretory protein by isolated porcine parathyroid cells, *Endocrinology*, 103, 2081, 1978.
65. **Mayer, G. P. and Hurst, J. G.,** Comparison of the effects of calcium and magnesium on parathyroid hormone secretion rate in calves, *Endocrinology*, 102, 1803, 1978.
66. **Reitz, R. E., Mayer, G. P., Deftos, L. J., and Potts, J. T.,** Endogenous parathyroid hormone response to thyrocalcitonin-induced hypocalcemia in the cow, *Endocrinology*, 89, 932, 1971.
67. **Blum, J. W., Fischer, J. A., Hunziker, W. H., Binswanger, U., Picotti, G. B., Da Prada, M., and Guillebeau, A.,** Parathyroid hormone responses to catecholamines and to changes of extracellular calcium in cows, *J. Clin. Invest.*, 61, 1113, 1978.
68. **Mayer, G. P., Habener, J. F., and Potts, J. T.,** Parathyroid hormone secretion in vivo: demonstration of a calcium-independent, nonsuppressible component of secretion, *J. Clin. Invest.*, 56, 678, 1976.
69. **Habener, J. F., Powell, D., Murray, T. M., Mayer, G. P., and Potts, J. T.,** Parathyroid hormone: secretion and metabolism in vivo, *Proc. Natl. Acad. Sci.*, 68, 2986, 1971.
70. **Neuman, W. F., Neuman, M. W., Sammon, P. J., Simon, W., and Lane, K.,** The metabolism of labelled parathyroid hormone. V. Collected biologic studies, *Calcif. Tissue Res.*, 18, 271, 1975.
71. **Habener, J. F. and Potts, J. T.,** Biosynthesis of parathyroid hormone, *N. Engl. J. Med.*, 299, 580, 635, 1978.
72. **Sammon, P. J., Stacey, R. E., and Bronner, F.,** Role of parathyroid hormone in calcium homeostasis in metabolism, *Am. J. Physiol.*, 218, 479, 1970.
73. **Clark, I. and Rivera-Cordero, F.,** Effects of endogenous parathyroid hormone on calcium, magnesium and phosphate metabolism in rats, *Endocrinology*, 92, 62, 1973.
74. **Mallette, L. E., Sode, J. E., Marx, S. J., Georges, L. P., and Aurbach, G. D.,** Total body retention of orally administered 47-Calcium in primary hyperparathyroidism, *J. Clin. Endocrinol. Metab.*, 40, 582, 1975.
75. **Mazzuoli, G. F., D'Erasmo, E., Scarda, A., Minisola, S., Mancini, D., and Malaguti Aliberti, L.,** Significance of early increase in stable and radioactive plasma calcium after parathyroidectomy in primary hyperparathyroidism, *Calcif. Tissue Int.*, 29, 185, 1979.
76. **Norman, A. W.,** *Vitamin D: The Calcium Homeostatic Steroid Hormone*, Academic Press, New York, 1979.
77. **DeLuca, H. F.,** *Vitamin D: Metabolism and Function*, Springer-Verlag, Berlin, 1979.
78. **Guggenheim, K. Y. and Wolinsky, I.,** *Nutrition and Nutritional Diseases: The Evolution of Concepts*, Collamore Press, D.C. Heath, Lexington, Mass., 1981.
79. **Holick, M. F., MacLaughlin, J. A., Clark, M. B., Holick, S. A., Potts, J. T., Anderson, R. R., Blank, I. H., Parrish, I. H., and Elias, P.,** Photosynthesis of previtamin D_3 in human skin and the physiologic consequences, *Science*, 210, 203, 1981.
80. **Holick, M. F., MacLaughlin, J. A., and Doppelt, S. H.,** Regulation of cutaneous previtamin D_3 photosynthesis in man: skin pigment is not an essential regulator, *Science*, 211, 590, 1980.
81. **Horsting, M. and DeLuca, H. F.,** In vitro production of 25-hydroxycholecalciferol, *Biochem. Biophys. Res. Commun.*, 36, 251, 1969.
82. **Bhattacharyya, M. and DeLuca, H. F.,** Subcellular location of rat liver calciferol-25-hydroxylase, *Arch. Biochem. Biophys.*, 160, 58, 1974.
83. **Gray, R. W., Omdahl, J. L., Ghazarian, J. G., and DeLuca, H. F.,** 25-Hydroxycholecalciferol-1-hydroxylase: subcellular location and properties, *J. Biol. Chem.*, 247, 7528, 1972.
84. **Ghazarian, J. F., Jefcoate, C. R., Knutson, J. C., Orma-Johnson, W. H., and DeLuca, H. F.,** Mitochondrial cytochrome P_{450}: a component of chick kidney 25-hydroxycholecalciferol-1-α-hydroxylase, *J. Biol. Chem.*, 249, 3026, 1974.
85. **Holick, M. F., Schnoes, H. K., DeLuca, H. F., Gray, R. W., Boyle, I. T., and Suda, T.,** Isolation and identification of 24,25-dihydroxycholecalciferol: a metabolite of vitamin D_3 made in the kidney, *Biochemistry*, 11, 4251, 1972.
86. **Henry, H. L., Midgett, R. J., and Norman, A. W.,** Regulation of 25-hydroxyvitamin D_3-1-hydroxylase in vivo, *J. Biol. Chem.*, 249, 7584, 1974.
87. **Garabedian, M., Holick, M. F., DeLuca, H. F., and Boyle, I. T.,** Control of 25-hydroxy-cholecalciferol metabolism by the parathyroid glands, *Proc. Natl. Acad. Sci.*, 69, 1673, 1972.
88. **Suda, T., DeLuca, H. F., Schnoes, H. K., Tanaka, Y., and Holick, M. F.,** 25,26-Dihydroxycholecalciferol, a metabolite of vitamin D_3 with intestinal calcium transport activity, *Biochemistry*, 9, 4776, 1970.
89. **Holick, M. F., Kleiner-Bossaller, A., Schnoes, H. K., Kasten, P. M., Boyle, I. T., and DeLuca, H. F.,** 1,24,25-Trihydroxyvitamin D_3: a metabolite of vitamin D_3 effective on intestine, *J. Biol. Chem.*, 248, 6691, 1973.

90. **Wasserman, R. H., Kallfelz, F. A., and Comar, C. L.,** Active transport of calcium by rat duodenum in vivo, *Science*, 133, 833, 1961.
91. **Walling, M. W.,** Intestinal calcium and phosphate transport: differential responses to vitamin D_3 metabolites, *Am. J. Physiol.*, 233, E488, 1977.
92. **Stern, P. H., Trummel, C. L., Schnoes, H. K., and DeLuca, H. F.,** Bone resorbing activity of vitamin D metabolites and congeners in vitro: influence of hydroxyl substituents in the A ring, *Endocrinology*, 97, 1552, 1975.
93. **Bronner, F., Sammon, P. J., Stacey, R. E., and Shah, B. G.,** Role of thyrocalcitonin in the regulation of the blood calcium level, *Biochem. Med.*, 1, 261, 1967.
94. **Sammon, P. J., Stacey, R. E., and Bronner, F.,** Further studies on the role of thyrocalcitonin in calcium homeostasis and metabolism, *Biochem. Med.*, 3, 252, 1969.
95. **Talmage, R. V., Grubb, S. A., Norimatsu, H., and Vanderwiel, C. J.,** Evidence for an important physiological role for calcitonin, *Proc. Natl. Acad. Sci.*, 77, 609, 1980.
96. **Binstock, M. L. and Mundy, G. R.,** Effect of calcitonin and glucocorticoids in combination on the hypercalcemia of malignancy, *Ann. Intern. Med.*, 93, 269, 1980.
97. **Barlet, J. P.,** Inhibition by calcitonin of hypercalcaemia induced by 1,25-dihydroxycholecalciferol, *J. Endocrinol.*, 85, 63, 1980.
98. **Deftos, L. J., Powell, D., Parthemore, J. G., and Potts, J. T.,** Secretion of calcitonin in hypocalcemic states in man, *J. Clin. Invest.*, 52, 3109, 1973.
99. **Metz, S. A., Deftos, L. J., Baylink, D. J., and Robertson, R. P.,** Neuroendocrine modulation of calcitonin and parathyroid hormone in man, *J. Clin. Endocrinol. Metab.*, 47, 151, 1978.
100. **Vora, N. M., Williams, G. A., Hargis, G. K., Bowser, E. N., Kawahara, W., Jackson, B. L., Henderson, W. J., and Kukreja, S. C.,** Comparative effect of calcium and of the andrenergic system on calcitonin secretion in man, *J. Clin. Endocrinol. Metab.*, 46, 567, 1978.
101. **Moriuchi, S., Yamanouchi, T., and Hosoya, N.,** Demonstration of two different vitamin D-dependent calcium-binding proteins in rat intestinal mucosa, *J. Nutr. Sci. Vitaminol.*, 21, 251, 1975.
102. **Hitchman, A. J. W. and Harrison, J. E.,** Calcium binding proteins in the duodenal mucosa of the chick, rat, pig, and human, *Can. J. Biochem.*, 50, 758, 1972.
103. **Morrissey, R. L., Zolock, D. T., Bucci, T. J., and Bikle, D. D.,** Immunoperoxidase localization of vitamin D dependent calcium binding protein, *J. Histochem. Cytochem.*, 26, 628, 1978.
104. **Corradiono, R. A. and Wasserman, R. H.,** Vitamin D: induction of calcium-binding protein in embryonic chick intestine in vitro, *Science*, 172, 731, 1971.
105. **Hermsdorf, C. L. and Bronner, F.,** Vitamin D-dependent calcium-binding protein from rat kidney, *Biochim. Biophys. Acta*, 379, 553, 1975.
106. **Taylor, A. N. and Wasserman, R. H.,** Vitamin D-induced calcium-binding protein: comparative aspects in kidney and intestine, *Am. J. Physiol.*, 223, 110, 1972.
107. **Stenflo, J., Fernlund, P., Egan, W., and Roepstorff, P.,** Vitamin K-dependent modification of glutamic acid residues in prothrombin, *Proc. Natl. Acad. Sci.*, 71, 2730, 1974.
108. **Stenflo, J.,** A new vitamin K-dependent protein, *J. Biol. Chem.*, 251, 355, 1976.
109. **Hauschka, P. V., Lian, J. B., and Gallop, P. M.,** Direct identification of the calcium-binding amino acid γ-carboxyglutamate, in mineralized tissue, *Proc. Natl. Acad. Sci.*, 72, 3925, 1975.
110. **Hauschka, P. V.,** Osteocalcin in developing bone systems, in *Vitamin K Metabolism and Vitamin K-Dependent Proteins*, Suttie, J. W., Ed., University Park Press, Baltimore, 1979, 227.
111. **Lian, J. B., Reit, R., Roufosse, A. H., Glimcher, M. J., and Gallop, P. M.,** Osteocalcin content in normal, rachitic, and vitamin K antagonized bone, in *Vitamin K Metabolism and Vitamin K-Dependent Proteins*, Suttie, J. W., Ed., University Park Press, Baltimore, 1979, 237.
112. **Kenny, A. D.,** Gastrointestinal absorption of alkaline earth ions, in *Pharmacology of Intestinal Absorption: Gastrointestinl Absorption of Drugs*, Section 39b, Encyclopedia of Pharmacology and Therapeutics, Rummel, W. and Farth, W., Eds., Pergamon Press, London, 1975, 507.
113. **Malm, O. J.,** On phosphates and phosphoric acid as dietary factors in the calcium balance of man, *Scand. J. Clin. Lab. Invest.*, 5, 75, 1953.
114. **Laflamme, G. H. and Jowsey, J.,** Bone and soft tissue changes with oral phosphate supplements, *Mod. Trends Orthop. Surg.*, 52, 151, 1975.
115. **Spencer, H., Menczel, J., Lewin, I., and Samachson, J.,** Effect of high phosphorus intake on calcium and phosphorus metabolism in man, *J. Nutr.*, 86, 125, 1965.
116. **Wilkinson, R.,** Absorption of calcium, phosphorus and magnesium, in *Calcium Phosphorus and Magnesium Metabolism*, Nordin, B. E. C., Ed., Churchill Livingstone, New York, 1976, 36.
117. **Spencer, H., Kramer, L., Osis, D., and Norris, C.,**Effect of phosphorus on the absorption of calcium and the calcium balance in man, *J. Nutr.*, 108, 447, 1978.
118. **Clark, I.,** Importance of dietary Ca:PO_4 ratios on skeletal Ca, Mg and PO_4 metabolism, *Am. J. Physiol.*, 217, 865, 1969.

119. **Anderson, G. H. and Draper, H. H.,** Effect of dietary phosphorus on calcium metabolism in intact and parathyroidectomized adult rats, *J. Nutr.*, 102, 1123, 1972.
120. **Howe, J. C. and Beecher, G. R.,** Effect of dietary protein and phosphorus levels on calcium and phosphorus metabolism of the young, fast growing rat, *J. Nutr.*, 111, 708, 1981.
121. **Howsey, J. and Balasubramanian, P.,** Effect of phosphate supplements on soft tissue calcification and bone turnover, *Clin. Sci.*, 62, 289, 1972.
122. **Shah, B. G., Krishnarao, G. V. G., and Draper, H. H.,**The relationship of Ca and P nutrition during adult life and osteoporosis aged mice, *J. Nutr.*, 92, 30, 1967.
123. **Schaafsma, G. and Visser, R.,** Nutritional interrelationships between calcium, phosphorus and lactose in rats, *J. Nutr.*, 110, 1101, 1980.
124. **Bell, R. R., Tzeng, D. Y., and Draper, H. H.,** Long-term effects of calcium, phosphorus and forced exercise on the bones of mature mice, *J. Nutr.*, 110, 1161, 1980.
125. **Ornoy, A., Wolinsky, I., and Guggenheim, K.,** Structure of long bones of rats and mice fed a low calcium diet, *Calcif. Tissue Res.*, 15, 71, 1974.
126. **Steenbock, H. and Hertig, D. C.,** Vitamin D and bone, *J. Nutr.*, 57, 449, 1955.
127. **Kemm, J. R.,** Effect of previous dietary history of calcium intake on the skeleton and calcium absorption in the rat, *J. Physiol.*, 230, 643, 1973.
128. **Guggenheim, K., Simkin, A., and Wolinsky, I.,** The effect of fluoride on bone of rats fed diets deficient in calcium or phosphorus, *Calcif. Tissue Res.*, 22, 9, 1976.
129. **Doige, C. E., Owen, B. D., and Mills, H. L.,** Influence of calcium and phosphorus on growth and skeletal development of growing swine, *Can. J. Anim. Sci.*, 55, 147, 1975.
130. **Draper, H. H., Sie, T. L., and Bergan, J. G.,** Osteoporosis in aging rats induced by high phosphorus diets, *J. Nutr.*, 102, 1133, 1972.
131. **Krishnarao, G. V. G. and Draper, H. H.,** Influence of dietary phosphate on bone resorption in senescent mice, *J. Nutr.*, 102, 1143, 1972.
132. **Jowsey, J.,** Why is mineral nutrition important in osteoporosis?, *Geriatrics*, 33, 39, 1978.
133. **Draper, H. H. and Bell, R. R.,** Dietary phosphate and the homeostasis of adult bone, *Baroda J. Nutr.*, 2, 167, 1975.
134. **Krook, L., Lutwak, L., Whalen, J. P., Henrikson, P. A., Lesser, G. V., and Uris, R.,** Human periodontal disease. Morphology and response to calcium therapy, *Cornell Vet.*, 62, 32, 1972.
135. **Anderson, M. P., Hunt, R. D., Griffiths, H. J., McIntyre, K. W., and Zimmerman, R. E.,** Long-term effect of low dietary calcium: phosphate ratio on the skeleton of *Cebus albifrons* monkeys, *J. Nutr.*, 107, 834, 1977.
136. **Page, L. and Friend, B.,** The changing United States diet, *Bioscience*, 28, 192, 1978.
137. **Ellinger, R. H.,** *Phosphates as Food Ingredients*, CRC Press, Boca Raton, Fla., 1972.
138. **Lutwak, L.,** Periodontal disease, in *Nutrition and Aging*, Winick, M., Ed., John Wiley & Sons, New York, 1972, 145.
139. **Hintz, H. F.,** Calcium, cola, calamity, *Cornell Vet.*, 70, 3, 1980.
140. **Chu, J. Y., Margen, S., and Costa, F. M.,** Studies in calcium metabolism. II. Effects of low calcium and variable protein intake on human calcium metabolism, *Am. J. Clin. Nutr.*, 28, 1028, 1975.
141. **Cummings, J. H., Hill, M. J., Jivraj, T., Houston, H., Branch, W. J., and Jenkins, D. J. A.,** The effect of meat protein and dietary fiber on colonic function and metabolism. I. Changes in bowel habit, bile acid excretion, and calcium absorption, *Am. J. Clin. Nutr.*, 32, 2086, 1979.
142. **Johnson, N. E., Alcantara, E. N., and Linkswiler, H.,** Effect of level of protein intake on urinary and fecal calcium and calcium retention of young adult males, *J. Nutr.*, 100, 1425, 1970.
143. **Linkswiler, H. M., Joyce, C. L., and Anand, C. R.,** Calcium retention of young adult males as affected by level of protein and of calcium intake, *Trans. N.Y. Acad. Sci.*, 36, 333, 1974.
144. **Walker, R. M. and Linkswiler, H. M.,** Calcium retention in the adult human male as affected by protein intake, *J. Nutr.*, 102, 1297, 1972.
145. **Kim, Y. and Linkswiler, H. M.,** Effect of level of protein intake on calcium metabolism and on parathyroid and renal function in the adult human male, *J. Nutr.*, 109, 1399, 1979.
146. **Anand, C. R. and Linkswiler, H. M.,** Effect of protein intake on calcium balance of young men given 500 mg of calcium daily, *J. Nutr.*, 104, 695, 1974.
147. **Hegsted, M., Schuette, S. A., Zemel, M. B., and Linkswiler, H. M.,** Urinary calcium and calcium balance in young men as affected by level of protein and phosphorus intake, *J. Nutr.*, 111, 553, 1981.
148. **Zemel, M. B., Schuette, S. A., Hegsted, M., and Linkswiler, H. M.,** Role of the sulfur-containing amino acids in protein-induced hypercalciuria in man, *J. Nutr.*, 111, 545, 1981.
149. **Hegsted, M. and Linkswiler, H. M.,** Long-term effects of level of protein intake on calcium metabolism in young adult women, *J. Nutr.*, 111, 244, 1981.
150. **Margen, S., Chu, J. Y., Kaufmann, N. A., and Calloway, D. H.,** Studies in calcium metabolism. I. The calciuretic effect of dietary protein, *Am. J. Clin. Nutr.*, 27, 584, 1974.

151. **Schwartz, R., Woodcock, N. A., Blakely, J. D., and MacKellar, I.,** Metabolic responses of adolescent boys to two levels of dietary magnesium and protein. II. Effect of magnesium and protein level on calcium balance, *Am. J. Clin. Nutr.*, 25, 519, 1973.
152. **Allen, L. H., Barlett, R. S., and Block, G. D.,** Reduction of renal calcium reabsorption in man by consumption of dietary protein, *J. Nutr.*, 109, 1345, 1979.
153. **Chu, J. Y., Margen, S., Calloway, D. H., and Costa, F. M.,** Integumentary loss of calcium, *Am. J. Clin. Nutr.*, 32, 1699, 1979.
154. **McCance, R. A., Widdowson, E. M., and Lehmann, H.,** The effect of protein intake on the absorption of calcium and magnesium, *Biochem. J.*, 36, 686, 1942.
155. **Schenk, W. G. and Moylan, J. A.,** Effect of parenteral amino acid solution administration on calcium balance following trauma, *Surg. Forum*, 28, 73, 1977.
156. **Spencer, H., Kramer, L., Osis, D., and Norris, C.,** Effect of a high protein (meat) intake on calcium metabolism in man, *Am. J. Clin. Nutr.*, 31, 2167, 1978.
157. **Schuette, S. A., Zemel, M. B., and Linkswiler, H. M.,** Studies on the mechanism of protein-induced hypercalciuria in older men and women, *J. Nutr.*, 110, 305, 1980.
158. **Block, G. D., Wood, R. J., and Allen, L. H.,** A comparison of the effects of feeding sulfur amino acids and protein on urine calcium in man, *Am. J. Clin. Nutr.*, 33, 2128, 1980.
159. **Allen, L. H., Block, G. D., Wood, R. J., and Bryce, G. F.,** The role of insulin and parathyroid hormone in the protein-induced calciuria of man, *Nutr. Res.*, 1, 3, 1981.
160. **Bell, R. R., Engelmann, D. T., Sie, T. L., and Draper, H. H.,** Effect of a high protein intake on calcium metabolism in the rat, *J. Nutr.*, 105, 475, 1975.
161. **Allen, L. H. and Hall, T. E.,** Calcium metabolism, intestinal calcium-binding protein, and bone growth of rats fed high protein diets, *J. Nutr.*, 108, 967, 1978.
162. **Denis, G., Kuczerpa, A., and Nikolaiczuk, N.,** Stimulation of bone resorption by increasing dietary protein intake in rats fed diets low in phosphorus and calcium, *Can. J. Physiol. Pharmacol.*, 51, 539, 1973.
163. **Whiting, S. J. and Draper, H. H.,** Effect of chronic high protein feeding on bone composition in the adult rat, *J. Nutr.*, 111, 178, 1981.
164. **Graves, K. L. and Wolinsky, I.,** Calcium and phosphorus metabolism in pregnant rats ingesting a high protein diet, *J. Nutr.*, 110, 2420, 1980.
165. **Lee, C. J., Cheaney, O. M., Smith, C. A., Marlatt, A. L., Skerski, G. M., and Packett, L. V.,** Effects of dietary levels of protein and calcium on utilization of milk protein and minerals in rats, *Nutr. Rep. Int.*, 6, 267, 1972.
166. **Weiss, R. E., Gorn, A., Dux, S., and Nimni, M. E.,** Influence of high protein diets on cartilage and bone formation in rats, *J. Nutr.*, 111, 804, 1981.
167. **Whiting, S. J. and Draper, H. H.,** The role of sulfate in the calciuria of high-protein diets in adult rats, *J. Nutr.*, 110, 212, 1980.
168. **Moyer, G. L., Wilson, H. D., and Schedl, H. P.,** Effects of dietary protein on body weight, mucosal growth, and calcium absorption in the rat, *Am. J. Dig. Dis.*, 23, 545, 1978.
169. National Dairy Council, The role of lactose in the diet, *Dairy Counc. Dig.*, 45, 25, 1974.
170. **Ali, R. and Evans, J. L.,** Lactose and calcium metabolism: a review, *J. Agric. Univ. Puerto Rico*, 57, 149, 1973.
171. **Armbrecht, H. J. and Wasserman, R. H.,** Enhancement of Ca^{++} uptake by lactose in the rat small intestine, *J. Nutr.*, 106, 1265, 1976.
172. **Condon, J. R., Nassim, J. R., Millard, F. J. C., Hilbe, A., and Stainthorpe, E. M.,** Calcium and phosphorus metabolism in relation to lactose tolerance, *Lancet*, 1, 1027, 1970.
173. **Kobayashi, A., Kawai, S., Ohbe, Y., and Nagashima, Y.,** Effects of dietary lactose and lactase preparation on the intestinal absorption of calcium and magnesium in normal infants, *Am. J. Clin. Nutr.*, 28, 681, 1975.
174. **Kocian, J., Skala, I., and Bakos, K.,** Calcium absorption from milk and lactose-free milk in healthy subjects and patients with lactose intolerance, *Digestion*, 9, 317, 1973.
175. **Greenwald, E., Samachson, J., and Spencer, H.,** Effect of lactose on calcium metabolism in man, *J. Nutr.*, 79, 531, 1963.
176. **Behar, J. and Kerstein, M. D.,** Intestinal calcium absorption: differences in transport between duodenum and ileum, *Am. J. Physiol.*, 230, 1255, 1976.
177. **Wasserman, R. H.,** Intestinal absorption of calcium and phosphorus, *Fed. Proc.*, 40, 68, 1981.
178. **Lengemann, F. W.,** The site of action of lactose in the enhancement of calcium utilization, *J. Nutr.*, 69, 23, 1959.
179. **Chang, Y. and Hegsted, D. M.,** Lactose and calcium transport in gut sacs, *J. Nutr.*, 82, 297, 1964.
180. **Lengemann, F. W., Wasserman, R. H., and Comar, C. L.,** Studies on the enhancement of radiocalcium and radiostrontium absorption by lactose in the rat, *J. Nutr.*, 68, 443, 1959.
181. **Wasserman, R. H.,** Lactose stimulated intestinal absorption of calcium: a theory, *Nature (London)*, 201, 997, 1964.

182. **Dupuis, Y., Digaud, A., and Fontaine, N.,** Further observations in favour of the role of alkaline phosphatase in calcium absorption, *Calcif. Tissue Res.*, Suppl. 22, 556, 1977.
183. **Anon.,** Metabolic bone disease as a result of lactase deficiency, *Nutr. Rev.*, 37, 72, 1979.
184. **Woodruff, C. W.,** Milk intolerances, *Nutr. Rev.*, 34, 33, 1976.
185. **Fasset, D. W.,** Oxalates, in National Research Council Publ. 1354, National Academy of Science-National Research Council, Washington, D.C., 1966, 257.
186. Council on Foods, The nutritional value of spinach, *JAMA*, 109, 1907, 1937.
187. **Pingle, U. and Ramasastri, B. V.,** Absorption of calcium from a leafy vegetable rich in oxalates, *Br. J. Nutr.*, 39, 119, 1975.
188. **Pingle, U. and Ramasastri, B. V.,** Effect of water-soluble oxalates in *Amaranthus* spp. leaves on the absorption of milk calcium, *Br. J. Nutr.*, 40, 591, 1978.
189. **Hodkinson, A. and Zarembski, P. M.,** Oxalic acid metabolism in man: a review, *Calcif. Tissue Res.*, 2, 115, 1968.
190. **Mackenzie, C. G. and McCollum, E. V.,** Some effects of dietary oxalate on the rat, *Am. J. Hyg.*, 25, 1, 1937.
191. **Tarjan, R., Sandi, E., and Denes, A.,** Einfluss der Ernährung auf das Knochensystem. I. Einfluss der mit der Nährung Zugeführten löslichen oxalate auf die Entwicklung des Knochensystems weisser Ratten, *Acta Physiol. Acad. Sci. Hung.*, 5, 313, 1954.
192. **Tarjan, R., Sandi, E., and Denes, A.,** Einfluss der Ernährung auf das Knochensystem. II. Wirkung des in der Nährung vorhandenen Oxalates auf die Entwicklung des Knochensystems weisser Ratten, *Acta Physiol. Acad. Sci. Hung.*, 5, 463, 1954.
193. **Shirly, E. K. and Schmidt-Nielsen, K.,** Oxalate metabolism in the pack rat, sand rat, hamster and white rat, *J. Nutr.*, 91, 496, 1967.
194. Committee on Food Protection, Toxicants Occurring Naturally in Foods, Publ. No. 73: 8968, 2nd ed., National Academy of Science - National Research Council, Washington, D.C., 1973, 346.
195. **Kimura, K. K.,** *The Nutritional Significance of Dietary Fiber*, Life Sciences Research Office, Federation of American Society of Experimental Biology, Bethesda, Md., 1977.
196. **Spiller, G. A. and Shipley, E. A.,** Perspectives in dietary fiber in human nutrition, *World Rev. Nutr. Diet.*, 27, 105, 1977.
197. **Kelsay, J. L.,** A review of research on effects of fiber intake on man, *Am. J. Clin. Nutr.*, 31, 142, 1978.
198. Symposium on Role of Dietary Fiber in Health, *Am. J. Clin. Nutr. Suppl.*, 31, S1, 1978.
199. **James, W. P. T., Branch, W. J., and Southgate, D. A. T.,** Calcium binding by dietary fiber, *Lancet*, 1, 638, 1978.
200. **Harmuth-Hoene, A. E. and Schelenz, R.,** Effect of dietary fiber on mineral absorption in growing rats, *J. Nutr.*, 110, 1774, 1980.
201. **Ismail-Beigi, F., Reinhold, J. G., Faradji, B., and Abadi, P.,** Effects of cellulose added to diets of low and high fiber content upon the metabolism of calcium, magnesium, zinc and phosphorus by man, *J. Nutr.*, 107, 510, 1977.
202. **Kelsay, J. L., Behall, K. M., and Prather, E. S.,** Effect of fiber from fruits and vegetables on metabolic responses of human subjects. II. Calcium, magnesium, iron, and silicon balances, *Am. J. Clin. Nutr.*, 32, 1876, 1979.
203. **Reinhold, J. G., Ismail-Beigi, F., and Faradji, B.,** Fibre vs. phytate as determinant of the availability of calcium, zinc and iron of breadstuffs, *Nutr. Rep. Int.*, 12, 75, 1975.
204. **Reinhold, J. G., Faradji, B., Abadi, P., and Ismail-Beigi, F.,** Decreased absorption of calcium, magnesium, zinc and phosphorus by humans due to increased fiber and phosphorus consumption as wheat bread, *J. Nutr.*, 106, 493, 1976.
205. **McCance, R. A. and Widdowson, E. M.,** Mineral metabolism on dephytinized bread, *J. Physiol.*, 101, 304, 1942.
206. **Godora, R., Kaul, A. P. and Bhat, C. M.,** Effect of cellulose incorporation in a low fiber diet on fecal excretion and serum levels of calcium, phosphorus, and iron in adolescent girls, *Am. J. Clin. Nutr.*, 34, 1083, 1981.
207. **Kay, R. M.,** Kinds and functions of plant fiber, in *Nutrition in Transition*, Proc. West. Hemisphere Nutr. Congr. V, White, P. L., and Selvey, N., Eds., American Medical Association, Chicago, 1978.
208. **Reinhold, J. G., Abadi, P., Faradji, B., Ismail-Beigi, F., and Russell, R. M.,** Depletion of calcium, magnesium, zinc and phosphorus associated with consumption by humans of wholemeal wheaten bread rich in fiber and phytate, *Baroda J. Nutr.*, 7, 55, 1980.
209. **Oberleas, D.,** Phytates, in Toxicants Occurring Naturally in Foods, 2nd ed., National Academy of Science-National Research Council, Washington, D.C., 1973, 363.
210. **Bruce, H. M. and Callow, R. K.,** Cereal and rickets. The role of inositol hexaphosphoric acid, *Biochem. J.*, 28, 517, 1934.
211. **Harris, R.,** Phytic acid and its importance in human nutrition, *Nutr. Rev.*, 13, 257, 1955.

212. **Erdman, J. W. and Forbes, R. M.,** Mineral bioavailability from phytate-containing foods, *Food Prod. Dev.*, 11, 46, 1977.
213. **Reinhold, J. G., Nasr, K., Lahimgarzadeh, A., and Hedayati, H.,** Effects of purified phytate and phytate-rich bread upon metabolism of zinc, calcium, phosphorus, and nitrogen in man, *Lancet*, 1, 283, 1973.
214. **Mellanby, E.,** *A Story of Nutritional Research. The Effect of Some Dietary Factors on Bone and the Nervous System*, Williams & Wilkins, Baltimore, 1950.
215. **Van Den Berg, C. J., Hill, L. F., and Stanbury, S. W.,** Inositol phosphates and phytic acid as inhibitors of biological calcification in the rat, *Clin. Sci.*, 43, 377, 1972.
216. **Thomas, W. C., Jr. and Tilden, M. T.,** Inhibition of mineralization by hydrolysates of phytic acid, *Johns Hopkins Med. J.*, 131, 133, 1972.
217. **Walker, A. R. P., Fox, F. W., and Irving, J. T.,** The effect of bread rich in phytate phosphorus on the metabolism of certain mineral salts with special reference to calcium, *Biochem. J.*, 42, 452, 1948.
218. **Bronner, F., Harris, R. S., Meletskos, C. J., and Benda, C. E.,** Studies in calcium metabolism. Effect of food phytates on calcium^{45} uptake in children on low-calcium breakfasts, *J. Nutr.*, 54, 523, 1954.
219. **Reinhold, J. G.,** Phytate destruction by yeast fermentation in whole wheat meals, *J. Am. Diet. Assoc.*, 66, 38, 1975.
220. **Wills, M. R., Day, R. C., Phillips, J. B., and Bateman, E. C.,** Phytic acid and nutritional rickets in immigrants, *Lancet*, 1, 771, 1970.
221. **Ford, J. A., Colhoun, E. M., McIntosh, W. D., and Dunnigan, M. G.,** Biochemical response of late rickets and osteomalacia to a chupatti-free diet, *Br. Med. J.*, 3, 446, 1972.
222. **Berlyne, G. M., Ari, J. B., Nord, E., and Shainkin, R.,** Bedouin osteomalacia due to calcium deprivation caused by high phytic acid content of unleavened bread, *Am. J. Clin. Nutr.*, 26, 910, 1973.
223. **Walker, A. R. P.,** The human requirement of calcium: should low intakes be supplemented?, *Am. J. Nutr.*, 25, 518, 1972.
224. **Reinhold, J. G.,** Phytate concentrations of leavened and unleavened Iranian breads, *Ecol. Food. Nutr.*, 1, 187, 1972.
225. **Brown, W. E. and Konig, K. G.,Eds.,** Cariostatic mechanisms of fluorides, *Caries Res.*, 11 (Suppl. 1), 1, 1977.
226. **Richmond, V. L.,** Health effects associated with water fluoridation, *J. Nutr. Ed.*, 11, 62, 1980.
227. Royal College of Physicians, *Fluoride, Teeth and Health*, Pitman Medical, London, 1976.
228. **Riggs, B. L., Hodgson, S. F., Hoffman, D. L., Kelly, P. J., Johnson, K. A., and Taves, D.,** Treatment of primary osteoporosis with fluoride and calcium. Clinical tolerance and fracture occurrence, *JAMA*, 243, 446, 1980.
229. **Baylink, D. J. and Ivey, J. L.,** Sodium fluoride for osteoporosis — some unanswered questions, *JAMA*, 243, 463, 1980.
230. **Guggenheim, K. Y. and Wolinsky, I.,** The effect of fluoride on the mechanical strength of bone, *Baroda J. Nutr.*, 3, 177, 1976.
231. **Wolinsky, I. and Guggenheim, K.,** Effect of low calcium diet on bone and calcium metabolism in rats and mice — a differential species response, *Comp. Biochem. Physiol.*, 49a, 183, 1974.
232. **Spencer, H., Osis, D., Kramer, L., Wiatrowski, E., and Norris, C.,** Effect of calcium and phosphorus on fluoride metabolism in man, *J. Nutr.*, 105, 733, 1975.
233. **Havivi, E.,** Effect of calcium and vitamin D on fluoride metabolism in the rat, *Nutr. Metab.*, 14, 257, 1972.
234. **Wagner, M. J. and Muhler, J. C.,** The effect of calcium and phosphorus on fluoride absorption, *J. Dent. Res.*, 39, 49, 1960.
235. **Ekstrand, J. and Ehrnebo, M.,** Influence of milk products on fluoride bioavailability in man, *Eur. J. Clin. Pharmacol.*, 16, 211, 1979.
236. **Chang, Y. O., Pan, M., and Varnell, T.,**The effect of fluoride on calcium absorption in rats, *Nutr. Rep. Int.*, 16, 539, 1977.
237. **Spencer, H., Lewin, I., Fowler, J., and Samachson, J.,** Effect of sodium fluoride on calcium absorption and balance in man, *Am. J. Clin. Nutr.*, 22, 381, 1969.
238. **Cohn, S. H., Dombrowski, C. S., Hauser, W., and Atkins, H. L.,** Effects of fluoride on calcium metabolism in osteoporosis, *Am. J. Clin. Nutr.*, 24, 20, 1971.
239. **Thiébaud, M., Zender, R., Courvoisier, B., Buad, C. A., and Jacot, C.,** The action of flouride on diffuse bone atrophies, in *Fluoride in Medicine*, Vischer, L., Ed., Hans Huber, Bern, 1970, 136.
240. **Spencer, H., Kramer, L., Norris, C., and Wiatrowski, E.,** Effect of calcium on fluoride metabolism of patients with chronic renal failure, *Nutr. Rep. Int.*, 21, 429, 1980.
241. **Hegsted, D. M.,** Calcium and phosphorus, in *Modern Nutrition in Health and Disease*, 5th ed., Goodhart, R. S. and Shils, M. E., Eds., Lea & Febiger, Philadelphia, 1973, 268.
242. **Avioli, L. V.,** Calcium and phosphorus, in *Modern Nutrition in Health and Disease*, 6th ed., Goodhart, R. S. and Shils, M. E., Eds., Lea & Febiger, Philadelphia, 1980, 294.

243. **FAO/WHO,** Calcium Requirements, Tech. Rep. Ser. No. 230, World Health Organization/Food and Agriculture Organization, Rome, 1962.
244. **Irwin, M. I. and Kienholz, E. W.,** A conspectus of research on calcium requirments of man, *J. Nutr.,* 103, 1019, 1973.
245. National Research Council, Recommended Dietary Allowances, 9th ed., Washington, D.C., 1980, 125.
246. **Widdowson, E. M.,** Nutritional requirements and its assessment, with special reference to energy, protein and calcium, *Bibl. Nutr. Dieta,* 28, 148, 1974.
247. **Paterson, C. R.,** Calcium requirements in man: a critical review, *Postgrad. Med. J.,* 54, 244, 1978.
248. **International Union Nutritional Sciences,** Report of the Committee on International Dietary Allowances of the International Union of Nutritional Sciences, *Nutr. Abstr. Rev.,* 45, 89, 1975.
249. **McBean, L. D. and Speckmann, E. W.,** A recognition of the interrelationship of calcium with various dietary components, *Am. J. Clin. Nutr.,* 27, 603, 1974.
250. **Bullamore, J. R., Gallagher, J. C., Wilkinson, R., Nordin, B. E. C., and Marshall, D. H.,** The effect of age on calcium absorption, *Lancet,* 2, 535, 1970.

PHOSPHORUS

G. M. Siu and H. H. Draper

INTRODUCTION

Phosphorus (P) is involved in more discrete biochemical processes than any other mineral element. Although 85% of the P in the body is located in the skeleton, that portion in the soft tissues serves as an essential component of many compounds of major biological importance, including phospholipids, phosphoproteins, nucleic acids, and intermediates of energy metabolism. The exchange of P among various body pools is not as rigidly controlled as it is for calcium (Ca), but P metabolism is nevertheless regulated by a complex hormonal system which includes parathyroid hormone, vitamin D, and calcitonin.

DIETARY PHOSPHORUS

Average daily intakes of P by adults consuming a mixed diet have been estimated to vary from 900 to 1600 mg depending mainly upon caloric intakes.[1] Individual intakes from diets high in protein or phosphate food additives may substantially exceed this range. Because of the ubiquitous occurrence of P in foods and its efficient absorption, P deficiency is rare. Consequently, there is little information on the P requirement of human subjects. Recommended intakes by adults are commonly set at parity with Ca, but if the recommendations for Ca are met, the recommendations for P are invariably exceeded.

The possibility that a large excess of dietary P relative to Ca may have adverse long-term effects on adult bone has been the subject of much recent research. Longitudinal experiments on several species of adult animals have demonstrated that high P diets increase aging bone loss and have led to speculation that the high P content of the mixed Western diet may be a factor contributing to osteoporosis, a prevalent disease of older women. It has also been proposed that osteopenia induced by excess dietary P may contribute to alveolar bone resorption and periodontal disease.[2]

The osteopenia induced by high P diets is mediated by a mild, chronic depression of plasmic Ca caused by the flux of excess P through the blood. The consequent increase in the synthesis of parathyroid hormone leads to activation of mechanisms for the maintenance of plasma Ca homeostasis including increased bone resorption, increased reabsorption of Ca from the renal tubules, and (by a mechanism involving the increased synthesis of 1,25-dihydroxyvitamin D_3) an increase in the efficiency of Ca absorption from the small intestine. At moderate excesses of dietary P, this system successfully maintains body Ca homeostasis, the increase in bone resorption evidently being fully coupled to an increase in the rate of bone formation.[3] Large excesses of dietary P, however, lead to net bone loss with an increase in endogenous fecal Ca.

The demonstration that these effects of excess P in experimental animals can, within limits, be offset by an increase in dietary Ca has led to the concept of an optimal Ca to P ratio, and this concept is firmly implanted in the formulation of diets for laboratory and farm animals. Laboratory rodents fed diets which contain a liberal amount of Ca (e.g., 0.6%) at Ca to P ratios of 1:2 or wider exhibit an increase in aging osteopenia. However, the effective ratio depends upon the absolute intake of both elements, since the efficiency of Ca absorption decreases markedly at high intakes whereas the efficiency of P absorption remains essentially unchanged. Hence, the ratio of absorbed Ca to absorbed P shifts in favor of P as the intake of both elements is increased proportionately. Over a wide range of intakes plasma Ca is more sensitive to dietary P than to dietary Ca. Unlike plasma P, plasma Ca homeostasis is protected against the influence of excess dietary Ca by a physiological vitamin

D-dependent mechanism that regulates the amount of Ca absorbed. There is no comparable mechanism that protects against the absorption of excess P. In evaluating the effect of dietary P on bone metabolism, it is therefore necessary to consider the total P content of the diet as well as the Ca to P ratio.

The applicability of these observations on experimental animals to human diets is unclear. Most Western diets contain an excess of P because of the natural association of P with protein and the widespread use of phosphate additives in processed foods.[4] Estimates of the Ca to P ratio of the average diet of adults in the U.S. and Canada are in the range of 1:1.5 to 1:1.8.[1] The diets of children and teenagers have a narrower ratio owing mainly to an increased consumption of Ca in dairy products.

A recent evaluation of the influence of various nutrients on bone metabolism in the U.S. led to the conclusion that excess dietary P is not a cause for concern among the general adult population. Ca balance studies on human adults have indicated that, although excess P consumption is associated with parathyroid stimulation, Ca balance can be maintained over a wide range of P intakes, apparently as the result of a stimulus to bone formation which offsets the parathyroid hormone-induced increase in bone resorption.[5] In addition, the hypercalciuria and negative Ca balance caused by excess dietary protein is offset, within limits, by the hypocalciuric effect of excess dietary P, indicating that there may be counterbalancing effects of these two naturally associated nutrients on body Ca homeostasis.[6] It must be considered, however, that statistics for the average consumption of Ca, P, and protein conceal the fact that there are many individuals who consume diets which are low in Ca (500/mg or less per day), high in protein (over 100 g/day), and high in P ($\sim$2000 mg/day including several hundred milligrams from additives). Whether Ca balance is maintained under these conditions is still unclear.

ABSORPTION

P is readily absorbed from the small intestine. Based on balance studies conducted under conditions of normal Ca intake, the amount of P absorbed appears to increase linearly with intake over a range of 4 to 40 mg/kg/day with an efficiency of $\sim$70%.[7,8] This linearity does not hold, however, at extremely high or low intakes. At intakes of <2 mg/kg/day, the efficiency of P absorption can increase to $\sim$90%.[8] P in most foodstuffs is absorbed with an efficiency of 70 to 75%, but cereal P, which is present mainly in phytin, is less efficiently absorbed owing to incomplete digestion.[9]

Many studies have been carried out on the factors influencing intestinal P absorption and the mechanism involved. P absorption can be described in terms of a nonsaturable diffusable component which is important in vivo at normal dietary P intakes and a saturable active component which is of primary importance at low intakes.[8] In experiments utilizing everted rat gut sacs, it has been determined that glucose and sodium are required for phosphate transport against a concentration gradient. Potassium enhances, while cyanide and anaerobiosis inhibit, Pi transport against a concentration gradient, indicating the presence of an active transport mechanism.[10,11]

Using ^{32}P and an *in situ* ligated gut loop technique in chicks, Wasserman and Taylor[12] showed that:

1. ^{32}P is rapidly absorbed and translocated across all segments of the small intestine.
2. The process has a saturable component, as indicated by decreased ^{32}P absorption with increasing Pi concentration
3. The rate-limiting step in phosphate translocation appears to be from serosa to blood.
4. The absorption of ^{32}P is independent of ethylene glycol tetraacetic acid (EGTA), suggesting that Pi is not transferred as a co-ion to Ca^{++}.

5. Absorbed ^{32}P is present in a form not distinguishable from orthophosphate (although the authors did not rule out the possibility that ^{32}P may be absorbed as a labile complex).
6. Vitamin D-stimulated ^{32}P translocation occurs in all segments of chick intestine.

The vitamin D enhancement effect was first reported by Harrison and Harrison,[13] who observed that the presence of Ca was required for vitamin D-stimulated P absorption and who therefore postulated a link between Ca and P absorption. This view was supported by the observations of Chen et al.[14] using everted gut sacs from vitamin D-deficient rats. They demonstrated an increase in Pi absorption from the jejunum following in vivo vitamin D repletion in the absence of dietary Ca, although there was a greater increase when Ca was present. These authors proposed that the active vitamin D metabolite responsible for enhanced Pi transport is 1,25-dihydroxyvitamin D_3 [1,25$(OH)_2D_3$] since nephrectomy prevented a response to 25-hydroxyvitamin D_3 (25-OH-D_3), but not to 1,25$(OH)_2D_3$, in vivo.

Subsequent studies on chick intestine *in situ* and in vitro showed, however, that vitamin D increased P absorption in the absence of Ca ions. Ouabain decreased phosphate transport but had no effect on Ca transport.[15] In addition, the vitamin D effect on Pi absorption was observed to be maximal in the jejunum while the maximal effect of vitamin D on Ca absorption occurred in the duodenum.[16] Intraperitoneally injected actinomycin D and cyclohexamide, which decreased the rate of ^{32}P absorption in vivo, did not affect the rate of ^{45}Ca absorption, while corticosteroids blocked both Ca and P absorption.[17]

Based on existing literature, Parfitt and Kleerekoper[18] have suggested that there are probably three components to normal P absorption:

1. Ca coupled, vitamin D dependent
2. Non-Ca coupled, vitamin D dependent
3. Non-Ca coupled, vitamin D independent

P absorption increases with increasing pH up to ~8.0.[18,19] No optimal pH has been reported, but at alkaline pH, precipitation of phosphate salts may be promoted, resulting in decreased phosphate absorption. In ligated chick intestinal loops, ammonium chloride has been shown to interfere with phosphate absorption.[20] Oral administration of nonabsorbable Mg and Al compounds as antacids impairs P absorption as a result of binding of Pi by these ions.[21]

Parathyroid hormone increases P absorption in vivo and in vitro. Postmenopausal women with primary hyperparathyroidism exhibit increased phosphate absorption.[22] Using everted rat duodenal loops, PTH has been shown to increase P influx from the mucosa to the serosa.[23] Parathyroidectomy (PTX) results in decreased phosphate absorption from rat gut sacs *in situ*, which can be reversed by administration of PTH. The mechanism of the PTH effect is believed to involve enhanced conversion of 25-OHD_3 to 1,25$(OH)_2D_3$.[24] This view is supported by experiments which demonstrated that the decreased phosphate absorption which follows thyroparathyroidectomy (TPTX) can be corrected by injection of either PTH or 1,25-$(OH)_2D_3$.[25]

EXCRETION

Renal excretion is the primary means of maintaining phosphate homeostasis. Under normal physiological conditions, the daily urinary phosphate excretion of ~450 μmol/kg of body weight per day approximates net intestinal phosphate absorption.[18] The tubular reabsorption of phosphate normally exceeds 85% of the filtered load.[26] Renal reabsorption of phosphate is responsive to changes in dietary phosphorus. In animals fed high P diets, renal reabsorption of phosphate is decreased, while in animals fed low P diets, phosphate reabsorption can increase to as much as 100% of the filtered load.[27]

The results of micropuncture studies indicate that renal phosphate reabsorption occurs primarily in the proximal tubules via a saturable active transport mechanism requiring Na ions,[28,29] with the monovalent $H_2PO_4^-$ ion as the predominant transported species.[30] The proximal tubules normally reabsorb ~75% of the filtered load with a further reabsorption of 5 to 20% in the distal tubule or collecting ducts.[31,32]

Net renal phosphate excretion relative to filtered load can be described by the threshold model relating maximal tubular reabsorption to glomerular filtration rate (Tm/GFR). Since plasma phosphate levels normally exceed the Tm/GFR, urine generally contains some phosphate. The reabsorption mechanism is responsive to hormonal and ionic influences as well as to changes in GFR and dietary phosphorus.

From free-flow micropuncture[33] and in vitro studies with renal cortical brush border membrane vesicles,[34,35] there is evidence that rats adapted to low P diets (0.1 to 0.2%) have a lower fractional excretion of P (FEPi) than rats fed normal P diets (0.5 to 0.8%) under conditions of equal filtered loads. This decrease in FEPi may be related to the increased rate of phosphate transport in the renal cortical brush border vesicles of rats adapted to a low P diet. In rats adapted to high P diets (1.3 to 1.8%), there is an increased FEPi as well as a decreased rate of phosphate transport by the renal cortical brush border vesicles. The rate of phosphate uptake in these vesicles is positively correlated with alkaline phosphatase activity.

Parathyroid hormone (PTH) produces phosphaturia in normal subjects and in patients with hypoparathyroidism.[36,37] This response is also seen in patients with primary hyperparathyroidism, in whom phosphaturia results in hypophosphatemia.[38] Using intrarenal arterial PTH infusion techniques, the hypophosphatemia has been shown to be the result of a decreased renal threshold for phosphate.[39,40] The site of action of PTH in the nephron has been investigated in many studies. Using free-flow micropuncture techniques,[41,42] decreased Pi reabsorption in response to PTH was demonstrated in the proximal tubules, which were found to be an important site (if not the only site) of PTH action. This view was confirmed by the results of in vivo studies on PTX animals which revealed increased phosphate reabsorption in the proximal tubules (~25%),[43-45] as well as by results of microinjection studies with radioactive tracers[46] and of microperfusion analysis.[47]

There is evidence that parathyroid hormone can also inhibit Pi reabsorption at sites distal to the proximal tubules.[32,48] On the basis of changes in Pi efflux rate in the presence and absence of PTH, it has been suggested that the pars recta, which is located between the late proximal tubule and the early distal tubule where it is inaccessible to micropuncture, may be involved in PTH-responsive phosphate reabsorption.[49] In TPTX rats, a PTH-induced decrease in Pi reabsorption has been demonstrated in the distal convuluted tubules.[50] There is also evidence from micropuncture studies that phosphate reabsorption in this segment of the nephron is inhibited by PTH.[51,52] Thus, it appears that PTH can act at all nephron sites to inhibit phosphate reabsorption.

There are several possible mechanisms for the inhibition of phosphate reabsorption by PTH. The phosphaturic response is accompanied by increased urinary cAMP in vivo and in isolated tubules.[53] Dibutyryl cyclic adenosine nucleotide mimics the physiologic effect of PTH on phosphate excretion.[54,55] PTH interacts with specific receptors on the contraluminal membrane resulting in activation of adenyl cyclase and increased intracellular cAMP.[56,57] The increase in urinary cAMP which follows PTH administration precedes the phosphaturic effect of PTH. Also, clinical conditions associated with inhibition of adenyl cyclase (e.g., hypercalcemia, hypomagnesemia, acute metabolic acidosis, or pseudohypoparathyroidism) result in blunting or ablation of the effects of exogenous PTH with little or no phosphaturia.[18]

Other proposed mechanisms for the phosphaturic action of PTH[58] include the suggestion that PTH effects on phosphate reabsorption are secondary to the PTH inhibition of Na transport. Phosphate reabsorption in the proximal tubule is coupled to Na reabsorption.

However, from micropuncture studies,[41,59] it appears that phosphate and Na transport in the distal tubules are dissociated. Hence, while PTH-stimulated Na transport may mediate Pi reabsorption in the proximal tubules, changes in Na transport do not entirely account for PTH-induced phosphaturia.

A further proposal is that PTH decreases carbonic anhydrase activity[60] and hence HCO_3^- reabsorption,[61,62] resulting in alkalinization of luminal fluids favoring formation of the HPO_4^{-2} ion in place of the $H_2PO_4^-$ ion. The increased ionic charge on HPO_4^{-2} decreases its transport across the luminal membrane, thus decreasing phosphate reabsorption.[32] However, conflicting evidence exists regarding an effect of luminal alkalinization of PTH-induced phosphaturia. From microperfusion studies, Bank and Aynedjian[63] concluded that acid phosphate is preferentially reabsorbed, whereas *in situ* perfusion experiments have indicated that Pi reabsorption is increased with alkalinization of the perfusate.[64,65] In any event, it appears that PTH-induced phosphaturia is mediated by cAMP and is related to the effect of PTH on Na and bicarbonate transport.

Vitamin D and its metabolites are antiphosphaturic,[66] but the increase in phosphate reabsorption can be inhibited by high doses of PTH, cAMP, or thyrocalcitonin,[67] as well as by Ca infusion. PTH, cAMP, and thyrocalcitonin increase phosphate excretion in the absence of vitamin D.[68,69] The phosphaturic agents are also capable of elevating cytosolic Ca,[70] suggesting that inhibition of the antiphosphaturic effect of vitamin D may be related to increased intracellular Ca[66] which may be an important modulator of electrolyte transport.[71] The antiphosphaturic effect of vitamin D reportedly requires the presence of PTH in the rat[72] but not in the dog.[68]

These observations support the contention that vitamin D metabolites have a direct effect on renal phosphate reabsorption. However, the evidence that PTH is required for this effect is inconclusive.[66] Growth hormone and insulin cause acute decreases in urinary phosphate excretion, whereas antidiuretic hormone, thyrotropic hormone, calcitonin, and glucagon are phosphaturic.[73]

BONE

Besides providing structural support, bone is a major reservoir of many of the body's inorganic elements (99% of the Ca, 80% of the P, and substantial amounts of Mg and Na) as well as of bicarbonate. Bone resorption, which involves removal of both mineral and matrix, is of primary importance in the maintenance of serum and tissue Ca homeostasis. Phosphates and bicarbonates derived from bone also provide buffering capacity for acid-base balance.

Bone formation or growth is influenced by parathyroid hormone, thyroxine, cortisol, and growth hormone, as well as by nutritional status with respect to vitamin D, Ca, P, Mg, and Na.[18] Bone resorption is increased by PTH, osteoclast activating factor, prostaglandin E, and vitamin D and is inhibited by calcitonin and estrogen. Using cultures of ^{45}Ca-labeled fetal calvaria, Raisz and Niemann[74] showed that increasing the phosphate concentration in the medium resulted in decreased spontaneous as well as PTH-stimulated Ca and ^{3}H-hydroxyproline release. Collagen synthesis was increased at high phosphate concentrations while low phosphate concentrations inhibited collagen synthesis and bone formation[75,76] and increased bone resorption.[77]

Phosphate deficiency in human subjects is associated with osteomalacia, increased resorption of skeletal Ca and P, and bone pain.[78] In rats prelabeled with ^{45}Ca, acute P depletion (0.04% diet P) results in a rise in the concentration and specific activity of serum Ca after 48 hr, but no increase in the number of osteoclasts until after 72 hr, indicating that the rise in serum Ca is due to decreased bone formation and/or increased resorptive activity of preexisting osteoclasts.[77] Also, the increase in specific activity of serum Ca is more rapid in TPTX rats, implying that the increased resorption is inhibited by calcitonin. The explanation proposed for the increase in serum Ca is that phosphate depletion results in a decrease in Ca deposition in bone which is followed by an increase in the resorptive activity of preexisting osteoclasts and finally an increase in the numbers of osteoclasts.

Phosphate inhibits PTH-induced bone resorption in vitro, but has no effect on PTH-induced bone resorption in vivo.[79] Phosphate supplements have been suggested as a treatment for osteoporotic bone loss on the basis of in vitro studies. In clinical trials involving post-menopausal osteoporotic women,[80] phosphate supplements resulted in improved Ca balance, but bone-forming surfaces decreased while bone resorbing surfaces increased. Feeding high P diets to rats,[81] mice,[82] dogs,[5] and pigs[83] results in increased bone resorption as measured by increased loss of ^{45}Ca from prelabeled bone,[81,84] decreased cortical thickness and mineral content,[85] and increased bone porosity.[86,87] Parathyroidectomy prevents the increase in bone resorption caused by feeding a high P diet to rats. Dietary phosphate-induced bone resorption is associated with increased serum P,[88] decreased serum Ca,[89] increased urinary cAMP,[90] and increased serum immunoreactive parathyroid hormone (iPTH) (secondary hyperparathyroidism). An increase in parathyroid activity is a stimulus for bone resorption in vitro and in vivo.[91,92]

Rasmussen and Bordier[70] have proposed a hypothesis to explain the action of PTH on target tissues such as bone and kidney. Parathyroid hormone interacts with membrane receptor sites, activating adenyl cyclase and causing a consequent increase in the intracellular concentration of cAMP. Cyclic AMP increases membrane permeability to Ca and activates a protein kinase which results in release of Ca from the mitochondria to the cytosol, with subsequent efflux of Ca from the cell to the extracellular fluids. Feedback regulation of intracellular cAMP concentration is provided via increased cytosolic Ca which activates phosphodiesterase to increase cAMP degradation while inhibiting adenyl cyclase.

Vitamin D increases the mobilization of Ca and P from bone to blood. In tissue cultures, $1,25(OH)_2D_3$ has been shown to increase bone resorption at physiological concentrations.[93] The increased resorption is enhanced by the presence of PTH but at high concentrations of $1,25(OH)_2D_3$ bone resorption can occur in the absence of PTH. Also, $1,25(OH)_2D_3$-induced resorption is not associated with the increase in cAMP[94,95] which is characteristic of PTH-induced bone resorption. In vivo, the effects of $1,25(OH)_2D_3$ on bone resorption and phosphate homeostasis are dependent on PTH secretion. In animals with low plasma P resulting from inadequate dietary P, $1,25(OH)_2D_3$ synthesis is enhanced,[96] while PTH levels remain low. Under these conditions the P released from one is conserved by the kidneys while the Ca released is excreted in the urine, resulting in a relatively greater increase in plasma P than in plasma Ca. $1,25(OH)_2D_3$ levels in the plasma are also increased at high P intakes as a result of the associated hypocalcemia which stimulates PTH secretion.[97]

Calcitonin administration in vivo produces decreases in plasma Ca and P. These changes are due to inhibition of bone resorption as well as increased Ca and P accumulation in bone. In tissue cultures, calcitonin results in decreased ^{45}Ca release from bone[98] as well as inactivation and reduction of osteoclasts.[98,99] Calcitonin also promotes phosphate uptake by soft tissues.[100]

A reduction in bone ash content[101,102] and cortical thickness[103] has been reported in ovariectomized animals. The presumption that these changes are related to the decrease in circulating estrogen is supported by clinical trials which have shown that estrogen replacement therapy in postmenopausal women protects against peripheral bone loss and decreases the incidence of vertebral compression[104] as well as hip fractures.[105]

DENTAL ASPECTS

P metabolism can affect dental development and maintenance as well as the integrity of the supporting structures. In animal studies, interglobular dentin, which is characterized by microscopic evidence of globular areas of incomplete mineralization in dentin, can be produced experimentally by feeding young animals vitamin D-deficient diets with distorted Ca to P ratios.[106] In humans, interglobular dentin has been reported in patients with rickets.[107] In a clinical study on children with various abnormalities in Ca and P metabolism, namely vitamin D-dependency rickets (VDDR), X-linked hypophosphatemia (XLH), and hypoparathyroidism, Nikiforuk and Fraser[108] observed that XLH patients exhibited interglobular dentin in the presence of hypophosphatemia and normolcalcemia, while VDDR patients exhibited enamel hypoplasia and interglobular dentin in the presence of hypophosphatemia and hypocalcemia. Patients with hypoparathyroidism were hypocalcemic and hyperphosphatemic; clinically, dentin development was normal and only enamel hypoplasia was observed. These authors therefore concluded that enamel hypoplasia is related to hypocalcemia, while interglobular dentin is caused by hypophosphatemia.

Phosphorus levels in saliva or oral fluids have been implicated in caries activity. In animal studies, addition of metaphosphates to the diet resulted in a significant reduction in caries.[109] In humans, Shannon[110] reported a negative correlation between decayed, missing, or filled surfaces (DMFS) rates and P concentration in the parotid fluids of young adults. However, oral administration of phosphate did not alter salivary phosphate levels significantly,[111] indicating that the cariostatic effect of phosphates is likely to be local and that their effectiveness depends on the duration and site of retention in plaque or saliva. This view is supported by studies[112] which showed that phosphates given by intubation exerted little or no caries reduction. In contrast to animal models, phosphate administration in humans has proved to be largely ineffective,[113] although in some studies[110,114] decreases in DMFS were dissociated with increased phosphate in the diet or parotid fluid. The discrepancy between animal and human data may be related to the higher phosphate levels used in animal experiments.[115] Also, the phosphate level in rat saliva (1.6 mg/dℓ) is much lower than in human saliva (12.1 mg/dℓ). Hence, rats may be more sensitive to small increases in phosphate concentration in the oral fluids.

Current theories on the etiology of periodontal disease favor local factors as the primary causative agents. It is recognized that the response to local irritants may be influenced by nutritional status,[117] but periodontal disease is prevalent in North America among patients considered to be in excellent nutritional health.

While most studies on nutritional secondary hyperparathyroidism have focused on osteoporotic bone changes, extensive alveolar bone loss and tooth mobility have been reported in dogs fed a Ca-deficient diet with a Ca to P ratio of 1:10.[106] This alveolar bone loss and periodontal disease could be prevented by feeding a Ca adequate diet (Ca to P ratio 1.3:1). On the basis of necropsy findings, Krook et al.[2] suggested that osteopenia occurs in the alveolar bone before clinical manifestation of bone loss in the vertebrae or long bones. These authors reported an association between clinical periodontal disease and spinal osteopenia. In their study, patients with clinical periodontal disease showed radiographic changes indicative of osteoporosis which were greatly increased for their chronological age. However, the study lacked age-matched controls and proper statistical analysis.

Thus, it appears that while oral hygiene and local factors are the primary agents responsible for the initiation of periodontal disease, imbalances in Ca and P homeostasis may be important in modulating the severity of bone loss.

HYPOPHOSPHATEMIA

Body P consists of inorganic and organic phosphates present intracellularly and extracellularly. Intracellular phosphates are primarily organic phosphates, such as phospholipids, phosphoglycerides, nucleic acids, and phosphoproteins, with small amounts of inorgaic phosphates which are important in the maintenance of electrical neutrality, as well as serving as substrates in the synthesis of organo-phosphates.

Although dietary deficiencies of P are rare, P depletion may occur as a result of starvation, ingestion of excessive nonabsorbable antacids,[118] or consumption of experimental diets selectively low in phosphorus. In short-term phosphorus depletion, there is tissue catabolism and plasma P can be maintained. However, if severe P restriction is continued, biochemical as well as pathological changes can occur in animals and in humans.[119, 120]

Lotz and co-workers[121] reported a P depletion syndrome produced by prolonged antacid treatment characterized by hypophosphatemia, decreased tissue phosphate, decreased ATP, hypophosphaturia, increased intestinal absorption of Ca, hypercalcemia, increased bone resorption, anorexia, malaise, weakness, and bone pain. In rats, similar changes were shown to occur independently of PTH, vitamin D,[122] and calcitonin,[123] while hypophysectomy or cessation of growth tend to blunt the effects of experimental P depletion.[119, 124]

Clinically, hypophosphatemia occurs in 10 to 15% of hospitalized patients.[125] Causes include P redistribution under the influence of glucose, insulin infusion, or respiratory alkalosis. Hypophosphatemia can also occur in cases of renal tubular disorders such as Faconi syndrome, vitamin D-resistant rickets, and hyperparathyroidism.

Under stimulation from respiratory alkalosis or glucose infusion, phosphates shift from extracellular to intracellular space due to increased cellular glucose utilization which increases the intracellular inorganic phosphate demand. Critical levels of phosphates are necessary for the production of ATP and 2,3-diphosphoglycerate (2,3-DPT) which is important in regulating hemoglobin-O_2 dissociation. Correlations between serum [Pi] and erythrocyte ATP and 2,3-DPG levels have been demonstrated.[126] Erythrocytes from patients with hypophosphatemia show increased fragility as well as changes in hemoglobin-O_2 dissociation resulting in tissue hypoxia.[127] In hospitalized alcoholic patients exhibiting hypophosphatemia following treatment with hyperalimentation, muscle weakness, rhabdomyolysis, increased creatine phosphokinase, and myoglobinuria have been reported.[128, 129] The mechanism of hypophosphatemia in alcoholics is not clear. Contributory factors include low dietary P intake, GI disturbances, metabolic acidosis, and hyperventilation associated with liver cirrhosis. Also, ethanol has been shown to stimulate release of calcitonin[130] which can inhibit P absorption and increase phosphaturia.[131, 132]

In experimental P depletion induced by feeding a P-deficient diet, liver and kidney Pi and ATP concentrations did not decrease significantly in rats until the 2nd and 6th week, respectively. The tissue response to depletion is in contrast to that of erythrocytes, which contain no mitochondria for storage of inorganic phosphorous. Vitamin D metabolites may also play a role in the maintenance of tissue ATP during the initial period of P depletion. At low plasma [Pi], $1,25(OH)_2D_3$ synthesis has been shown to be enhanced while renal cell [Pi] is decreased.[96] It has been suggested that this increase in $1,25(OH)_2D_3$ mediates the increased intestinal absorption of Ca observed in P depletion. Birge and Haddad[133] showed that $25\text{-}OHD_3$ can facilitate Pi entry into muscle cells, resulting in increased tissue ATP. Thus, severe P depletion is necessary to effect falls in cellular Pi and ATP.

HYPERPHOSPHATEMIA

The diagnosis of hyperphosphatemia must take into account the considerable variations in plasma Pi concentrations observed in normal subjects and differences associated with age and sex. Round[134] found average values of about 4.5 mg/dℓ at 7 to 10 years of age, decreasing to a mean of 3.5 mg/dℓ at 17 years, the values for girls being slightly lower than those for boys. Levels in younger children also average approximately 4.5 mg/dℓ, although values as high as 8 mg/dℓ have been reported for normal children. Adult values generally fall between 2.5 and 4.5 mg/dℓ.[135] The higher plasma Pi values in children are associated with a higher TmPi/GFR, but it is not clear to what extent this difference is attributable to their higher plasma growth hormone level.[73]

Because of the dominant role of the kidney in P homeostasis, most instances of hyperphosphatemia are the result of diseases of this organ or of abnormal renal handling of Pi arising from aberrations in parathyroid function. Clinical hyperphosphatemia is a feature of hypoparathyroidism, pseudohypoparathyroidism, acromegaly, and acute renal failure. The first three conditions are associated with an increased TmPi/GFR and the last with a fall in GFR.

Hyperphosphatemia resulting from a rise in TmPi is elicited in a logarithmic dose-responsive manner by administration of growth hormone to experimental animals[136] and is seen in states of excess growth hormone secretion such as acromegaly. The response persists in TPTX dogs, indicating that it is not mediated by inhibition of the secretion or peripheral action of PTH. Thyroxin has a permissive effect on the response to growth hormone, and it is of interest in this regard that serum phosphate is elevated in hyperthyroidism.

The decline in GFR which accompanies early progressive renal failure leads to a state of secondary hyperparathyroidism in which hyperphosphatemia is prevented by a compensating decrease in the tubular reabsorption of phosphate (TRP).[137] In the advanced stages of the disease, however, when the loss of functional nephrons reduces the GFR to 25% of normal or less, TRP no longer responds to further increases in PTH and plasma Pi accumulates.[138] Although PTH apparently plays an important role in modulation of tubular phosphate transport in renal disease, this condition also may inhibit the renal synthesis of vitamin D metabolites which affect the excretion of phosphate.

Phosphate may accumulate in the plasma as a result of PTH insufficiency arising from idiopathic or familial hypoparathyroidism or from other diseases of the parathyroid. Hyperphosphatemia is accompanied by hypocalcemia and hypophosphaturia. An increase in phosphate excretion has been observed by some investigators following Ca infusion, possibly in response to increased secretion of calcitonin, which has been reported to be hypophosphatemic and phosphaturic in hypoparathyroid subjects.

Pseudohypoparathyroidism (PHP) is a condition in which there is end organ resistance to PTH. There are several variants, distinguishable by differences in renal vs. skeletal resistance to PTH, $1,25(OH)_2D_3$ or cAMP, or by the presence or absence of facial abnormalities.[139] Hyperphosphatemia is accompanied by hypocalcemia, hypophosphaturia, and elevated serum levels of immunoreactive PTH. PHP patients generally fail to show a normal rise in blood or urine cAMP after PTH administration. Some patients respond to cAMP administration, indicating an abnormality in the adenyl cyclase system, but in other patients the pathophysiology is obscure.

Protracted hyperphosphatemia entails a risk of deposition of Ca salts in the soft tissues with serious clinical consequences. Nephrocalcinosis is readily produced in adult laboratory rats and mice by chronic feeding of a high P diet. Although the incidence of kidney calcification in rodents is increased if the diet is also low in Ca, some investigators have observed nephrocalcinosis in growing rats fed diets adequate (0.5%) or high (1.0%) in Ca and not markedly excessive in P (1.0%).[140]

The risk of soft tissue calcification can be related to the product of the concentrations of Ca and Pi in the plasma. This product tends toward constancy and any marked deviation arising from an excess of either element denotes a calcinotic state. The maximal value for the Ca $\times$ P product which is consistent with prevention of calcinosis (the concentration of each element being expressed in milligrams per deciliter) has been estimated by Nordin[141] to be approximately 70. This value may be exceeded in acute hypercalcemias, but is seldom exceeded as a result of hyperphosphatemia except in advanced renal disease.

REFERENCES

1. U.S. Department of Agriculture, Food and Nutrient Intakes of Individuals in One Day in the United States, Spring 1977, Nationwide Food Consumption Survey 1977—1978, Preliminary Report No. 2, Science and Education Administration, Hyattsville, Md., 1980.
2. **Krook, L., Whalen, J. P., Lesser, G. V., and Berens, D. L.,** Experimental studies on osteoporosis, in *Methods and Achievements in Experimental Pathology*, Vol. 7, Jasmin, G. and Cantin, M., Eds., S. Karger, Basel, 1975, 72.
3. **Draper, H. H. and Scythes, C. A.,** Calcium, phosphorus, and osteoporosis, *Fed. Proc.*, 40, 2434, 1981.
4. **Ellinger, R. H.,** Phosphates in food processing, in *Handbook of Food Additives*, 2nd ed., CRC Press, Boca Raton, Fla., 1968, 617.
5. **Laflamme, G. H. and Jowsey, J.,** Bone and soft tissue changes with oral phosphate supplement, *J. Clin. Invest.*, 51, 2834, 1972.
6. **Goldsmith, R. S., Jowsey, J., Dube, W. J., Riggs, B. L., Arnaud, C. D., and Kelly, P. J.,** Effects of phosphorus supplementation on serum parathyroid hormone, *J. Clin. Endocrinol. Metab.*, 43, 523, 1976.
7. **Krane, S. M.,** Calcium, phosphate and magnesium, in *The International Encyclopedia of Pharmacology and Therapeutics*, Rasmussen, H., Ed., Pergamon Press, London, 1970, 19.
8. **Wilkinson, R.,** Absorption of calcium, phosphorus and magnesium, in *Calcium, Phosphate and Magnesium Metabolism*, Nordin, B. E. C., Ed., Churchill Livingstone, Edinburgh, 1976, 37.
9. **Parfitt, A. M., Higgins, B., Nassium, J. R., Collins, J., and Hilb, A.,** Metabolic studies in hypercalciuria, *Clin. Sci.*, 27, 463, 1964.
10. **Glorieux, F. H., Morin, C. L., Tracers, R., Delvin, E. E,, and Poirier, R.,**Intestinal phosphate transport in familial hypophosphatemic rickets, *Pediatr. Res.*, 10, 691, 1976.
11. **Harrison, H. C. and Harrison, H. E.,** Intestinal transport of phosphate: action of vitamin D, calcium and potassium, *Am. J. Physiol.*, 201, 1007, 1961.
12. **Wasserman, R. H. and Taylor, A. N.,** Intestinal absorption of phosphate in the chick: effect of vitamin D_3 and other parameters, *J. Nutr.*, 103, 586, 1973.
13. **Harrison, H. E. and Harrison, H. C.,** Sodium, potassium and intestinal transport of glucose, L-tyrosine, phosphate and calcium, *Am. J. Physiol.*, 205, 107, 1963.
14. **Chen, T. C., Castillo, L., Korycka-Dahl, M., and DeLuca, H. F.,** Role of vitamin D metabolites in phosphate transport of rat intestine, *J. Nutr.*, 104, 1056, 1974.
15. **Taylor, A. N.,** *In vitro* phosphate transport in chick ileum: effect of cholecalciferol, calcium, sodium and metabolic inhibitors, *J. Nutr.*, 104, 489, 1974.
16. **Hurwitz, S. and Bar, A.,** Site of vitamin D action in chick intestine, *Am. J. Physiol.*, 222, 761, 1972.
17. **Ferraro, C., Ladizesky, M., Cabrejas, M., Montoreano, R., and Mautalen, C.,** Intestinal absorption of phosphate: action of protein synthesis inhibitors and glucocorticoids in the rat, *J. Nutr.*, 106, 1752, 1976.
18. **Parfitt, A. M. and Kleerekoper, M.,** The divalent ion homeostatic system: physiology and metabolism, in *Clinical Disorders of Fluid and Electrolyte Metabolism*, 3rd ed., Maxwell, M. H. and Kleeman, C. R., Eds., McGraw-Hill, New York, 1980, 269.
19. **Berner, W., Kinne, R., and Murer, H.,** Phosphate transport into brush-border membrane vesicles isolated from rat small intestine, *Biochem. J.*, 160, 467, 1976.
20. **Chow, K. W., Pond, W. G., and Lengemann, F. W.,** Effects of ammonia on the absorption and accumulation of glucose, ^{45}Ca and ^{32}P from ligated intestinal loop of the chick, *J. Nutr.*, 102, 1513, 1977.
21. **Clarkson, E. M., Luck, V. A., Hynson, W. V., Bailey, R. R., Eastwood, J. B., Woodhead, J. S., Clements, V. R., O'Riordan, V. L. H., and deWandener, H. E.,** The effect of aluminum hydroxide on calcium, phosphorus and aluminum balances, the serum parathyroid hormone concentration and the aluminum content of bone in patients with chronic renal failure, *Clin. Sci.*, 43, 519, 1972.
22. **Gallagher, J. C. and Wilkinson, R.,** The effect of ethinyloestradiol on calcium and phosphorus metabolism of post-menopausal women with primary hyperparathyroidism, *Clin. Sci.*, 45, 785, 1973.
23. **Borle, A. B., Keutmann, H. T., and Newman, W. F.,** Role of parathyroid hormone in phosphate transport across rat duodenum, *Am. J. Physiol.*, 204, 705, 1963.
24. **DeLuca, H. F. and Schnoes, H. K.,** Metabolism and mechanism of action of vitamin D, *Annu. Rev. Biochem.*, 45, 631, 1976.
25. **Rizzoli, R., Fleisch, H., and Bonjour, J. P.,** Role of 1,25-dihydroxyvitamin D_3 on intestinal inorganic phosphorus absorption in rats with normal vitamin D supply, *Calcif. Tissue Res. Suppl.*, 22, 561, 1977.
26. **Aurbach, G. D. and Potts, J. T.,** The parathyroids, *Adv. Metab. Dis.*, 1, 45, 1964.
27. **Steele, T. H., Underwood, J. L., Stromberg, B. A,, and Lamore, C. A.,** Renal resistance to parathyroid hormone during phosphorus deprivation, *J. Clin Invest.*, 58, 1461, 1976.
28. **Strickler, J. C., Thompson, O. D., Klose, R. M., and Giebisch, G.,** Micropuncture study of inorganic phosphate excretion in the rat, *J. Clin. Invest.*, 43, 1596, 1964.

136. **Durand, D., Prelot, M., and Raoul, Y.,** Effets precoces de l'hormone de croissance sur les phosphates seriques et urinaires du rat, *Experientia*, 32, 120, 1976.
137. **Goldman, R. and Bassett, S. H.,** Phosphate excretion in renal failure, *J. Clin. Invest.*, 33, 1623, 1954.
138. **Massry, S. G. and Brautbar, N.,** Renal handling of phosphate in renal failure, in *Renal Handling of Phosphate*, Massry, S. G. and Fleisch, H., Eds., Plenum Medical, New York, 1980, 307.
139. **Chesney, R. W.,** Tubular defects in phosphate reabsorption in clinical medicine, in *Renal Handling of Phosphate*, Massry, S. G. and Fleisch, H., Eds., Plenum Medical, New York, 1980, 349.
140. **Hitchman, A. J., Hasany, S. A., Hitchman, A., Harrison, J. E., and Tam, C.,** Phosphate-induced renal calcification in the rat, *Can. J. Physiol. Pharmacol.*, 57, 92, 1979.
141. **Nordin, B. E. C.,** Primary and secondary hyperparathyroidism, *Arch. Intern. Med.*, 9, 81, 1958.

Pulp

PULP COMPOSITION

Taro Hayakawa

SAMPLE PREPARATION

Dissection of Dental Pulp from Matured Teeth

Pincus[1] simply cracked open human central incisors with light hammer blows and removed the pulpal tissue with tweezers. However, histological examination showed that tiny fragments of dentin were often embedded in the pulpal tissue. These pulps were carefully examined under stereoscopic microscope in order to remove all fragments of enamel and dentin.[2]

The technique developed by Engström and Öhman[3] has offered the possibility of an easy approach to obtain pulpal tissues with a minimum of trauma. With the aid of a diamond wheel, a groove was ground into the desired cleavage plane of the enamel and cementum of the tooth. The tooth was then split with a vice provided with two pairs of sharp edges, made especially for the purpose. Whereas incisors, cuspids, and premolars were most conveniently split from the apical end, molars generally proved easier to deal with when a large portion of the crown was first split off by cleaving the tooth at right angles to its long axis and then cracking open each root separately.

Orlowski[4] lyophilized the sectioned blocks of both bovine and porcine jaws before cracking open the teeth and suggested that the drying facilitated the handling of pulps for chemical analysis.

Isolation of Odontoblasts

Gotjamanos[5] has shown that the odontoblasts of the rat incisor adhere to the pulp proper when dissecting it out, thus providing an opportunity to study the morphological interrelationship between the odontoblasts and subodontoblast layer in a dentinogenically active tooth. Linde[6] has also described a dissection technique to obtain the basal two thirds of rat maxillary incisor pulp and has found that the major part of the odontoblasts adhere to the pulpal connective tissue and the remaining odontoblasts, adhering to the dentin, could be scraped out with an excavator. On the other hand, several investigators, using teeth from several different species,[7-10] have shown that the odontoblasts tend to remain attached to the dentin when the pulp is dissected out.

Karjalainen[10] has histologically revealed that if the coronal pulp is gently removed, whole odontoblasts remain attached to the predentin surface where reparative dentin is being formed. Samples of the odontoblast layer and predentin from such areas could be obtained by scraping with a dental excavator. Kramer[8] also introduced a detailed isolation technique of dental pulps and odontoblasts from human premolars. Small groups of odontoblasts were isolated manually from the surface of human pulp with the aid of a dissection microscope.[11] Odontoblasts were isolated adhering to a sheet of membrane filter which was applied to the surface of the dental papilla of bovine mandibular third incisor and cuspid.[12] Kira[13] used a "freeze-shaving" technique to separate the odontoblast layer with a glass knife from frozen rabbit pulpal tissue.

Dental Pulp Fluid

"Dental pulp fluid" was extracted from dog teeth by drilling a hole into the pulp chamber.[14] This pulp fluid seems to correspond to the water-rich phase of the extracellular matrix.

TOTAL NITROGEN

Total nitrogen (N) contents in human, bovine, monkey, sheep, pig, goat, and dog dental pulps were determined by the micro-Kjeldahl method and no significant species differences were observed (mean value: 0.117 mg/mg of dry tissue).[1,2,15] Total N content in bovine dental pulp was also determined by the indophenol method.[16] The total N content (micro-Kjeldahl) of bovine incisor papilla showed a developmental increase.[17]

PROTEINS

The total protein contents in human permanent tooth pulps[18] and the odontoblast-predentin region of human third molars (50 intact and 80 carious)[19] were determined by the α-amino N method, and the total protein content in developing rabbit tooth germs was determined by the Folin-Ciocalteu method.[20]

Collagen

Based on hydroxyproline analysis, collagen contents in the pulps of several different species[4,18,20-23] were determined at different developmental stages. Generally speaking, the dental pulp contains relatively small amounts of collagen compared with other soft connective tissues. The collagen concentration increased significantly during the development of rabbit tooth germ[20] and human dental pulp.[18] The difference in collagen concentration between molar and incisor pulps of rabbit was statistically significant.[22] The solubility of collagen in the bovine molar pulp decreased as the development proceeded; however, this was not true in the rabbit incisor erupting continually.[23]

The collagen content of the odontoblast-predentin area in intact teeth matched that of teeth with advanced dentin caries.[19] The collagen concentrations of in vivo and xenografted pulps of 19-day fetal rat molars were compared.[21]

The presence of type III collagen besides type I collagen was initially demonstrated in bovine dental pulp[24] by detecting α_1(III)CB3 peptide. Two different types of collagen (I and III) were also identified in embryonic mouse tooth buds by CM-cellulose chromatography of pepsin-solubilized collagens and SDS-polyacrylamide gel electrophoresis of the CNBr peptides.[25]

Type I and III collagens were isolated from dental papilla and pulp of bovine teeth by pepsin digestion and differential NaCl precipitation, and were further purified by molecular sieve and ion-exchange chromatography prior to amino acid analyses.[23]

The content of type III collagen in bovine pulp was found to increase with age.[26] A small increase in the amount of type III was also observed between two stages of the development, papilla and pulp, using two different techniques; interrupted gel electrophoresis and direct estimation of α_1(III)CB3 peptide.[27] Recently, Hayakawa et al.[28] confirmed the consistent increase of type III/type I ratio during the development of bovine premolar pulp. Cultured odontogenic cells from rabbit tooth germ (RP cell) synthesized a significant amount of type III collagen comparable to type I collagen.[29]

Both dihydroxylysinonorleucine (DHLNL) and hydroxylysinonorleucine (HLNL) contents in bovine dental pulps decreased with age, however, the ratio of these reducible cross-links, DHLNL/HLNL, increased with age.[26]

Hydroxylysyl glycoside (GlcGalHyl and GalHyl) contents of bovine pulp collagen were determined by ion exchange chromatography using orcinol-H_2SO_4 reagent.[30] Imino acid contents in human pulp[31] and in the collagens extracted with hot trichloroacetate from bovine[32] and dog[33] pulps were determined.

Other Proteins

Steelmon and co-workers[34] analyzed the albumin and globulin fractions of dental pulp fluid by paper chromatography and electrophoresis. The average protein concentration in the pulp fluid was about one fifth that of venous blood plasma.[14] Relative concentrations of protein fractions, i.e., serum albumin, α_1-, α_2-, β-, and γ- globulins, and fibrinogen in these two body fluids was analyzed by electrophoresis.[35] There was but little difference in the relative amounts of the various fractions in the two fluids. The localization of these plasma proteins in human pulp was studied using fluorescent antibody technique.[36] Total protein in dog pulp fluid increased in the pulps inflamed with croton oil or turpentine oil.[37] Phosphoprotein was identified by polyacrylamide gel electrophoresis in the insoluble fraction of bovine unerupted molar pulp.[38]

LIPIDS

The total content and the gross composition of pulp lipids were studied with human[39] and young porcine[40] pulps. Recently Rabinowitz and Rossman[41] carried out an intensive study on the lipid composition of human pulp and revealed that the total lipids were 1.3% of wet weight, phospholipids were 57.4% of the total lipids, and neutral lipids were 42.6%. Cholesterol was the major lipid in the neutral lipids, much of it in free form. The cholesterol contents in human and bovine pulps increased slightly with age.[42]

Qualitative and quantitative analyses of phospholipid fractions using pulp from several different species[43-49] were made mainly by means of single- and two-dimensional thin-layer chromatography. Phosphatidylcholine was the most abundant phospholipid in all the pulps examined.

With increasing age of pulps, there was no obvious qualitative or quantitative change in the phospholipids isolated.[44,45] Also, no significant difference was found between bovine incisor and molar pulps.[45] The overall profile of pulp phospholipids was similar in all species (rat, rabbit, and cattle) examined, regardless of the rate of dentin formation or the stage of development.[50]

Fatty acid compositions of phospholipids were determined in dental pulp from different species and developmental stages.[50,51] The fatty acid compositions of purified pulp phospholipids are markedly different from those of dentin lipids which have been reported to contain little or no arachidonic acid and docosapolyenoic acids.[52,53] The subcellular distribution of phospholipids in rat incisor pulp were studied using subcellular fractionation techniques.[48]

CARBOHYDRATES

Glucose content in bovine dental papilla was determined by Somogyi-Nelson's method.[54] The glucose concentration in dog dental pulp fluid was higher than that in venous blood plasma and at the same level as in arterial plasma.[14] Pentoses were identified in bovine dental pulp and their content was determined.[55] The presence of mannose and xylose was also identified. Glycogen contents in human[56] and rat[57] pulps and bovine tooth papilla[54] were determined.

Total carbohydrate measured by the anthrone reaction accounted for 7.8% (dry weight) of porcine and 5.93% of bovine pulp.[4] The pulp, compared to other tissues, contains relatively large amounts of carbohydrates.

Glycosaminoglycans

Glycosaminoglycans (GAGs) of dental pulp from several different species were separated and quantitated using two different methods, cetylpyridinium chloride (CPC)-cellulose column technique[58-60] and a two-dimensional electrophoresis on cellulose-acetate strip.[61,62] Chondroitin sulfates and hyaluronic acid (HA) were two major constituents in all the pulps examined by several investigators[4,63-67] except for human pulp. Linde[65] showed that the major component of human dental pulp was HA, while Embery[68] reported that both chondroitin-

4-sulfate (C4S) and dermatan sulfate (DS) were major ones. Recently Sakamoto et al.[67] has revealed that DS and C4S exist in human dental pulp as the major components.

By CPC-cellulose column technique, chondroitin-6-sulfate (C6S) was detected in porcine,[63] rat incisor,[64] and human[68] pulps, and keratan sulfate (KS) was detected in bovine,[69] rat, and rabbit incisor[64,65] pulps. Neither C6S nor KS was, however, detected in bovine,[66,67] rabbit incisor and first molar,[67] and human[67] pulp by two-dimensional electrophoresis. These discrepancies seem to be attributed to the difference of analytical methodology. The absence of 3-*O*-(β-D-Δ-4,5-glucuronopyranosyl)-*N*-acetylgalactosamine-6-sulfate in the products of chondroitinase AC digestion leads to an inevitable conclusion that bovine, rabbit, and human pulp GAGs do not contain C6S.[66,67] C4S content in bovine first molar pulp increased during the eruption of the teeth and decreased sharply after the root apex was completed, whereas such a marked decrease was not obvious in the nonrooted rabbit dental pulp which was constantly rebuilding teeth.[67] These results are consistent with the possibly important role of C4S in maintaining the high concentration of Ca ion in the calcifying tissues.[70] The GAGs in the odontoblast-predentin layer, dissected out from dentinogenically active permanent porcine teeth, were isolated and analyzed.[71] Slavkin and Bavetta[21] compared the hexosamine concentrations of in vivo and xenografted pulps of 19-day fetal rat molars.

NUCLEOTIDES AND NUCLEIC ACIDS

Acid soluble nucleotides (AMP, ADP, ATP, and UDP) were identified in rabbit dental pulp using Dowex® column and paper chromatography.[72] ATP content in both normal and diseased human pulps were determined with the firefly luciferin-luciferase assay system.[73] ATP levels of developing teeth were statistically higher than that of fully formed teeth. Cyclic AMP and cyclic GMP content in normal and pulpitic human pulps were determined.[74] A statistically positive correlation was observed between pulpitic pulps and cyclic GMP concentrations.

DNA and RNA contents in developing bovine incisor[17] and rabbit tooth[20] germs and human pulp[41] were determined. DNA concentration in bovine premolar pulp was quite constant during development and showed a significant increase when attrition began.[75] DNA content of the human pulp below advanced caries was higher than pulp elsewhere.[19]

INORGANIC COMPOSITION

Dry tissue weights and water contents of young porcine,[40] bovine,[76] rabbit,[13] and human[77] pulps were determined. Water contents in bovine and human pulps decreased with age.

The ash content of young porcine[40] and human[1,77] pulp was reported. The ash content in human pulp increased approximately tenfold with age over the span of less than 20 years to over 50 years.[77] Calcium (Ca) and phosphorus (P) contents in young bovine[40] and human[1,77] pulp ash were determined. The percentages of Ca and P in human pulp ash increased with advancing age,[77] and the Ca/P ratio increased from 1.6 to 2.0. Ca content in embryonic, neonatal, and early postnatal rabbit tooth germs was determined using a Ca electrode.[20] Very rapid increase in Ca was observed between birth and fifth postnatal day.

The contents of total and inorganic phosphate and pyrophosphate in rabbit incisor pulp were analyzed and gave 1.6 to 3.5 times higher values than those in serum.[78] Total P content in bovine pulp was also determined by Fiske-Subbarow's method after the materials were hydrolyzed with $HClO_4$, indicating the presence of 0.11% (wet weight) P.[16] This value agrees fairly well with the value (0.10%) for rabbit incisor pulp,[78] but considerably lower than that (0.28 to 2.36%) reported for human pulp.[77]

Fluoride content in the ash of young porcine pulp[40] was measured by the gravimetric analysis of its Ca salt. The amount of fluoride in human pulp was analyzed by colorimetry[79] and a titration method.[77] The value in the pulp was many times greater than has been reported for other human soft tissues, and almost without exception there was an increase in the concentration of fluoride with age.[77]

Motomura[40] measured the content of potassium, sodium, magnesium, sulfur silica, and chloride in the ash of young porcine pulp. Potassium and sodium contents of human dental pulp were determined by flame photometry.[80] Spectrographic analysis of dental pulp fluid showed the presence of Mg, Fe, Ca, P, Cu, and Na.[14]

OTHERS

Histamine in adult rat, rabbit, and human pulps was separated by gel electrophoresis and determined by spectrophotofluorometry (*O*-phthalaldehyde-induced fluorescence).[81]

Rat pulp contained an appreciable amount of prostaglandins, but sex- and age-dependent differences in their concentrations were not observed.[82] The concentrations of dopamine and norepinephrine in bovine dental pulps were determined simultaneously by trihydroxyindole fluorometry.[83] The ion exchange chromatography and the spectrophotofluorometric technique were used for the quantitative determination of norepinephrine content in cat canine pulp.[84]

Presence of substance P was demonstrated by an immunofluorescence method in cat pulp.[85] Polyamines (putrescine, spermidine, and spermine) in rat incisor pulp were separated and the amounts were measured fluorometrically.[86] Pulpal putrescine concentration may change rhythmically in 24 hr.

By paper chromatography, amino acids were analyzed qualitatively in bovine dental pulps[87] and quantitatively in normal and inflamed human dental pulps.[88] In the water extract of bovine dental pulp, the following free amino acids were detected: lysine, arginine, histidine, glutamine, aspartic acid, serine, threonine, proline, valine, phenylalanine, isoleucine, and α- and γ-aminobutyric acids.[89]

Vitamin C contents in different stages of dog, rabbit, guinea pig,[90,91] and bovine[92] pulps were determined by the 2,6-dichlorophenol-indophenol method. Vitamin C content in tooth germs was generally higher than that in young pulps.[91] Lactic acid content of bovine tooth papilla were determined.[54]

REFERENCES

1. **Pincus, P.,** Some physioloical data on the human dental pulp, *Br. Dent. J.*, 89, 143, 1950.
2. **Riedesel, M. L. and Fisher, A. K.,** Nitrogen content of dental pulp, *J. Dent. Res.*, 33, 632, 1954.
3. **Engström, H. and Öhman, A.,** Studies on the innervation of human teeth, *J. Dent. Res.*, 39, 799, 1960.
4. **Orlowski, W. A.,** Analysis of collagen, glycoproteins and acid mucopolysaccharides in the bovine and porcine dental pulp, *Arch. Oral Biol.*, 19, 255, 1974.
5. **Gotjamanos, T.,** A method for isolating an intact dental pulp from rat dentine, *Arch. Oral Biol.*, 14, 729, 1969.
6. **Linde, A.,** A method for the biochemical study of enzymes in the rat odontoblast layer during dentinogenesis, *Arch. Oral Biol.*, 17, 1209, 1972.
7. **Flieder, D. E. and Fisher, A. K.,** The rate of endogenous oxygen consumption in bovine dental pulp, *J. Dent. Res.*, 34, 921, 1955.
8. **Kramer, I. R. H.,** The isolation and examination of odontoblasts in the fresh unfixed state, *Proc. R. Soc. Med.*, 49, 545, 1956.
9. **Zerlotti, E.,** Histochemical study of the connective tissue of the dental pulp, *Arch. Oral Biol.*, 9, 149, 1964.
10. **Karjalainen, S.,** A quantitative study of the *N*-L-Arginyl-2-naphthylamidase activity in the odontoblast-predentine region of intact and carious human teeth, *Arch. Oral Biol.*, 22, 13, 1977.
11. **Röckert, H.,** Methods for isolation of odontoblasts and determination of intracellular potassium, *Acta Odontol. Scand.*, 22, 373, 1964.

12. **Sasaki, S.**, Biochemical studies on the odontoblast (Japanese with English abstract), in *Kososhiki-kenkyu*, Araya, S. et al., Eds., Ishiyaku, Tokyo, 1969, 49.
13. **Kira, K.**, Studies on the succinic oxidase system of the dental pulp tissue (Japanese with English abstract), *Mon. Bull. Dent. Res.*, 35, 309, 1961.
14. **Haldi, J., Wynn, W., and Culpepper, W. D.**,Dental pulp fluid. I. Relationship between dental pulp fluid and blood plasma in protein, glucose and inorganic element content, *Arch. Oral Biol.*, 3, 201, 1961.
15. **Fisher, A. K., Wait, D. M., Chalup, M. J., and Nash, L. L.**, Further studies on the nitrogen content of dental pulp, *J. Dent. Res.*, 37, 214, 1958.
16. **Harada, M., Matsuoka, K., and Fukasawa, K.**, Phosphate and insoluble protein interaction in bovine dental pulp, *Arch. Oral Biol.*, 20, 849, 1975.
17. **Sakamoto, S., Sasaki, S., and Araya, S.**, Regional changes in nitrogen and nucleic acid levels in bovine incisors during development, *Arch. Oral Biol.*, 12, 683, 1967.
18. **Uitto, V.-J. and Ranta, R.**, Protocollagen proline hydroxylase activity in human dental pulp at various stages of development, *Proc. Finn. Dent. Soc.*, 69, 254, 1973.
19. **Karjalainen, S., Söderling, E., and Kuutti-Savolainen, E.-R.**, Predentinal collagen and pulpal prolyl hydroxylase of intact and carious human teeth, *Arch. Oral Biol.*, 24, 555, 1979.
20. **Slavkin, H. C. and Bavetta, L. A.**, Total protein, collagen, nucleic acids and calcium ions during rabbit odontogenesis, *Arch. Oral Biol.*, 13, 1465, 1968.
21. **Slavkin, H. C. and Bavetta, L. A.**, Odontogenesis *in vivo* and in xenografts on chick chorio-allantois. I. Collagen and hexosamine biosynthesis, *Arch. Oral Biol.*, 13, 145, 1968.
22. **Uitto, V.-J. and Antila, R.**, Characterization of collagen biosynthesis in rabbit dental pulp *in vitro*, *Acta Odontol. Scand.*, 29, 609, 1971.
23. **Shuttleworth, C. A., Ward, J. L., and Hirschmann, P. N.**, Extraction of collagen fractions from bovine and rabbit dental follicle, papilla and pulp, *Arch. Oral Biol.*, 23, 235, 1978.
24. **Takagi, T.**, Analysis of collagen cross-links and detection of Type III collagen in bovine periodontium, gingiva and dental pulp (Japanese), *J. Stomatol. Soc. Jpn.*, 42, 42, 1975.
25. **Lesot, H. and Rush, J. V.**, Analyse des types de collagène synthétisés par l'ébauche dentaire et ses constituants dissociés chez l'embryon de Souris (French with English abstract), *Biol. Cell.*, 34, 23, 1979.
26. **Takagi, T., Saito, S., Kuboki, Y., and Sasaki, S.**, Age-related changes of the collagen of periodontium, gingiva and dental pulp (Japanese), *Jpn. J. Oral Biol.*, 17, 432, 1975.
27. **Shuttleworth, C. A., Ward, J. L., and Hirschmann, P. N.**, The presence of Type III collagen in the developing tooth, *Biochim. Biophys. Acta*, 535, 348, 1978.
28. **Hayakawa, T., Hashimoto, Y., and Myokei, Y.**, Changes in collagen types during the development of bovine dental pulp, *J. Dent. Res.*, 58, 2279, 1979.
29. **Gotoh, Y., Saito, S., and Sato, A.**, Synthesis of procollagen by odontogenic cells of rabbit tooth germ, *Biochim. Biophys. Acta*, 587, 253, 1979.
30. **Pearson, C. H. and Ainsworth, L.**, Variations in the hydroxylysyl glycoside contents of collagens in bovine periodontal ligament and dental pulp, *J. Dent. Res.*, 57, 894, 1978.
31. **Trentalancia, M.**, L'idrossiprolina nelle polpe dentali (Italian), *Dent. Cadmos*, 35, 1497, 1967.
32. **Iwayama, Y., Kawakami, H., Ishihara, H., Okamoto, H., and Yokomizo, I.**, Imino acids contents in collagenous protein of the bovine dental pulp (Japanese), *J. Osaka Univ. Dent. Soc.*, 13, 305, 1968.
33. **Okamoto, H. and Shima, Y.**, Imino acids contents in collagenous protein of the dog dental pulp (Japanese), *J. Hiroshima Univ. Dent. Soc.*, 4, 50, 1972.
34. **Steelmon, G., Haldi, J., and Wynn, W.**, Protein fractions in dental pulp fluid and in blood plasma, *J. Dent. Res.*, 37, 66, 1958.
35. **Haldi, J. and Wynn, W.**, Protein fractions of the blood plasma and dental-pulp fluid of the dog, *J. Dent. Res.*, 42, 1217, 1963.
36. **Honjo, H., Tsubakimoto, K., Utsumi, N., and Tsutsui, M.**, Localization of plasma proteins in the human dental pulp, *J. Dent. Res.*, 49, 888, 1970.
37. **Aono, M., Yoshida, M., Munemoto, K., Okada, H., and Yokomizo, I.**, Physiopathological studies on dental pulp (Japanese with English abstract), *Osaka Univ. Dent. Soc.*, 9, 53, 1964.
38. **Fukasawa, K., Fukasawa, K. M., Hiraoka, B. Y., and Harada, M.**, Identification of the phosphoprotein in the insoluble fraction of bovine dental pulp, *J. Dent. Res.*, 55, 702, 1976.
39. **Hodge, H. C.**, Lipids of the tooth pulp, *Proc. Soc. Exp. Biol. Med.*, 35, 53, 1936.
40. **Motomura, S.**, Chemische Untersuchung der Zahnpulpa (German), *Hoppe Seylers Z. Physiol. Chem.*, 270, 33, 1941.
41. **Rabinowitz, J. L. and Rossman, S.**, Lipid composition of human dental pulp, *Arch. Oral Biol.*, 24, 477, 1979.
42. **Fisher, A. L. and Stickley, J. J.**, The cholesterol content of some bovine and human oral tissues, *J. Dent. Res.*, 39, 1037, 1960.
43. **D'Alberton, A., Miclavez, N., and Zucchi, F.**, I Fosfolipidi della Polpa Dentaria (Italian), *Minerva Stomatol.*, 13, 89, 1964.

44. **Shapiro, I. M. and Wuthier, R. E.**, A study of the phospholipids of bovine dental tissues. II. Developing bovine foetal dental pulp, *Arch. Oral Biol.*, 11, 513, 1966.
45. **Manzoli, F. A. and Gelli, M.**, Quantitative determination of lipids in the dental pulp of *Bos taurus* during development, *Arch. Oral Biol.*, 13, 705, 1968.
46. **Mule, V.**, Phospholipids in the dental pulp and periodontium of rats (Russian with English abstract), *Stomatologiia*, 51, 18, 1972.
47. **Ellingson, J. S. and Smith, M.**, Phospholipid compositions of rat, rabbit and bovine dental pulp, *Arch. Oral Biol.*, 20, 731, 1975.
48. **Mühle, W. and Doronin, P. P.**,Dünnschichtchromatographische Analyse der Phospholipide aus Pulpagewebe von Schneidezähnen der Ratte (German with English abstract), *Zahn Mund Kieferheilkd.*, 64, 561, 1976.
49. **Rabinowitz, J. L.**, Composition of human dental pulp: lipid differences in age, pathologic condition, and tooth type, *J. Endod.*, 6, 556, 1980.
50. **Ellingson, J. S. and Gibson, B. J.**, Fatty acid compositions of phospholipids in rat, rabbit and bovine dental pulp, *Arch. Oral Biol.*, 20, 735, 1975.
51. **Larson, L. R. and Ellingson, J. S.**, Fatty acid profiles of phospholipids in rabbit and bovine dental pulp, *Biochim. Biophys. Acta*, 486, 437, 1977.
52. **Prout, R. E. S. and Atkin (née Shutt), E. R.**, Effect of diet deficient in essential fatty acid on fatty acid composition of enamel and dentine of the rat, *Arch. Oral Biol.*, 18, 583, 1973.
53. **Odutuga, A. A. and Prout, R. E. S.**, Fatty acid composition of carious molar enamel and dentine from rats deficient in essential fatty acids, *Arch. Oral Biol.*, 20, 49, 1975.
54. **Sasaki, S.**, Glycolytic activity of the tooth germ, *J. Dent. Res.*, 46, 347, 1967.
55. **Graziano, V.**, Presenza e Determinazione dei Pentosi nella Polpa di Denti di Bue (Italian with English abstract) *Ann. Stomatol.*, 17, 1005, 1968.
56. **Tempestini, E. and Messina, G.**, Sulla Presenza di Glicogeno nella Polpa Dentaria (Italian), *Boll. Soc. Ital. Biol. Sper.*, 31, 915, 1955.
57. **Galluzzo, F., Frumento, F. and Siervo, R.**, Determinazione Biochimica del Glicogeno Pulpare (Italian), *Minerva Stomatol.*, 11, 496, 1962.
58. **Antonopoulos, C. A., Borelius, E., Gardell, S., Hamnström, B., and Scott, J. E.**, The precipitation of polyanions by long-chain aliphatic ammonium compounds. IV. Elution in salt solutions of mucopolysaccharide-quaternary ammonium complexes adsorbed on a support, *Biochim. Biophys. Acta*, 54, 213, 1961.
59. **Antonopoulos, C. A., Gardell, S., Szirmai, J. A., and De Tyssonsk, E. R.**, Determination of glycosaminoglycans (mucopolysaccharides) from tissues on the microgram scale, *Biochim. Biophys. Acta*, 83, 1, 1964.
60. **Anseth, A., Antonopoulos, C. A., Bjelle, A., and Fransson, L.-Å.**, Fractionation and quantitative determination of keratan sulfate using cetylpyridinium chloride and ECTEOLA-cellulose, *Biochim. Biophys. Acta*, 215, 522, 1970.
61. **Hata, R. and Nagai, Y.**, A rapid and micro method for separation of acidic glycosaminoglycans by two-dimensional electrophoresis, *Anal. Biochem.*, 45, 462, 1972.
62. **Hata, R. and Nagai, Y.**, A micro colorimetric determination of acidic glycosaminolycans by two dimensional electrophoresis on a cellulose acetate strip, *Anal. Biochem.*, 52, 652, 1973.
63. **Linde, A.**, Glycoaminoglycans (mucopolysaccharides) of the porcine dental pulp, *Arch. Oral Biol.*, 15, 1035, 1970.
64. **Linde, A.**, Glycosaminoglycans of the rat incisor pulp, *Biochim. Biophys. Acta*, 279, 446, 1972.
65. **Linde, A.**, A study of the dental pulp glycosaminoglycans from permanent human teeth and rat and rabbit incisors, *Arch. Oral Biol.*, 18, 49, 1973.
66. **Murakawa, A.**, Acid mucopolysaccharides in the bovine dental pulp, *J. Osaka Dent. Univ.*, 8, 19, 1974.
67. **Sakamoto, N., Okamoto, H., and Okuda, K.**, Qualitative and quantitative analyses of bovine, rabbit and human dental pulp glycosaminoglycans, *J. Dent. Res.*, 58, 646, 1979.
68. **Embery, G.**, Glycosaminoglycans of human dental pulp, *J. Biol. Buccale*, 4, 229, 1976.
69. **Graziano, V.**, Sul Contenuto di Mucopolisaccharidi Acidi nella Polpa di Denti di Bue (Italian), *Boll. Soc. Ital. Biol. Sper.*, 42, 324, 1966.
70. **Woodward, C. and Davidson, E. A.**, Structure-function relationships of protein polysaccharide complexes: specific ion-binding properties, *Proc. Natl. Acad. Sci. U.S.A.*, 60, 201, 1968.
71. **Linde, A.**, Glycosaminoglycans of the odontoblast-predentine layer in dentinogenically active porcine teeth, *Calcif. Tissue Res.*, 12, 281, 1973.
72. **Suzuki, K., Ito, H., Tanimoto, M., and Ishizu, Y.**, Acid soluble nucleotides of rabbit dental pulp (Japanese), *Nihon Univ. Dent. J.*, 38, 338, 1964.
73. **Le Bell, Y. and Larmas, M.**, Adenosine-5′-triphosphate levels of the human tooth pulp during health and disease, *Arch. Oral Biol.*, 24, 313, 1979.

74. **Sproles, A. C., Schilder, H., and Schaffer, L. D.,** Cyclic AMP and cyclic GMP concentrations in normal and pulpitic human dental pulps, *J. Dent. Res.*, 58, 2369, 1979.
75. **Hayakawa, T., Numata, Y., and Iijima, K.,** Changes in prolyl hydroxylase activity during the development of the bovine dental pulp, *Arch. Oral Biol.*, 23, 831, 1978.
76. **Toyama, S.,**The changes of succinic dehydrogenase in dental pulp of cows (Japanese with English abstract), *J. Tokyo Med. Coll.*, 17, 1531, 1959.
77. **Yoon, S. H., Brudevold, F., Smith, F. A., and Gardner, D. E.,** Fluoride, calcium, phosphate, ash, and water content of human dental pulps, *J. Dent. Res.*, 44, 696, 1965.
78. **Fujimoto, H., Kohno, J., and Nishida, M.,** Kato-zenshibu-shizui no rin-bunkaku ni tsuite (Japanese), *J. Kyoto Pref. Med. Univ.*, 66, 268, 1959.
79. **Police, P.,** Il Fluoro nella Polpa Dentale Umana (Italian with English abstract), *Arch. Stomatol.*, 2, 377, 1962.
80. **Messina, G.,** Potassio nella Polpa Dentaria (Italian), *Boll. Soc. Ital. Biol. Sper.*, 38, 170, 1962.
81. **Pohto, P. and Antila, R.,** Assay of histamine in dental pulps, *Acta Odontol. Scand.*, 28, 691, 1970.
82. **Hirafuji, M., Satoh, S., and Ogura, Y.,** Prostaglandins in rat pulp tissue, *J. Dent. Res.*, 59, 1535, 1980.
83. **Nakano, G., Kuzuya, H., and Nagatsu, T.,** Catecholamines in the dental pulp, *J. Dent. Res.*, 49, 1549, 1970.
84. **Pohto, P., Antila, R., and Klinge, E.,** Quantitation of transmitter noradrenaline in the pulps of cat canine teeth, *Proc. Finn. Dent. Soc.*, 68, 20, 1972.
85. **Olgart, L., Hökfelt, T., and Nilsson, G.,** Localization of substance P in the dental pulp, in *Pain in the Trigeminal Region,* Anderson, D. J. and Matthews, B., Eds., Elsevier/North-Holland Biomedical Press, Amsterdam, 1977, 49.
86. **Furuta, H.,** Separation of polyamines in rat incisor pulp and drug effects, *J. Dent. Res.*, 58, 1413, 1979.
87. **Spinella, D. J. and Fisher, A. K.,** Qualitative chromatographic analysis of amino acids of bovine dental pulp, *J. Dent. Res.*, 37, 65, 1958.
88. **Andreescu, C., Verzea, N., and Stănciulescu, A.,** Quantitative determination of aminoacids from the-human dental pulp in the normal subject and in various inflammatory disturbances, with the aid of chromatography (Rumanian with English abstract), *Stomatologia,* 20, 97, 1973.
89. **Motomura, S.,** Weitere Beitraege zu "Chemischen Untersuchungen der Zahnpulpa" (German), *J. Biochem.*, 50, 167, 1961.
90. **Aoki, C.,** Shizui no Vitamin C (Ascorbin-san) Ganryo ni kansuru Kenkyu (Japanese), *J. Nippon Dent. Assoc.*, 29, 279, 1936.
91. **Aoki, C.,** Shiga oyobi Yojaku-shizui no Vitamin C Ganryo ni tsuite (Japanese), *J. Nippon Dent. Assoc.*, 29, 832, 1936.
92. **Shirane, Y.** Studies on vitamin C in the bovine pulp in relation to dental development (Japanese with English abstract), *Tokyo Idai Z.*, 14, 121, 1956.

METABOLISM OF DENTAL PULP

M. O. Hutchins and M. L. Mailman

INTRODUCTION

Biochemically, the dental pulp remains the least well characterized of the tooth tissues. For purposes of this review, we distinguished between intracellular metabolism and extracellular matrix metabolism.

Studies of intracellular metabolism has been in two major areas: (1) respiration and (2) injury response. Respiration and ATP production have been studied chiefly because the pulp chamber is traditionally reviewed as a confined chamber with a limiting blood (and therefore oxygen) supply; such studies have been directed to understanding the metabolic response of the tissue to anoxia, which is thought to result from an inflammatory response. Other studies of cellular activity have been directed to defining the extent of the inflammatory response, using a variety of lysosomal enzyme as markers for host responses.

Many of the studies concerning pulp extracellular matrix metabolism seek to define the differentiative changes that occur in the tooth during development, aging, and disease. Both the macromolecules of the extracellular matrix and enzymes which modify them provide convenient molecular markers for changes in phenotype expression.

Interest in dental pulp has traditionally been based on a view of this tissue as functioning primarily as a reservoir of stem cells for primary or secondary dentinogenesis. Throughout the literature confusion exists in defining the role of the odontoblast. In some studies, a view that odontoblasts are terminally differentiated pulp cells prevails, and odontoblasts are considered to be a pulp cell type. In other studies, odontoblasts are clearly viewed as the cellular component of dentin. The authors of this review incline to the latter view since connective and skeletal tissues are generally defined by both their cellular and extracellular components. We have attempted to distinguish pulp and dentin and their respective secretory cells, the fibroblast and odontoblast, throughout this review.

AEROBIC RESPIRATORY RATE

Only a few respiratory studies in the dental pulp have been performed and experimental data describing the relationship between various metabolic pathways are fragmentary, but the extant data indicates that pulp metabolism proceeds via both aerobic and anaerobic pathways. Direct measurements by the Warburg method of oxygen utilization in vitro by pulp from rat incisors and bovine molars indicated that QO_2 of pulp varies considerably with age and region and is low compared to kidney cortex.[1] The QO_2 was higher in pulps of dentinogenically active teeth than their more mature counterparts and, regionally, apical pulp had a higher QO_2 than coronal tissue. In a comparative study of pulp from dentinogenically active teeth from 13 mammalian species ranging in mass from 19 g to 800 kg, the QO_2 of pulp was inversely related to body mass.[2] Pooled pulp from the tundra voles (body weight of 18 g) had a QO_2 of 12.69 compared to 1.34 for pulp from humans (body weight of about 70 kg).

Oxygen utilization in pulp appears to vary in response to environmental factors. Local anesthetics commonly used in clinical dentistry were shown to depress pulp respiratory activity in vitro at dose levels ranging from 5 to 20% but not at a dose level of 2.5%.[3] The anoxic depression of pulp respiratory activity was shown to be readily reversible using respirometric techniques in vitro;[4] the QO_2 of pulp exposed to N_2 and subsequently to O_2 was not significantly different than pulp exposed to O_2 alone over the same incubation period. When aerobic respiratory rates were measured as $^{14}CO_2$ release from pulps incubated in vitro in ^{14}C-sodium succinate, orthodontically prestressed rat incisor pulp exhibited a significantly higher respiratory rate than the nonstressed control.[5] Such changes may reflect stress-induced changes in pulpal blood flow and/or cellular metabolism. When blood flow (measured by the radioactive microsphere technique) and O_2 utilization (measured by the Warburg method) were compared in exposed pulp with and without steroid treatment, neither parameter was affected by steroid treatment.[6]

ANAEROBIC METABOLISM

The dental pulps of bovine, porcine, and human teeth appear to have a high capacity for anaerobic glucose utilization, measured either as glycolytic enzyme activity, lactic acid production in vitro, or as lactate dehydrogenase (LDH) activity in pulp homogenates.[7-9] Of particular significance is the electrophoretic identification of both the M (muscle-type) and H (heart-type) subunits in LDH isoenzymes from pulp, which indicates that normal dental pulp can metabolize glucose by either aerobic or anaerobic pathways.[9] The relative activity of different pathways of glucose utilization in pulp of developing deciduous porcine molars was assessed from the relative activities of marker enzymes for glycogenolysis, glycolysis, anaerobic glycolysis, and the pentose shunt.[8] This study suggested that glycogenolysis and aerobic glycolysis were higher in pulps from erupting and exfoliating deciduous teeth than their occlusal counterpart, but that the pentose shunt was relatively higher in the occlusal tooth and was reduced during exfoliation. Verifying studies using isotopic techniques have not been performed.

Electrophoretic comparison of LDH isoenzymes in homogenates of healthy vs. inflamed human pulp indicated total LDH activity and LDH isoenzymes containing the M subunits were elevated in severely inflamed pulp.[10] Therefore, the potential for anaerobic glycolysis appears to be increased in dental pulp in response to carious lesions. However, LDH activity, measured by Na-pyruvate reduction, increased in gangrenous pulp but not in dental pulp from normal, developing, dentin caries, or restored teeth.[11] Inflammation-induced changes in pulp enzyme activities appear to be selective since aspartate transaminase activity resembled LDH activity while malate dehydrogenase activity was unaffected. Aminoacyl peptidase activity was elevated in inflamed pulp and decreased markedly in necrotic or gangrenous tissue.[12] The metabolic significance of changes in transaminase or peptidase activity in inflamed pulp is not clear and may be related to the release of lysosomal enzymes. Alternatively, the elevation in aminoacyl peptidase in less severely inflamed pulp may be related to the release of inflammatory mediators such as bradykinin or derived from infiltrating leukocytes. Changes in LDH, cytochrome oxidase, and lysosomal enzymes were reported in incisal odontoblasts from hypocalcemic vitamin D-deficient rats,[13] in which the odontoblast-predentin layer was prepared by a specific method.[14] Although an increase in cytochrome oxidase and LDH activity was found in the affected odontoblasts, the LDH subunit distribution was unchanged indicating that a shift in anaerobic glycolysis had not occurred. In the latter study, the increase in lysosomal enzymes (acid phosphatase, cathepsin D, and hyaluronidase) as well as LDH and cytochrone oxidase activity may be related to parathyroid hormone secretion or hypocalcemia.

ADENOSINE-5′-TRIPHOSPHATE

The production of adenosine-5′-triphosphate (ATP) by anaerobic and/or aerobic pathways provides on index of pulp metabolic activity which may be related to energy requirements during pulpal development, calcification, and response to injury or infection.

A bioluminescence method was used to measure ATP levels in extracts of pulp from normal, developing, diseased, and restored human teeth and to compare ATP levels in pulp vs. dentin of carious teeth.[15,16] In healthy teeth, ATP levels in developing pulp were higher than in mature teeth [15] in correspondence with the higher O_2 utilization reported in developing teeth.[1] Additionally, the ATP levels in the pulp were 100 to 200 times higher than in dentin.[16] In diseased teeth, ATP levels were similar in inflamed and normal pulp, but were sharply decreased in gangrenous pulp.[15] In carious teeth, ATP levels were elevated in dentin but not in corresponding pulp.[16] The level of ATP in pulp during development, disease, and health appears to be related to the activity of ATP hydrolases.[17] Ca^{2+} and Mg^{2+} activated ATPase parallels the tissue levels of ATP, with high levels in pulp collected from developing teeth and decreases at the completion of root morphogenesis. In diseased pulp, both Mg^{2+}- and Ca^{2+}-stimulated ATPase activities were elevate. Since pulpal ATPase activities are increased by detergent solubilization, both appear to be membrane-associated enzymes.

CYCLIC AMP AND GMP

The cyclic nucleotides play a key regulatory role not only in receptor-mediated hormonal responses but also in the release of inflammatory mediators from cells of the affected tissues. Both cyclic nucleotides have been identified in normal and pulpitic human dental pulp.[18] In inflamed pulp cAMP levels were at normal levels but cGMP levels were significantly elevated, probably in conjunction with the cellular inflammatory response. Immunohistochemical studies indicate that cAMP is present both in odontoblasts and fibroblasts and is uniformly distributed throughout the pulp tissue during feline tooth eruption.[19] In contrast, alkaline phosphatase in the same tissues is localized in the odontoblastic-subodontoblastic zone, suggesting that the latter enzyme may play a specific role in dentinogenesis.[20]

PHOSPHOESTERASES OF PULP

Although the metabolic role of alkaline phosphatase (APase) is not clearly defined, increased levels of this enzyme are frequently found in mineralizing tissues and it is generally accepted as a histochemical enzyme marker for calcifying tissues.[21] In rodent maxillary incisors, APase was localized in the odontoblast[22] by the hydrolysis of the substrate para-nitrophenylphosphate (pNPP) and its properties were similar to bone APase. Biochemically, two forms of APase have been identified in odontoblasts and their properties have recently been reviewed.[23] The nonspecific APase, assayed with pNPP as a substrate, has been extracted from the odontoblast with Triton® X-100, indicating that it is a membrane-associated protein. Nonspecific APase is inhibited by levamisole, R 8321, 1-P-bromotetramisole and slightly by Ca^{2+} ions. The second alkaline phosphatase does not hydrolyze pNPP or inorganic pyrophosphate (PP_i), is active against ATP at alklaine pH, is unaffected by levamisole, and is activated by Ca^{2+} and Mg^{2+}.

High levels of APase appear to be always present in calcifying tissues; however, this enzyme is ubiquitious and elevated activity may also be associated with noncalcifying processes. Odontoblasts from vitamin D-deficient rats, where calcification may be impaired, exhibited a significant increase in nonspecific APase activity.[24] In a developmental study of mouse tooth germs,[25] nonspecific APase showed a progressive increase in mesenchymal cells during the bell stage and decreased thereafter to a steady state level at stages preceding the stages of odontoblast differentiation and dentinogenesis.

Both alkaline and acid phosphatases in the dental pulp of several species have been identified and partially characterized.[26-28] Ion exchange chromatography of homogenized rabbit pulp identified three forms of APase with similar properties.[26] Inhibition and electrophoretic studies of the dental pulp of monkey teeth[27] demonstrated two forms of both alkaline and acid phosphatases in the odontoblastic, subodontoblastic, and pulpal areas. Phosphatase activities were identified in normal homogenized human dental pulp with pNPP as a substrate at both acid and alkaline pH;[27] acid phosphatase, which is generally considered a marker enzyme for lyosomal activity, was located in the some subcellular fraction as APase. In healthy pulp, APase activity was higher than the acid phosphatase, but its activity was decreased in severely diseased pulp, suggesting that pulp cellular function is depressed under these conditions. Acid phosphatase was elevated in a Triton® X-100 extract of homogenized human dental pulp that was inflamed or necrotic compared to pulpal tissue from normal, developing, or restored human teeth.[29] Elevation of this enzyme was similar to that reported for another lysosomal enzyme, aminoacyl peptidase.[12]

Pyrophosphatase (PPase) is postulated to play an important role in calcification since its substrate is inorganic pyrophosphosphate, an inhibitor of mineralization. PPase has been identified in homogenates of dental pulp.[30] Its location in the supernatant fraction is distinct from that APase which is in the microsomal fraction. PPase is sensitive to Mg^{+2}, but unlike APase is not inhibited by CN^-. Pulp PPase appears to have a narrow specificity for PP_i since it does not hydrolyze GCP, ATP, PEP, or phosphocreatine.

CALCIUM-STIMULATED ATPase

Adenosine triphosphatase (ATPase) activities have been identified in dental pulp collected from the teeth of fetal calves,[31] rabbits,[32,33] rats,[34-36] and humans.[17] In the pulp, ATPase(s) have been implicated in several activities, including (1) hydrolysis of pulpal ATP to provide metabolic energy, (2) regulation of intracellular transport of Ca^{2+}, and (3) provision of either energy or phosphate for dentin mineralization (based on similar properties of pulpal APase and Ca^{2+}- and MG^{2+}-ATPase). The distribution of Ca^{2+}- and Mg^{2+}-activated ATPase in bovine dental pulp is elevated in the odontoblast region relative to the central region.[31] Pulpal ATPase appears to be a membrane-bound enzyme with properties that distinguish it from mitochondrial ATPase, from membrane-bound ATPases of other tissues, and from alkaline phosphatase.[32-36] In the rabbit, pulp ATPase is a microsomal enzyme,[32,33] while in the rat pulp, ATPase is associated with the plasma membrane.[34-36] Pulpal ATPase from rat exhibits a pH optimum of 7.9 to 8.1, is activated by either Ca^{2+} or Mg^{2+}, is not inhibited by ouabain or fluoride, and hydrolyzes the phosphorylated nucleotides. The biochemical characteristics and distribution profile of pulp ATPase are consistent with a role in dentinogenesis and/or matrix vesicle-mediated calcification, but definitive evidence based on protein purification is lacking. Alternatively, the presence of Ca^{2+}-activated ATPase throughout the pulp suggests that it may also have other functions such as the regulation of intracellular Ca^{2+} transport (the calcium pump). To this point, when mineralization rates were experimentally impaired in vitamin D-deficient rat incisors, ATPase was not concomitantly depressed.[37]

SUMMARY — CELLULAR METABOLISM

A review of the literature indicates that cellular metabolism of the dental pulp has the general characteristics of other connective tissues and can function either aerobically or anaerobically. Anaerobic pathways may be utilized preferentially in severely inflamed pulp whereas a high rate of aerobic glycolysis is characteristic of developing pulp.

The metabolic activity of several enzymes differs in the pulp proper and in the odontoblast-subodontoblast region in normal and diseased teeth. In healthy teeth, increased cGMP, APase, and a Ca^{2+}- and Mg^{2+}-dependent ATPase appear to be specifically associated with odontoblast differentiation and/or dentinogenesis. In diseased teeth, lysosomal enzyme activity is increased in both pulp and dentin. It is not clear whether tissue cells or inflammatory cells are the source of these enzymes.

PROTEIN AND COLLAGEN SYNTHESIS AND TURNOVER

Biochemical and autoradiographic tracer studies in vivo indicate that prolyl-labeled collagen and noncollagen proteins are synthesized and turned over rapidly in the dental pulp.[38-42] There is a discrepancy in comparative studies directed toward estimating the protein synthetic capacity of the pulp relative to gingival connective tissue (GCT) and periodontal ligament (PL). Biochemical studies of proline and hydroxyproline specific activity in rat oral tissues radiolabeled continuously in vivo for 24 hr indicated that net incorporation of proline into pulp was three to ten times higher than in gingiva which was in turn twice as high as PL.[38] A quantitative autoradiographic study comparing the kinetics of incorporation of proline over a 24-hr labeling period indicated that the *initial* rate of proline incorporation (in the first hour of labeling) into the PL was two to three times higher than in GCT and four to five times higher than pulp.[42] In the latter study, the relative biosynthetic activities of PL and GCT agree closely with those measured in a biochemical microassay in which pulp incorporation was not studied.[43] Although total incorporation of proline (grains per unit surface area) into dental pulp was less than into GCT, the protein synthetic capacity (grains per cell) was greater in pulp than in GCT. In the autoradiographic study, the kinetic patterns of tissue labeling were similar to those obtained by Sodek[43] for insoluble collagen of skin, GCT, and PL in that a steady state with respect to incorporated proline was achieved by 1 to 6 hr after administration of isotope with no subsequent loss or gain for 18 to 23 hr. This type of kinetic pattern indicates that active protein turnover and efficient proline reutilization as well as active biosynthesis occur in the dental pulp. The net incorporation of radiolabeled precursor in a 24-hr period is a sum of both rates of synthesis and rates of degradation and may also reflect reutilization of labeled precursor; therefore, a valid comparison of relative biosynthetic capacity must be based on the kinetics of incorporation as well as tissue differences in precursor specific activity over the course of the experiment.

Estimates of collagen turnover in rat dental pulp and in other oral connective tissues have been made by microchemical methods directed toward establishing the kinetics of disappearance of radioactive hydroxyproline from pulp prelabeled in vivo.[39,40] Half-life estimates of 6 and 13.5 days were obtained for collagen from incisor and molar pulp, respectively. In a similar study, the half-life of collagen in PL was estimated to be 9.5 days.[44]

In both rats and rabbits radiolabeled in vivo, higher rates of pulpal collagen synthesis and turnover appear to be associated with eruption. Incorporation of proline into rat incisor collagen was three times higher than into rat molar collagen.[38] The largest fraction of proline incorporation in both teeth was into noncollagen proteins with 34% as collagen in the incisor and 16% as collagen in the molar. Similar results were obtained in rabbits of several ages radiolabeled in vivo with proline where a higher hydroxyproline specific activity in all pulp collagen fractions (soluble and insoluble) was observed in the incisor compared to the molar.[41] Comparable differences were not obtained when rabbit incisor and molar pulps were labeled in vitro.[45] However, in human premolars labeled in vitro rates of pulpal collagen synthesis were higher in teeth with incomplete root development than in fully erupted teeth.[46] As noted above, the turnover of collagen in erupting incisors is nearly twice as fast as molar collagen in the rat. Thus, higher rates of collagen metabolism generally appear to be associated with tooth eruption and morphogenesis both in vivo and in vitro (except for the rabbit incisor).

Pulpal protocollagen prolyl hydroxylase (PPH) activity generally parallels patterns of pulp collagen biosynthesis. In human and bovine teeth, pulp from all developmental stages contained considerable PPH activity compared to skin and gingiva.[47] Although species and tooth variation in PPH activity was evident, the highest levels of activity were obtained in dentinogenically active, erupting teeth where crown formation was completed, and enzyme activity decreased by about 50% following root morphogenesis.[47,48]

Pulp collagen metabolism responds to injury and to a variety of therapeutic agents used clinically. Pulp exposure induces increased pulpal collagen synthesis. In human premolar pulps exposed in vivo, capped, and subsequently removed 2 to 4 weeks after exposure, collagen synthesis, measured in vitro using tracer techniques, was increased in exposed vs. intact pulp.[46] The activity of PPH was similarly increased in exposed vs. intact pulp, but in human teeth with advanced caries pulpal PPH activity was significantly depressed.[46,49] The latter suggest that pulp fibroblast synthetic function is impaired in advanced caries, since PPH activity is not itself rate-limiting for collagen synthesis.

Corticosteroids selectively inhibit pulp collagen synthesis in vitro, both in human premolar pulps previously exposed to steroid-containing capping agents in vivo and also in rabbit pulps exposed to steroids in vitro.[46] In human pulp, glucocorticoid inhibition of collagen synthesis appeared partly to result from increased degradation of newly synthesized collagen, since the specific activity of dialyzable ^{14}C-hydroxyproline was increased in pulps treated with hydrocortisone. Similarly tetracycline inhibited pulpal collagen synthesis in vitro at dose levels lower than those required to inhibit general protein synthesis.[46] PPH activity was alsoinhibited by tetracycline but was restored by the addition of excess ferrous ions; therefore, the selective inhibitory effect of tetracycline on collagen biosynthesis may be due to chelation of ferrous ions, a required cofactor for collagen prolyl hydroxylation. Several metals which are used in dental alloys also inhibit pulpal collagen biosynthesis, either by a selective effect on prolyl hydroxylation or by a general effect on protein synthesis.[51,52]

Changes in the rate of pulpal collagen biosynthesis during pulp development do not parallel patterns of net collagen accumulation in the pulp where a differential increase in collagen concentration occurs after occlusal contact has been established.[53, 54] The disparity between decreasing rates of pulp collagen synthesis and increasing net concentration of pulp collagen during development and eruption suggests that rates of collagen maturation and/or degradation play a significant role in regulating pulp collagen content.

COLLAGEN SOLUBILITY AND CROSS-LINKS

Sequential extraction of pulp indicated that the bulk of collagen is in the insoluble fibers of the extracellular matrix. In both rabbit and bovine teeth, only a small fraction of the total pulp collagen could be extracted in neutral salt or acetic acid solutions,[53,54] while the largest fraction was solubilized by limited pepsin digestion.[53] In bovine molars, pulp collagen solubility decreases as a function of developmental stage. The pepsin-susceptibility of pulp collagen is consonant with the types of reducible cross-link and the distribution of collagen types in the pulp tissue. As in other soft oral and nonoral connective tissues, the predominant reducible cross-links are dihydroxylysinonorleucine and hydroxylysinonorleucine, the latter of which is predominant. Hydroxymerodesmosine and an unidentified cross-link unique to pulp were also present.[55] The age-related decrease in pulp collagen solubility may be related to the preferential increase in Type III collagen deposition, a reduction in the rate of collagen turnover, in increase in the rate of collagen maturation, or to other factors such as variations in proteoglycans or in collagen posttranslational modifications. Developmental differences have been demonstrated in the distribution of glycosylated and unglycosylated hydroxylsyl residues in pulp collagen with higher levels of glycosylated collagen present in pulps of unerupted teeth compared to fully erupted teeth.[56]

Tracer studies in vivo indicate that pulp collagen insolubility results from rapid polymerization and cross-linking of newly synthesized collagen into the fibers of the extracellular matrix by a pathway that differs from predentin collagen biosynthesis.[53,57,58] When rabbits of various ages were radiolabeled with proline for 24 hr in vivo, significant radioactivity was present in both newly synthesized salt-soluble collagen and also in insoluble collagen fractions, indicating that significant conversion of soluble to insoluble collagen occurred during the labeling period in both molars and incisors at all ages.[53] These studies suggest that some developmental variation may occur in the rates of collagen synthesis and/or maturation since the specific activity of soluble collagen fractions was higher in incisors vs. molars at all stages, but the specific activity of the pepsin-insoluble collagen fraction varied with age.

COLLAGEN TYPES

A more detailed in vivo tracer study of the kinetics of procollagen processing in the rat incisor pulp, using SDS-PAGE fluorographic analysis of sequentially extracted collagen, indicates that the secretion and maturation of Type I and Type III procollagens follow somewhat different pathways in the pulp.[57] Type I procollagen was preferentially cleaved at the carboxyterminus leaving prominent p_N intermediates which appeared to be cross-linked. In contrast, Type III collagen was only recovered in the pepsin-solubilized fraction 1 hr after administration of isotope, suggesting an unusual mechanism for the secretion or processing of this species in the pulp. Comparison of these results from in vivo studies of rat incisor pulp with similar studies of rat incisor predentin indicates tissue specificity not only with respect to collagen type, but also in the biosynthetic pathway, i.e., in predentin Type I and Type I-trimer, but not Type III are synthesized and procollagen processing occurs via cleavage of the aminoterminal propeptide leaving transient p_C intermediates.[58]

Unlike skin and other connective tissues which contain a lesser proportion of Type III collagen in mature tissue than in fetal tissue, the proportion of Type III to Type I collagen in the dental pulp increases during development.[59,60] Interrupted gel electrophoresis and direct measurement of $\alpha 1$ (III)-CB3 from developing bovine dental pulp indicated that Type III collagen is 18% of dental follicle collagen, 20 to 24% of papilla collagen, and 28 to 31% of pulp,[59] with even higher values of 46 to 50% observed in pulps of mature bovine teeth with attrition.[60] Because Type III collagen is a major constituent of the vascular wall and the dental pulp becomes increasingly vascularized during development, the significance of increasing Type III collagen concentrations in the pulp is not clear.

The relative synthesis of collagen types in the pulp and its precursor the dental papilla have been studied using tracer techniques and immunochemical techniques in vivo and in vitro.[57,61-66] In vivo Type III procollagen constitutes a significant portion of total collagen synthesis in the rat incisor pulp but the cellular source of Type III collagen synthesis cannot be identified using biochemical techniques.[57] Immunochemical localization studies in developing mouse tooth germs, using antibodies directed against Type I collagen and Type III procollagen, indicate that Type III collagen is synthesized by the undifferentiated mesenchymal cells of the dental papilla, by preodontoblasts, and by the endothelial cells of the blood vessels, but not by terminally differentiated odontoblasts.[65,66] Similar studies have not been performed in more mature teeth. A detailed biochemical analysis of collagen synthesis in embryonic mouse tooth germs in organ culture demonstrated that at least two collagen types are synthesized, one of which is Type I and the other of which is unidentified.[64] In contrast to the dental papilla and the dental pulp, tracer studies of collagen synthesis in the odontoblast-predentin region in vivo and in vitro indicate that Type I collagen and Type I-trimer, but not Type III collagen, are synthesized by odontoblasts and secreted into the predentin matrix in approximately equal amounts in murine teeth[58,67-69] and in a 40:1 ratio in bovine molars.[70]

The distribution of collagen types synthesized by three cell lines derived from unmineralized dental tissues has been described and does not appear to correlate precisely with the in vivo distribution of collagen types in the tissue of origin. A cell line derived from rabbit incisor tooth germs synthesized 33% of total protein as collagen with Type I and Type III collagen in a ratio of 40:60, respectively, while older cultures decreased the proportion of

Type III collagen and also synthesized Type I-trimer.[62] Although the young cultures approximated the 1:1 ratio of Type III to Type I collagen observed in rat pulp,[57] the older cultures expressed a collagen phenotype more characteristic of rat odontoblasts.[58,67] Two fibroblastic cell lines derived from the pulp of rabbit incisors or premolars synthesized and secreted collagen as the predominant protein product, but only about 15% of the collagen synthesized by these cells was Type III, considerably less than in the tissue of origin.[61,63] In one study, extensive analysis of the Type I collagen fraction indicated that two forms of Type I procollagen were produced in significant amounts, which the authors concluded were the intact Type I procollagen and also the p_C-Type I intermediate; however, the absence of the α2-CB8 peptide in cyanogen bromide digest of the putative p_C species leaves open the possibility that this species may be Type I-trimer.[63] The relationship between the in vivo and in vitro biosynthesis of collagens in the pulp cells remains obscure.

DIFFERENTIATIVE CHANGES IN MATRIX CONSTITUENTS

Synthesis of collagens and other extracellular constituents, particularly those of the basement membrane, is intimately involved in odontogenic processes.[66,71-73] In the absence of ascorbate, mouse tooth germs exhibit a 65% depression in proline incorporation and fail to differentiate in organ culture, and, although the relative rate of collagen synthesis and the proportional synthesis of collagen types is maintained, the collagen synthesized in ascorbate-deficient cultures is underhydroxylated and preferentially degraded.[64,66] Transfilter labeling experiments with cultured rat or mouse tooth germs indicated that ameloblast differentiation involves synthesis and transfilter migration of collagen from the mesenchymal component of the tooth germ[74] and that odontoblast differentiation depends upon the formation of the basement membrane at the epithelio-mesenchymal interface.[75] In several experiments where basement membrane was destroyed or its formation inhibited, subsequent cellular differentiation in tooth germs failed to occur.[66,72]

Immunochemical studies of mouse tooth germs identified Type IV collagen, basement membrane proteoglycan, laminin, and fibronectin as constituents of the basal lamina both at the epithelial-mesenchymal junction and in the blood vessels investing the dental papilla.[66,76] During the predifferentiative bud and cap stages of odontogenesis the junctional basement membrane also reacts with anti-Type III procollagen,[65] but at later stages where differentiated odontoblasts are apposed to preameloblasts, Type III collagen no longer appears to be a component of the basement membrane.[66] A consideration of the temporospatial sequence of distribution of the basement membrane components as well as the effects of its experimental disruption has led Lesot et al.[66] to propose that mediated binding of the odontoblast to the basement membrane may be required for its terminal differentiation and polarization. The observation that each component of the basement membrane shows a distinct temporospatial pattern of distribution and disappearance from the epithelio-mesenchymal junction during odontoblast differentiation and dentinogenesis has led Thesleff[76] to propose that ameloblast differentiation may be inhibited in the presence of the interfacing basement membrane and permitted by its disappearance.

At odontogenic stages before and during odontoblast differentiation, indirect immunofluorescent studies indicate that there is a transition in the phenotypic expression of the mesenchymal cells.[65,66] Undifferentiated mesenchymal cells of the dental papilla appear to synthesize and secrete fibronectin as well as Types I and III collagen in a nonpolarized fashion. However, differentiated odontoblasts do not appear to synthesize Type III collagen and the secretion of fibronectin is localized to the apical region near the epithelial-mesenchymal interface. These patterns confirm biochemical studies indicating that pulp and papilla cells synthesized Types I and III collagen while odontoblasts synthesized Type I and Type I-trimer. Immunochemically a significant portion of Type III procollagen and Type IV collagen is located in the blood vessels of the papilla and the traces of A and B collagens[57] are probably likewise derived from the vascular endothelium. Laminin appears to be always located just at the epithelial perimeter of the dental epithelium but not within it, suggesting that the peripheral epithelial cells may be the source of its synthesis.

COLLAGENASE AND COLLAGENOLYSIS

Collagen turnover in the pulp appears to be mediated predominantly by an extracellular pathway involving collagenase. Collagenase has been identified in serum-free culture medium of bovine dental pulp extracts[77] and in inflamed human pulp.[78] Biochemically the pulp enzyme is similar to other vertebrate collagenases in that it is about 45,000 daltons, is inhibited by EDTA and by serum, has activity toward native collagen under physiologic conditions, and produces the α^A and α^B fragments characteristic of specific collagenase. The majority of collagenase produced by dental pulp in vitro is in a latent form and can be activated by treatment with trypsin, thiol-blocking agents, or chaotropic agents.[77] Since bovine pulp explants in serum-free culture medium secrete a factor which inhibits exogenous collagenase, pulp collagenase latency appears to result at least in part from inhibitor complex with endogenous enzyme.[79] Production of active vs. latent collagenase by the dental pulp appears to vary developmentally. Although pulp from erupted or unerupted teeth with fully formed crowns produce only latent collagenase,[77,80] explants of embryonic dental sacs secrete mostly active collagenase in culture.[81]

The cellular source of dental pulp collagenase has not been identified. In other connective tissues, biochemical and immunochemical techniques have demonstrated that the fibroblast synthesizes both collagenase and its inhibitor; the similarity between pulp and gingiva with respect to collagen synthesis and turnover invites speculation that pulp fibroblasts are the resident cells whose activity regulates pulp collagen metabolism under physiologic conditions. Cultured human pulp fibroblasts have been observed to phagocytose and digest collagen intracellularly in vitro and thus have the potential for mediating collagenolysis via an intracellular pathway.[82] However, among the extensive electronmicroscopic studies of human pulps,[83-86] only a single example of fibroblast-mediated intracellular collagenolysis has been reported, which was in an inflammed region subjacent to a carious lesion.[87]

In inflamed pulp high levels of collagenase activity have been reported in homogenates of pulps showing extensive pulpolysis.[78] However, in healthy pulp and in less severely diseased pulp, little or no collagenase activity was present. Several features of the latter study indicate that the collagenase was probably released by infiltrating inflammatory cells rather than fibroblasts. These are

1. Techniques used to obtain collagenase in this study include freeze-thawing and direct extraction, both of which have previously been effective in releasing active collagenase from neutrophils, but not connective tissues.
2. The parallel activity profiles for elastase and collagenase also indicate that the enzymes may be from a common cellular source, the most likely candidate being the neutrophil granules where elastase and collagenase are localized.
3. Histologically a significant inflammatory infiltrate of polymorphonuclear leukocytes and macrophages (both of which may produce collagenase during an inflammatory response) were observed in pulp samples which produce collagenase.

GLYCOSAMINOGLYCAN METABOLISM

The dental pulp matrix is rich in glycosaminoglycans (GAGS), but detailed biochemical studies of these have been limited to a few species in which the GAG composition has been

Table 1
GLYCOSAMINOGLYCAN CONTENT AND DISTRIBUTION IN SEVERAL SPECIES AT DIFFERENT AGES

Species	Total GAG (μg/100 μg dry pulp)	Total GAG (%)				Ref.
		HA	CS[a]	DS	CS-ABC[b]	
Human						
Fetal tooth germs	1.78[c]	66	33	0	33	92
Unerupted — apex open	Not done	60	12	28	40	94
Unerupted — apex closed	Not done	13	41	35	76	96
Erupted	0.50	2	60	34	94	93
Porcine (premolars)						
Unerupted	0.86[c]	10	81	6	87	95
Erupted	0.48	25	67	2	69	
Bovine (molar)						
2 weeks	1.44[d]	37	50	9	59	96
1 year	1.38[d]	36	51	10	61	
4 years	0.46[d]	34	46	17	63	
10 years	0.06[d]	32	20	28	48	
Rabbit						
Premolar						
3 weeks	1.62[d]	31	52	11	63	96
1 year	1.52[d]	30	50	14	64	
3 years	1.18[d]	29	47	16	63	
Incisor						
3 weeks	1.66[d]	29	53	13	66	96
1 year	1.28[d]	28	52	14	66	
3 years	1.26[d]	24	53	13	66	
Adult	Not done	24	72	0	72	94

[a] CS = chondroitin-4-sulfate + chondroitin-6-sulfate (CS-A and C).
[b] Chondroitin-4-sulfate + chondroitin-6-sulfate + dermatan sulfate.
[c] Recalculated as two times the value measured as micrograms hexosamine per 100 μg dry pulp.
[d] Recalculated as two times the value measured as micrograms uronic acid per 100 μg dry pulp.

analyzed. There are no reports on the proteoglycan composition of the dental pulp using dissociative techniques so the physical state of the GAGs in vivo is not known. As mentioned above, basement membrane proteoglycan has been localized in the vasculature of the pulp and at the interface between epithelium and mesenchyme in tooth germs, from which it disappears during dentinogenesis. In the species which have been analyzed biochemically there appears to be a consistent pattern of a decrease in total pulp GAG concentration during tooth development and eruption, confirming earlier histochemical observation.[88-91]

AGE-DEPENDENT CHANGES IN GAG

As summarized in Table 1, there is an age-dependent decrease in total glycosaminoglycan content of the pulp, measured biochemically either as hexosamine or as uronic acid after tissue solubilization and isolation of the GAG fraction(s). This decrease is most pronounced in species which contain rooted teeth and in which eruption is not continuous. In humans, total pulp GAGS decrease from nearly 2% of the tissue dry weight in fetal tooth germs[92] to about 0.5% in mature teeth[93] with intermediate values for teeth with incomplete apexification.[94] Similarly in porcine premolars, pulp GAG concentration (estimated as twice the uronic acid concentration) decreases from nearly 0.9% in unerupted teeth to about 0.5% in erupted teeth.[95] In bovine mandibular molars, pulp GAG concentration (estimated as twice the uronic acid content per 100 μg dry weight) decreased progressively from about 1.4% in 2-week calves to about 0.1% in 10-year-old adults, with a precipitous threefold decrease occurring between 1 and 4 years of life.[86] There is a decrease in pulpal GAG concentration in developing rabbit premolars from about 1.6% at 3 weeks to about 1.1% at 3 years, while in the continuously erupting rabbit incisor, a similar decrease in GAG, from 1.7 to 1.3%, was noted between 3 weeks and 3 years. The age-related decrease in GAG concentration has been interpreted to be associated with dentinogenesis and eruption.[97] However, the biochemical data from the bovine premolar show that GAGs decrease continuously with age long after the tooth has matured. Additionally, the GAG concentrations at all ages are similar in both the rabbit molar, which is nonerupting, and the incisor, which is continuously erupting. These data suggest that age or other function-related factors may also play a role in determining GAG concentration.

Age-related changes in the distribution of GAG classes have also been observed in several species. Since a variety of methods have been used by different investigators, a comprehensive comparison of developmental differences in the distribution of all GAG subclasses cannot be made, but there is sufficient data to compare the distribution of sulfated vs. unsulfated GAG. Since keratan sulfate and heparan sulfate are minor components of pulp which do not exhibit age-related changes the major sulfated GAG are the chondroitin sulfates (CS) A, B, and C. Hyaluronic acid (HA) is the only unsulfated GAG which has been identified in the dental pulp. For comparative purposes, in Table 1 a distinction has not been made between chondroitin sulfate A and C isomers (chondroitin sulfate 4 and 6, respectively), since these were not always clearly distinguished in every study. However, dermatan sulfate (chondroitin sulfate B) appears to exhibit some significant variations in some species and has therefore been included as a separate category. In human pulp unsulfated GAG represents 66% of the total in fetal tooth germs[92] and thereafter decreases from 60 to 13% during root morphogenesis[94,96] and constitutes only 2% in erupted molars.[93] In porcine premolars, the converse pattern was reported; the fraction of total pulp GAG as hyaluronate increased from 10 to 35% during eruption while sulfated GAG decreased from 87 to 69%.[95] In bovine molars hyaluronate constituted about 1/3 of total GAG throughout life while the chondroitin sulfates decreased from 61% at 1 year to 49% at 10 years.[96] In rabbits of varying ages from 3 weeks to 3 years, both incisor and premolar pulp exhibited a pattern similar to bovine molars; hyaluronic acid constituted about 30% of GAG throughout life and the chondroitin sulfates accounted for 60 to 66%.[96] It is clear that there is considerable species variation in the age-related distribution of GAG classes.

Age-related changes in the distribution of dermatan sulfate relative to the 4 and 6 isomers of chondroitin sulfate were also observed in some species, but a general pattern associated with dentinogenesis or root apexification is not apparent. In most species, dermatan sulfate is a relatively minor component and the chondroitin sulfate isomers constitute the major sulfated GAG. In humans, no dermatan sulfate was observed in tooth germs, but it constituted 30 to 35% of total GAGs in pulps of teeth with crowns during and after root morphogenesis.[92-94,96] Similarly dermatan sulfate represented an increasing fraction of total GAG in bovine molars constituting about 10% of total GAG in the 2-week calf and about 30% in the 10-year-old adult. In bovine pulp there appears to be an age-related reciprocal relationship in the distribution of dermatan sulfate vs. chondroitin sulfates since the sum of these moieties varies only slightly with age, i.e., the age-dependent decrease in chondroitin sulfates 4 and 6 is paralleled by an increase in dermatan sulfate. In the porcine premolar, the chondroitin sulfates are the major sulfated GAG constituting 81 to 67% of total GAGs in erupted and unerupted teeth, respectively; all the sulfated GAG exhibit a decrease during eruption.[95] In the pulps of rabbit premolars and incisors, no significant age-related variations were observed; about 50% of total GAG was the chondroitin sulfates and 13% was dermatan sulfate.[96]

GAG BIOSYNTHESIS AND TURNOVER

Tracer studies in vivo and in vitro indicate the glycoasminoglycans are actively synthesized and turned over in the dental pulp.[98,99] The rate of disappearance of $^{35}SO_4$ in tissue digests and GAG fractions of rat incisors prelabeled in vivo was used to estimate half-lives of 4.5 days for total incisor GAG, 4.1 days for chondroitin-4-sulfate, 5.2 days for chondroitin-6-sulfate, and 6.8 days for the fraction containing keratan sulfate and other glycoproteins.[98] These data indicate that GAG turnover time in rat dental pulp is short compared to other rat connective tissues.

The relative rates of synthesis of sulfated GAGs were estimated from in vitro incorporation of $^{35}SO_4$ into rat incisor pulp to be in the order chondroitin-6-sulfate $>$ chondroitin-4-sulfate $>$ keratan sulfate $+$ glycoproteins.[98] When relative rates of GAG synthesis were similarly estimated from ^{14}C-acetate incorporation, which is also incorporated into hyaluronic acid, the relative incorporation of acetate into the chondroitin sulfates isomers was the same as for sulfate incorporation and the specific activity of hyaluronate was lower than that of chondroitin-4-sulfate. A higher relative specific activity of the glycoprotein-containing fraction with respect to acetate vs. sulfate incorporation probably reflects the higher level of acetylation and the lower level of sulfation of the carbohydrate residues in glycoproteins. The high levels of acetate uptake suggest that considerable synthesis of glycoproteins occurs in the dental pulp.

A developmental gradient of activity with respect to sulfated GAG biosynthesis was observed when regionally defined segments of calf molars were labeled in vitro.[99] Temporally, the highest levels of sulfate incorporation were observed in pulp from embryonic third molars with lower levels in the partially erupted second molar and the fully erupted first molar. Spatially, the developing molars exhibited an increasing gradient of incorporation along the coronal to apical axis, suggesting that the highest cellular activity with respect to sulfated GAG synthesis is in the morphogenetically active apical region.

Proteoglycan degradation in vivo is thought to proceed via an initial proteolytic cleavage of the core protein and the resultant products are then probably degraded by lysosomal glycosidases and sulfatases with the required specificity for the glycosaminoglycan constituents.[97,100] Pulp contains at least one enzyme with activity against hyaluronic acid and the chondroitin sulfates.[101,105] The hyaluronidase activity in homogenates of rabbit dental pulp[105] appears to be similar to the mucopolysaccharidase activity identified histochemically in human dental pulp, dental sac, and periodontal ligament during tooth exfoliation and permanent tooth eruption.[101-104] Both activities exhibit a pH optimum between 3.5 and 5.0 and have similar substrate specificity in that both enzymes degrade chondroitin-4-sulfate and chondroitin-6-sulfate as well as hyaluronic acid in vitro. The dental pulp enzyme is a typical mammalian lysosomal hyaluronidase in that it is inhibited by iodoacetamide and ferrous ions and activated by polylysine and protamine.[105] The enzyme has properties of an endoglycosidase; the final products appear to be tetrasaccharides with a reducing terminal hexosamine.[105] Comparative studies of enzyme activity during development or disease have not been performed.

REFERENCES

1. **Fisher, A.,** Respiratory variations within the normal dental pulp, *J. Dent. Res.*, 46, 424, 1967.
2. **Fisher, A.,** Relation of body mass to oxygen consumption rate in dental pulp, *J. Dent. Res.*, 56, 78, 1977.
3. **Rockert, H.,** Effects of local anesthetics on the respiratory activity *in vitro* of cells in the dental pulp, *Scand. J. Dent. Res.*, 86, 415, 1978.

4. **Rockert, H. and Ortendal, T.,** Recovery of aerobic respiratory activity of the dental pulp after anoxic periods of two to five hours, *IRCS Med. Sci; Bioch; Cell and Mem. Biol; Dent. and Oral Biol; Path.*, 8, 99, 1980.
5. **Labart, W., Taintor, J., Dyer, J., and Weimer, D.,** The effect of orthodontic forces on pulp respiration in the rat incisor, *J. Endo.*, 6, 724, 1980.
6. **De Leon, R., Path, M., and Meyer, M.,** Blood flow and oxygen consumption in steroid-treated dental pulps, *Oral Surg.*, 45, 784, 1978.
7. **Pejrone, C.,** Anaerobic glycolysis in dental pulp, *J. Dent. Res.*, 44, 521, 1965.
8. **Siquara-Da-Rocha, M. and Nicolau, J.,** Metabolic activity of deciduous porcine dental pulp in different phases of root development, *J. Dent. Res.*, 59, 762, 1980.
9. **Linde, A. and Ljunggren, A.,** Lactate dehydrogenase isoenzyme pattern of human dental pulp, *J. Dent. Res.*, 49, 1469, 1970.
10. **Messelt, E., Skogedal, O., and Eriksen, H.,** Lactate dehydrogenase (LDH) isoenzyme pattern in normal and inflamed human dental pulp, *Acta Odontol. Scand.*, 36, 345, 1978.
11. **Le Bell, Y. and Larmas, M.,** A quantitative study of lactate and malate dehydrogenase and aspartate transaminase activities in the human dental pulp, *Arch. Oral Biol.*, 23, 925, 1978.
12. **Le Bell, Y. and Larmas, M.,** A quantitative study of α aminoacylpeptide hydrolase activity in the human dental pulp, *Arch. Oral Biol.*, 21, 195, 1976.
13. **Linde, A.,** A method for the biochemical study of enzymes in the rat odontoblast layer during dentinogenesis, *Arch. Oral Biol.*, 17, 1209, 1972.
14. **Engstrom, C.,** Odontoblast metabolism in rats deficient in vitamin D and calcium. IV. Lysosomal and energy metabolic enzymes, *J. Oral Pathol.*, 9, 246, 1980.
15. **Le Bell, Y. and Larmas, M.,** Adenosine-5′-triphosphate levels of the human tooth pulp during health and disease, *Arch. Oral Biol.*, 24, 313, 1979.
16. **Laikko, I. and Larmas, M.,** Adenosine triphosphate in normal and carious human dentine, *Arch. Oral Biol.*, 24, 15, 1979.
17. **Le Bell, Y.,**Ca^{2+}- and Mg^{2+}-activated ATP hydrolysis in human tooth pulp, *J. Dent. Res.*, 60, 128, 1981.
18. **Sproles, A., Schilder, H., and Schaffer, D.,** Cyclic AMP and cyclic GMP concentrations in normal and pulpitic human dental pulps, *J. Dent. Res.*, 58, 2369, 1979.
19. **Gustafson, G., Echerdol, O., Leever, D., Shanfeld, J., Montgomery, P., and Davidovitch, Z.,** Cyclic AMP in dental and periodontal tissues during tooth eruption in kittens, *J. Dent. Res.*, 56, 407, 1977.
20. **Choby, M., Shanfeld, J., Montgomery, P., and Davidovitch, Z.,** Simultaneous localization of cyclic AMP and alkaline phosphatase in feline dental pulp, *J. Dent. Res.*, 60 (Abstr.), 422, 1981.
21. **Linde, A. and Maagnussom, B.,** Inhibition studies of alkaline phosphatases in hard tissue-forming cells, *J. Histochem. Cytochem.*, 23, 342, 1975.
22. **Granstrom, G. and Linde, A.,** A biochemical study of alkaline phosphatases in isolated rat incisor odontoblasts, *Arch. Oral Biol.*, 17, 1213, 1972.
23. **Linde, A. and Granstrom, G.,** Odontoblast alkaline phosphatases and Ca^{2+} transport, *J. Biol. Buccale*, 6, 293, 1978.
24. **Engstrom, C., Granstrom, G., and Linde, A.,** Odontoblast metabolism in rats deficient in vitamin D and calcium. II. Changes in activities of alkaline phosphatases, *J. Oral Pathol.*, 6, 367, 1977.
25. **Larmas, M. and Thesleff, I.,** Biochemical study of changes in non-specific alkaline phosphomonoesterase activity during mouse tooth ontogeny, *Arch. Oral Biol.*, 25, 791, 1980.
26. **Fugiwara, T., Noda, H., and Suzuki, K.,** Properties of alkaline phosphatase fractions in rabbit dental pulp, *J. Dent. Res.*, 57 (Abstr.), 458, 1978.
27. **Franzen, A. and Hasselgren, G.,** Electrophoretic separation of alkaline and acid phosphatase isoenzymes from the pulp of monkey teeth, *Acta Odontol. Scand.*, 36, 371, 1978.
28. **Masi, P. and Viti, M.,** A research of human dental pulp metabolism, *Bull. Group Int. Rech. Sc. Stomat. et Odont.*, 21, 129, 1978.
29. **Le Bell, Y. and Larmas, M.,** Changes of non-specific phosphomonoesterase activity of the human dental pulp in health and disease, *Arch. Oral Biol.*, 23, 921, 1978.
30. **Furuyama, S., Mitsuma, M., Doi, N., Sugiya, H., and Oyama, S.,** Properties and distributions of inorganic pyrophosphatase in rabbit dental pulp, *J. Dent. Res.*, 56, 1239, 1977.
31. **Messer, H., Tate, G., and Reade, P.,** Distribution of Ca^{2+}-, Mg^{2+}- ATPase activity in bovine dental pulp, *J. Dent. Res.*, 57 (Abstr.), 865, 1978.
32. **Abiko, Y., Yokoyama, I., Asanuma, T., Ishido, T., and Takiguchi, H.,** Ca^{2+}-stimulated adenosine triphosphatase in dental pulp of albino rabbit, *J. Dent. Res.*, 56, 91, 1976.
33. **Abiko, Y.,** Studies on calcium-stimulated adenosine triphosphatase in the albino rabbit dental pulp. Its subcellular distribution and properties, *J. Dent. Res.*, 56, 1558, 1977.
34. **Guo, M. and Messer, H.,** Properties of Ca^{2+}-, Mg^{2+}-activated adenosine triphosphatase from rat incisor pulp, *Arch. Oral Biol.*, 21, 637, 1976.

35. **Guo, M. and Messer, H.,** A comparison of Ca^{2+}-, Mg^{2+}-ATPase and alkaline phosphatase activities of rat incisor pulp, *Calcif. Tissue Res.*, 26, 33, 1978.
36. **Guo, M. and Messer, H.,** Influence of calcium deficiency on Ca^{2+}-, Mg^{2+}-ATPase activity of rat incisor pulp, *J. Dent. Res.*, 56, 1084, 1977.
37. **Messer, H. and Guo, M.,** Lack of relation of pulp Ca^{2+}-,Mg^{2+}-ATPase to mineralization rate in rat incisor dentine in response to vitamin D and calcium intake, *Arch. Oral Biol.*, 24, 271, 1979.
38. **Orlowski, W. and Doyle, J.,** Collagen metabolism in the pulps of rat teeth, *Arch. Oral Biol.*, 24, 391, 1976.
39. **Orlowski, W.,** The turnover of collagen in the dental pulp of rat incisors, *J. Dent. Res.*, 56, 437, 1977.
40. **Orlowski, W.,** A potential for high collagen turnover in the molar pulp independent of eruption, *J. Dent. Res.*, 56, 1488, 1977.
41. **Shuttleworth, C., Ward, J., and Hirschmann, P.,** *In vivo* incorporation of ^{3}H-proline into rabbit dental pulp collagen, *Arch. Oral Biol.*, 24, 613, 1979.
42. **Kameyama, Y.,** Autoradiographic study of ^{3}H-proline incorporation by rat periodontal ligament, gingival connective tissue, and dental pulp, *J. Periodontal Res.*, 10, 88, 1975.
43. **Sodek, J.,** A new approach to assessing collagen turnover by using a microassay, *Biochem. J.*, 160, 243, 1976.
44. **Orlowski, W.,** Biochemical study of collagen turnover in rat incisor periodontal ligament, *Arch. Oral Biol.*, 23, 1163, 1978.
45. **Uitto, V. and Antila, R.,** Characterization of collagen biosynthesis in rabbit dental pulp *in vitro*, *Acta Odontol. Scand.*, 29, 609, 1971.
46. **Uitto, V., Antila, R., and Ranta, R.,** Effects of topical glucocorticoids medication on collagen biosynthesis, *Acta Odontol. Scand.*, 33, 287, 1975.
47. **Uitto, V. and Ranta, R.,** Protocollagen prolyl hydroxylase activity in human dental pulp at various stages of development, *Proc. Finn. Dent. Soc.*, 69, 254, 1973.
48. **Hayakawa, T., Numata, Y., and Iigimo, K.,** Changes in prolyl hydroxylase activity during the development of the bovine dental pulp, *Arch. Oral Biol.*, 23, 831, 1978.
49. **Karjalainen, S., Soderling, E., and Kuutti-Savolainen, E.,** Predentinal collagen and pulpal prolyly hydroxylase of intact and carious human teeth, *Arch. Oral Biol.*, 24, 555, 1979.
50. **Uitto, V.,** Effect of tetracycline on collagen biosynthesis in the dental pulp, *Acta Odontol. Scand.*, 33, 279, 1975.
51. **Helgeland, K.,** Effect of some heavy metals on protein and collagen biosynthesis in rabbit dental pulp *In vitro*, *Scand. J. Dent. Res.*, 85, 266, 1977.
52. **Helgeland, K.,** pH and the effect of fluoride and zinc on protein and collagen biosynthesis in rabbit dental pulp *in vitro*, *Scand. J. Dent. Res.*, 85, 407, 1977.
53. **Shuttleworth, C., Ward, J., and Hirschmann, P.,** Extraction of collagen fractions from bovine and rabbit dental follicle, papilla, and pulp, *Arch. Oral Biol.*, 23, 235, 1978.
54. **Hutchins, M., Mailman, M., Patten, J., and Turner, S.,** Development changes in collagen and non-collagen protein synthesis in bovine pulp, *J. Dent. Res.*, 58 (Abstr.), 247, 1979.
55. **Kuboki, Y., Takagi, T., Sasaki, S., Saito, S., and Mechanic, G.,** Comparative biochemistry of bovine periodontium, gingiva, and dental pulp, *J. Dent. Res.*, 60, 159, 1981.
56. **Pearson, C. and Ainsworth, L.,** Variations in the hydroxylysyl glycoside contents of collagens in bovine periodontal ligament and dental pulp, *J. Dent. Res.*, 57, 894, 1978.
57. **Sodek, J. and Mandell, S.,** Procollagen processing and collagen maturation in the rat incisor *in vivo*, *J. Dent. Res.*, 59 (Abst.), 549, 1980.
58. **Sodek, J. and Mandell, S.,** Formation of the collagenous matrix in rat incisor predentine, *J. Dent. Res.*, 60 (Abstr.), 637, 1981.
59. **Shuttleworth, C., Ward, J., and Hirschmann, P.,** The presence of Type III collagen in the developing tooth, *Biochim. Biophys. Acta*, 535, 348, 1978.
60. **Hayakawa, T., Hashimoto, Y., and Myokei, Y.,** Changes in collagen types during development of the dental pulp, *J. Dent. Res.*, 58 (Abstr.), 2279, 1979.
61. **Shuttleworth, C., Berry, L., and Wilson, N.,** Collagen synthesis in rabbit dental pulp fibroblast cultures, *Arch. Oral Biol.*, 25, 201, 1980.
62. **Gotoh, Y., Saito, S., and Sato, A.,** Synthesis of procollagen by odontogenic cells of rabbit tooth germs, *Biochim. Biophys. Acta*, 587, 253, 1979.
63. **Gotoh, Y., Saito, S., and Sato, A.,** Synthesis of procollagen by dental pulp cells from incisor *in vitro*, *J. Biochem.*, 86, 1037, 1979.
64. **Schiltz, J., Rosenbloom, J., and Levenson, G.,** The effects of ascorbic acid deficiency on collagen synthesis by mouse molar tooth germs in organ culture, *J. Embryol. Exp. Morphol.*, 37, 49, 1977.
65. **Thesleff, I., Stenman, S., Vaheri, A., and Timpl, R.,** Changes in matrix proteins, fibronectin, and collagen during differentiation of mouse tooth germs, *Dev. Biol.*, 70, 116, 1979.

66. **Lesot, H., Osman, M., and Ruch, V.,** Immunofluorescent localization of collagens, fibronectin, and laminin during terminal differentiation of odontoblasts, *Dev. Biol.*, 82, 371, 1981.
67. **Munksgaard, E., Rhodes, M., Mayne, R., and Butler, W.,** Collagen synthesis and secretion by rat incisor odontoblasts in organ culture, *Eur. J. Biochem.*, 82, 600, 1978.
68. **Wohllebe, M. and Carmichael, D.,** Type I trimer and Type I collagen in neutral salt-soluble lathyritic rat dentine, *Eur. J. Biochem.*, 92, 183, 1978.
69. **Wohllebe, M. and Carmichael, D.,** Biochemical characterization of guanidinium chloride-soluble dentine collagen from lathyritic rat incisors, *Biochem. J.*, 181, 667, 1979.
70. **Munksgaard, E. and Moe, D.,** Types of collagen in an extract of odontoblasts and dentine from developing teeth, *Arch. Oral Biol.*, 25, 485, 1980.
71. **Levenson, G. and Schiltz, J.,** The role of ascorbic acid on the structural integrity of developing tooth germs, *J. Biol. Buccale*, 7, 137, 1979.
72. **Thesleff, I. and Pratt, R.,** Tunicamycin-induced alterations in basement membrane formation during odontoblast differentiation, *Dev. Biol.*, 80, 175, 1980.
73. **Thesleff, I.,** Role of the basement membrane in odontoblast differentiation, *J. Biol. Buccale*, 6, 241, 1978.
74. **Wolters, J.,** The transfilter transmission of ^{3}H-proline labeled material in cultured rat tooth germs, *Arch. Oral Biol.*, 23, 51, 1978.
75. **Thesleff, I., Lehtonen, E., and Saxen, L.,** Basement membrane formation in transfilter tooth culture and its relation to odontoblast differentiation, *Differentiation*, 10, 71, 1978.
76. **Thesleff, I., Barrach, H., Foidart, J., Vaheri, A., Pratt, R., and Martin, G.,** Changes in the distribution of Type IV collagen, laminin, proteoglycan and fibronectin during mouse tooth development, *Dev. Biol.*, 81, 182, 1981.
77. **Kishi, J., Iijima, K., and Hayakawa, T.,** Dental pulp collagenase: initial demonstration and characterization, *Biochim. Biophys. Res. Commun.*, 86, 27, 1979.
78. **Morand, M., Schilder, H., Blondin, J., Stone, P., and Franzblau, C.,** Collagenolytic and elastinolytic activities from diseased dental pulps, *J. Endo.*, 7, 156, 1981.
79. **Mailman, M.,** unpublished observation 1980.
80. **Cunningham, K.,** A biochemical Study of the Effects of Glutaraldehyde and Formocresol on Bovine Pulp Tissues, Master of Science thesis, University of Texas Dental Branch, Houston, 1979.
81. **Iijima, K., Kishi, J., Fuyamada, H., Nagatsu, T., and Hayakawa, T.,** Dental sac collagenase: demonstration and initial characterization, *J. Dent. Res.*, 57, 724, 1978.
82. **Magliore, H. and Dumont, J.,** Etude Ultrastructurale de Cellules Pulpaires Humaines Cultivees *"in Vivo"*, *J. Biol. Buccale*, 4, 3, 1976.
83. **Torneck, C.,** Changes in the fine structure of the dental pulp in human caries pulpitis, *J. Oral Pathol.*, 3, 71, 1974.
84. **Harris, R. and Griffin, C.,** Histogenesis and fibroblasts in the human dental pulp, *Arch. Oral Biol.*, 12, 459, 1967.
85. **Harris, R. and Griffin, C.,** The fine structure of odontoblasts and the cell rich zone of the human dental pulp, *Aust. Dent. J.*, 3, 168, 1969.
86. **Griffin, C. and Harris, R.,** Ultrastructure of collagen fibrils and fibroblasts of the developing human dental pulp, *Arch. Oral. Biol.*, 11, 659, 1966.
87. **Torneck, C.,** Intracellular destruction of collagen in the human dental pulp, *Arch. Oral. Biol.*, 23, 745, 1978.
88. **Wislocki, G. and Sognnaes, R.,** Histochemical reactions of normal teeth, *Am. J. Anat.*, 87, 239, 1950.
89. **Fullmer, H.,** Histochemical studies of the acid mucopolysaccharides, *Oral Surg. Oral Med. Oral Pathol.*, 33, 976, 1972.
90. **Fullmer, H. and Alpher, N.,** Histochemical polysaccharide reactions in human developing teeth, *Lab. Invest.*, 7, 163, 1958.
91. **Quintarelli, G. and Dellovo, M.,** Mucopolysaccharide histochemistry of rat tooth germs, *Histochemistry*, 3, 195, 1963.
92. **Matthiessen, M. and Helin, G.,** Acidic glycosaminoglycuronoglycans of human tooth germs, *Scand. J. Dent. Res.*, 81, 171, 1973.
93. **Embery, G.,** Glycosaminoglycans of human dental pulp, *J. Biol. Buccale*, 4, 229, 1976.
94. **Linde, A.,** A study of the dental pulp glycosaminoglycans from permanent human teeth and rat and rabbit incisors, *Arch. Oral Biol.*, 18, 49, 1973.
95. **Linde, A.,** Glycosaminoglycans (mucopolysaccharides) of the porcine dental pulp, *Arch. Oral Biol.*, 15, 1035, 1970.
96. **Sakamoto, N., Okamoto, H., and Okuda, K.,** Qualitative and quantitative analysis of bovine, rabbit, and human dental pulp glycosaminoglycans, *J. Dent. Res.*, 58, 646, 1979.
97. **Linde, A.,** Glycosaminoglycans of the dental pulp. A biochemical study, *Scand. J. Dent. Res.*, 81, 177, 1973.

98. **Linde, A.,** Glycosaminoglycan turnover and synthesis in the rat incisor pulp, *Scand. J. Dent. Res.*, 81, 145, 1973.
99. **Snyder, J., Patten, J., Turner, S., Hutchins, M., and Heath, G.,** Uptake of sulfate in developing bovine dental pulp: a connective tissue model, *J. Endod.*, 5, 305, 1979.
100. **Dingle, J., Barrett, H., and Weston, P.,** Cathepsin D; characteristics of immuno-inhibition and confirmation of a role in cartilage breakdown, *Biochem. J.*, 123, 1, 1971.
101. **Alexander, S., Swerdloff, M., and Caputo, L.,** Dental pulp mucopolysaccharidase. Identification and role in root resorption, *J. Dent. Res.*, 59, 1404, 1980.
102. **Alexander, S. and Swerdloff, M.,** Partial characterization of deciduous dental pulp mucopolysaccharidase, *J. Dent. Res.*, 59, 1991, 1980.
103. **Alexander, S. and Swerdloff, M.,** Mucopolysaccharidase activity and glycosaminoglycan content in traumatized resorbing deciduous teeth, *J. Dent. Res.*, 59, 766, 1980.
104. **Alexander, S., Swerdloff, M., Ceen, R., and Bertolami, C.,** Hyaluronidase activity in human premolar and third molar dental sacs, *Arch. Oral Biol.*, 25, 207, 1980.
105. **Sakamoto, M., Nakajima, T., Ikunaga, K., Shidahara, H., Okamota, H., and Okuda, K.,** Identification of hyaluronidase activity in rabbit dental pulp, *J. Dent. Res.*, 60, 850, 1981.

Dentin

DENTIN COMPOSITION

Arthur Veis

DENTIN COMPOSITION

Introduction

Dentin is not uniform in composition, particularly with respect to mineral content. Interglobular and peritubular dentin are hypo- and hypermineralized, respectively, in comparison with the overall average mineral content. Of the organic matrix, 90% or more is collagen and the remainder is a mixture of noncollagenous proteins, glycoproteins, and phosphoproteins, with small amounts of proteoglycans, and low molecular weight polypeptides. It is likely that predentin and dentin differ in their content of noncollagenous protein constituents.

Serious study of the chemistry of the nonmineralized portion of dentin began almost simultaneously in the laboratories of Eastoe, Leaver, and Veis in the early 1960s. Several thorough reviews cover this early work.[1-3] It can be quite fairly said that the modern era of research on dentin, and the paths it has taken, have derived from the approach taken by Schlueter in the author's laboratory. In his dissertation research,[4-8] the detailed physical and chemical characterization of dentin collagen was begun, the presence of unique, phosphorylated noncollagenous components was described, and the interactions between collagenous and noncollagenous components was pointed to as an important aspect of dentin biochemistry. Leaver and colleagues[9-12] attacked the problem differently, initially emphasizing the presence of acidic peptides abundant in dentin. They later paid special attention to the nature of components released only after collagenase degradation of the matrix.

Proteoglycans and glycosaminoglycans may be important structural components of the dentin matrix but are likely to prove to have even more important regulatory roles in the predentin. Engfeldt and Hjerpe[15] and Jones and Leaver[12] led the way in the study of these tissue components.

The crucial ultrastructural studies of Weinstock and Leblond[13,14] and their visualization of the time dependence of the incorporation of proline, serine, and organic phosphate into the extracellular matrix opened up the study of the biosynthesis and secretory pathways of the matrix components. As hard-working cells engaged in extensive secretory activity during their most active phase, odontoblasts must contain a full armamentarium of intracellular enzymes. In addition, the predentin, particularly in the region of the mineralization front, is an enzymically active zone. The enzymes have not been described very well, but it is evident that degradative enzymes such as lysozyme, cathepsin D, glycosidases, and neutral proteases are present in dentin as extracellular components. Alkaline phosphatases and a calcium-dependent ATPase are also present. Where these tissue constituents are located is an open, but vital, question.[16]

Sample Preparation

The inorganic phase of human dentin constitutes, on a dry weight basis, up to about 75% of the tooth. The majority is in the form of an imperfect calcium hydroxypaptite but with a Ca/P ratio close to 2. However, there are significant amounts of Mg^{2+}, Sr^{2+}, Na^{+}, and K^{+} as cations and citrate, HCO_3^{-}, Cl^{-}, and F^{-} as anions. The mineral crystals play a crucial role in determining the physical properties of the dentin. However, to the biochemist they are an impediment. In order to study the organic matrix components the first task is to remove the mineral phase since the majority of the protein components cannot be extracted prior to dissolution of the mineral phase.

The method selected for dissolution of the hydroxyapatite depends very critically on ones ultimate purpose. However, because of the uncertain content and location of the variety of degradative enzymes listed in the introduction, unless one is purposefully intending to study these enzymes in active form, the first step in any preparative scheme is to stabilize the system against such degradation. As illustrated in Figure 1, all useful fractionation schema begin with tissue stabilization. The first step after extraction of a tooth is to cool it on ice and wash the tooth with some mix of proteolytic enzyme inhibitors. Following such a wash, mechanical cleaning of adherent tissue is desirable, and where the extracellular matrix is the desired product, scraping of the pulp cavity to disrupt the odontoblast layer is useful. Isolation of cellular constituents, such as the odonblast mRNA, requires very careful removal of the pulp to preserve odontoblast integrity. Since the odontoblastic processes pervade the entire circumpulpal dentin it is impossible to mechanically isolate dentin free from cellular contaminants.

Extraction time may be reduced if the teeth are crushed after washing, however, care must be taken to avoid local heating, and reduction to very small particle size is not required. The dentin matrix is very porous and readily accessible to extractants. In fact, it is the author's practice to reduce the teeth only to thin shards after freezing the teeth in liquid N_2.

As depicted in Figure 1, one may next proceed to dissociative extraction, essentially as practiced for proteoglycan subunit dissociative extraction in 4 *M* guanidine·HCl to produce fraction DS2, those components extractable without demineralization; DS3, the mineral bound, EDTA, guanidine soluble components; and DI3, an insoluble matrix residue. Any cellular or enamel matrix proteins would appear in DS2 along with some dentin specific proteins, but by far the largest portion of dentin specific noncollagenous proteins (NCP) appear in DS3, all of course, in a denatured state. DI3 is principally collagenous, but important NCP constituents remain.

Nondegradative extraction can be easily achieved under associative conditions as well. The pathway labeled A2-1, utilizing 0.6 *N* HCl to dissolve the hydroxyapatite, is rapid and surprisingly gentle. This is, in fact, the first step in preparing dentin matrix to retain the property of bone induction.[17] Fraction AS-2 in this case contains the major dentin phosphoprotein, plus a number of low molecular weight polypeptides.[18] The residue AI2-1 contains mainly collagen plus some constituents soluble only at neutral to slightly basic pH, producing AS3-1,2. The final insoluble residue AI3-1,2 is mainly, but not entirely collagen. AI3-1,2 is quite similar in composition to DI3, the product of dissociative extraction. Acetic acid demineralization, step A2-2, is not recommended. It is slower than HCl demineralization, and at the higher pH, several degradative reactions appear to take place. At the higher pH, phosphoproteins are not extracted so that the components of AS2-1 and AS2-2 are not identical. After extraction of AI2-2 with NaCl-Tris at neutral pH the final residue AI3-1,2 is virtually identical in composition with that from the HCl procedure, except that bound phosphoproteins might be partially degraded.

The mildest and by far most commonly preferred associative method is A2-3, utilizing 0.5 *M* EDTA at pH7.4 to dissolve the hydroxyapatite and extract the soluble NCP. AS2-3 contains a mix of components comparable to those in DS2 and DS3. The residue AI2-3 is virtually identical to DI3 if the enzyme inhibitors have been retained at all stages of extraction and washing. Procedure A2-3 does have some disadvantages. The amount of EDTA used is large, multiple extractions are required, and the volumes of solution of soluble NCP are large. To obviate this latter difficulty, the dentin may be kept in dialysis bags, but in this case extraction is slower and frequent changes of the external solution are required. Removal of the EDTA is a problem. Except under dissociative conditions the NCP appear to bind EDTA avidly. Nondissociative gel filtration will not remove the EDTA. The only nondissociative procedures for removal of the EDTA are electrophoresis or chromatography on DEAE-cellulose. Choice of dialysis membrane also affects the content of all AS2 fractions. There are low molecular weight NCP components which appear to pass the usual M_r 10,000 cut-off membranes but are retained by M_r 3500 cut-off membranes.

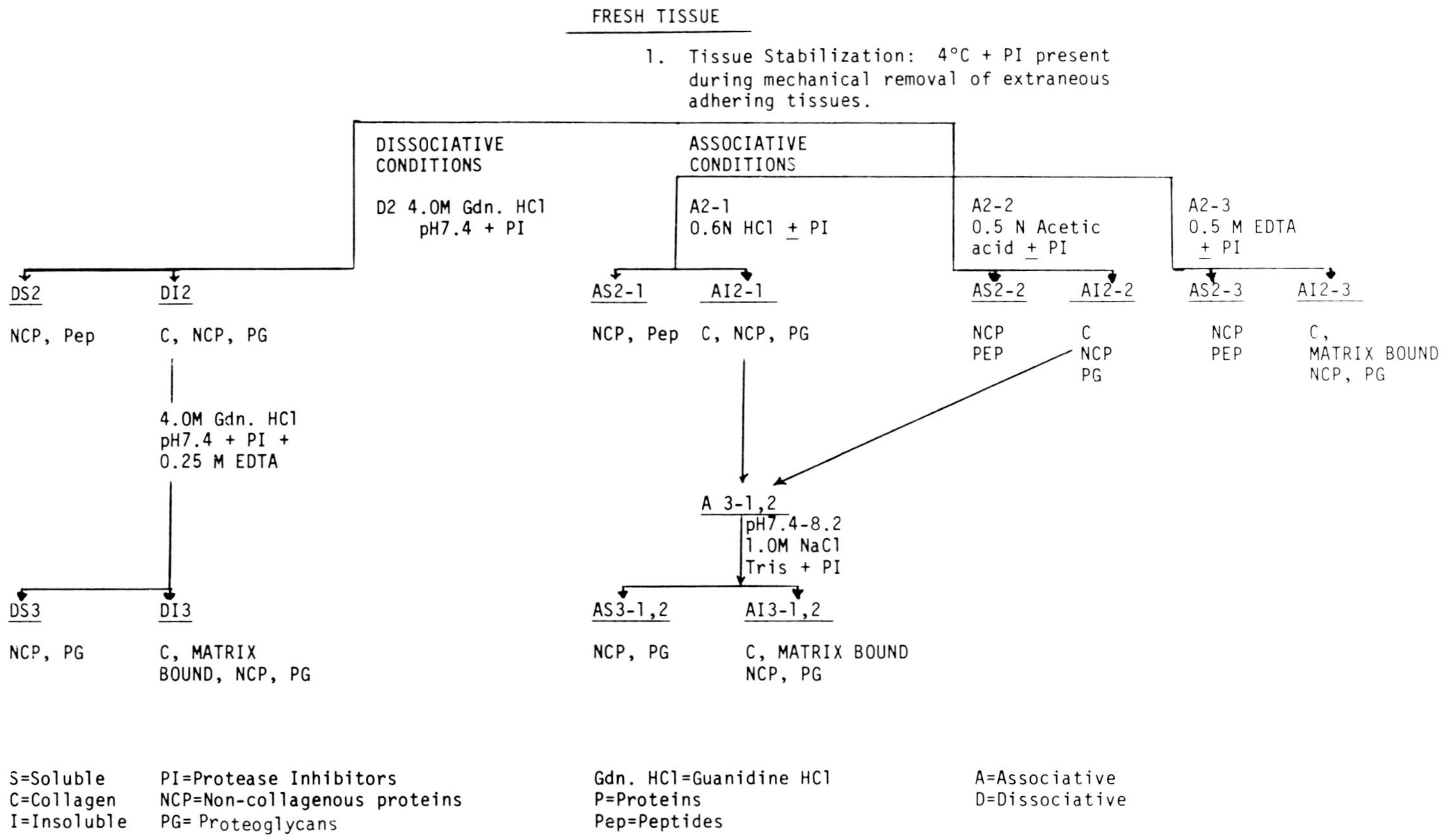

FIGURE 1. Flow diagram for the stabilization, demineralization, and extraction of teeth. All possible combinations of associative (A) or dissociative (D) methods may be used. The use of appropriate protease inhibitors through all steps is crucial.

From any of the procedures outlined in Figure 1, one is left with one or more extracts which contain all of the soluble components of the matrix in native or denatured form and in various potential stages of degradation. On a weight basis the insoluble residue, which is principally collagenous, but clearly contains several nonextractable NCP constituents, is the major constituent. In the balance of this section, we shall deal with the nature of each of the components in turn.

Collagen

Nature and Composition

Except in fetal bovine teeth[19] extracted under dissociative conditions, virtually all the collagen of dentin is found in the insoluble AI3 or DI3 fractions. This residue is exceedingly stable, resists swelling at acid pH,[7] and dissolution by pepsin and other proteases generally used to dissolve collagen fibers.[20] All the collagen is essentially Type I,[21,22] although Volpin and Veis[21] found some evidence for the presence of a small amount of Type III. Wohlebe and Carmichael[23] report the presence of some Type I trimer, $[\alpha 1(I)]_3$, as a constituent of mature dentin. The presence of such a trimer, manifest as a slight excess in α1(I) chain content, would have been impossible to detect in the CNBr degradation studies carried out by Volpin and Veis[21] and Scott and Veis.[22]

At the ultrastructural level dentin collagen fibers can be distinguished from other soft-tissue Type I collagen fibers only in terms of gross fibrillar interweaving. The individual fibers of dentin show the typical 690 nm cross-striations and periodicity characteristic of other collagen fibers at the electron microscopic level.[14] They have no peculiar staining properties. However, Katz and Li[24,25] have interpreted density and X-ray diffraction data to indicate that the pore spaces between dentin collagen fibrils are larger than in the soft tissue collagens, permitting access of phosphate ions into the fibril interior.

Detailed analyses of the corresponding CNBr peptides do show that dentin and skin collagens have some chemical differences, related to posttranslational and postsecretional modifications. Bovine skin and dentin CNBr peptides from equivalent AI2-3 preparations contained, respectively, 221 and 220 residues per 1000 of (Pro + Hyp). Hydroxylation in the dentin was greater, however, with 101 residues per 1000 Hyp in dentin compared to 95 in the skin collagen. Lysine hydroxylation was more marked, with 10 residues per 1000 Hyl in dentin compared with 6 in skin, again with the same sum of (Lys + Hyl) in both tissues, 33 residues per 1000.[21] In bovine insoluble skin, the principal sites of glycosylation were in α1(I)CB5 and α2(I)CB4, with indications of partial glycosylation in α1(I)CB6 and α2(I)CB3-5. The main derivatives in α1CB5 and α2CB4 were the dissaccharide, glucosyl-galactosyl hydroxylysine. In dentin α1(I)CB5 was also a principal site for glycosylation in dissaccharide form, but in the α2 chain, both α2(I)CB4 and α2(I)CB3-5 were glycosylated to an equivalent extent and, in this case were principally modified to monosaccharide form.[21]

Purified bovine dentin α1(I)CB6, α2(I)CB3-5, and α2(I)CB4 each contained one residue of covalently bound phosphate per peptide.[21] The α2(I)CB4 peptides from both skin and dentin were heterogenous, i.e., upon chromatography, two closely related peptides were isolated, differing in only a few residues. One peptide, true α2(I)CB4, had the requisite one third glycine content. The second peptide, α2(I)CB4′, was about 15 amino acid residues larger but had only one additional glycine. It was suggested that, in both cases, α2(I)CB4 was the site of some covalent peptide attachment. Recent work[26,27] points more to the involvement of α2(I)CB4 in intermolecular cross-linking, particularly in dentin,[26] and that the extra peptide material on α2(I)CB4 could have originated in fragments of the amino or carboxyl telopeptide. At the time of the studies of Volpin and Veis,[21] the compositions of the telopeptides were not known. The cross-linking problem will be addressed below.

A comment on the CNBr technique is in order. Application of the standard 70% formic acid digestion for 4 hr at 26°C using in excess of a 200/1 to CNBr/methionyl residue ratio,

Table 1
COMPOSITIONS OF COLLAGEN-NCP CONJUGATES ISOLATED FROM CNBr DIGESTS OF EDTA-EXTRACTED (AI2-3) INSOLUBLE BOVINE DENTIN MATRIX[a]

	Hydroxylapatite chromatography (30)	DEAE-Chromatography (31)
4-OH Pro	30	64
Asp	228	192
Thr	16	14
Ser + Ser (P)[b]	282	180
Glu	51	58
Pro	53	74
Gly	151	213
Ala	54	65
Half-Cys	5	—
Val	13	16
Met[c]	1	—
Ile	7	9
Leu	16	19
Tyr	4	4
Phe	8	10
OH Lys	4	5
Lys	34	33
His	6	8
Arg	25	33
Homoser	T[d]	2
Hexosamine	Present	N.D.[d]
Phosphorus (w/w) (%)	3.9	N.D.

[a] Results in residues per 1000 amino acid residues
[b] Sum of serine and phosphoserine. No corrections made for degradation on hydrolysis.
[c] Present as methionine sulfoxide.
[d] T, trace amount; N.D., not determined.

completely converts soluble Type I collagen to 0.1 *M* acetic acid soluble peptides. This same digestion does not convert dentin AI2-3 or DI3 completely to acetic acid soluble peptides.[21,28-30] The acetic acid insoluble digestion residue can be solubilized at pH 7.4 to 8.3. This fraction is partly collagenous but also contains additional NCP components. Chromatography on DEAE-cellulose[29] or hydroxylapatite columns[30,31] yields fractions containing collagen and NCP which can be isolated as homogeneous single components. Table 1 shows compositions determined for these components from bovine teeth. Both Curley-Joseph and Veis[30] and Fujisawa et al.[31] took pains to assure that neither free collagen CNBr peptides nor free NCP were present in the fractions whose compositions are listed. These two sets of data obviously differ in terms of the ratio of collagen to NCP, possibly as the result of incomplete cleavage of the NCP portion by CNBr in the work of Curley-Joseph and Veis (note the presence of one residue Met) or degradation of the NCP in the work of Fujisawa et al. because of a chromatography step carried out under acidic conditions at 40°C. The two sets of data are in excellent agreement, however, pointing to an NCP component rich in equimolar amounts of Asp and Ser, with most of the Ser being present as phosphoserine. Butler and colleagues have,[32,33] until recently,[34,35] disputed the presence of such conjugates in dentin, as have Jontell et al.[36] In both cases, we attribute their failure to isolate these conjugates to overzealous and degradative isolation practices.

To summarize this section, all the collagen in mature dentin is highly insoluble, appearing in DI3 or AI2-3. It is Type I, with some indications that $[\alpha 1(I)]_3$, the $\alpha 1(I)$-trimer, may be a normal, minor constituent. The Type I is slightly more highly hydroxylated in proline and

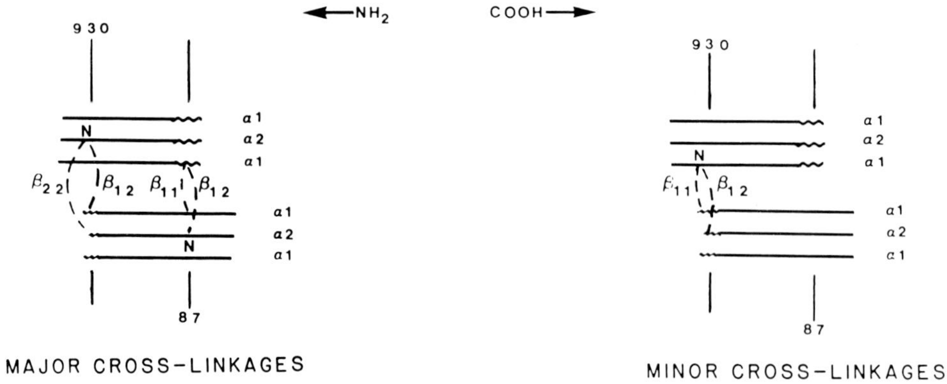

FIGURE 2. The cross-linkages of dentin collagen. The symbol N denotes the position of the amine donor for the Schiffs base formation which is the first step in each cross-linking reaction. In every case, the aldehyde group is on the telopeptides. The numbers 930 and 87 refer to the helix sequence positions of the amine donor lysyl or hydroxylysyl residues. All of these linkages are between a pair of 4D-shifted molecules. Other cross-linkages are possible, but not yet verified in detail.

much more hydroxylated with respect to lysine. Dentin and skin collagens have equivalent sites which are glycosylated, but dentin has more monosaccharide than disaccharide modifications. There is evidence that selected sites on the collagen backbone may be phosphorylated. Dentin also contains a small quantity of conjugate of collagen and an NCP, and this may also be phosphorylated.

Cross-Linking

The insolubility and structural stability of dentin collagen has been ascribed to an unusual degree of cross-linking,[7,20] but it seems now that the number of cross-linkages may be about the same as in insoluble skin or tendon and that it is the placement of these linkages which may be unique.[22,26,27,37] Scott, who has made the study of dentin cross-linking one of his major activities, concludes that the collagen fibers are stabilized by two or more intermolecular cross-linkages between 4D-shifted monomers, as depicted in Figure 2. Both α1 and α2 chains may participate in these linkages. One type cross-link involves an aldehyde of the telopeptide region of α1CB6, the carboxyl terminal end of one molecule, with the lysine or hydroxylysine on either α1CB5 or α2CB4 at residue 87, i.e., β_{11} CB6 × CB5 or β_{12} CB6 × CB5. The second and very common linkage in dentin[26] involves hydroxylysine residue 930 in the α2CB3,5 peptide, and the amino telopeptide aldehyde of an α1 or α2 chain, i.e., β_{22} CB1 × CB3,5 or β_{12} CB0,1 × CB3,5. The α1 chain helical region cross-linkages, at residue 930, β_{11} CB0,1 × CB6 and β_{12} CB6 × CB1, are also present in dentin, but in lesser amount. Scott and Edwards[37] contend that only 10 to 20% of the cross-linkages in dentin are recovered as reducible peptides, e.g., dihydroxylysinonorleucine and that stabilization through nonreducible bonds must be more prevalent. The in vivo conversion of dehydrohydroxylysinohydroxynorleucyl cross-linkages to the stable hydroxylysino-5-ketonorleucyl form is probably a major event in dentin cross-link maturation. The pyridinoline linkage described by Fujimoto et al.[38] in other tissues may be present but no studies on dentin have been carried out.

Noncollagenous Proteins (NCP)

Phosphophoryns

The principal high molecular weight NCP are a class unique to dentin, serine rich-aspartate rich proteins in which almost all seryl residues are phosphorylated. These proteins were first described in detail by Veis et al.[39] and later named phosphophoryns, phosphate-carrier

Table 2
THE COMPOSITIONS OF PHOSPHOPHORYNS FROM VARIOUS SPECIES[a]

	Fetal bovine molars[b]	Unerupted bovine molars[c]	Erupted human molars[d]	Rat incisor[e] α	Rat incisor[e] β	Rabbit incisor[f]
Asp	384	335	360	363	360	276
Thr	6	13	20	15	19	20
Ser + PSer	430	439	387	489	447	500
Glu	29	27	49	36	65	56
Pro	8	18	49	8.5	20	9
Gly	39	35	46	39	71	49
Ala	14	14	6.8	14	29	13
1/2 Cys	T[g]	1.4	15	0.6	—	—
Val	7	9	8.5	2.4	6.7	5.3
Met	T	4.2	—	0.9	2.5	—
Isoleu	6	5.6	3.6	2.5	4.4	4.0
Leu	8	13	8.2	2.4	5.7	5.9
Tyr	2	5.4	4.4	3.5	4.7	2.8
Phe	2	5.3	4.9	1.4	3.9	5.2
Lys	54	53	13.4	10.1	7.9	43
His	7	7.4	14.7	9.5	10.2	4.4
Arg	5	8.8	8.4	1.0	3.9	6.7

[a] Residues per 1000 amino acid residues.
[b] Termine et al.[19]
[c] Stetler-Stevenson and Veis.[61]
[d] Takagi et al.[62]
[e] Dimuzio and Veis.[40]
[f] Richardson et al.[63]
[g] T, trace.

proteins, by Dimuzio and Veis.[40] Table 2 shows the compositions of phosphophoryns isolated from various species and by various laboratories. The data cited are those derived from minimally degraded preparations. There has been a debate over the composition of the phosphophoryns and it has been claimed that "pure" phosphophoryns contain, for example, no aromatic amino acids.[41] These arguments have been laid to rest by the isolation of high molecular weight phosphophoryns, the use of protease inhibited extractants, and prevention of the degradation by acid hydrolysis.[19,35,36,40]

Almost all of the phosphorus is present as orthophosphate esters of seryl residues.[42,43] A maximum of ~1% of the phosphate in rat incisor dentin could be present as *O*-phosphothreonine.[43] Upon hydrolysis in 6 *N* HCl in preparation for amino acid analysis some serine is destroyed but, more importantly, a substantial fraction of the phosphoseryl residues are hydrolyzed to serine. The appropriate correction factors are described by Richardson et al.[43] All the phosphate may be released by β-elimination, using a reaction mixture of 0.2 *M* NaOH and 0.3 *M* $NaBH_4$ at 37° for 12 to 24 hr.[43]

The soluble phosphophoryns vary in amount in dentins from different species. The organic matrix of rat incisor dentin contains 10 to 11% by weight, bovine molars ~4%, and normal human molars contain 2.5%. The majority of the other soluble components of fractions DS2, AS2-1, and AS2-3 are also acidic, but can be separated to a large extent by chromatography on DEAE-cellulose.[39,40] At pH 8.2 in 0.05 *M* Tris HCl and 40°C, the phosphophoryns characteristically elute between 0.26 and 0.30 *M* NaCl, whereas the other glycoproteins elute at salt concentrations below 0.26 *M*. Proteoglycans, on the other hand, elute at about 0.55 *M* NaCl in this system. Dimuzio and Veis[40] devised a high resolution system for isolation of phosphophoryns on this basis.

Unfortunately, the molecular weight degradation is frequently evident because the lability of the phosphophoryn leads to fragments of similar character. All of these fragments elute on DEAE-cellulose at the same position. Molecular weight fractionation is therefore very necessary. Further, some nonphosphophoryn components bind and coelute with phosphophoryn unless dissociative conditions are used. Because of their random chain conformation and high charge density, the phosphophoryns frequently chromatograph anomalously. The system of Termine et al.[19] using Sephacryl® S300 with 4.0 *M* guanidine.HCl provides a very clean high molecular weight preparation. Sephacryl® S300 with a 0.34 *M* glycine, 0.05 *M* Tris.HCl, in 0.1% SDS at pH 8.3[44] is also useful, particularly since the eluting buffer is also a good gel electrophoresis buffer and chromtographic fractions can be analyzed directly. As noted very early,[39] phosphophoryn does not stain with the usual protein stains. Coomassie® blue does not stain phosphophoryn.[33] Alcian® blue[33,39] is effective, and even more so, Stains-all is effective because it gives a distinctive blue color for phosphorylated proteins and a reddish-blue for more usual proteins.[19,44] The relative molecular weight of bovine phosphophoryn is 158,000 based on gel electrophoresis. Recently Stetler-Stevenson and Veis[45] have used gel filtration on high-pressure liquid chromatography Biosil® TSK400 columns in 4.0 *M* guanidine hydrochloride to obtain very good resolution of bovine and rat phosphophoryns.

Proteoglycans

Dentin contains some proteoglycans. The glycosaminoglycan components comprise about 2.5 to 5.0% of the noncollagenous components[2, 15] and less than 0.05% of the entire organic matrix.[46] Chondroitin-4-sulfate, chondroitin-6-sulfate, dermatan sulfate, and keratan sulfate have all been identified in the organic matrix. Of these, chondroitin-4-sulfate is present in the largest amount.[15,46] Chondroitin-6-sulfate may be the major constituent in predentin, degraded as the mineralization front passes through a predentin zone. Only chondroitin-4-sulfate is found in DS3, i.e., in the fraction where mineral has been dissolved, while chondroitin-6-sulfate has been found in DS2.[47] One recent report[48] suggests that heparan sulfate may be present in large amount. Linde[49] has demonstrated the presence of large amounts of keratan sulfate in the predentin of porcine teeth.

There has been a resurgence of interest in the function of the glycosaminoglycan components of dentin and the techniques for their analysis have been improved, along with further insight into the nature of the proteoglycans in other tissues. It is likely that studies now in progress in several laboratories will provide more specific information on the glycosaminoglycan composition of dentin and predentin. The most certain statement that can be made at this time is the general observation that these two regions do differ in glycosaminoglycan content and composition.

The core protein of the dentin matrix proteoglycan has a high content of seryl and aspartyl residues[12] (Table 3) and comprises about 6% by weight of the proteoglycan moiety.

Glycoproteins

Dissociative extract DS3[19] and associative extract AS2-3[50] are very similar in composition and contain a number of glycoproteins ranging in molecular weight from greater than 100,000 to less than 14,000. The phosphophoryns, as already indicated, are the major constituents, comprising about 60% of the fraction.[35] Gel electrophoresis of AS2-3 or DS3 show more than 20 Coomassie® blue staining bands attributable to dentin and representing all the noncollagenous, nonphosphophoryn proteins. Butler et al.[35] used the fractionation scheme shown in Figure 3 to separate the soluble components. Fraction AS2-3-I contains the phosphophoryns, fraction AS2-3-II from rat incisor is similar to phosphophoryn in composition except that it has only 73 residues PSer per 1000 residues. This lower degree of phosphorylation appears to make this component not precipitable by 1.0 *M* $CaCl_2$. Chro-

Table 3
REPRESENTATIVE AMINO ACID COMPONENTS OF NONCOLLAGENOUS PROTEINS FROM DENTINS OF VARIOUS SOURCES[a]

	Human[b]			Rat incisor[c]		
	Proteoglycan core protein	Molars (20K)	AS2-2,II (14K)	AS2-2,II	AS2-3,III (95K)	AS2-3,IV Gla-Prot
4-OH Pro	—	—	—	—	—	10
Asp	206	257	254	253	165	161
Thr	83	42	32	37	65	47
Ser	154	180	190	284[d]	111	73
Glu	154	138	180	140	180	149
Pro	45	49	51	37	43	79
Gly	167	102	126	65	156	101
Ala	61	47	34	36	66	70
Val	37	37	32	20	34	38
1/2 Cys	—	—	—	—	—	23
Met	T[e]	T	T	T	T	5
Isoleu	14	18	23	12	26	29
Leu	19	35	39	23	36	67
Tyr	2	12	36	16	6	32
Phe	10	12	13	12	8	19
OH Lys	—	—	—	—	—	—
Lys	27	31	12	25	43	30
His	23	9	10	14	33	34
Arg	17	27	18	27	28	35

[a] Residues per 1000 amino acid residues.
[b] Jones and Leaver.[12]
[c] Butler et al.[35]
[d] Sum of serine and phosphoserine.
[e] T, trace amount.

matography on Sephadex® G-50 separates the remaining proteins by molecular weight into two groups. Within the G-50 void volume the most interesting components are proteins with weights 95K, 20K, and 14K. Table 3 shows the compositions of representative rat incisor and human molar proteins. All have characteristically high Asp contents but, in contrast to the phosphophoryns, have high Glu contents as well. The composition of the core protein of the dentin proteoglycan fits this same pattern. The 95K glycoprotein is 33 wt % carbohydrate, bound in at least two types of linkage to the core protein. In one case the polysaccharide chains are linked to asparaginyl residues by *N*-acetylglucosamine. The other chains are linked to seryl or threonyl residues via *O*-glycosidic bonds.

Gamma-Carboxyglutamic Acid Containing Protein

A low molecular weight protein, in fraction AS2-3-IV, and well included in Sephadex® G-50, contains γ-carboxyglutamic acid. Its composition is shown in Table 3. While its sequence has not been determined it is probable that this Gla-protein is similar in sequence to the bone Gla-protein[51] and that it will prove to contain 49 to 51 amino acid residues homologous with the bone protein. However, the AS2-3-IV fraction as isolated by Butler et al.[35] has three closely related components with weights ranging from molecular weight 10,000 to 15,000. Perhaps the dentin Gla-protein is produced in a precursor form which is processed differently from the bone Gla-protein. As in other mineralized tissues the function of the Gla-protein, which accounts for ~5% of the NCP,[35] is unknown.

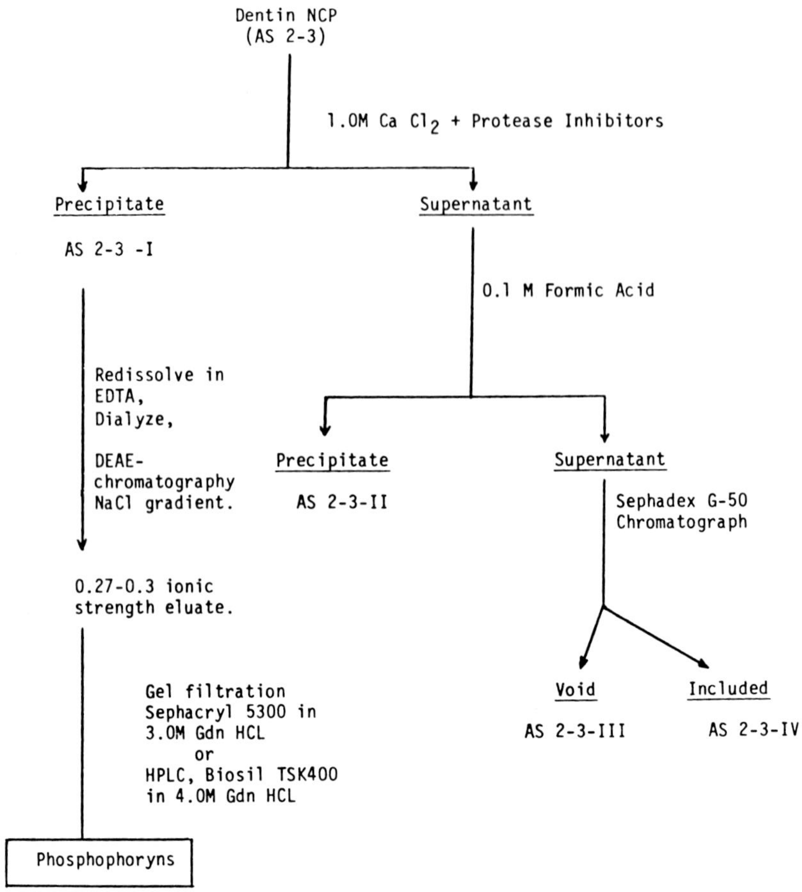

FIGURE 3. The Butler et al.[35] scheme for fractionation of the soluble proteins obtained during the demineralization of dentin.

Exogenous Proteins

All of the proteins mentioned above are endogenous to the dentin and presumably are synthesized directly by the odontoblasts. Transferrin, immunoglobulins, albumin, and α2-HS-glycoprotein are all present and can be readily identified.[35] Kinoshita[52] has demonstrated convincingly that albumin from the circulation is directly deposited in mineralizing dentin without being processed through the odontoblasts, i.e., the albumin bypasses the odontoblast interior entirely. It is likely that this undefined pathway of entry is used by the other serum protein constituents. Colchicine also enters the extracellular matrix by this route.[53] The retention of all these components is related to the process of active mineralization.

Extracellular Enzymes

Fetal bovine dentin has been extracted in EDTA to yield the equivalent of AS2-3, but without the use of protease inhibitors. The extract, after dialysis to remove the EDTA, was chromatographed on DEAE-cellulose and each fraction was checked for proteolytic and glycosidic activity.[54] Neutral protease, trypsin inhibitor, and lysozyme activities were all determined to be present. Lysozyme was detected in many of the DEAE-fractions, apparently complexed to the numerous acidic glycoproteins. Cathepsin D was also present extracellularly in predentin[55] as well as within the odontoblasts and odontoblastic processes.

Alkaline phosphatases are present in all mineralizing tissues, including dentin. The rat incisor phosphatase has a molecular weight of 79,000,[56] smaller than that in cartilage.

Alkaline phosphatases, monoester phosphohydrolases (EC 3.1.3.1), have an optimum pH range from 8.5 to 10.5 in vitro and are zinc requiring.

Lipids

Lipids are present in dentin and predentin to the extent of 2 to 6% of the organic matrix.[57,58] They are primarily acidic phospholipids, of which phosphatidyl serine and phosphatidyl inositol are the principal constituents. While they are always present it is difficult to say that the phospholipids are intrinsic components of the extracellular matrix. They appear to arise as fragments of cell membranes or budded off vesicle membranes. As such they are cellular components and are complexed with other cell membrane proteins to form proteolipids. A unique feature of dentin is the presence of calcium-acidic phospholipid-phosphate complexes.[58] These are present in the predentin and appear to accumulate, as lipid staining material, in the vicinity of the mineralization front.[59] Boskey[60] has recently reviewed this subject in detail and has concluded that the calcium-acidic phospholipid-phosphate complexes, and the proteolipids may have important roles in the mineralization process. In particular, the earliest stages of mineral deposition in dentinogenesis appear to depend upon membrane related lipid components.

Summary and Concluding Thoughts on the Composition of Dentin Matrix

The dentin matrix is a typical connective tissue containing collagen, proteoglycans, glycoproteins, various enzymes, and some lipid. The unique features of the collagen are that it contains only Type I collagen, prominently cross-linked via the $\alpha 2$ chains, and that the collagen itself may be phosphorylated. In addition, the major noncollagenous proteins are also unique in having very high contents of aspartyl residues. The principal NCP, phosphophoryn, contains about 35 residue percent Asp and 40 residue % Ser, with almost every seryl hydroxyl group phosphorylated. The Asp-rich proteins are very labile and relatively easily degraded by exposure to acids. They also appear to be easily degraded by proteases. Both of these features seriously complicate isolation and purification of these components. Finally, dentin matrix contains endogenous proteases which are difficult to inactivate adding to the problems of tissue stabilization prior to fractionation.

All dentins appear to have very similar compositions, but molars and continuously erupting incisors do not have identical distributions of the NCP and there may be wide species differences. Human dentin may be among the more unusual forms in terms of composition.

REFERENCES

1. **Veis, A., Spector, A. R., and Carmichael, D. J.,** Organization and polymerization of bone and dentin collagens, *Clin. Orthopaed. Rel. Res.*, 66, 188, 1969.
2. **Leaver, A. G., Triffitt, J. T., and Holbrook, I. B.,** Newer knowledge of non-collagenous proteins in dentin and cortical bone matrix, *Clin. Orthopaed. Rel. Res.*, 110, 269, 1975.
3. **Eastoe, J. E.,** Chemical organization of the organic matrix of dentine, in *Structural and Chemical Organization of Teeth*, Vol. 2, Miles, A. E. W., Ed., Academic Press, New York, 1967, 270.
4. **Schlueter, R. J.,** Macromolecular Organization of Dentine Collagen, Ph.D. dissertation, Northwestern University, Evanston, Ill., 1963.
5. **Veis, A. and Schlueter, R. J.,** Presence of phosphate mediated cross-linkages in hard tissue collagens, *Nature (London)*, 197, 1204, 1963.
6. **Schlueter, R. J. and Veis, A.,** Phosphorous containing proteins of teeth, *Fed. Proc.*, 22, 478, 1963.
7. **Veis, A. Schlueter, R. J.,** Macromolecular organization of dentine matrix collagen. I. Characterization of dentine collagen, *Biochemistry*, 3, 1650, 1964.
8. **Schlueter, R. J. and Veis, A.,** Macromolecular organization of dentine matrix collagen. II. Periodate degradation and carbohydrate cross-linking, *Biochemistry*, 3, 1657, 1964.

9. **Leaver, A. G. and Shuttleworth, C. A.,** The isolation from human dentine and ox bone of phosphate containing peptides, *Arch. Oral Biol.*, 11, 1209, 1966.
10. **Leaver, A. G. and Shuttleworth, C. A.,** Fractionation of the acid-soluble nitrogen of bone and dentin, *Arch. Oral Biol.*, 12, 947, 1967.
11. **Leaver, A. G.,** Studies on certain peptide fractions isolated from human dentine, *Arch. Oral Biol.*, 14, 503, 1969.
12. **Jones, I. L. and Leaver, A. G.,** Studies on the minor components of the organic matrix of human dentine, *Arch. Oral Biol.*, 19, 371, 1974.
13. **Weinstock, M. and Leblond, C. P.,** Radioautographic visualization of the deposition of a phosphoprotein at the mineralization front in the dentin of the rat incisor, *J. Cell Biol.*, 56, 838, 1973.
14. **Weinstock, M. and Leblond, C. P.,** Synthesis, migration, and release of precursor collagen by odontoblasts as visualized by radioautography after [^{3}H] proline administration, *J. Cell Biol.*, 60, 92, 1974.
15. **Engfeldt, B. and Hjerpe, A.,** Glycosaminoglycans of dentine and predentine, *Calcif. Tissue Res.*, 10, 152, 1972.
16. **Linde, A.,** On enzymes associated with biological calcification, in *The Chemistry and Biology of Mineralized Connective Tissues,* Veis, A., Ed., Elsevier/North Holland, New York, 1981, 559.
17. **Urist, M. R., Iwata, H., Ceccotti, P. W. L., Dorfman, R. L., Boyd, S. D., McDowell, R. M., and Chien, C.,** Bone morphogenesis in implants of insoluble bone gelatin, *Proc. Natl. Acad. Sci. U.S.A.*, 70, 3511, 1973.
18. **Takagi, Y. and Veis, A.,** Matrix protein difference between human normal and dentinogenesis imperfecta dentin, in *The Chemistry and Biology of Mineralized Connective Tissues,* Veis, A., Ed., Elsevier/North-Holland, New York, 1981, 233.
19. **Termine, J. D., Belcourt, A. B., Miyamoto, M. S., and Conn, K. M.,** Properties of dissociatively extracted fetal tooth matrix proteins. II. Separation and purification of fetal bovine dentin phosphoprotein, *J. Biol. Chem.*, 255, 9769, 1980.
20. **Carmichael, D. J., Dodd, C. M., and Veis, A.,** Solubilization of bone and dentin collagens by pepsin. Effect of cross-linkages and non-collagen components, *Biochim. Biophys. Acta,* 491, 177, 1977.
21. **Volpin, D. and Veis, A.,** Cyanogen bromide peptides from insoluble skin and dentin bovine collagens, *Biochemistry,* 12, 1452, 1973.
22. **Scott, P. G. and Veis, A.,** Cyanogen bromide peptides of bovine soluble and insoluble collagens. II. Tissue specific cross-linked peptides of insoluble skin and dentin collagen, *Connect. Tissue Res.*, 4, 117, 1976.
23. **Wohlebe, M. and Carmichael, D. J.,** Type I collagen in neutral-salt-soluble lathyritic rat dentine, *Eur. J. Biochem.*, 92, 183, 1978.
24. **Katz, E. P. and Li, S.-T.,** The packing of collagen in mineralized and non-mineralized tissues, *Biochem. Biophys. Res. Commun.*, 46, 1368, 1972.
25. **Katz, E. P. and Li, S.-T.,** Structure and function of bone collagen fibrils, *J. Mol. Biol.*, 80, 1, 1973.
26. **Scott, P. G.,** A major intermolecular cross-linking site in bovine dentine collagen involving the α 2 chain and stabilizing the 4D overlap, *Biochemistry,* 19, 6118, 1980.
27. **Scott, P. G., Veis, A., and Mechanic, G.,** Identity of a cyanogen bromide fragment of bovine dentin collagen containing the site of an intermolecular cross-link, *Biochemistry,* 15, 3191, 1976.
28. **Volpin, D. and Veis, A.,** Differences between CNBr peptides of soluble and insoluble bovine collagens, *Biochem. Biophys. Res. Commun.*, 44, 804, 1971.
29. **Dickson, I. R., Dimuzio, M. T., Volpin, D., Ananthanarayanan, S., and Veis, A.,** The extraction of phosphoproteins from bovine dentin, *Calcif. Tissue Res.*, 19, 51, 1975.
30. **Curley-Joseph, J. and Veis, A.,** The nature of covalent complexes of phosphoproteins with collagen in the bovine dentin matrix, *J. Dent. Res.*, 58, 1625, 1979.
31. **Fujisawa, R., Takagi, T., Kuboki, Y., and Sasaki, S.,** The characterization of collagen-phosphophoryn complex in bovine dentin, in *The Chemistry and Biology of Mineralized Connective Tissues,* Veis, A., Ed., Elsevier/North-Holland, New York, 1981, 483.
32. **Butler, W. T., Finch, J. E., Jr., and DeSteno, C. V.,** Chemical character of proteins in rat incisors, *Biochim. Biophys. Acta,* 257, 167, 1972.
33. **Linde, A., Bhown, M., and Butler, W. T.,** Non-collagenous proteins of rat dentin. Evidence that phosphoprotein is not covalently bound to collagen, *Biochim. Biophys. Acta,* 667, 341, 1981.
34. **Butler, W. T., Mikulski, A., and Urist, M. R.,** Non-collagenous proteins of a rat dentin matrix possessing bone morphogenetic activity, *J. Dent. Res.*, 56, 228, 1977.
35. **Butler, W. T., Bhown, M., Dimuzio, M. T., and Linde, A.,** Noncollagenous proteins of dentin. Isolation and partial characterization of rat dentin proteins and proteoglycans using a three-step preparative procedure, *Coll. Relat. Res.*, 1, 187, 1981.
36. **Jontell, M., Linde, A., and Lundvik, L.,** Comparative studies of phosphoprotein preparations from rat incisor dentin, *Prep. Biochem.*, 10, 235, 1980.

37. **Scott, P. G. and Edwards, H.,** Qualitative and quantitative studies on intermolecular cross linking in bovine dentine collagen, in *The Chemistry and Biology of Mineralized Connective Tissues,* Veis, A., Ed., Elsevier/North-Holland, New York, 1981, 45.
38. **Fujimoto, D., Akiba, K., and Nakamura, N.,** Isolation and characterization of a fluorescent material in bovine achilles tendon collagen, *Biochem. Biophys. Res. Commun.,* 76, 1124, 1977.
39. **Veis, A., Spector, A. H., and Zamoscianyk, H.,** Isolation of an EDTA-soluble phosphoprotein from mineralizing bovine dentin, *Biochim. Biophys. Acta,* 257, 404, 1972.
40. **Dimuzio, M. T. and Veis, A.,** Phosphophoryns — major non-collagenous proteins of rat incisor dentin, *Calcif. Tissue Res.,* 225, 169, 1978.
41. **Munksgaard, E. C., Butler, W. T., and Richardson, W. S.,** Phosphoprotein from dentin: new approaches to achieve and assess purity, *Prep. Biochem.,* 7, 321, 1977.
42. **Lee, S. L., Veis, A., and Glonek, T.,** Dentin phosphoprotein: an extracellular calcium binding protein, *Biochemistry,* 16, 2971, 1977.
43. **Richardson, W. S., Munksgaard, E. C., and Butler, W. T.,** Rat incisor phosphoprotein: the nature of the phosphate and quantitation of the phosphoserine, *J. Biol. Chem.,* 253, 8042, 1978.
44. **Veis, A., Stetler-Stevenson, W. G., Takagi, Y., Sabsay, B., and Fullerton, R.,** The nature and localization of the phosphorylated proteins of mineralized dentin, in *The Chemistry and Biology of Mineralized Connective Tissues,* Veis, A., Ed., Elsevier/North-Holland, New York, 1981, 377.
45. **Stetler-Stevenson, W. G. and Veis, A.,** Characterization of undegraded phosphophoryn of bovine dentin, *Fed. Proc.,* 41(4), 1438, 1982.
46. **Hjerpe, A. and Engfeldt, B.,** Proteoglycans of dentine and predentine, *Calcif. Tissue Res.,* 22, 173, 1976.
47. **Rahemtulla, F., Prince, C. W., Caterson, B., Christner, J. E., Baker, J. R., and Butler, W. T.,** Studies on the chemistry and metabolism of dentin proteoglycans using an *in vivo* labelling technique, in *The Chemistry and Biology of Mineralize Connective Tissues,* Veis, A., Ed., Elsevier/North-Holland, New York, 1981, 389.
48. **Branford-White, C. J.,** Molecular organization of heparan sulphate proteoglycan from human dentine, *Arch. Oral Biol.,* 23, 1141, 1978.
49. **Linde, A.,** Glycosaminoglycans of the odontoblast-predentine layer in dentinogenically active porcine teeth, *Calcif. Tissue Res.,* 12, 281, 1973.
50. **Veis, A., Sharkey, M., and Dickson, I.,** Non-collagenous proteins of bone and dentin extracellular matrix and their role in organized mineral deposition, in *Calcium Binding Proteins and Calcium Function,* Wasserman, et al., Eds., Elsevier/North-Holland, New York, 1977, 409.
51. **Price, P. A., Poser, J. W., and Roman, N.,** Primary structure of the γ-carboxylglutamic acid-containing protein from bovine bone, *Proc. Natl. Acad. Sci. U.S.A.,* 73, 3374, 1976.
52. **Kinoshita, Y.,** Incorporation of serum albumin into the developing dentine and enamel matrix in the rabbit incisor, *Calcif. Tissue Int.,* 29, 41, 1979.
53. **Cho, M. I. and Garant, P. R.,** Autoradiographic demonstration of colchicine binding at the mineralizing front of bone and dentin, in *The Chemistry and Biology of Mineralized Connective Tissues,* Veis, A., Ed., Elsevier/North-Holland, New York, 1981, 303.
54. **Sharkey, M. A.,** Isolation and identification of a nucleoside conjugated to dentin phosphoprotein, Ph.D. dissertation, Northwestern University, Evanston, Ill., 1976.
55. **Nygren, H., Persliden, B., Hansson, H-A., and Linde, A.,** Cathepsin D: ultra-immunohistochemical localization in dentinogenesis, *Calcif. Tissue Int.,* 29, 251, 1979.
56. **Granstrom, G., Jontell, M., and Linde, A.,** Separation of odontoblast Ca^{2+}-ATPase and alkaline phosphatase, *Calcif. Tissue Int.,* 27, 211, 1979.
57. **Odutuga, A. A. and Prout, R. E. S.,** Lipid analysis of human enamel and dentine, *Arch. Oral Biol.,* 19, 729, 1974.
58. **Boskey, A. L., Goldberg, M. R., and Posner, A. S.,** Calcium-phospholipid-phosphate complexes in mineralizing tissue, *Proc. Soc. Exp. Biol. Med.,* 157, 588, 1977.
59. **Irving, J. T. and Wuthier R. E.,** Histochemistry and biochemistry of calcification with special reference to the role of lipids, *Clin. Orthopaed. Rel. Res.,* 56, 237, 1968.
60. **Boskey, A. L.,** Current concepts of the physiology and biochemistry of calcification, *Clin. Orthop. Relat. Res.,* 157, 225, 1981.
61. **Stetler-Stevenson, W. G. and Veis, A.,** unpublished results.
62. **Takagi, Y., Veis, A., and Sauk, J. J.,** Mineralization defects in collagen matrices may be related to noncollagen protein components, *Clin. Orthopaed. Rel. Res.,* in press.
63. **Richardson, W. S., Beegle, W. F., Butler, W. T., and Munksgaard, E. C.,** The phosphoprotein of rabbit incisors, *J. Dent. Res.,* 56, 233, 1977.

ODONTOBLASTIC ACTIVITIES IN DENTINOGENESIS

Michael T. DiMuzio and William T. Butler

Dentinogenesis is comprised of three biological processes: matrix protein synthesis, accumulation and maturation, and mineralization. The evidence to date is that the first two of these processes are associated with cellular activity. Mineralization is a multifold result of systemic and local controls, which has not been linked to actual cellular direction. The present review will relate odontoblastic activity to the formation of a mature and functional dentin matrix.

ODONTOBLASTS

The odontoblasts are the cells responsible for the formation of the extracellular matrix of dentin. As late as 1968, dentin was still conceived to be the result of a coordinated effort between the odontoblasts and those subodontoblastic fibroblasts located in the pulp.[1] This belief stems from the data dealing with the origin of the mantle dentin (the first layer of dentin) which is synthesized by fibroblasts prior to the total differentiation of the odontoblasts.[2,3] These coarse fibers of collagen become the layer upon which the total dentin of the tooth is secreted. Their function is unknown, and interestingly, their differential staining compared to the collagen of odontoblast origin is visible throughout the lifetime of the tooth.

As the odontoblasts differentiate, their increased contents of ribonucleic acid have been easily demonstrated histochemically.[1] In addition, increased amounts of intracellular vacuoles and hydrolytic enzyme activity are measurable.[4] The appearance of alkaline phosphatase (APase) within the odontoblast at this time can also be compared to the osteoblasts of bone which characterizes their synthetic capacities.[5,6] No specific function has been determined for this APase other than it appears to be a marker for the secretory cells of mineralizing connective tissues. More will be said about APase below.

The continual synthetic requirements of the odontoblast make the findings by Ten Cate[7] of high levels of intracellular oxidative enzyme activity unsurprising. He showed that anerobic and aerobic metabolism as well as pentose phosphate shunt activity increases with increasing differentiation of the odontoblasts.

A number of morphologists[8-10] have shown that developing odontoblasts contain prominent Golgi apparati, secretory vesicles, localized mitochondria, lysosomal-like bodies, and regions of rough endoplasmic reticulum. These features are typical of secretory cells and their presence expected, but the polar nature of the odontoblast and the segregation of these intracellular compartments has been used to great advantage by those interested in the detailed study of extracellular protein synthesis. Therefore, regarding odontoblastic activity, the prime questions become how, what, and where do these cells secrete their products destined for the extracellular space.

Larsson and Bloom,[10-12] in a definitive group of studies of the fine structure of developing odontoblasts, outlined the sequence of events dealing with synthetic activities. Following the earliest stages of mesenchymal cell differentiation, odontoblasts begin to organize for the synthesis of extracellular matrix: rough endoplasmic reticulum and Golgi and the initial secretion and organization of the extracellular space. As noted above, the presence of prominent endoplasmic reticulum and Golgi complexes underline the secretory capacity of the odontoblasts, but it is the increasing number of cytoplasmic organelles involved in secretion which is the most conspicuous feature of the developing odontoblasts. Prior to mineralization of the matrix, increased amounts of collagenous fibers appear between the epithelial-mesenchymal membrane. Upon this thin layer of collagenous mantle dentin is secreted a layer of predentin, of odontoblastic origin, concomitant with the differentiation of epithelial cells to ameloblasts.[11] This widening predentin width is interspersed with globular electron-dense structures very similar to the matrix vesicles of early calcification in cartilage[13,14] and bone.[15] The needle-like structures within these globules are thought to be crystallites and form the initial sites of mineral salt precipitation within the ever-widening predentin matrix. However, these matrix vesicular structures have not been identified in later stages of dentinogenesis and therefore may only relate to the initiation, and not to the continual deposition and growth, of hydroxyapatite.[16]

The continual differentiation to odontoblasts, matrix synthesis, secretion, and calcification proceed in an ordered sequence that constitutes the process of dentinogenesis. The static results of fine structure analyses, although valuable in the identification and localization of structures, offer little information regarding the synthetic activities of the odontoblasts; however, the organization of the tissue has proven advantageous for a number of other approaches. The unique acellular matrix and the lining odontoblasts have provided an ideal model for autoradiographic studies[17-20] which describe the intracellular translocation, secretion, and deposition of matrix components. These odontoblastic activities will now be discussed.

DENTIN COLLAGEN

The major extracellular dentin component, the fibrous protein collagen,[21] has been the only macromolecule to be carefully followed from assembly to exocytosis.[22] In an elegant combination of electron microscopy, histochemistry, and autoradiography, Weinstock and Leblond[22] were able to trace the steps of dentin collagen processing within the odontoblasts. Fine structure analysis of mature mouse odontoblasts suggested that collagen-like material was restricted to cylindrical Golgi regions and secretory granules in the apex of the odontoblasts.[22] For these studies, silver methenamine (which stains collagen fibrils specifically) and low pH-phosphotungstic acid were used to accentuate the contents of these regions. That the secretory granules could be observed in fusion with the plasmalaemma of the odontoblastic process indicate their transitory nature. This seqence was quantitated by autoradiographs after injection of ^{3}H-proline. In short, presumptive collagen was found at 2 min over the rough endoplasmic reticulum, at 10 min over the spherical portions of the Golgi, at 20 min over the cylindrical portion, and at 30 min within secretory granules and predentin. This sequence of events indicated that a proline-labeled protein (collagen precursor) was moving through the odontoblast toward the predentin.

Collagen fibrils in distal predentin are thicker and more numerous than the newly formed fibers in the proximal predentin.[11] Presumably, as new molecules are added to proximal predentin, the existent extracellular fibrils enlarge by their accretion of additional molecules and gradually become part of the distal predentin, due to the advancement of the mineralization front. Radioautographs point toward the continuing secretion of labeled secretory granules throughout the length of the odontoblastic process which transverses the entire predentin width.[20, 22]

Bone collagen synthesis is a combination of osteoblastic activity and extracellular modifications of the secreted molecule.[23,24] This extracellular handling in dentin was explored by immunohistochemical means in order to characterize the collagen presursor and its eventual fate[25] using specific anti-pro type I collagen antibodies. These studies allowed Karim et al.[25] to trace in a similar pathway the procollagen molecule through the odontoblast rough ER, Golgi, secretory granules, and into the predentin. However, the specificity of the antibody to the proportion of type I collagen led to some unexpected results. Immunostaining was rare in the simple lysosomes, but frequent in multivesicular bodies. In odontoblasts, these bodies are known to contain hydrolytic enzymes which might be involved in procollagen breakdown (known to occur in such collagen-rich tissues) or with pinocytosis of amino- and carboxy-terminal peptides released in the course of procollagen transformation into collagen.[22] Cournil et al.[26] showed that procollagen type I was present in predentin, but antigenically decreased at distances away from the odontoblast bordering membrane. As noted above, it is known that collagen fibers increase in number and size as the dentin mineralization front approaches.[10] Karim et al.[25] showed that the reactivity with anti-pro type I collagen is next to the odontoblasts and suggested that the transformation took place in the extracellular space just before the collagen is polymerized into fibrils. The absence of reaction in the distal third of the predentin indicated that procollagen synthesis, secretion, and conversion was completed before the approach of the mineralization front.

NONCOLLAGENOUS PROTEINS

It was only after investigations began to focus on the noncollagenous proteins (NCPs) of dentin that researchers became aware that odontoblasts were not simply secreting a collagenous organic layer which became mineralized with time. On the contrary, the NCPs studied have been shown to have an activity in the extracellular matrix that is quite different from dentin collagen. These NCPs have an ''intimate'' relationship with the expanding mineralization front, whereas collagen is apparently passively engulfed by the ever-expanding mineralization front of mineral crystal growth. Carniero and Leblond,[17] using autoradiography in the rat incisor, showed that odontoblasts synthesized and secreted collagen into the predentin by 4 hr and that only after 35 hr was the label (^{3}H-glycine) found within the mineralized dentin. This kinetic work was repeated at the electron microscope level of resolution using ^{3}H-proline as a precursor.[22]

The continual growth and eruption of the rat incisor requires the rapid mineralization of the extracellular matrix. When investigations focused on two groups of NCPs, namely, glycoproteins and phosphoproteins, their involvements were shown to be quite different from the activity of collagen. Weinstock et al.[18] in a light microscopic study used ^{3}H-fucose as a precursor for glycoprotein biosynthesis. The principle advantage of this precursor is that it is utilized in tissues mainly for protein glycosylation without significant conversion into other substances.[27,28] After i.v. injection of ^{3}H-fucose into young rats, the label was found within two distinct zones after only 4 hr: one zone was in the proximal dentin close to the odontoblast and a second distinct band of activity was at the predentin-dentin junction. Recall that this rapid migration far exceeds the movement of dentin collagen at 4 hr postinjection, which has just reached the predentin next to the odontoblast.[20,22]

Further experiments on the major NCP of rat incisor dentin, the phosphoproteins, added credence to the belief that the NCPs were actively involved in the mineralization process.[20,29] Weinstock and Leblond,[20] using three different tracers, ^{3}H-proline, ^{3}H-serine, and ^{33}P-phosphate in three separate investigations, dramatically showed that the phosphoproteins were associated with the mineralization front soon after their synthesis and secretion by odontoblasts. Their results elegantly showed that dentin collagen (^{3}H-proline) label required at least 30 hr to become engulfed in the mineralization front whereas the phosphoprotein (^{3}H-serine and ^{33}P-phosphorus) label, apparently without delay, migrated rapidly to the mineralization front as early as 90 min postinjection. As a compliment to these autoradiographic studies, biochemical studies of the rat incisor phosphoproteins and dentin collagen biosynthesis in vivo not only pointed directly toward the different secretion times but the fates of these two proteins.[29] Collagen was shown to be insolubilized within the matrix as early as 45 to 60 min, whereas phosphoproteins reached a maximal extracellular specific activity at 60 min. However, after 60 min a slow gradual decrease occurred up to 30 hr in phosphoprotein specific activity, in contrast to a steady accumulation of dentin collagen specific activity within the continually erupting incisor tooth. This slow turnover of phosphoproteins may be a requirement toward decreasing their local concentration at the mineralization front, since it has been shown that phosphoproteins have a concentration-dependent inhibition toward the growth of hydroxyapatite crystals in vitro.[30,31]

Another NCP of interest, although in much lower quantities in dentin than the phosphoproteins, are the glycosaminoglycan (GAG)-containing proteins, the proteolgycans. These sulfated glycoproteins, because of their affinities for calcium ion, have been thought to participate in the mineralization processes in all mineralizing tissues (i.e., bone, dentin, epiphyseal growth plate).[19,32-35] The loss of hexosamine and of noncollagenous nitrogen has been shown to occur during the formation of fully mineralized bone tissue.[36] In the tooth, using a single large dose of ^{35}S-sulfate, Sundstrom[37] showed in a kinetic autoradiographic study that a similar loss of labeled components occurred during the process of dentinogenesis in the rat.

In a recent study, Takagi et al.[38] showed, using a specific stain for glycoconjugates, that a 50% decrease in GAG content of the matrix occurred in moving from the predentin into the dentin. These studies compliment and are consistent with numerous studies, both biochemical and ultrastructural that document an abrupt change in the appearance of these acidic complex macromolecules.[32,33,37,39] Another in vivo biochemical study on the identification and kinetics of ^{35}S-sulfate labeled proteolgycans have pointed toward a more specific metabolism in regards to the transformation of predentin to dentin.[40] Presumptive predentin-extracted proteolgycans contained both chondroitin-4-sulfate (C-4-S) and C-6-S, whereas dentin-derived proteoglycans contained only C-4-S. In addition, the high molecular weight proteoglycan extracted from predentin was absent from dentin-derived proteoglycans. More recent observations[108] have shown that proteoglycans are not degraded in a wholesale manner during mineralization. After injections of ^{35}S-sulfate, the label could be traced from the predentin proteoglycan pool to the dentin pool with no apparent losses, as opposed to what was indicated earlier.[40]

Although no direct evidence exists, these apparent enzymic modifications of proteoglycans and phosphoproteins may be a requirement in order to compensate for the documented effects of these NCPs on in vitro mineralization systems.[30,35] The extracellular enzymes of the dentin matrix may be responsible for these turnover activities of the NCPs. Cathepsin D has been shown to degrade cartilage matrix.[41] Linde and Persliden[42] showed a high cathepsin D activity in isolated odontoblasts. In addition, other glycosidases necessary for total proteoglycan degradation are present in the odontoblast-predentin area: hyaluronidase, β-glucuronidase, and β-*N*-acetylglucosaminidase.[43-46] Another ubiquitous proteoglycan degradative enzyme, lysozyme, is known to exist in bone,[47] although its activity within the dentin matrix has not been substantiated.

It is attractive to focus the activity of dentin alkaline phosphatase (APase) as being directed toward the phosphoproteins and their apparent turnover however, no such documentation exists. Purified calf intestinal APase has been shown via in vitro incubations to dephosphorylate the phosphoserine residues of the native protein.[48] The diversity of hydrolytic functions of APase towards phosphate esters leaves open its role in dentinogenic activities. Inorganic pyrophosphate, β-glycerolphosphate, and adenosine triphosphate (ATP) all provide attractive and acceptable roles within the dentin matrix. Interestingly, the calcium-activated ATPase in dentin[49] may be related to a similar activity found in enamel[50] and calcifying cartilage.[51] The odontoblasts, although their role in calcium ion movement is unclear,[52] may utilize the ATPase as a membrane-bound enzyme within the odontoblastic process[53] as a means to concentrate or localize calcium ions. Unfortunately, attempts to link these functions to the actual mineralization process and/or dentinogenesis have been inconclusive.

CALCIUM

Autoradiography using ^{45}calcium suggests two routes for the transfer of calcium to the organic matrix of dentin, a direct transfer from the capillaries of the dental pulp via intercellular spaces[54] or both between and through the odontoblasts.[55] Appleton and Morris[52] used potassium pyroantimonate osmium staining for calcium in odontoblasts and the predentin matrix. Electron microscopic studies using this technique indicated the presence of calcium within the odontoblast[56] and its process.[57] In addition, previous investigators had suggested that insoluble calcium phosphate was found intracellularly and was secreted directly into the predentin matrix.[58] Appleton and Morris showed distinct staining within vesicular Golgi structures and the odontoblastic processes[52] which appeared to be bound to organic macromolecules. However, the dynamics of the calcium release mechanism could only be implied in their study.

ENZYMES AND MATRIX VESICLES

The role of the phosphatases (alkaline and acid) in mineralization processes has been controversial. As early as 1923, Robison[59] showed that developing bone contained phosphatases active on various substrates, but primarily on hexose and protein-bound phosphates. He reported that actively calcifying and hypertrophic cartilages had high activities, while resting cartilage had no demonstrable activity. Robison suggested that a phosphatase catalyzed the accumulation of mineral forming salt in the ossification process.

Other investigators have since confirmed the role of phosphatases in ossification but have suggested that their role is in matrix formation[60-62] and in the energy release from ATP.[63] Phosphatases hydrolyze the phosphate ester from poly- and pyro-phosphates which are known to inhibit crystal growth in vivo and in vitro.[64] It has also been proposed that the presence of specific phosphatases might be necessary to permit hydroxyapatite formation.[65-67] In the physiological process of mineralization, pyrophosphate (PP_i) might need to be destroyed before mineral deposition can proceed.[64] A neutral pyrophosphatase appears in costal cartilage at the onset of calcification.[68] Alkaline phosphatase, which possesses pyrophosphatase activity,[69] has long been in the forefront in discussions of mineralization.[59] Ali et al.[70] demonstrated that the major part of the alkaline phosphatase and pyrophosphatase activities were located within the matrix vesicles of epiphyseal cartilage.

Certain studies have suggested that PP_i might have a role in the process of calcification.[66] PP_i inhibits the precipitation of calcium phosphate in vitro both in the presence and absence of collagen, and it prolonged the time required for the conversion of amorphous calcium phosphate (ACP) to crystalline hydroxyapatite.[65,71,72]

Wuthier et al.[73] evaluated the role of PP_i in biological mineralization by measuring the ratios of PP_i to inorganic phosphate in the growth plate and other tissues. Their investigation showed that the PP_i/P_i ratio decreased in going from the resting zone to the calcified cartilage (210 to 13.4 mg P per gram P). These values are compared to 3.3 to 4.3 mg P per gram P in compact, periosteal, and cancellous bone (no values are known for the dentin matrix). Either new mineral is depositing with a lower ratio of PP_i/P_i or PP_i is being destroyed. These same investigators stressed that until the local concentrations and rates of synthesis of PP_i in calcifying tissues are determined, the role of PP_i will be uncertain.[73] The major drawback to implicating PP_i in mineralization theories lies in its ubiquitous production in the body which is obligatory in synthetic reactions involving trinucleotides. Similarly, alkaline and pyrophosphatases are also ubiquitous so as to allow amino acid and fatty acid activation to proceed by hydrolyzing PP_i.

Matrix vesicles and the phenomena of mineralization have always shared a somewhat troubled spotlight. Felix and Fleisch[74] showed that in isolated vesicles that an active calcium transport was absent. However, the vesicles did have a significant positive effect on the nucleation of calcium phosphate precipitation, whose nucleation was delayed in the presence of PP_i. These results led the authors to propose a dual role of in vivo action: nucleation of mineral precipitation and local destruction of PP_i. Furthermore, Ali[75] proposed that the hydrolysis of phosphorus-containing compounds was probably essential in the act of increasing local concentrations of calcium and phosphorus. In bone and dentin these phorphorus containing compounds may be the NCPs.[48,76-78] In addition, in teeth and bones, the initiation and maintenance of mineralization appear as two separate processes. Matrix vesicles appear to be necessary as a factor in initiation, whereas other factors must be concerned with the transport and localization of ions to maintain mineral balance. Since collagen can be eliminated as a factor, the NCPs of the matrix must be considered as prime candidates.

EFFECTS ON DENTINOGENESIS

Dentinogenic activity involves those processes of tissue formation and maturation which provide a suitable environment for the deposition and growth of mineral crystals. These coordinated activities are difficult to define in their individual roles and relative importance in mineralization because of their interdependence. However, studies which disrupt or inhibit this normal synergism of events have provided some insights into individual mechanisms involved in dentinogenesis.

Larsson and Larsson[79,80] used a potent inhibitor of mineralization (EHDP: ethylene-1-hydroxy-1,1-diphosphonate) and showed that matrix effects were dose dependent. At low doses (0.5 to 2.5 mg/kg) EHDP interfered with the crystal growth in early dentin formation in that rat.[79] At high doses (7.5 to 10 mg/kg), total inhibition of apatite crystal formation was attained as well as collagen related effects. These latter effects provided some clues into the mechanisms surrounding the secretion and maturation of the dentin collagen matrix. At high doses, an increased predentin width was observed which contained grossly disturbed collagen fibrils. Apparently EHDP was interfering at the level of extracellular aggregation. Larsson proposed that EHDP, a known calcium chelator, might be disrupting the calcium co-factor for procollagen-peptidase, which has not been isolated within the predentin matrix but has been found in calcifying cartilage.[80] Thus, EHDP may be disrupting the calcium ion distribution within predentin inhibiting the peptidase, resulting in a disordered polymerization of collagen, which is known to exist in certain collagen disorders lacking the available peptidase altogether.[81] Agents such as EHDP may become useful probes with which to delineate the extracellular activities regarding the production and maturation of a functional dentin matrix.

Vitamin D appears to be required for skeletal growth and mineralization, but is paradoxical in that the active form of the vitamin, 1α,25 dihydroxy-vitamin D_3 is a potent stimulator of resorption and inhibitor of collagen synthesis in bone.[82,83] Dentin in the continually erupting rodent incisor provides an excellent mineralizing tissue with which to access the formative role of vitamin D. Since dentin, in contrast to bone, is not usually resorbed, it is an ideal tissue in which to study growth. Mineralization of rachitic dentin is poor and the width of predentin larger than in normal animals.[84-86] Administration of vitamin D does restore mineralization but its direct involvement has not been experimentally demonstrated.[87] In vitro experiments on bone slices from animals deficient in phosphate and vitamin D were able to remineralize in a normal serum, whereas they failed to remineralize in serum from rachitic animals.[88] Yoshiki and co-workers[89] showed that dentin made rachitic by a low calcium and vitamin D-deficient diet could be restored by recovery of normal serum calcium levels. It appeared that vitamin D was only an indirect factor in the process of mineralization. However, under normal conditions, it is still unknown as to how vitamin D in its various metabolic forms relates to the activity of the odontoblast; such probes may be best made using in vitro organ cultures where direct effects can be monitored.

ORGAN CULTURES

Organ cultures have provided some valuable insight into the formative activities taking place during dentinogenesis. Early studies using tooth organ culture emphasized the viability of the systems via ultrastructural criteria.[90,91] More recent experiments have used biochemical criteria to compare in vitro synthesized products with those chemically characterized from preparative studies.[92-94] Munksgaard et al.[92] used rat incisor dentin fragments incubated in the presence of ^{14}C-proline as a collagen precursor and recovered type I collagen in all compartments of the culture (odontoblasts, dentin, and medium) as well as a type I trimer. Type I trimer is a molecular species of collagen shown to be a product of certain cells in culture[95-97] and in cultured mouse tooth germs,[98] although there is evidence of its presence in vivo.[99] DiMuzio et al.[94] have used rat incisor fragments with the pulp intact and shown that not only was type I collagen secreted and insolubilized, but also phosphoproteins and γ-carboxyglutamic acid (Gla)-containing proteins were synthesized, secreted, and accummulated in the dentin matrix. These latter organ cultures were performed in the presence of a well-defined medium containing heterologous serum and established, collagen-enhancing factors. However, in these systems the fact that no direct evidence of matrix mineralization has been shown is a major drawback in considering them as adequate models of dentinogenesis.

Tooth germs from the mouse and rat have provided models for developmental processes as well as tools toward understanding the activities of the dentin matrix.[91,100,101] Tooth germs have been used successfully as a means to follow the order of mesenchymal and epithelial tissue interactions in terms of morphogenesis and cytodifferentiation.[102,103] Recent advances by Yamada et al.[104] have led to defined conditions where total cytodifferentiation takes place in the absence of exogenous serum. They showed quite dramatically that a "normal" matrix was produced but did not mineralize in the absence of serum.

Such systems as the mouse tooth germ have permitted the probing of odontoblast activity with a variety of agents. Thesleff and Pratt[105] used tunicamycin (an inhibitor of protein glycosylation) and found a total inhibition of morphogenesis and differentiation of odontoblasts in the molar tooth germ in vitro. These results were suggested to be the result of alterations in the extracellular matrix components required for the full differentiation of the odontoblast. Further studies[106] with vitamin A showed that an excess of the vitamin prior to odontoblast differentiation prevented tooth morphogenesis. Vitamin A is known to interfere, at high doses, with the glycosylation of cell-surface related proteins; however in contrast to tunicamycin effects, vitamin A had no effect if added after the onset of odontoblast differentiation. These studies serve to highlight the importance of glycoproteins in normal dentin development and maturation of a viable matrix.

Recently, Thesleff et al.[107] used an immunohistochemical approach to study the specific changes that occur during dentinogenesis in the mouse tooth germ culture. Using specific antibodies to type IV collagen, proteoglycan, fibronectin, and laminin, they documented the following metabolic changes. The basement membrane produced by epithelial and mesenchymal cells contained all of these glycoproteins and the proteoglycan prior to the secretion of predentin. As predentin accumulated, the basement membrane began to be degraded, with fibronectin being the first component removed, followed *en bloc* by type IV collagen, laminin, and proteoglycan. This removal was timed to the polarization of the epithelial cells into ameloblasts. The observed changes in these collagenous and noncollagenous components appeared to be related to odontoblast differentiation and matrix formation in the tooth. Their studies attempted to delineate the early changes in metabolism in the process of dentinogenesis and provide an excellent framework with which to further investigate the later steps of cytodifferentiation, protein synthesis, secretion, and interaction in more detail.

CONCLUSION

The goal in studying the biochemistry of dentinogenesis is to understand the role of the odontoblast and its products relative to tissue formation and growth. As discussed in this review, the questions and investigations into these processes are becoming more refined and directed. The near future promises an assured excitement and advancement toward that ultimate goal.

REFERENCES

1. **Ten Cate, A. R.,** Current concepts and problems of dentinogenesis, in *Dentin and Pulp,* Symous, N. B. B., Ed., Churchill Livingstone, London, 1968, 9.
2. **von Korff, K.,** Die Entwicklung der Zahnbeingrundsubstanz der Saǔgetiere, *Arch. Mikr. Anat.,* 67, 1, 1906.
3. **Nylen, M. V. and Scott, D. B.,** Electron microscope studies of odontogenesis, *J. Indian St. Dent. Assoc.,* 39, 406, 1960.
4. **Katchburian, E., Katchburian, A. V., and Pearse, A. G.,** Histochemistry of lysosomal enzymes in developing teeth of albino rats, *J. Anat.,* 101, 783, 1967.
5. **Ten Cate, A. R.,** The distribution of alkaline phosphatase in the human tooth germ, *Arch. Oral Biol.,* 7, 195, 1962.
6. **Wong, G. L. and Cohn, D. V.,** The effect of parathormone on the synthesis of collagenous matrix by isolated bone cells in *Mechanisms of Localized Bone Cells, in Mechanisms of Bone Loss,* Horton, J. E., Tarpley, T. M., and Davis, W. F., Eds., Information Retrieval, Arlington, Va., 1978, 47.
7. **Ten Cate, A. R.,** A histochemical study of the human odontoblast, *Arch. Oral Biol.,* 12, 963, 1967.
8. **Takuma, S.,** Ultrastructure of dentinogenesis, in *Structural and Chemical Organization of Teeth,* Vol. 1, Miles, A. E. W., Ed., Academic Press, New York, 1967, 325.
9. **Weinstock, A.,** Matrix development in mineralizing tissues as shown by radioautography: formation of enamel and dentin, in *Developmental Aspects of Oral Biology,* Slavkin, H. C. and Bavetta, L., Eds., Academic Press, New York, 1972, 202.
10. **Larsson, Å. and Bloom, G. D.,** Studies on dentinogenesis in the rat, *Z. Anat. Entwicklungsgesch.,* 39, 227, 1973.
11. **Larsson, Å.,** Studies on dentinogenesis in the rat, *Z. Anat. Entwicklungsgesch.,* 142, 103, 1973.
12. **Larsson, Å.,** Studies on dentinogenesis in the rat, *Calcif. Tissue Res.,* 16, 93, 1974.
13. **Anderson, H. C.,** Vesicles associated with calcification in the matrix of epiphyseal cartilage, *J. Cell Biol.,* 41, 59, 1969.
14. **Bonucci, E.,** Fine structure and histochemistry of "calcifying globules' in epiphyseal cartilage, *Z. Zellforsch.,* 103, 192, 1970.
15. **Bernard, G. W. and Pease, D. C.,** An electron microscopic study of initial intramembraneous osteogenesis, *Am. J. Anat.,* 125, 271, 1969.
16. **Slavkin, H. C.,** Intercellular communication during odontogenesis, in *Developmental Aspects of Oral Biology,* Slavkin, H. C. and Bavetta, L. A., Ed., Academic Press, New York, 1972, 165.
17. **Carniero, J. and Leblond, C. P.,** Role of the osteoblasts and odontoblasts in secreting the collagen of bone and dentin, as shown by radioautography in mice given tritium-labeled glycine, *Exp. Cell Res.,* 18, 291, 1959.
18. **Weinstock, A., Weinstock, M., and Leblond, C. P.,** Autoradiographic detection of ^{3}H-fucose incorporation into glycoprotein by odontoblasts and its deposition at the site of the calcification front in dentin, *Calcif. Tissue Res.,* 8, 181, 1972.
19. **Leblond, C. P. and Weinstock, M.,** Comparative study of dentin and bone formation, in *The Biochemistry and Physiology of Bone,* Vol. 4, 2nd ed., Bourne, G. H., Ed., Academic Press, New York, 1976, 518.
20. **Weinstock, M. and Leblond, C. P.,** Radioautographic visualization of the deposition of a phosphoprotein at the mineralization front in the dentin of the rat incisor, *J. Cell Biol.,* 56, 838, 1973.
21. **Butler, W. T., Finch, J. E., and Desteno, C. V.,** Chemical character of proteins in rat incisors, *Biochim. Biophys. Acta,* 257, 167, 1972.
22. **Weinstock, M. and Leblond, C. P.,** Synthesis, migration and release of precursor collagen by odontoblasts as visualized by radioautography after ^{3}H-proline administration, *J. Cell Biol.,* 60, 92, 1974.
23. **Bornstein, P., Ehrlich, H. P., and Wyke, A. W.,** Procollagen: conversion of the precursor to collagen by a neutral protease, *Science,* 175, 544, 1972.
24. **Bornstein, P.,** The biosynthesis of collagen, *Annu. Rev. Biochem.,* 43, 567, 1974.
25. **Karim, A., Cournil, I., and Leblond, C. P.,** Immunohistochemical localization of procollagens. II. Electron microscope distribution of procollagen I antigenicity in the odontoblasts and predentin of rat incisor teeth by a direct method using peroxidase linked antibodies, *J. Histochem. Cytochem.,* 27, 1070, 1979.

26. **Cournil, I., Leblond, C. P., Pomponio, J., Hand, A. R., Sederlof, L., and Martin, G. R.,** Immunohistochemical localization of procollagens. I. Light microscopic distribution of procollagen I, III and IV antigenicity in the rat incisor tooth by the indirect peroxidase-anti-peroxidase method, *J. Histochem. Cytochem.*, 27, 1059, 1979.
27. **Coffey, J. W., Miller, O. N., and Sellinger, O. Z.,** The metabolism of L-fucose in the rat, *J. Biol. Chem.*, 239, 4011, 1964.
28. **Bekesi, J. G. and Winzler, R. J.,** The metabolism of plasma glycoproteins, *J. Biol. Chem.*, 242, 3873, 1967.
29. **DiMuzio, M. T. and Veis, A.,** The biosynthesis of phosphophoryns and dentin collagen, in the continuously erupting rat incisor, *J. Biol. Chem.*, 253, 6845, 1978.
30. **Termine, J. D. and Conn, K. M.,** Inhibition of apatite formation by phosphorylated metabolites and macromolecules, *Calcif. Tissue Res.*, 22, 149, 1976.
31. **Nawrot, C. F., Campbell, D. J., Schroeder, J. K., and Van Valkenburg, M.,** Dental phosphoprotein-induced formation of hydroxylapatite during *in vitro* synthesis of amorphous calcium phosphate, *Biochemistry*, 15, 3445, 1976.
32. **Engfeldt, B. and Hjerpe, A.,** Glycosaminoglycans of dentine and predentin, *Calcif. Tissue Res.*, 10, 152, 1972.
33. **Hjerpe, A. and Engfeldt, B.,** Proteoglycans of dentine and predentin, *Calcif. Tissue Res.*, 22, 173, 1976.
34. **Linde, A.,** Glycosaminoglycans of the odontoblast-predentin layer in dentinogenically-active porcine teeth, *Calcif. Tissue Res.*, 12, 281, 1973.
35. **Blumenthal, N. C., Posner, A. S., Silverman, L. D., and Rosenberg, L. C.,** Inhibition of hydroxyapatite precipitation by cartilage proteoglycans, *Calcif. Tiss. Int.*, 27, 75, 1979.
36. **Pugliarello, M. C., Vittur, F., DeBernard, B., Bonucci, E., and Ascenzi, A.,** Chemical modifications in osteones during calcification, *Calcif. Tissue Res.*, 5, 108, 1970.
37. **Sundström, B.,** New aspects on the utilization of inorganic sulfate during dentin formation, *Histochemie*, 26, 61, 1971.
38. **Takagi, M., Parmley, R. T., and Denys, F. R.,** Ultrastructural localization of complex carbohydrates in odontoblasts predentin and dentin, *J. Histochem. Cytochem.*, 29, 747, 1981.
39. **Leblond, C. P., Belanger, L. F., and Greulich, R. C.,** Formation of bones and teeth as visualized by radioautography, *Ann. N.Y. Acad. Sci.*, 60, 630, 1955.
40. **Rahemtulla, F., Prince, C. W., Caterson, B., Christner, J. E., Baker, J. R., and Butler, W. T.,** Studies on the chemistry and metabolism of dentin proteoglycans using *in vivo* labeling technique, in *The Chemistry and Biology of Mineralized Connective Tissues*, Veis, A., Ed., Elsevier, New York, 1981, 395.
41. **Dingle, J. T., Barrett, A. J., Poole, A. R., and Stovin, P.,** Inhibition by pepstatin of human cartilage degradation, *Biochem. J.*, 127, 443, 1972.
42. **Linde, A. and Persliden, B., Cathepsin D.,** Activity in isolated odontoblasts, *Calcif. Tissue Res.*, 23, 33, 1977.
43. **Buddecke, E. and Kresse, H.,** Enzymes involved in proteoglycan degradation, in *Connective Tissues*, Fricke, R. and Hartman, F., Eds., Springer-Verlag, Berlin, 1974, 131.
44. **Morrison, R. I. G.,** Glycosidases necessary for proteoglycan removal, in *Connective Tissues*, Fricke, R. and Hartman, F., Eds. Springer-Verlag, Berlin, 1974, 150.
45. **Engström, C., Linde, A., and Persliden, B.,** Glycosidases involved in glycosaminoglycan removal in dentin, *Arch. Oral Biol.*, 17, 1421, 1972.
46. **Engström, C., Linde, A., and Persliden, B.,** Histochemistry of specific glycosidases in dentin, *Scand. J. Dent. Res.*, 84, 76, 1976.
47. **Kuettner, K. and Sorgente, N.,** Lysozyme in preosseous cartilage, *Biochim. Biophys. Acta*, 372, 335, 1974.
48. **Linde, A., Bhown, M., and Butler, W. T.,** Noncollagenous proteins of dentin. A re-examination of proteins from rat incisor dentin utilizing techniques to avoid artifacts, *J. Biol. Chem.*, 255, 5931, 1980.
49. **Linde, A. and Magnussen, D. C.,** Inhibition studies of alkaline phosphatases in hard tissue forming cells, *J. Histochem. Cytochem.*, 23, 342, 1975.
50. **Mörnstad, H.,** Calcium-activated ATPase in enamel, *Scand. J. Dent. Res.*, 86, 1, 1978.
51. **Kahn, S. E. and Arsenis, C.,** Identification, characterization and localization of A($Ca^{+2}Mg^{+2}$) activated purine nucleoside triphosphate phosphohydrolase from calcifying cartilage, *Biochim. Biophys. Acta*, 569, 52, 1979.
52. **Appleton, J. and Morris, D. C.,** An ultrastructural investigation of the role of the odontoblast in matrix calcification using potassium pyroantimonate osmium method for calcium localization, *Arch. Oral Biol.*, 24, 467, 1979.
53. **Granström, G., Linde, A., and Nygren, H.,** Ultrastructural localization of alkaline phosphatases in rat incisor odontoblasts, *J. Histochem. Cytochem.*, 26, 359, 1978.
54. **Munhoz, C. O. G. and Leblond, C. P.,** Deposition of calcium phosphate into dentin and enamel as shown by radioautography of sections of incisor teeth following injection of ^{45}Ca into rats, *Calcif. Tissue Res.*, 15, 221, 1974.

55. **Nagai, N. and Frank, R. M.**, Electron microscopic autoradiography of ^{45}Ca during dentinogenesis, *Cell Tissue Res.*, 155, 513, 1974.
56. **Kashiwa, H. K. and Sigman, M. D.**, Calcium localized in odontogenic cells of rat mandibular teeth by the glyoxal-bis (2-hydroxyamil) method, *J. Dent. Res.*, 45, 1796, 1966.
57. **Appleton, J. and Morris, D. C.**, The use of potassium pyroantimonate osmium as a means of identifying and localizing calcium at the ultrastructural level in the cells of calcifying systems, *J. Histochem. Cytochem.*, 27, 676, 1979.
58. **Takuma, S., Yanagisawa, T., and Sin, W. L.**, Ultrastructural and microanalytical aspects of developing dentin in rat incisors, *Calcif. Tissue Res.*, 24, 215, 1977.
59. **Robison, R.**, The possible significance of hexosephosphoric esters in ossification, *Biochem. J.*, 17, 286, 1923.
60. **Göthlin, G. and Ericsson, J. L. E.**, Line structural localization of acid phosphomonoesterase in the osteoblasts and osteocytes of fracture callus, *Histochemie*, 35, 81, 1973.
61. **Silberman, M. and Frommer, J.**, Hydrolytic enzymaticity during endochondral ossification of secondary cartilage, *Am. J. Anat.*, 140, 309, 1974.
62. **Wergedal, J. E. and Baylink, D. J.**, Distribution of acid and alkaline phosphatase activity in undemineralized sections of the rat tibial diaphysis, *J. Histochem. Cytochem.*, 17, 799, 1969.
63. **Granstrøm, G. and Linde, A.**, A comparison of ATP-degrading enzyme activities in rat incisor odontoblasts, *J. Histochem. Cytochem.*, 6, 322, 1976.
64. **Russel, R. G. G., Bisaz, S., and Fleisch, H.**, Pyrophosphate and disphosphonates in calcium metabolism and their possible roll in renal failure, *Arch. Intern. Med.*, 124, 571, 1969.
65. **Fleisch, H. and Neuman, W. F.**, Mechanism of calcifications: role of collagen polyphosphatase and phosphatase, *Am. J. Physiol.*, 200, 1296, 1961.
66. **Fleisch, H.**, Role of nucleation and inhibition in calcification, *Clin. Orthop.*, 32, 170, 1964.
67. **Neuman, W. F. and Neuman, M. W.**, *The Chemical Dynamics of Bone Mineral*, University of Chicago Press, Chicago, 1958.
68. **Alcock, N. W. and Shils, M. E.**, Association of inorganic pyrophosphatase activity with normal calcification of rat costal cartilage, *Biochem. J.*, 112, 505, 1969.
69. **Fernley, H. N. and Walker, P. G.**, Studies on alkaline phosphatase, *Biochem. J.*, 104, 1011, 1967.
70. **Ali, S. Y., Sajdera, S. W., and Anderson, H. C.**, Isolation and characterization of calcifying matrix vesicles from epiphyseal cartilage, *Proc. Natl. Acad. Sci. U.S.A.*, 67, 1513, 1970.
71. **Francis, M. D.**, The inhibition of calcium hydroxyapatite crystal growth by polyphosphonates and polyphosphates, *Calcif. Tissue Res.*, 3, 151, 1969.
72. **Termine, J. D., Wuthier, R. E., and Posner, A. S.**, Amorphous-crystalline mineral changes during endochondial and periosteal bone formation, *Proc. Soc. Exp. Med.*, 125, 4, 1967.
73. **Wuthier, R. D., Bisax, S., Russell, R. G. G., and Fleisch, H.**, Relationship between pyrophosphate, amorphous calcium phosphate and other factors in the sequence of calcification, *Calcif. Tissue Res.*, 10, 198, 1972.
74. **Felix, R. and Fleisch, H.**, Pyrophosphatase and ATPase in isolated cartilage matrix vesicles, *Calcif. Tissue Res.*, 22, 1, 1976.
75. **Ali, S. Y.**, Analysis of matrix vesicles and their role in the calcification of epiphyseal cartilage, *Fed. Proc.*, 35, 135, 1976.
76. **DiMuzio, M. T. and Veis, A.**, Phosphophoryns — major non-collagenous proteins of rat incisor dentin, *Calcif. Tissue Res.*, 25, 169, 1978.
77. **Leaver, A. G., Triffitt, J. T., and Holbrook, I. B.**, Newer knowledge of noncollagenous protein in dentin and cortical bone matrix, *Clin. Orthop. Rel. Res.*, 110, 269, 1975.
78. **Lee, S. L. and Glimcher, M. J.**, Purification, composition and ^{31}O NMR spectroscopic properties of a non-collagenous phosphoprotein isolated from chicken bone matrix, *Calcif. Tissue Int.*, 33, 385, 1981.
79. **Larsson, Å. and Larsson, S. E.**, Ethane-1-hydroxy-1,1-diphosphonate (EHDP) inhibits crystal growth in predentin calcification, *Virchows Arch. Abt. A*, 360, 315, 1973.
80. **Larsson, Å.**, The short-term effects of high doses of ethylene-1-hydroxy-1,1-diphosphonate upon early dentin formation, *Calcif. Tissue Res.*, 16, 109, 1974.
81. **Lapiere, C. M., Lenaers, A., and Kohn, L. D.**, Procollagen peptidase: an enzyme excising the coordination peptides of procollagen, *Proc. Natl. Acad. Sci. U.S.A.*, 68, 3054, 1971.
82. **Raisz, L. G., Trummel, C. L., Holick, M. F., and Deluca, H. F.**, 1,25-Dihydroxycholecalciferol: a potent stimulator of bone resorption in tissue culture, *Science*, 175, 768, 1972.
83. **Raisz, L. G., Maina, D. M., Gworek, S. C., Dietrich, J. W., and Canalis, E. M.**, Hormonal control of bone collagen synthesis *in vitro:* inhibitory effect of 1-hydroxylated vitamin D metabolites, *Endocrinology*, 102, 731, 1978.
84. **Mellanby, M.**, The influence of diet on the structure of teeth, *Physiol. Rev.*, 8, 545, 1978.
85. **Weinmann, J. P. and Schour, I.**, Experimental studies in calcification. I. The effect of a rachitogenic diet on the dental tissues of the white rat, *Am. J. Pathol.*, 21, 821, 1945.

86. **Leaver, A. G.,** Inter-relationships of calcium, phosphorus and vitamin D in the bones and teeth of the rat, *Clin. Orthop.*, 78, 90, 1971.
87. **Irving, J. T.,** A comparison of the action of vitamin D on the teeth of rachitic rats with that of additional calcium or phosphorus added to rachitogenic diets, *J. Physiol.*, 104, 253, 1946.
88. **Lamm, M. and Neuman, W. F.,** On the role of vitamin D in calcification, *Arch. Pathol.*, 66, 204, 1958.
89. **Yoshiki, S., Yanagisawa, T., Suda, T., and Sasaki, S.,** The role of vitamin D in the mineralization of dentin in rats made rachitic by a diet low in calcium and deficient in vitamin D, *Calcif. Tissue Res.*, 15, 295, 1974.
90. **Glasstone, S.,** The development of tooth germs *in vitro*, *J. Anat.*, 70, 260, 1932.
91. **Koch, W. E.,** *In vitro* development of tooth rudiments of embryonic mice, *Anat. Rec.*, 152, 513, 1965.
92. **Munksgaard, E. C., Rhodes, M., Mayne, R., and Butler, W. T.,** Collagen synthesis and secretion by rat incisor odontoblasts in organ culture, *Eur. J. Biochem.*, 82, 609, 1978.
93. **Munksgaard, E. C., Richardson, W. L., III, and Butler, W. T.,** Biosynthesis of phosphoprotein by rat incisor odontoblasts in *in vitro* culture, *Arch. Oral Biol.*, 23, 583, 1978.
94. **DiMuzio, M. T., Bhown, M., and Butler, W. T.,** Odontoblast/dentin organ cultures: the biosynthesis of matrix proteins, in *The Chemistry and Biology of Mineralized Connective Tissues*, Veis, A., Ed., Elsevier, New York, 1981, 427.
95. **Mayne, R., Vail, M. S., Mayne, P. M., and Miller, E. J.,** Changes in type of collagen synthesized as clones of chick chondrocytes grow and eventually lost division capacity, *Proc. Natl. Acad. Sci. U.S.A.*, 73, 1674, 1976.
96. **Benya, P. D., Padilla, S. R., and Nimni, M. E.,** The progeny of rabbit articular chondrocytes synthesize collagen Types I and III and Type I trimer, but not Type III. Verifications by cyanogen bromide peptide analysis, *Biochemistry*, 16, 865, 1977.
97. **Little, C. D., Church, R. L., Miller, R. A., and Ruddle, F. H.,** Procollagen and collagen produced by a tetatocarcinoma-derived cell line, TSD4: evidence for a new molecular form of collagen, *Cell*, 10, 287, 1977.
98. **Schlitz, J. R., Rosenbloom, J., and Levenson, G. E.,** The effects of ascorbic acid deficiency on collagen synthesis by moose molar tooth germs in organ culture, *J. Embryol. Exp. Morphol.*, 37, 49, 1977.
99. **Wohllebe, M. and Carmichael, D. J.,** Type-I trimer and Type I collagen in neutral-salt-soluble lathyritic-rat dentin, *Eur. J. Biochem.*, 92, 183, 1978.
100. **Wigglesworth, D. J.,** Formation and mineralization of enamel and dentin by rat tooth germs *in vitro*, *Exp. Cell Res.*, 49, 211, 1968.
101. **Thesleff, I.,** Differentiation of odontogenic tissues in organ culture, *Scand. J. Dent. Res.*, 84, 353, 1976.
102. **Grobstein, C.,** Mechanisms of organogenetic tissue interaction, *Natl. Cancer Inst. Monogr.*, 26, 279, 1967.
103. **Koch, W. E.,** Tissue interactions during *in vitro* odontogenesis, in *Developmental Aspects of Oral Biology*, Slavkin, H. C. and Bavetta, L. A., Eds., Academic Press, New York, 1972, 151.
104. **Yamada, M., Bringas, P., Grodin, M., MacDougall, M., Cummings, E., Grimmitt, J., Weliky, B., and Slavkin, H. C.,** Chemically-defined organ culture of embryonic mouse tooth organs: morphogenesis, dentinogenesis and amelogenesis, *J. Biol. Buccale*, 8, 127, 1980.
105. **Thesleff, I. and Pratt, R. M.,** Tunicamycin inhibits mouse tooth morphogenesis and odontoblast differentiation *in vitro*, *J. Embryol. Exp. Morphol.*, 58, 195, 1980.
106. **Hurmerinta, K., Thesleff, I., and Sexen, L.,** *In vitro* inhibition of mouse odontoblast differentiation by vitamin A, *Arch. Oral Biol.*, 25, 385, 1980.
107. **Thesleff, I., Barrach, H. J., Foidart, J. M., Vaheri, A., Pratt, R. M., and Martin, G. R.,** Charges in the distribution of Type IV collagen, laminin, proteoglycan and libronectin during mouse tooth development, *Dev. Biol.*, 82, 182, 1981.
108. **Rahemtulla, F., Prince, C. W., Duong, H. L., and Butler, W. T.,** unpublished observation.

DENTIN CONDITIONS AND DISEASES

David H. Pashley

INTRODUCTION

Dentin is the secretory product of the pulp. Although most of the secretion takes place during tooth development, odontoblasts and other pulpal cells continue to modify dentin throughout life. The rate of modification depends upon the nature of the stimuli that are applied to dentin. The following discussion will cover the most common conditions or diseases that affect dentin. These include fractured, cut, and abraded dentin as examples of hypersensitive dentin; dentin exposed by attrition; dentin invaded by microorganisms (carious dentin); and finally, the interrelationship between dentin and the pulp. The emphasis will be on presenting a functional perspective of these conditions rather than providing an exhaustive review of their literature.

GENERAL

As long as dentin is covered by enamel on coronal surfaces and cementum on radicular surfaces, the dental pulp will generally remain healthy for life, unless the apical blood supply is disrupted by excessive orthodontic forces or acute trauma. Most pathologic pulpal conditions begin with removal of one or both of these protective barriers via dental caries, fractures, erosion, or abrasion. This results in the communication of pulpal soft tissues with the oral cavity by way of dentinal tubules. Under such conditions, thermal, osmotic, hydrostatic, and chemical insults can act upon a wide variety of pulpal cells such as odontoblasts, nerves, endothelial cells, and vascular smooth muscle. The initial stages of the pulpal reaction(s) begin with stimulation or irritation of these cells, but the process may proceed to inflammation and often to tissue destruction.

To understand how these processes may lead to pulpal damage, it is instructive to examine the pulpodentin complex in its separate forms, beginning with dentin. The variables regulating the rate at which substances move across dentin will be reviewed along with how these might affect the pulp. The mechanism(s) of dentin sensitivity will be covered briefly so that the reader will understand the lack of dentin sensitivity in some conditions. Consideration will be given to a wide variety of mechanisms, both passive and active, that may alter dentin sensitivity. Dentin will be further subdivided into the different types of dentin that are found under a wide variety of clinical conditions. Carious dentin will be examined from the perspective of how it responds to infection and how the pulp responds to caries. The mechanisms of bacterial invasion of dentin will be discussed along with the effects of bacterial products on the pulp. A discussion of the sources of nutrition for carious microorganisms will follow. A section titled "Speculations" is included to permit the author to bring together several observations that appear unrelated but which may provide a more unifying perspective of dentinogenesis. Finally, several speculative discussions of the functional implications of dentin permeability, dentin sensitivity, and the pulpal response to caries will complete this chapter.

DENTIN STRUCTURE

Dentin is a calcified connective tissue penetrated by thousands of tubules per square millimeter. Estimates of the tubular density vary from 40,000 to 70,000 tubules per square millimeter,[1,2] depending upon the distance from the pulp. The tubules are about 1 μm in diameter at their periphery to about 3 μm at their pulpal surface. The walls of the tubules are lined by a hypermineralized "peritubular" dentin throughout most of their length with the exception of that part of the tubule nearest the pulp.[3]

The dentin between the tubules is called "intertubular" dentin which is richer in organic matrix and somewhat less calcified than peritubular dentin. The tubules contain an extracellular-like fluid which cumulatively accounts for a significant fraction of the total volume of dentin (20 to 30%). If the fluid is contaminated with bacterial byproducts such as endotoxins and exotoxins, then it can be regarded as a reservoir of injurious agents that can permeate to the pulp and cause inflammation. This process of permeation of substances through dentin, while a passive process is determined in large part by the structure of dentin.

Permeability Characteristics of Dentin

Dentinal tubules converge as they approach the pulp chamber because the surface area of the pulp chamber is much smaller than the surface area of the dentinoenamel junction (DEJ). This results in a concentrating or focusing effect of permeating substances into a smaller area at the termination of the tubules than at their origin. One can calculate total tubular surface areas from Garberoglio and Brännström's data[1] to be about 1% at the DEJ, 10% halfway through dentin, to about as high as 45% at the pulp chamber. The clinical implications of this are that as dentin is removed, either mechanically by the dentist or pathologically by disease processes, the remaining dentin becomes more and more permeable because there are more and more tubules per unit surface area, each tubule is larger[1], and the dentin becomes thinner, thereby increasing diffusion gradients.[4] Thus, excavation of dentin, while necessary, increases the potential for chemical or bacteriologic irritation to the pulp. It provides strong evidence for the necessity of "basing" deep cavities prior to final restoration. The convergence of dentinal tubules is also responsible for the fall in dentin microhardness that has been observed in dentin as Knoop indentation measurements were made from the DEJ toward the pulp.[5,6] Dentin near the pulp has more tubules and less calcified intertubular matrix per unit area than more peripheral dentin. It is this relative softness of the dentin lining the pulp chamber that permits endodontic treatment. Another important variable affecting dentin permeability is the surface area of the exposed dentin. Obviously a leaking restoration on a full crown restoration will provide more diffusional surface area for bacterial products than a small occlusal restoration.[4] Large surface areas of dentin that are also reduced in thickness (i.e., crown preparation for a porcelain jacket crown) would work together to increase the rate at which injurious substances could diffuse across dentin from the mouth to the pulp.

The character of the dentin surface can also modify dentin permeability. Two extremes are possible, viz., tubules that are completely open (i.e., seen in freshly fractured or acid etched dentin) and tubules that appear, even at the scanning electron microscopic level, to be completely occluded with microcrystalline debris, commonly referred to as the "smear layer".[7] This forms on dentin surfaces whenever they are cut with either hand or rotary instruments.[8] Although the smear layer prevents bacteria from penetrating dentin,[9,10] there apparently are spaces between the microscopic particles making up the smear layer that permit molecules as large as albumin (molecular weight 68,000) to readily permeate such dentin, but at a reduced rate. By using a variety of radioactive molecules of differing size, a reciprocal relationship between dentin permeability and molecular size was found such that small molecules like water or glucose permeate dentin much faster than larger molecules such as dextrans and proteins. Removal of the "smear layer" by acid etching increases dentin permeability because the microcrystalline debris no longer restricts the diffusional surface area.[11] This procedure also permits bacteria to penetrate into dentin where they may migrate slowly toward the pulp by cell division or they may pass rapidly if aided by hydrostatic pressures such as those produced during the mastication of food. There are enough constrictions and irregularities within dentinal tubules, however, to remove 99.8% of the bacteria that enter the dentin surface.[10] These intratubular resistances to permeating substances result in the functional or physiologic tubular radius being only about one tenth of the anatomic or microscopic tubular radius[12] or about 0.05 μm, which is small enough to remove bacteria from saliva that may move across dentin. Thus, oral fluids teeming with microorganisms may be rendered sterile as they pass across dentin. This accounts for the fact that patients may complain of hypersensitive dentin for years without developing pulpal inflammation. Fluid movement is thought to occur in such conditions and is, in fact, the basic premise of the hydrodynamic theory of dentin pain.[13]

Pulpal Reaction to Permeating Substances

In the last example, if fluid can move across exposed dentin to cause dentinal pain or hypersensitivity, then other substances should also permeate dentin to reach the pulp. It is appropriate to consider what happens to permeating substances after they reach the pulp chamber. Although bacteria may not actually permeate dentin until very late in the caries process, carious dentin or products from dental plaque when placed on dentin have been shown to cause severe pulpal reactions.[14-16] The difference between these two extremes (no inflammation vs. abscess formation) is most likely due to differences in the pulpal concentration of these injurious substances. If the pulpal interstitial fluid concentration of these substances can be maintained at a vanishingly low concentration by the pulpal circulation, then even though exposed dentin may permit substances to permeate,[17] they will never reach a concentration high enough to trigger PMN chemotaxis, mast cell degranulation, or other components of the cascade of events that are associated with inflammation. The key element in maintaining the pulpal interstitial fluid concentration of harmful substances at a low level is the balance between the rate at which they permeate dentin relative to the rate at which they can be removed by the pulpal circulation.[18] As long as the rate of pulpal blood flow is normal, the microcirculation is extremely efficient at removing substances diffusing across dentin to the pulp chamber.[19] Meyer's group,[20-22] using radioactive microspheres, has reported that dog pulpal blood flow varies (with tooth type and with location within the pulp) from 0.4 to 1.0 $m\ell/min^{-1}/g^{-1}$ being higher in the crown than the root. These rates of blood flow are similar to those reported in gingivae by Kaplan et al.,[23] using radioactive microspheres, but are much higher than blood flows to bone or skeletal muscle. Pulpal blood flow rates are comparable to blood flow rates in liver, heart, and brain.[24] Thus, there is enough blood flowing through the pulp each minute to completely replace between 40 to 100% of the entire volume of the pulp. Since blood is confined to the vasculature which only comprises about 6.8% of the total pulpal volume,[25] the blood volume of the pulp is then replaced 5 to 14 times each minute. This high rate of blood turnover tends to keep the concentration of substances diffusing into the pulp from dentin at a very low level. Pulpal blood is in equilibration with pulpal interstitial fluid. Interstitial fluid is both extracellular and extravascular, but is in direct communication with both plasma and dentinal fluid.[26] If the pulpal interstitial fluid volume is like those of other organs,[27] it must occupy about 20% of the total pulpal volume. The high rate of pulpal blood flow keeps the interstitial fluid concentration of substances diffusing across dentin very low. This has recently been confirmed in studies on dog teeth which were prepared with chambers on both the buccal and lingual dentin. Isotopic tracers placed in the buccal chamber diffused rapidly into the pulp where they were removed by the circulation. Procedures which interfered with pulpal circulation permitted the tracers to appear in the lingual chamber.[19]

If, for any reason, pulpal blood flow is reduced,[20,28,29] then the rate of turnover or replacement of both interstitial fluid and pulpal blood will decrease, resulting in a rise in the interstitial fluid concentration of substances that had been permeating across dentin. Since vascular smooth muscle, capillary endothelia, and perivascular cells such as mast cells are all located in or bathed by interstitial fluid, as the concentration of injurious agents increases, concentrations sufficient to degranulate mast cells[30,31] may be achieved. This would result in the release of histamine[32,33] or substance P,[28,34] produce bradykinin[35,36] or prostaglandins,[37] activate plasma proteins including the complement series[38-40] which will initiate a host of reactions resulting in inflammation. The net result of the action of endogenous mediators of inflammation is arteriolar vasodilation, elevated capillary hydrostatic pressure, increased leaking of plasma proteins into the pulpal interstitium,[41] and increased pulpal tissue pressure.[42,43] These events lead to a further reduction in pulpal blood flow which allows the interstitial concentration of bacterial products to increase to even higher values which complete the loop of a positive feedback system.[19] This results in further decreases in pulpal blood flow. Pulpal blood flow will fall in direct proportion to any increase in pulpal tissue pressure. Van Hassel,[42] using chronic taps just into the pulp, reported that pulpal tissue pressure is elevated in pulpitis, but that the elevation is localized within specific regions of the pulp, being normal in noninflamed areas. Thus, total pulpal blood flow need not fall when a pulpal abscess begins since the elevation in tissue pressure may be localized, producing regional reductions in the distribution of pulpal blood flow. The localized reduction in pulpal blood flow, however, will allow the accumulation of mediators of inflammation which will cause a spread in the elevation of tissue pressure, reducing pulpal blood flow to a larger volume of pulp, etc.[19,42] The elevated pulpal tissue pressure causes the dull aching, poorly localized pulpal sensory nerve activity[44,45] which apparently is perceived with a much different quality than the brief, sharp, well-localized dentinal pain which is postulated to be due to fluid movement within dentin.[46] When the pulp chambers of teeth with elevated pulpal pressures are opened, the pain generally subsides rapidly as pulpal tissue pressures fall to normal values.

This difference in dental pain is probably due to several factors. The dull pain of pulpitis is associated with pulpal inflammation and elevated tissue pressures. The presence of algesic or pain producing substances like bradykinin,[36,47] substance P,[48] and perhaps prostaglandins must contribute to the activation of sensory nerves within the pulp. The reduction in pulpal blood flow brought about by elevations in pulpal tissue pressure prevents the circulation from removing these substances as fast as they are produced. Opening the tooth and allowing the abscess to drain lowers tissue pressure and resolves the pain. The brief, sharp pain of unanesthetized or exposed dentin may be the result of a different mechanism operating in a different (i.e., normal) pulpal environment.

THEORIES OF DENTIN SENSITIVITY

The sensitivity of dentin to a wide variety of stimuli is well known to clinicians but the mechanism underlying the sensitivity is not well understood. Early investigators assumed that dentin sensitivity was due to the presence of nerves in dentin. Light microscopists, using heavy metal stains, demonstrated what were thought to be nerves in the dental pulp and in dentin.[49] Arguing against the theory that dentin sensitivity was due to direct stimulation of nerves on exposed dentin was the observation by Fernhead[50,51] that newly erupted teeth had few nerves and yet were as sensitive as ''normal'' mature teeth. Anderson's group failed to cause pain in patients' dentin when agents known to cause nerve depolarization (potassium chloride, acetylcholine, etc.) were applied to fresh cavities (see Reference 52). Both Anderson[53] and Brännström in Sweden[54] were unable to anesthetize freshly cut dentin by applying local anesthetics topically. All of these observations argue against any mechanism of dentin sensitivity that is due to the direct stimulation of nerves in dentin by thermal or mechanical stimuli. This led others[55,56] to suggest that dentin sensitivity was mediated by odontoblasts. The proponents of this theory argued that nerves in dentin were in contact (synapse-like) with odontoblastic processes which, in turn, projected through dentinal tubules to their termination. Indeed, Arwill et al.[57,58] and Frank et al.[59] published electron-micrographs of unusual structural junctions between intratubular nerves and odontoblastic processes. This hypothesis had some merit in that it postulated that dentinal or pulpal nerves were stimulated indirectly, but careful examination of odontoblasts or their processes for synaptic vesicles[60] failed to demonstrate the morphologic requirements of such a theory. This theory was also weakened by recent electron microscopic studies designed to determine the extent of the odontoblastic process. These electron microscopic reports in animals[61-63] and humans[64-66] found that the peripheral one half to two thirds of the tubules are devoid of odontoblastic processes and nerves. Odontoblastic processes and nerves extend into the tubules about one third of their length, but the rest of the tubule is simply filled with an extracellular-like fluid.[67] Thus, there are no structures in peripheral dentin (no odontoblastic processes and no nerves) capable of being stimulated. This strengthens Brännström's "hydrodynamic theory" of dentin sensitivity.[13,65] In essence, the theory states that dynamic fluid movement within dentinal tubules is common to mechanical, thermal, and osmotic stimuli. This fluid movement in either direction stimulates nerves in dentin or pulp which then results in dentinal pain. This may be due to slight physical displacement of the odontoblast and/or adjacent nerves into the predentin tubule orifice, thereby mechanically stimulating them or causing them to release intracellular potassium from odontoblasts[68] which might depolarize nearby nerves to thresholds resulting in conducted nerve impulses. The theory does not rule out the possibility that fluid movement affects odontoblasts which, in turn, somehow stimulate pulpal nerves. Such a mechanism seems unlikely, however, in view of the experiments in which Brännström allowed sufficient time for injured odontoblasts to die and yet found the affected dentin as sensitive as it was before destroying the odontoblasts.[69]

Since fluid movement across dentin is fundamental to the "hydrodynamic theory", several studies of fluid movement across dentin have been done to clarify the influence of dentin structures. These show that normal dentin can be regarded as having three zones: a surface zone, an intratubular zone, and a pulpal zone (Figure 1). The resistance to fluid movement is different in each zone,[70] and each zone can be modified to increase its resistance to fluid movement, with resulting changes in both dentin sensitivity and dentin permeability. The surface zone, which extends ± 10 to 15 μm from the dentin surface, can either be free of debris, as in freshly fractured or acid-etched dentin, or the tubules can be occluded with a layer of of microcrystalline debris or "smear layer". This layer, when present, accounts for 86% of the total resistance to fluid movement across dentin in vitro.[70] Acid-etching dentin, even for a few seconds, removes this layer of debris, thereby increasing the permeability[11,71,72] and the sensitivity of the dentin.[73]

The middle zone of dentin (which extends from just beneath the surface zone to the most peripheral extent of the odontoblastic processes) can be altered by the deposition of mineral or organic substances within the tubule. Intratubular mineral deposition from saliva (see the section entitled "Precipitation of Salivary Calcium Phosphates in Tubules") or from odontoblasts (see the section entitled "Deposition of Intratubular Crystalline Material") has been reported in teeth affected by attrition, abrasion, and caries. The intratubular secretion of collagen and/or amorphous organic material may also occur (see the section entitled "Secretion of Protein Material into Tubules"). These materials would presumably decrease both dentin permeability and dentin sensitivity. Sclerosis of this zone in advance of dental caries presumably contributes to the lack of pain in spite of the presence of dentin exposed to the oral cavity.

Finally, the inner zone of dentin is perhaps the most dynamic since it is here where viable cells can respond to a wide variety of stimuli. The reaction in this zone may range from a narrowing of the width of predentin, to the formation of irritation dentin (see the section entitled "Aspiration of Cells"), to the destruction of odontoblasts.[14,74,75] Although no quantitative studies have been done on the hydraulic conductance (the ease of fluid flow) across dentin which has been occluded by odontoblastic nuclei, one would expect fluid flow to be severely limited.[76] Similarly, if the tubules of irritation dentin do not match up with those

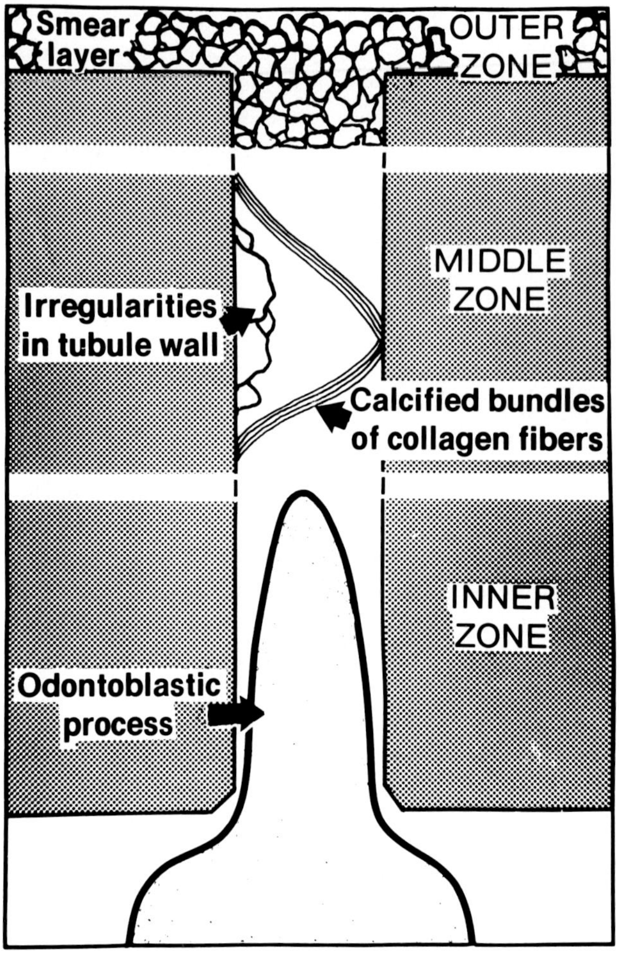

FIGURE 1. Dentin can be regarded as consisting of three zones: an outer zone that is created whenever dentin is cut with hand or rotary instruments; a middle zone devoid of odontoblastic processes or nerves but containing nonliving irregularities that can modify dentin permeability; an inner zone containing odontoblastic processes and nerves. Dimensional changes can occur independently in any zone which, in turn, can alter overall dentin permeability.

of primary dentin, fluid flow would be greatly restricted. Alternatively, if atubular dentin is produced it would greatly reduce dentin permeability. Thus, before one can understand the reaction of the pulpodentinal complex to abnormal stimuli, it is necessary to understand normal dentin structure and function. Investigators are at the threshold of understanding the properties of normal dentin, but much more work is required.

DENTIN CONDITIONS

Hypersensitive Dentin

Normally, dentin is covered with enamel or cementum and thus is not available for direct stimulation. When the peripheral ends of the tubules are exposed we speak of it as ''hypersensitive'' dentin when, in fact, it is no more sensitive than it should be. Many patients suffer from ''hypersensitive dentin'' following peridontal surgery, calculus removal, and/or root planning, but this often occurs spontaneously as ''normal'' gingival recession exposes root surfaces to the oral cavity. The thin (50 to 100 μm) covering of soft cementum is then slowly removed by tooth brushing, resulting in exposure of dentinal tubules to mechanical, osmotic, or thermal stimuli. What is puzzling is the fact that almost 40% of such patients who report initial dentin sensitivity show "spontaneous" loss of such sensitivity with time (days to weeks). If one believes that shifts in dentinal fluid are responsible for the sensitivity (the hydrodynamic theory), then the loss of sensitivity must be due to a loss of the ability of fluid to move through dentinal tubules. This must be due to the occlusion of tubules somewhere between the pulp and the surface. Several mechanisms are possible which might accomplish this result.

Passive Mechanisms of Decreasing Dentin Sensitivity

Precipitation of Salivary Calcium Phosphates in Tubules

Salivary calcium phosphate salts may precipitate in the tubules passively, as was recently reported by Brännström and Garberoglio.[77] These investigators compared the scanning electron microscopic appearance of dentinal tubules exposed to the oral cavity by enamel attrition vs. extracted young premolars mounted in a removable denture in which the cusp tips were ground down to expose dentin to simulate the in vivo condition. After "extraction", all teeth were fractured 1.5 mm below the exposed surface and examined by an SEM for tubule occlusion. They found that many, but not all, of the exposed tubules in the vital teeth were completely occluded with a material similar to peritubular dentin. The nonvital teeth showed similar occlusion of dentinal tubules indicating salivary mineral (the only route available in these pulpless teeth) may contribute to the development of dentinal sclerosis. Presumably, similar reactions could occur on root surfaces, since saliva is often supersaturated with respect to most forms of calcium phosphate. Saturation depends upon a host of variables such as salivary flow rate, pH, protein, and calcium and phosphate concentrations (see Reference 78 for review).

Adsorption of Plasma-Derived Proteins onto the Tubule Contents

Another possible explanation for spontaneous recovery from hypersensitive dentin is the fact that plasma per se or large plasma proteins such as fibrinogen and γ-globulins can decrease the ease with which fluid can move across dentin disks in vitro.[79] These proteins are present in dentinal fluid.[40,41] Since pulpal tissue pressure is higher than oral (atmospheric) pressure, there is a very slow filtration of dentinal fluid and its contents from the pulp to the surface. Large molecular weight proteins, lipoproteins, or glycoproteins may adsorb to the inside of dentinal tubules (which are essentially hydroxyapatite ion exchange beds), thereby reducing their radii. If this occurs at areas within the tubule which were already naturally constricted due to surface irregularities or to the presence of granular material within the tubule,[65] the functional radius[12] of the tubules would be reduced to a point where fluid flow would be significantly reduced.

Adsorption of Salivary Constituents (Glycoproteins) onto the Tubule Contents

While this is occurring from the inside (pulpal side) outward, salivary constituents may be permeating the tubules from the outside. Whole saliva (containing various salivary proteins, bacteria, shed epithelial cells, etc.) is very effective at reducing fluid flow through dentin disks using the in vitro techniques previously employed by Reeder et al.[71] It is not yet known which components of saliva are responsible, although it is known that bacteria can partially occlude dentinal tubules.[10] When submandibular saliva is passed through dentin disks in vitro, the saliva loses its apparent viscosity due to removal of glycoproteins by dentin. Protein determinations also show a loss of salivary protein across dentin. The adsorption of salivary proteins onto the hydroxyapatite of enamel to form an "acquired pellicle" has been extensively studied (see Reference 78 for review). Little work has been done on the adsorption of salivary proteins by dentin.

Active Mechanisms of Decreasing Dentin Sensitivity

Deposition of Intratubular Crystalline Material

In contrast to "passive" mechanisms of reducing fluid movement across dentin and thereby reducing dentin sensitivity, the activities of odontoblasts may also lead to tubular occlusion. Hiatt and Johansen[80] reported that patients who burnished their root surfaces with toothpicks (a form of mechanical stimulation) had hard, smooth, nonsensitive dentin. Electron microscopic examination of such dentin demonstrated that most of the peripheral portion of the tubules was occluded with crystalline material. Dreyfus et al.,[81] Van Huysen,[82] and Gwinett and Jendresen[83] have reported that exposed (irritated) dentinal tubules are often completely occluded by what appears to be a progressive growth of peritubular dentin. Others have reported the tubules are occluded by crystalline material that ranges from amorphous to fine, needle-shaped crystallites to large cuboidal crystals nearly the size of dentinal tubules.[84,85] Presumably, these mineralizations required active participation of the odontoblast and/or its process to modify the calcium or phosphate concentrations of dentinal fluid or otherwise influence mineralization mechanisms.

Secretion of Protein Material into Tubules

Odontoblasts that are irritated may not die but may secrete collagen very rapidly for a short period of time, which causes partial to total occlusion of the tubule for varying lengths. Tsatsas and Frank[64] found protein material within tubules near the DEJ using electron microscopy. This would effectively reduce dentin permeability and sensitivity. This extra material may only occupy 5 to 10 μm within dentinal tubules that are 2000 to 3000 μm (2 to 3 mm) long, but it would effectively "plug" the tubules. Garant[61] reported that the dentinal tubules of both normal and germ-free rats were often occluded by calcified deposits and/or wide collagen fibers. In such experiments, one cannot ascribe the reactions of dentin to bacterial products although they may still be due to the mechanical trauma of attrition. Furseth and Mjör[86] lined cavities cut in human teeth with a corticosteroid preparation. After 40 to 138 days, the teeth were extracted and examined by electron microscopy. The width of the predentin was reduced in keeping with the well-known effects of corticosteroids at reducing protein synthesis. Unexpectedly, more than normal collagen secretion was found in dentinal tubules, sometimes nearly occluding them. These two results seem contradictory and no explanation for them is known. Many clinicians use corticosteroid creams under temporary crowns to reduce dentin sensitivity. Their clinical success may be due to the occlusion of dentinal tubules by collagen and mineralized tissue, although no good clinical studies of the efficacy of corticosteroids as a dentin desensitizing treatment have been published.

Formation of Irritation Dentin

Normally, odontoblasts lay down dentin at a rate of 4 to 15 μm a day, depending upon tooth type, location within root canal, age, etc.[87] This has been determined by histologically measuring the distance between incremental fluorescent bands on dentin ground sections after sequential doses of tetracycline, a drug that deposits in calcifying structures.[88,89] If odontoblasts are irritated, this rate can increase greatly and the dentin formed is called "reparative" or "irritation" dentin. Although it is generally tubular in nature, it can be atubular.[90] Atubular dentin is formed by odontoblastic-like cells which often have minimal to absent terminal cell processes. This type of dentin is thought to be relatively impermeable since it lacks tubules. Dentin formed less rapidly, presumably in response to a lesser stimulation, shows much more tubularity, although the tubular density is less than that of primary (normal) dentin.[83,91-93]

The exact mechanism(s) which translate the irritational stimuli into accelerated rates of collagen secretion and matrix mineralization are not known. Perhaps fluid movement not only stimulates mechanoreceptors in the pulp which are perceived as pain (the hydrodynamic theory of dentin sensitivity), but also results in the production of chemical signals to the odontoblasts. This might take the form of changes in the distribution of ions across the cell membrane, changes in the intracellular levels of cyclic AMP or GMP, etc.

If the odontoblasts are injured or die, new odontoblasts differentiate from pulpal mesenchymal cells (measured autoradiographically after ^{3}H-thymidine administration) and begin formation of irritation dentin.[95] The probability of these new cells orienting their tubular lumina with those of the previous tubules is very low. If irritation dentin is laid down before all of the cellular debris from the dying odontoblasts can be phagocytized, there may be a cleft formed between primary and irritation dentin which permits communication between tubules even though they do not line up perfectly.[83] The resulting permeability would be dependent upon the number of tubules in the irritation dentin. If, on the other hand, the odontoblasts die, are completely removed, and irritation dentin formation proceeds on "sound" primary dentin, there would be even less chance for communication between tubules of the two forms of dentin.

Aspiration of Cells

Numerous investigators have reported the "aspiration" of a variety of cell types up into dentinal tubules. Langeland[14] reviewed the early literature and Eda and Saito[96] have reviewed the later literature on the subject. The movement of odontoblastic nuclei from their usual position in the basal end of the odontoblastic cell body up into the odontoblastic process at the level of predentin or even more peripherally, often follows cavity preparation with inadequate cooling, application of hypertonic solutions to dentin, prolonged desiccation of dentin, or application of force to teeth during extraction with forceps.

The term "aspiration" implies that the nucleus is sucked up the tubule by a negative pressure. While this may be the case under certain conditions, it is probably not a universal mechanism, and hence the term "aspiration" should be used selectively and perhaps replaced by the more general term "migration"[14] or "displacement". For instance, squeezing a crown with extraction forceps in vitro has been reported to cause movement of odontoblasti nuclei up into the tubules.[60] This maneuver probably creates a positive intrapulpal pressure which "pushes" the cell, rather than "pulling" it into the tubule. The osmotic pressure gradient produced across dentin by application of hypertonic solutions[14,97-101] would tend to draw cells up into dentinal tubules. Similar "aspiration" forces would be produced by air blast via capillary forces acting at the air-dentinal fluid interface of tubules.[102-104]

Langeland[14] suggested that injury to odontoblasts may alter their permeability characteristics and lead to a release of osmotically active solutes across their membranes thereby upsetting osmotic equilibrium leading to creation of a negative intratubular pressure which might "displace" the nucleus. It is interesting to note that the odontoblastic nucleus is the organelle most often referred to as being displaced. Most of this is probably due to the fact that it is large enough to be seen at the light microscopic level. Electron microscopic observations indicate generalized organelle movement with disruption of the Golgi apparatus, endoplasmic reticulum, and terminal web complex.[96,105]

Another possible mechanism of nuclear displacement is through the action of odontoblastic actin and myosin. These contractile proteins have been found in almost all cells including odontoblastic processes[106] It is possible that cell injury might cause an increase in the permeability of the odontoblast to extracellular calcium which could diffuse into the cell[107] and activate the contractile proteins[108] or induce other effects leading to changes in cell volume. Since the odontoblastic process has branches that tend to fix it in place in the peripheral dentin, cell shortening would lead to "aspiration" of the cell body up into the tubule. This writer is unaware of any experiments in which calcium iontophores like A23187[109] have been used to try to induce odontoblastic contraction in vivo or in vitro by increasing calcium permeability. Garant[61] reported that microtubules and microfilaments are present in odontoblasts along with side projections between them reminiscent of the cross bridges of muscle myosin. Although he was unsuccessful in attempting to disrupt the microtubules with a single dose of colchicine, more work should be done using higher doses of these drugs[110,111] as well as cytochalasin B, an agent known to inhibit microfilament function. If, after disrupting or blocking either cytoplasmic structure, aspiration of odontoblasts could no longer be induced, we would have a better understanding of the mechanism of "aspiration".

Other cell types have been found in dentinal tubules. Orban,[112] Langeland,[14] and Eda and Saito[96] have reported the presence of polymorphonuclear leukocytes (PMNs) in tubules. They presumably move up the tubules in response to chemotactic stimuli provided by injured or disintegrating odontoblasts. If the odontoblastic cell layer is destroyed or peeled away from dentin by edema fluid, PMN and other cell types may be passively forced into the tubules. These include red blood cells.[14,96,113] These cells are so deformable that even normal pulpal pressures (25 mmHg) are sufficient to deform them into the tubules. PMNs are less deformable than red blood cells;[114] however, no information is available on the pressures required to passively displace odontoblastic contents into dentinal tubules.

It should be pointed out that the movement of any cell type, including odontoblasts, up into the tubules will temporarily decrease dentin permeability and dentin sensitivity. This may be the mechanism of action of some agents used to desensitize hypersensitive dentin. Hodosh[115] reported that hypertonic solutions of KNO_3 desensitized dentin for several days to weeks. This may have caused displacement of odontoblasts up into the tubule so that no fluid could pass around this cellular "plug" until the cell disintegrated and/or until the PMNs attempting to clear the debris had also disintegrated.

Bacterial Invasion of Dentin

Mechanisms

Mumford[116] has discussed the mechanism of fluid movement through dentin whenever occlusal forces are applied through a self-sealing, incompressible object to a defect between a restoration and the wall of a cavity. Generally, any such defects will be filled with saliva containing 10^6 to 10^7 microorganisms per milliliter. The maximum occlusal forces generated by mastication vary from 9 kg[117] in anterior regions to ~200 kg in posterior segments.[118] As a first approximation, we can consider normal chewing forces to be about one tenth of these maximal values or about 1 to 10 kg. Since pressure is force per unit area, the hydrostatic pressures generated by such force are inversely related to the surface area of the defect. Table 1 shows a variety of pressures that would be available to force fluid and bacteria across dentin, depending upon the force applied and the area of the defect. Similar arguments can be made for carious lesions, although carious dentin must have a lower hydraulic conductance due to the presence of bacteria, intratubular mineralizations, irritation of dentin, etc. Clearly, the hydrostatic pressures available for filtration are enormous if, and only if, the force can be applied, without any leakage, to the defect. This is why patients occasionally experience brief, sharp pain during mastication on teeth that are carious or have defective restorations. Only osmotic or hydrostatic pressures can move fluid across dentin. Application of forces to dentin through hand instruments is ineffective because the force is resisted by mineralized dentin matrix.[119] Impression material, on the other hand, or anything like it, would permit much more force to be transmitted to the tubules.

Bacterial invasion of dentinal tubules would seem to be extremely easy if one views the tubules as long, straight, smooth-bore tubes. Actually, tubules have numerous constrictions and considerable material, both granular and fibrillar, projecting from the walls of the tubules, sometimes crossing to the other side. These structures impede fluid flow enormously as was demonstrated when the resistance to fluid movement of dentin in vitro was compared to glass microtubule-arrays of the same dimensions.[12] These intratubular resistances are also

Table 1
POTENTIAL HYDROSTATIC PRESSURES GENERATED BY MASTICATING FORCES[a]

Defect area (cm^2)	Force applied (kg)	Pressures generated		
		mmHg	p.s.i.	kg/cm^2
0.01	1	73556	1422	100
0.05	1	14711	284	20
0.10	1	7355	142	10
0.15	1	4904	95	7
0.20	1	3678	71	5
0.50	1	1471	28	2[b]

[a] The 1-kg force used in the above example is very conservative. Forces of 10 kg would generate 10 times higher pressures.

[b] Brännström reported patients experience dental pain at a threshold of 1 to 3 kg/cm^2.

responsible for trapping 99% of the bacteria that are forced through dentin. Thus, dentin acts much like a Millipore® filter in retaining bacteria. The bacteria retained in dentin, if viable, can "grow" through dentin by repeated cell division, even in the face of these obstacles. This was shown in a simple in vitro study in which a 1-mm-thick disk of dentin was clamped between two halves of a sterile split chamber device containing nutrient broth on both sides.[10] Streptococci were innoculated on the "enamel" side of the disk and the time required to detect these streptomycin-resistant organisms on the other side was determined by twice-daily sampling. Generally it took about 5 days for the organisms to "grow" through 1 mm (1000 μm) of normal dentin. Fracturing the dentin and examining it for bacteria using a scanning electron microscope revealed that only a few tubules had "permitted" bacterial penetration throughout the entire length of the dentin. While most of the tubules at the "enamel surface" contained bacteria, most of the tubules at the pulpal side of the dentin were free of bacteria. This suggests that dentin provides a significant barrier to bacterial permeation.

In the above in vitro study, there was no loss of peritubular dentin within 5 days. Other authors have reported that carious dentin in vivo is associated with a loss of peritubular dentin adjacent to invading bacteria.[85] Five days was probably insufficient time to demineralize the peritubular dentin in vitro.

Bacterial production of proteolytic enzymes such as collagenase, other proteases, and hyaluronidase[120] must contribute to organic matrix destruction at the same time that organic acids cause dissolution of calcified dentin matrix. These two events are generally separated both in space and time. Acidogenic bacteria seem to lead the process of carious destruction with proteolytic microorganisms following in a more superficial location. The advancing demineralizing front is generally free of bacteria.[121,122] The presence of bacteria within the tubules decreases the ease with which fluid can move across dentin.[10] This is one reason why exposed carious dentin is often far less sensitive and less permeable than a comparable area of exposed noncarious dentin.

Effects of Bacterial Products on the Pulp

Numerous investigators have reported that bacteria seldom gain direct access to pulpal soft tissue via tubule penetration, yet the pulpal soft tissue underlying carious dentin or even bacterially contaminated dentin are generally heavily inflamed.[14,123] This is due to bacterial products permeating through dentin via the tubules[15] to reach the pulpal interstitium.[124] These products move down their concentration gradients by way of diffusion rather than by filtration.[125] Thus, even though little fluid movement may occur through tubules contaminated by bacteria, bacterial products can still gain access to the pulp chamber by diffusion where they induce an inflammatory response.[15,124] A recent report[126] of the biological effects of degradation products of collagen indicates that such products may have kinin-like effects on the permeability of blood vessels. Thus, bacterial products and degradation products of collagen matrix may promote pulpal inflammation. Indeed, Bergenholtz has used the reaction of the pulp to microbial products as a functional indicator of changes in dentin permeability. Bergenholtz and Reit[16] treated cavities cut in monkey teeth with calcium hydroxide or sodium chloride for 14 or 60 days, followed by replacement of these agents with microbial test substance for 3 more days. The teeth were then removed and examined microscopically for pulpal reactions. In the sodium chloride treated teeth (control group) 13 of 22 teeth showed severe pulpal necrosis while only 1 of 24 teeth treated with $Ca(OH)_2$ showed a severe reaction. Their results indicate that $Ca(OH)_2$ treatment decreases permeability of individual tubules since there was no increase in dentin thickness. This would support the use of $Ca(OH)_2$-containing cavity liners so widely used in restorative dentistry. Although Furseth and Mjör[86] demonstrated by electron microscopy that $Ca(OH)_2$ treatment produced intratubular deposits, exactly how calcium hydroxide accomplishes this reaction is not known. Other bases may be as effective at elevating pH. Elevations in pH may cause physichemical reactions within the tubules and/or stimulate the odontoblasts to modify the local chemical composition of dentinal fluid which, in turn, promotes mineral deposition. Clearly, more work needs to be done in this important area.

Sources of Nutrition for Microorganisms Associated with Caries

Once bacteria have penetrated dentin, especially in lesions that have only a small access opening through the enamel into the oral cavity, they theoretically could derive some of their nutrition from pulpal (via dentinal tubules) fluids as well as oral fluids. Several investigators have suggested that pulpal or dentinal fluid may provide invading bacteria with critical growth-limiting nutrients, such as specific amino acids or vitamins. Brown and Wheatcroft[127] performed an in vitro caries study in which bacteria invaded exposed dentin in teeth in which growth-essential nutrients were present in the pulp chamber but absent in fluid around the crown. Replacement of the pulpal growth media with distilled water did not permit bacterial invasion of the exposed dentin. This supported earlier work that demonstrated that placement of essential nutrients on the pulpal side of dentin produced artificial caries which had a shape much like that of natural caries.[128] They argued that since dentinal fluid supports the carious process, then the incidence of caries should be lower in pulpotomized teeth. Their study demonstrating that result in rats[129] can be criticized on technical grounds (i.e., use of zinc oxide and eugenol, agents known to alter bacterial growth, in test groups, but not controls), but the idea is still intriguing.

Arguing against the clinical significance of pulpally derived nutrition in caries are recent studies in which investigators have purposely applied pit and fissure sealants to occlusal caries to see how long the carious organisms remain viable. The resin application would prevent oral fluids from permeating the lesions but would not prevent pulpal fluid from reaching the organisms through the dentinal tubules by diffusion, although it would prevent filtration of fluid to the surface. Handelman et al.[130] reported a 98% reduction in the number of viable organisms recovered from carious lesions 30 days after sealing. Later studies, done for up to 24 months, showed a marked reduction in viable organisms and a change in the character of the dentin from being soft and friable to being dry and leathery.[131] Similar findings were reported by Theilade et al.[132] and more recently by Mertz-Fairhurst et al.[133,134] They noted that sealed cavities showed no change in depth over a 1-year period while open lesions increased in depth. In the process of applying pit and fissure sealants to carious teeth, there is a first treatment with 37% phosphoric acid for 1 min to roughen the enamel. Jensen and Handelman[135] have recently shown that acid etching carious lesions results in an immediate 74% reduction in viable organisms. If this is followed by sealing, the reduction in viable counts falls 99.9% over a 1-year period. These data suggest that bacteria in carious lesions derive most of their nutrition from oral fluids. The flora of carious lesions in heterogeneous and their nutritional requirements are very complex. Some organisms may obtain their amino acids and carbon sources from saliva and yet derive a critical nutrient (such as a vitamin or one essential amino acid) from dentinal (pulpal) fluid. Blocking the salivary route of the lesion's nutrition with pit and fissure sealant effectively stops bacterial growth because the nutrients which had previously been provided by saliva, in abundance, are now no longer available. Dentinal fluid apparently cannot support these bacteria as a sole source of nutrition since the number of viable organisms recovered from sealed lesions rapidly drops off with time. These studies are complicated by fundamental unknown variables such as the permeability of dentin when the lesion was sealed. Massler[136] pointed out that rapidly progressing, active carious lesions are generally sensitive to air, thermal, osmotic, and chemical stimuli, as well as permeable to dyes.[137,138]

According to the hydrodynamic theory of dentin sensitivity, such dentin is also quite permeable to substances moving in either direction. If these rapidly progressing lesions were sealed with pit and fissure sealants, one wonders if their progress would be arrested. Perhaps under these conditions, pulpal fluids can provide more nutrition than under the more usual type of carious attack. Most carious lesions develop intermittently, with active periods alternating with less active or even arrested levels of activity. This type of activity permits odontoblasts to react to the invading microorganisms by producing "sclerotic dentin" and/or irritation dentin,[138] both of which greatly reduce dentin permeability[138,139] and dentin sensitivity. Thus, the probability that chronic dental caries receives significant nutrition from the pulp is much less than that of acute caries.

Sclerotic dentin found under caries lesions was originally identified as translucent dentin in ground sections. Microradiographic studies of increasing sophistication[139,140] demonstrated varying degrees of obturation of dentinal tubules. Most such studies exaggerate the degree of sclerosis by using thick sections (>10 μm) of dentin thereby permitting structural superimpositions of tangentially cut tubules over their lumina. This method lacked the resolution required to determine how complete the tubule occlusion was or what type of material obturated the tubules. This has been provided by numerous electron microscopic studies (see References 85 and 141 for recent reviews). These studies have shown that the development of the sclerotic zone under caries varies from deposition of small amorphous granular mineral to thickening of peritubular dentin until the tubules were completely occluded, to the presence of large intratubular crystals, packed loosely in some cases, but sometimes forming calcific plugs. Such dentin must be less permeable and less sensitive. Irritation dentin, being less tubular than primary dentin,[83,94] must, consequently, be less permeable, if not impermeable.[139]

Thus, the question of the relative contribution of pulpal vs. oral fluids to the nutrition of the bacterial flora of caries is complex. If the dentin is sclerotic, has dead tracts or irritation dentin, one would expect little opportunity for nutrients to diffuse from the pulp to the lesion by way of dentinal tubules. Conversely, if the tubules remain open and/or the affected odontoblasts die and the pulp cannot decrease the permeability characteristics of dentin, then one might expect some pulpal nutrients to reach the bacteria. The latter condition is probably the exception rather than the rule.

The traditional ideas that salivary buffers are required to buffer the organic acids produced by carious bacteria and that salivary calcium phosphate and fluoride are necessary for remineralization are still valid. The rampant caries seen in patients suffering from acute loss of salivary function provide dramatic testimony to the importance of saliva in maintaining normal homeostasis. The problem is one of rates of supply of salivary buffers and nutrients.[142] Bacteria can store carbon sources both intra- and extracellularly, but the body cannot store salivary buffers. Thus, presumably, during active fermentation, the rate of bacterial production of organic acids deep within carious dentin is much faster than the rate of diffusion of salivary buffers through carious dentin. Although this has never been measured directly, the rates of change of pH in carious lesions have been measured.[143] Active carious lesions sometimes convert to arrested lesions when undermined enamel breaks away, exposing more dentin surface area to salivary buffers. Sealants or well-placed restorations,[144,145] by blocking carious bacterial carbon and nitrogen sources from saliva, seem to "starve" the organisms into dormancy or death. More studies are required to determine if "sealed" carious dentin gains calcium and phosphate. No one has examined the spectra of bacterial products and/or enzymes of "active caries" vs. the "arrested caries" found under sealants.

Dentin Response to Caries

Pulpal Response

The capillaries and venules of the pulp permit a slow leakage of plasma proteins across their walls into the pulpal interstitial fluid. These proteins and the fluid that they are dissolved in will follow the path of least resistance in their exit from the pulp. Normally, this would be accomplished via pulpal lymphatics. If, however, enamel or cementum covering dentin is lost by attrition, caries, etc., a second pathway for fluid loss from the pulp is established. Since the pulpal tissue pressure (25 mmHg) is higher than that of the oral cavity (generally 0 mmHg, except during mastication or swallowing), interstitial fluid from the pulp will flow through dentin to the surface. All of the plasma proteins in interstitial fluid[41] will accompany that fluid. In fact, if bacterial products derived from saliva or carious dentin permeate to the pulp and cause the release of histamine or other endogenous mediators of inflammation[126] known to increase the leak rate of plasma proteins across the microcirculation, the concentration of plasma proteins in dentinal fluid will rise.[113] Okamura et al.[38-40] have shown the presence of both salivary and plasma proteins in human carious dentin, including immunoglobulins and portions of the complement series using immunohistological techniques. They demonstrated the presence of numerous plasma proteins, immunoglobulins, and complement fragments, but no salivary proteins, in dentin exposed by attrition or abrasion. Similar results were obtained in dentin beneath amalgam-filled cavities. They demonstrated that these plasma proteins pass through dentinal tubules and that they can be found on bacteria in human carious dentin. They speculate that their results demonstrate a physiological defense reaction of dentin against carious microorganisms. It would be very interesting to see if similar reactions occur beneath lesions sealed with pit and fissure sealant.

Zonal Classification of Caries

Miller and Massler[138] have contributed a useful classification of the carious lesions in dentin by distinguishing among several different zones: the "infected zone" is the most superficial zone and generally makes up most of the lesion. It is quite soft, varies in color from yellow to brown, and is heavily infected with a wide variety of microorganisms ($\sim 10^8$ cells per gram). It is in this zone where most of the proteolysis of dentin occurs. Deep to this zone is the "affected zone" of dentin which, though partially demineralized, nevertheless retains tubular organization. The number of viable microorganisms falls off from $\sim 10^7$ to almost 0 as one samples the dentin across this zone toward the pulp.[146] Most of the acidogenic organisms are in this zone and it is here where demineralization of dentin occurs which exposes collagen. If the collagen can be remineralized it will remain structurally sound; if it is hydrolyzed by bacterial collagenase or other proteases (after being denatured), it will be lost. Beyond the "affected zone" is a hypermineralized zone of dentin. The thickness of this zone depends upon the rate of progression of caries, being thicker in chronic (slowly progressive) caries than in acute caries. This hypermineralized zone has been demonstrated to be translucent on ground sections,[139] more radiopaque by microradiographic studies,[147] nonstaining in vivo, and harder according to Knoop microhardness tests.[6]

Formation of Sclerotic Dentin

The nature of this hypermineralization has recently been determined by SEM studies by Mendis and Darling.[85] They demonstrated that the hypermineralization beneath carious dentin was due to occlusion of the tubules by rod-like plugs which were not continuous with peritubular dentin. These "plugs" were often separated from the peritubular dentin by fibrillar material. The length of these plugs was quite long (i.e., in excess of 50 μm) and the authors speculated that they might extend the remaining distance to the pulp chamber. In some tubules, the "plugs" or intratubular "rods" seemed to have a granular substructure consisting of large (0.1 to 0.2 μm) granules of mineral occluding the tubules. They speculated that the large granules and plugs are composed of whitlockite (β-$Ca_3(PO_4)_2$),[148] which precipitates in advance of the carious attack from calcium and phosphate released from hydroxyapatite by bacterial acid production.[122] The separation of these rods or plugs from the peritubular dentin is probably due to the presence of a limiting membrane consisting of organic material including collagen fibrils within the peritubular dentin.[66]

The presence of this sclerotic zone of dentin containing occluded dentinal tubules must certainly decrease dentin permeability. This would tend to protect the pulp from injurious bacterial products and would also greatly restrict fluid movement. The latter would contribute to the lack of sensitivity of exposed carious dentin which is so characteristic of the disease process.

The mechanism of formation of these tubular plugs is not known. Although the pulpal ends of the tubules are generally within secondary dentin, these tubules may provide enough pulpal mineral salts for diffusion out into the hypermineralized zone. The redistribution of intratubular calcium phosphate from apatite of the peritubular dentin in the "affected zone" to whitlockite in the hypermineralized zone has been discussed above. One final source of mineral would be from saliva permeating the lesion. The contribution of each of these possibilities remains to be determined.

Formation of Irritation Dentin

The presence of irritation dentin, while generally found in carious dentin, is not unique to this disease process since it is often found under noninfected dentin exposed by attrition or abrasion.[85] The presence of irritation (secondary) dentin protects the pulp from bacterial insult in several ways. Such dentin generally has fewer tubules than "normal" dentin and hence would be less permeable. The production of this dentin tends to maintain a thick diffusion barrier between the source of bacterial products and the pulp. This dissipates the chemical concentrations of these products over a long diffusion distance (i.e., >50 to 100 μm is a long diffusion distance), thereby lowering their pulpal concentration.

Speculations

Speculations on the Mechanisms of Production of Sclerotic and Irritation Dentin

Sclerotic dentin is found not only under areas of dentin exposed by caries or occlusal attrition,[85] but also in root surfaces, especially apical radicular dentin, as a function of age.[78,139] In the latter case, no pathological conditions exist, yet the tubules become mineralized. Although more electron microscopy of such tissue seems warranted, even more important is the need to understand the mechanism(s) involved. Identification of the specific stimuli is required. Knowledge of the coupling between stimuli and the cellular response(s) is needed before we can understand why some tubules become sclerotic while adjacent tubules remain normal.

Irritation dentin obviously forms in response to some as yet unknown molecular signal. Such dentin is commonly found under caries, under occlusal attrition, and sometimes beneath abraded cervical dentin. It is not clear why some dentin responds to stimuli by forming intratubular deposits of mineral (sclerosis), while other dentin forms more new dentin (irritation dentin) under similar stimuli. Perhaps we have not yet clearly identified the stimuli. It may be that fluid movement or minute movements of odontoblasts are required. There may be chemical substances in saliva or specific products of bacterial metabolism that are responsible. Avery's group[60,149] has performed numerous important experiments in which the sensory or autonomic nerves to teeth were sectioned and cavities cut into dentin. Cutting the mandibular nerve led to accelerated rates of dentinogenesis in response to test cavities in these animals, sometimes leading to near obliteration of the pulp chamber. The response was not limited to the tubules under the cavity, but was a generalized reaction circumferentially. They speculate that these nerves exert a trophic restraining influence on dentinogenesis that is released when the nerves are severed. This is a very interesting hypothesis that may explain several seemingly unrelated observations. Bernick[150] reported that the density of subodontoblastic nerves falls dramatically under carious dentin. Perhaps bacterial products or enzymes released from inflammatory cells injure the subodontoblastic nerves. The loss of nerves may release intrinsic tendencies of odontoblasts and pulpal mesenchymal cells from the trophic influence of nerves[35] to form calcified structures. Is this the reason why irritation dentin forms? Fernhead[50] pointed out that the number of nerves innervating root dentin is far less than that of coronal dentin. If the formation of sclerotic dentin (i.e., calcification of dentinal tubules) is also under nervous control, this may be why radicular dentin scleroses more extensively, even in the absence of external stimuli, than coronal dentin.

Primary teeth are less well innervated than permanent teeth.[151] They also seem to produce irritation dentin at faster rates than permanent teeth.

Alternate mechanisms for the production of irritation dentin are possible. Avery[149] reported that sectioning the inferior alveolar nerve also resulted in a decrease in the number of odontoblasts. This decrease in the number of odontoblasts (and decrease in the number of nerve fibers) was associated with a generalized increase in primitive atubular dentin. In Baume's recent review,[90] he points out that atubular dentin is formed in the absence of odontoblasts. Indeed, Gottlieb[152] noted that local or total odontoblastic atrophy is accompanied by accelerated atubular dentin formation. He inferred that odontoblasts actually slow down dentinogenesis. Under this mechanism, odontoblasts restrain and control dentinogenesis. Anything that interferes with the metabolism of odontoblasts obviously will modify dentin formation. This could be accomplished by the action of agents directly on odontoblasts or indirectly on subodontoblastic nerves or capillaries.[153] If nerves exert a trophic influence on odontoblasts, then nerve degeneration might lead to odontoblast degeneration which would, in turn, "permit" the production of irritation dentin. Finally, if odontoblasts not only inhibit dentiogenesis, but also tubular sclerosis, then the observation that apical root dentin becomes sclerotic more frequently than cervical root dentin correlates well with the fewer number of odontoblasts in the apical region relative to midroot or crown dentin.

Speculations about Dentin Permeability

Little effort has been made to exploit the permeability characteristics of dentin to therapeutic advantage. Acid etching of dentin is regarded as heresy, yet there are theoretical reasons for such action if it is done with very dilute, isotonic acid solutions for very brief (10 to 20 sec) times. This might be employed to provide qualitative information to the clinician about the permeability of that dentin. If, after such treatment, the dentin remains moist even after several attempts to dry it with absorbent paper, then one can conclude that the tubules are open and permeable. If it does not become moist (glistening to reflected light), one can conclude that the tubules are occluded with something (i.e., intratubular mineral deposits in sclerotic dentin-aspirated odontoblasts, presence of irritation dentin) or that there is poor or absent circulation beneath such dentin. Studies should be done to correlate these clinical observations with histologic examination of these teeth. If the tubules are "open" one has the option of using topical agents to medicate the pulp. This might vary from parathyroid hormone used to resorb pulp stones to permit endodontic therapy to calcitonin and/or 1,25-dihydroxycholecalciferol to arrest root resorption during orthodontic treatment or internal resorption. Perhaps local anesthetics with vasoconstrictors could be filtered across such acid-etched dentin by filling an occlusal, acid-etched cavity with the solution and having the patient bite on a piece of rubber base impression material. If the dentin is permeable (i.e., fluid can move across dentin), then the clinician would be justified in applying topical medicaments to the dentin with reasonable assurance that they would reach the pulp chamber. If the pulp is inflamed or "hyperemic", one could place a dry cotton pellet in the cavity under a temporary restoration to permit excessive fluid to "drain" from the pulp, through the tubules, into the cotton to relieve elevations in pulpal tissue pressure. However, if fluid cannot be seen to move across dentin then these measures would probably not be effective.

If restorative materials are ever developed that will bond to dentin, there is a high probability that such dentin will have to be cleansed of its "smear layer". If the material is hydrophobic, the presence of dentinal fluid may "contaminate" the dentin surface and prevent proper bonding. This might be minimized by applying topical epinephrine (1:1000) or felypressin to the acid-etched dentin to purposely vasoconstrict subodontoblastic vessels and temporarily slow or prevent dentin fluid production. Vasoconstrictors might also be used to permit higher pulpal concentrations of topically applied agents to be achieved.[19,154]

Speculations Regarding Hypersensitive Dentin

The hydrodynamic theory of dentin sensitivity assumes that painful stimuli are being transmitted to pulpal nerves via fluid movement through tubules. More clinical work needs to be done to support this theory. For instance, patients with teeth scheduled for extraction should be screened for hypersensitive areas which should be "mapped" very carefully for areas of hypersensitivity as well as areas of insensitivity. If such patients were given fluorescein dye by capsule or intravenously, the investigator may be able to see yellow-green fluorescein dye (under UV light) permeating the "hypersensitive areas" but not coming across insensitive areas of exposed dentin. This could be correlated by SEM examination of these same teeth after extraction to examine the correlation between the presence of open tubules at the surface, with the distribution of the "hypersensitivity map" of that tooth and the appearance of fluorescein dye in the affected dentin. Although some tubules may be open at the dentin surface (as seen by SEM examination), vital dye may not penetrate these tubules due to the presence of irritation dentin or sclerotic dentin on the pulpal side of the dentin. Such dentin should also be insensitive. Thus, each piece of information would complement the other and contribute to a more accurate interpretation of the data. Similar studies should be done after treating hypersensitive areas with substances[155] or procedures[156] designed to reduce or eliminate dentin hypersensitivity. Such studies would be very helpful in testing the "hydrodynamic theory" of dentin sensitivity (see the section entitled "Theories of Dentin Sensitivity").

Speculations Regarding Pulpal Reaction to Caries

If we knew the molecular mechanisms controlling the production of sclerotic dentin, or irritation dentin, or atubular dentin, then we might be able to reproducibly stimulate one or more of the above as a means of therapy. For instance, those patients who have persistent hypersensitive dentin are apparently unable to produce either sclerotic or irritation dentin.

Are there substances that can trigger odontoblasts to accelerate peritubular matrix production? Our understanding of odontoblast metabolism is rudimentary at present.

If we could selectively stimulate pulpal cells to form either tubular or atubular irritation dentin, we might choose to stimulate the production of 1 to 2 mm of irritation dentin to replace that dentin which was destroyed by caries and then stimulate production of about 1 mm of atubular dentin to seal off the cavity from the pulp. We will have to learn much more about the regulation of dentinogenesis by odontoblasts and other pulp cells before we can even begin to plan such experiments.

The material covered above attempts to unify observations and hypotheses from many seemingly unrelated areas in an effort to stimulate thought in pulpal biology. Baume, in his recent scholarly review,[90] quoted Piaget's important notion, "The true object of interdisciplinary research, therefore, is to reshape or reorganize the fields of knowledge by means of exchanges which are in fact constructive recombinations."

ACKNOWLEDGMENTS

The author would like to thank Dr. Jackie Weatherred and Dr. Gary M. Whitford for reviewing this manuscript and for their helpful suggestions. This work was supported, in part, by NIDR Grant #DE 06427.

REFERENCES

1. **Garberoglio, R. and Brännström, M.,** Scanning electron microscopic investigation of human dentin tubules, *Arch. Oral Biol.*, 21, 355, 1962.
2. **Forssell-Ahlberg, K., Brännström, M., and Edwall, L.,** The diameter and number of dentinal tubules in rat, cat, dog and monkey, *Acta Odontol. Scand.*, 33, 243, 1975.
3. **Fiore-Donno, G. and Baume, L. J.,** Etude histochemique de la dentinogenese humaine, *Helv. Odontol. Acta*, 10 (Suppl. 47), 141, 1966.
4. **Outhwaite, W., Livingston, M., and Pashley, D.,** Effects of changes in surface area, thickness, temperature and post-extraction time on dentin permeability, *Arch. Oral Biol.*, 21, 599, 1976.
5. **Craig, R. G., Gehring, K., and Hosoda, H.,** Relation of structure to the microhardness of dentin, *J. Dent. Res.*, 38, 624, 1959.
6. **Fusayama, T., Okuse, K., and Hosoda, H.,** Relationship between microhardness, discoloration and microbial invasion in carious dentin, *J. Dent. Res.*, 4, 1033, 1966.
7. **Pashley, D. H., Michelich, V., and Kehl, T.,** Dentin permeability: effect of smear layer removal, *J. Prosth. Dent.*, 46, 531, 1981.
8. **Brännström, M., Glantz, P. O., and Nordenvall, K. J.,** The effect of some cleaning solutions on the morphology of dentin prepared in different ways: an *in vivo* study, *J. Dent. Child.*, 46, 291, 1979
9. **Olgart, L., Brännström, M., and Johnson, G.,** Invsion of bacteria into dentinal tubules. Experiments *in vivo* and *in vitro*, *Acta Odontol. Scand.*, 32, 61, 1974.
10. **Michelich, V., Schuster, G. S., and Pashley, D. H.,** Bacterial penetration of human dentin, *in vitro*, *J. Dent. Res.*, 59(8), 1398, 1980.
11. **Pashley, D. H., Livingston, M. J., Reeder, O. W., and Horner, J.,** Effects of the degree of tubule occlusion on the permeability of human dentin *in vitro*, *Arch. Oral Biol.*, 23, 1127, 1978.
12. **Michelich, V., Pashley, D. H., and Whitford, G. M.,** Dentin permeability: a comparison of functional versus antomical tubular radii, *J. Dent. Res.*, 57(11-12), 1019, 1978.
13. **Brännström, M., Linden, L. A., and Aström, A.,** The hydrodynamics of the dentinal tubule and of pulp fluid. A discussion of its significance in relation to dentinal sensitivity, *Caries Res.*, 1, 310, 1967.
14. **Langeland, K.,** Tissue changes in the dental pulp, *Odontol. Tidskr.*, 65, 239, 1957.
15. **Mjör, I. A. and Tronstad, L.,** Experimentally induced pulpitis, *Oral Surg.*, 34, 102, 1972.
16. **Bergenholtz, G. and Reit, C.,** Reactions of the dental pulp to microbial provocation of calcium hydroxide treated dentin, *Scand. J. Dent. Res.*, 88, 187, 1980.
17. **Dahlen, G. and Bergenholtz, G.,** Endotoxic activity in teeth with necrotic pulps, *J. Dent. Res.*, 59(6), 1033, 1980.

18. **Katz, M. A.,** Interstitial space — the forgotten organ, *Med. Hypothesis,* 6, 885, 1980.
19. **Pashley, D. H.,** The influence of dentin permeability and pulpal blood flow on pulpal solute concentrations, *J. Endodont.,* 5(12), 355, 1979.
20. **Meyer, M., Weiner, D., and Grim, E.,** Blood flow in the dental pulp of the dog, *Proc. Soc. Exp. Biol. Med.,* 116, 1038, 1964.
21. **Meyer, M. W. and Path, M. G.,** Blood flow in the dental pulp of dogs determined by hydrogen polarography and radioactive microsphere methods, *Arch. Oral Pathol.,* 24, 601, 1979.
22. **Path, M. G. and Meyer, M.,** Heterogeneity of blood flow in the canine tooth of the dog, *Arch. Oral Biol.,* 25(2), 83, 1980.
23. **Kaplan, M. L., Davis, M. A., and Goldhaber, P.,** Blood flow measurements in selected oral tissues in dogs using radiolabelled microspheres and rubidium-86, *Arch. Oral Biol.,* 23, 281, 1978.
24. **Bard, P., Ed.,** *Medical Physiology,* 11th ed., C. V. Mosby, St. Louis, 1961.
25. **Kraintz, L., Poon, P., and Reitzik, M.,** Blood volume determination of human dental pulp, *J. Dent. Res.,* 59 (Special Issue A), 544, 1980.
26. **Tanaka, T.,** The origin and localization of dentinal fluid in developing rat molar teeth studies with lanthanum as a tracer, *Arch. Oral Biol.,* 25, 153, 1980.
27. **Gamble, J. L.,** *Chemical Anatomy, Physiology and Pathology of the Extracellular Fluid,* Harvard University Press, Cambridge, Mass., 1950.
28. **Olgart, L. and Gazelius, B.,** Effects of adrenalin and felypressin (octapressin) on blood flow and sensory nerve activity in the tooth, *Acta Odontol. Scand.,* 35, 69, 1977.
29. **Edwall, L.,** Regulation of pulpal blood flow, *J. Endodont.,* 6(2), 434, 1980.
30. **Zacharisson, B. U.,** Mast cells in human dental pulp, *Arch. Oral Biol.,* 9, 149, 1964.
31. **Miller, G. S., Sternberg, R. N., Piliero, S. J., and Rosenberg, P. A.,** Histologic identification of mast cells in human dental pulp, *Oral Surg.,* 46(4), 559, 1978.
32. **Pohto, P. and Antila, R.,** Innervation of blood vessels in the dental pulp, *Int. Dent. J.,* 22, 228, 1972.
33. **Delbalso, A. M., Nishimura, R. S., and Setterstrom, J. A.,** The effects of thermal and electrical injury on pulpal histamine levels, *Oral Surg.,* 41(1), 110, 1976.
34. **Olgart, L., Gazelius, B., Brodin, E., and Nilsson, G.,** Release of substance P-like immunoreactivity from the dental pulp, *Acta Physiol. Scand.,* 101, 510, 1977.
35. **Kroeger, D. C.,** Possible role of neurohumeral substances in the pulp, in *Biology of the Dental Pulp Organ: A Symposium,* Finn, S. B., Ed., University of Alabama Press, Tuscaloosa, 1968, 333.
36. **Inoki, R., Toyoda, T., and Yamamoto, I.,** Elaboration of a bradykinin-like substance in dog's canine pulp during electrical stimulation and its inhibition by nacrotic and non-narcotic analgesics, *Naunyn-Schmiedebergs Arch. Pharmacol.,* 279, 387, 1973.
87. **Hirafuji, M., Satoh, S., and Ogura, Y.,** Prostaglandins in rat pulp tissue, *J. Dent. Res.,* 59(9), 1535, 1980.
38. **Okamura, K., Tanaka, A., Kahehi, A., Maeda, M., and Tsutsui, M.,** Plasma components in deep lesions of human carious dentin, *J. Dent. Res.,* 58(10), 2010, 1979.
39. **Okamura, K., Tsubakimota, K., Uobe, K., Nishida, K., and Tsutsui, M.,** Serum proteins and secretory component in human carious dentin, *J. Dent. Res.,* 58(3), 1127, 1979.
40. **Okamura, K., Meada, M., Nishikawa, T., and Tsutsui, M.,** Dentinal response against carious invasion: localization of antibodies in odontoblastic body and process, *J. Dent. Res.,* 59(8), 1368, 1980.
41. **Haldi, J. and Wynn, W.,** Protein fractions of the blood plasma and dental-pulp fluid of the dog, *J. Dent. Res.,* 42(5), 1217, 1963.
42. **Van Hassel, H. J.,** Physiology of the human dental pulp, *Oral Surg.,* 32, 126, 1971.
43. **Stenvik, A., Iversen, J., and Mjör, I. A.,** Tissue pressure and histology of normal and inflamed tooth pulps in macaque monkeys, *Arch. Oral Biol.,* 17, 1501, 1972.
44. **Hakumäki, M. O. K. and Narhi, M. V. O.,** Effect of intrapulpal pressure stimulation on the activity of sensory nerves of dental pulp, *Acta Physiol. Scand.,* 88, 584, 1973.
45. **Edwall, L. and Olgart, L.,** A new technique for recording of intradental sensory nerve activity in man, *Pain,* 3, 121, 1977.
46. **Brännström, J. and Aström, A.,** The hydrodynamics of the dentin: its possible relationship to dentinal pain, *Int. Dent. J.,* 22, 219, 1972.
47. **Inoki, R., Matsumoto, K., Kudo, T., Kotani, Y., and Oka, M.,** Bradykinin as an algesic (pain producing) substance in the pulp, *Naunyn-Schmiedebergs Arch. Pharmacol.,* 306, 29, 1979.
48. **Olgart, L.,** Local mechanisms in dental pain, in *Mechanisms of Pain and Analgesic Compounds,* Miles Symp. Ser. No. 11, Beers, R. F. and Bassett, E. G., Eds., Raven Press, New York, 1979.
49. **Arwill, T.,** Some morphological aspects of the dentinal innervation, in *Sensory Mechanisms In Dentin,* Anderson, D. J., Ed., Pergamon Press, Oxford, 1963, 3.
50. **Fernhead, R. W.,** Histological evidence for the innervation of human dentin, *J. Anat.,* 91, 267, 1957.
51. **Fernhead, R. W.,** The neurohistology of human dentin, *Proc. R. Soc. Med.,* 54, 877, 1961.

52. **Anderson, D. J., Hananm, A. G., and Matthews, B.,** Sensory mechanisms in mammalian teeth and their supporting structures, *Physiol. Rev.*, 50, 171, 1970.
53. **Anderson, D. J., Curwen, M. P., and Howard, L. V.,** The sensitivity of human dentin, *J. Dent. Res.*, 37, 669, 1958.
54. **Brännström, M.,** Sensitivity of dentin, *Oral Surg.*, 21(4), 517, 1966.
55. **Avery, J. K. and Rapp, R.,** Investigation of the mechanism of neural impulse transmission in human teeth, *Oral Surg.*, 12, 190, 1959.
56. **Bock, O.,** Über die Innervation des Dentins, *Dtsch. Zahnaerztl. Z.*, 15, 838, 889, 1960.
57. **Arwill, T.,** Studies on the ultrastructures of dental tissue. II. The predentin-pulpal border zone, *Odontol. Rev.*, 18, 191, 208, 1967.
58. **Arwill, T., Edwall, L., Lilja, J., Olgart, L., and Svensson, S. E.,** Ultrastructure of nerves in the dentinal-pulp border zone after sensory and autonomic nerve transsection in the cat, *Acta Odontol. Scand.*, 31, 273, 1973.
59. **Frank, R. M., Sauvage, C., and Frank, P.,** Morphological basis of dentin sensitivity, *Int. Dent. J.*, 22, 1, 1972.
60. **Avery, J. K.,** Anatomic considerations in the mechanism of pain and sensitivity in the teeth and supporting structures, in *Mechanisms of Pain and Sensitivity in the Teeth and Supporting Structures*, Chasens, A. I. and Kaslick, R. S., Eds., Fairleigh Dickinson University, Rutherford, N.J., 1974, 16.
61. **Garant, P. R.,** The organization of microtubules within rat odontoblast processes revealed by perfusion fixation with glutaraldehyde, *Arch. Oral Biol.*, 17, 1047, 1972.
62. **Holland, G. R.,** The dentinal tubules and odontoblast process in the cat, *J. Anat.*, 120, 169, 1975.
63. **Holland, G. R.,** The extent of the odontoblast process in the cat, *J. Anat.*, 121, 133, 1976.
64. **Tsatsas, B. G. and Frank, R. M.,** Ultrastructure of the dentinal tubular substances near the dentino-enamel junction, *Calcif. Tissue Res.*, 9, 238, 1972.
65. **Brännström, M. and Garberoglio, R.,** The dentinal tubules and the odontoblastic processes. A scanning electron microscopic study, *Acta Odontol. Scand.*, 30, 291, 1972.
66. **Thomas, H. F.,** The extent of the odontoblastic process in human dentin, *J. Dent. Res.*, 58, 2207, 1979.
67. **Coffey, C. T., Ingram, M. J., and Bjorndall, A.,** Analysis of human dentinal fluid, *Oral Surg.*, 30(6), 835, 1970.
68. **Scott, D., Jr.,** Physiologic considerations in the arousal of pain in the teeth: structure and function of existing mechanisms and a hypothesis of their transducer action, in *Mechanisms of Pain and Sensitivity in the Teeth and Supporting Tissue*, Chasen, A. L., Kaslick, R. S., and Hackensack, N. J., Eds., Fairleigh Dickinson University, Rutherford, N. J., 1974.
69. **Brännström, M. and Aström, A.,** A study on the mechanism of pain elicited from the dentin, *J. Dent. Res.*, 43, 619, 1964.
70. **Pashley, D. H., Livingston, M. J., and Greenhill, J. D.,** Regional resistances to fluid flow in human dentin, *in vitro*, *Arch. Oral Biol.*, 23, 807, 1978.
71. **Reeder, O. W., Walton, R. E., Livinston, M. J., and Pashley, D. H.,** Dentin permeability: determinants of hydraulic conductance, *J. Dent. Res.*, 57(2), 187, 1978.
72. **Pashley, D. H., Nelson, R., and Pashley, E.,** Fluid movement across dog dentin, *in vitro*, *Arch. Oral Biol.*, 26, 707, 1981.
73. **Johnson, G. and Brännström, M.,** The sensitivity of dentin: changes in relation to conditions at exposed tubule apertures, *Acta Odontol. Scand.*, 32, 29, 1974.
74. **Stanley, H. R., Jr. and Swerdlow, H.,** Aspiration of cells into dentinal tubules, *Oral Surg.*, 11, 1007, 1958.
75. **Brännström, M.,** Dentin sensitivity and aspiration of odontoblasts, *J. Am. Dent. Assoc.*, 66, 366, 1963.
76. **Stevenson, T. S.,** Fluid movement in human dentin, *Arch. Oral Biol.*, 10, 935, 1965.
77. **Brännström, M. and Garberoglio, R.,** Occlusion of dentinal tubules under superficial attrited dentin, *Swed. Dent. J.*, 4, 87, 1980.
78. **Jenkins, G. N.,** *The Physiology and Biochemistry of the Mouth*, 4th ed., Blackwell Scientific, Oxford, 1978, 100.
79. **Pashley, D. H., Nelson, R., and Kepler, E. E.,** The effects of plasma in salivary constituents on dentin permeability, *J. Dent. Res.*, 61, 978, 1982.
80. **Hiatt, W. H. and Johansen, E.,** Root preparation. I. Obturation of dentinal tubules in treatment of root hypersensitivity, *J. Periodontol.*, 43, 373, 1972.
81. **Dreyfus, F., Frank, R. M., and Gutman, B.,** *La Selerose Dentinaire*, *G. I. R. S.*, 7, 207, 1964.
82. **Van Huysen, G.,** The microstructure of normal and sclerosed dentin, *J. Prosth. Dent.*, 10, 976, 1960.
83. **Gwinett, A. J. and Jendresen, M. D.,** Micromorphologic features of cervical erosion after acid conditioning and its relation with composite resin, *J. Dent. Res.*, 57(4), 543, 1978.
84. **Mendis, B. R. R. N. and Darling, A. I.,** Distribution with age and attrition of peritubular dentin in the crowns of human teeth, *Arch. Oral Biol.*, 24, 131, 1979.

85. **Mendis, B. R. R. N. and Darling, A. I.,** A scanning electron microscope and microradiographic study of human coronal dentinal tubules related to occlusal attrition and caries, *Arch. Oral Biol.*, 24, 725, 1979.
86. **Furseth, R. and Mjör, I. A.,** The fine structure of corticosteroid covered, human dentin, *Arch. Oral Biol.*, 17, 719, 1972.
87. **Johannssen, L. B.,** Dentin apposition in the mandibular first molars of albino rats, *Arch. Oral Biol.*, 3, 81, 1961.
88. **Kawasaki, K., Tanaka, S., and Ishikawa, T.,** On the incremental lines in human dentine as revealed by tetracycline labelling, *J. Anat.*, 123, 427, 1977.
89. **Yilmaz, S., Newman, H. N., and Poole, E. F. G.,** Diurnal periodicity of von Ebner growth lines in pig dentin, *Arch Oral Biol.*, 22, 511, 1977.
90. **Baume, L. J.,** *The biology of pulp and dentine*, Vol. 8, Monographs In Oral Science, Myers, H., Ed., S. Karger, New York, 1980.
91. **Nalbandian, A., Gonzales, F., and Sognnaes, R. F.,** Sclerotic changes in root dentin of human teeth as observed by optical, electron and X-ray microscopy, *J Dent. Res.*, 39, 598, 1960.
92. **Sognnaes, R. F.,** Dental hard tissue destruction with special reference to idiopathic erosions, in *Mechanisms of Hard Tissue Destruction*, Washington, D.C., American Association for the Advancement of Science, 75, 91, 1963.
93. **Schmidt, W. J. and Keil, A.,** *Polarizing Microscopy Of Dental Tissue*, 1st ed., Pergamon Press, New York, 1971, 512.
94. **Scott, J. N. and Weber, D. F.,** Microscopy of the junctional region between human coronal primary and secondary dentin, *J. Morphol.*, 154(1), 133, 1977.
95. **Fitzgerald, M.,** Cellular mechanics of dentin bridge repair using ^{3}H-thymidine, *J. Dent. Res.*, 59(D), 2198, 1979.
96. **Eda, S. and Saito, T.,** Electron microscopy of cells displaced into the dentinal tubules due to dry cavity preparation, *J. Oral Pathol.*, 7, 326, 1978.
97. **Anderson, D. J. and Ronning, G. A.,** Dye diffusion in human dentine, *Arch. Oral Biol.*, 7, 505, 1962.
98. **Linden, L. and Brännström, M.,** Fluid movement in dentine and pulp, *Odontol. Rev.*, 18(3), 227, 1967.
99. **Horiuchi, H. and Matthews, B.,** In vitro observation on fluid flow through dentin caused by pain producing stimuli, *Arch. Oral Biol.*, 18, 275, 1973.
100. **Pashley, D. H., Livingston, M. J. and Whitford, G. M.,** The effect of molecular size on reflection coefficients in human dentin, *Arch. Oral Biol.*, 24, 455, 1979.
101. **Pashley, D. H. and Whitford, G. M.,** Permeability of human dentin in vitro interpreted from reflection coefficients, *Arch. Oral Biol.*, 35, 141, 1980.
102. **Berrgren, G. and Brännström, M.,** The rate of flow in dentinal tubules due to capillary attraction, *J. Dent. Res.*, 44(2), 408, 1965.
103. **O'Brien, W. J.,** Capillary action around dental structures, *J. Dent. Res.*, 52(3), 544, 1973.
104. **Brännström, M. and Johnson, G.,** The sensory mechanism in human dentin as revealed by evaporation and mechanical removal of dentin, *J. Dent. Res.*, 57(1), 49, 1978.
105. **Furseth, R. and Mjör, I. A.,** Electron microscopy of human coronal dentin. A methodological study with emphasis on the "aspiration" of odontoblast nuclei, *Acta Odontol. Scand.*, 27, 577, 1969.
106. **Leonard, L. A. and Sharawy, M.,** Histochemical identification of myosin-like substance in cells of the human dental pulp, *J. Dent. Res.*, 55(4), 628, 1976.
107. **Schanne, F. A. X., Kane, A. B., Young, E. E., and Farber, J. L.,** Calcium dependence of toxic cell death: a final common pathway, *Science*, 9(206), 700, 1979.
108. **Weber, A. and Murray, J. M.,** Molecular control mechanisms in muscle contraction, *Physiol. Rev.*, 53(3), 612, 1973.
109. **Stygles, V. G., Bonhaus, D. W. and Hook, J. B.,** A method for identifying Ca ionophores using a unicellular organism, *J. Pharmacol. Meth.*, 3, 243, 1980.
110. **Mikkelsen, H. B.,** Acute and protracted effect of vinblastine on odontoblasts and dentinogenesis in rat incisors, *Scand. J. Dent. Res.*, 86, 313, 1978.
111. **Stene, T.,** Effect of vincristine on odontoblasts in rat incisor, *Scand. J. Dent. Res.*, 86, 346, 1978.
112. **Orban, B.,** Migration of leukocytes into the dentinal tubules, *J. Am. Dent. Assoc.*, 27, 239, 1940.
113. **Pashley, D. H., Nelson, R., Williams, E. C., and Kepler, E.,** Use of dentine fluid protein concentrations to measure pulpal capillary reflection coefficients, *Arch. Oral Biol.*, 26, 703, 1981.
114. **Schmid-Schöbein, G. W., Stalak, R., Usuemi, S. and Chien, S.,** Cell distribution in capillary networks, *Microvasc. Res.*, 19, 18, 1980.
115. **Hodosh, M.,** A superior desensitizer. Potassium nitrate, *J. Am. Dent. Assoc.*, 88, 831, 1974.
116. **Mumford, J. M.,** Stimulus evoked pain in teeth, in *Frontiers in Oral Physiology*, Vol. 2, Kawamura, Y., Ed., S. Karger, Basel, 1976, 51.
117. **Hannan, A. G.,** The regulation of the jaw bite force in man, *Arch. Oral Biol.*, 21, 641, 1976.

118. **Van Steenberghe, D. and DeVries, J. H.,** The development of maximal clenching force between two antagonistic teeth, *J. Periodontal Res.*, 13, 91, 1978.
119. **Bender, I. G., Seltzer, S. and Kaufman, I. J.,** Infectability of the dental pulp by way of dentinal tubules, *J. Am. Dent. Assoc.*, 59, 466, 1959.
120. **Toto, P. D. and Predergast, R. C.,** Hyaluronidase-producing microorganisms in carious dentin, *J. Dent. Res.*, 47, 173, 1968.
121. **Parikh, S. R., Massler, M., and Bahn, A.,** Microorganisms in active and arrested carious lesions of dentin, *N.Y. Dent. J.*, 29, 347, 1963.
122. **Sarnat, H. and Massler, M.,** Microstructure of active and arrested dentinal caries, *J. Dent. Res.*, 44, 1389, 1965.
123. **Brännström, M. and Nordenvall, K. J.,** Bacterial penetration, pulpal reaction and the inner surface of concise enamel bond. Composite fillings in etched and unetched cavities, *J. Dent. Res.*, 57(1), 3, 1978.
124. **Bergenholtz, G.,** Effect of bacterial products on inflammatory reactions in the dental pulp, *Scand. J. Dent. Res.*, 85, 122, 1977.
125. **Merchant, V. A., Livingston, M. J., and Pashley, D. H.,** Dentin permeation: comparison of diffusion with filtration, *J. Dent. Res.*, 56(10), 1161, 1977.
126. **Buczko, W., Dziaczkowski, J., Kopec, M., Moniuszko-Jakonink, J., Sopeta, I., Wisniewski, K., Wize, J., and Wojtecka-Lukasik, E.,** Biological effects of degradation products of collagen by bacterial collagenase, *Br. J. Pharmacol.*, 69, 551, 1980.
127. **Brown, L. R. and Wheatcroft, M.,** Effect of the diffusion of microbial growth factors through tooth substance on production of carious lesions *in vitro*, *J. Dent. Res.*, 45(3), 830, 1966.
128. **Wachtel, L. W. and Brown, L. R.,** In vitro caries: factors influencing the shape of the developing lesion, *Arch. Oral Biol.*, 8, 99, 1963.
129. **Brown, L. R. and Lefkowitz, W.,** Influences of dentinal fluids on experimental caries, *J. Dent. Res.*, 45(5), 1493, 1966.
130. **Handelman, S. L., Buonocore, M. G., and Heseck, D. J.,** A preliminary report on the effect of fissure sealant on bacteria in dental caries, *J. Prosth. Dent.*, 27, 390, 1972.
131. **Handelman, S. L., Washburn, F., and Wopperer, P.,** Two year report of sealant effect on bacteria in dental caries, *J. Am. Dent. Assoc.*, 93, 967, 1976.
132. **Theilade, E., Fejerskov, O., Migasena, K., and Prachyabrued, W.,** Effect of fissure sealant on the microflora in occlusal fissures of human teeth, *Arch. Oral Biol.*, 22, 251, 1977.
133. **Mertz-Fairhurst, E. J., Schuster, G. S., Williams, J. E., and Fairhurst, C. W.,** Clinical progress of sealed and unsealed caries, I. Depth changes and bacterial counts, *J. Prosth. Dent.*, 42(5), 521, 1979.
134. **Mertz-Fairhurst, E. J., Schuster, G. S., Williams, J. E., and Fairhurst, C. W.,** Clinical progress of sealed and unsealed caries. II. Standardized radiographs and clinical observations, *J. Prost. Dent.*, 42(6), 633, 1979.
135. **Jensen, O. E. and Handelman, S.,** Effect of autopolymerizing sealant on viability of microflora in occlusal dental caries, *Scand. J. Dent. Res.*, 88, 382, 1980.
136. **Massler, M.,** Pulpal reactions to dental caries, *Int. Dent. J.*, 11, 363, 1967.
137. **Miller, W. A.,** Layering of Dental Caries as Demonstrated by Localization of Dyes, Master's thesis, University of Illinois, Urbana, 1959.
138. **Miller, W. A. and Massler, M.,** Permeability and staining of active and arrested carious lesions in dentin, *Br. Dent. J.*, 112, 187, 1962.
139. **Blake, G. C.,** An experimental investigation into the permeability of enamel and dentin with reference to its relation to dental caries, *Proc. R. Soc. Med.*, 51, 678, 1958.
140. **Weber, D. F.,** Human dentin sclerosis: a microradiographic survey, *Arch. Oral Biol.*, 19, 163, 1974.
141. **Daclusi, G., Kerebel, B., Le Cabellec, M. T., and Kerebel, L. M.,** Qualitative and quantitative data on arrested caries in dentin, *Caries Res.*, 13, 190, 1979.
142. **Critchley, P.,** Effect of foods on bacterial metabolic processes, *J. Dent. Res.*, 49(6), 1283, 1970.
143. **Dirksen, T. R., Little, M. F., Bibby, B. G., and Crump, S. L.,** The pH of carious cavities. I. The effect of glucose and phosphate buffer on cavity pH, *Arch. Oral Biol.*, 7, 49, 1962.
144. **Fisher, F. F.,** The viability of microorganisms in carious dentin beneath amalgam restoration, *Br. Dent. J.*, 121, 413, 1966.
145. **Fisher, F. J.,** The effect of a calcium hydroxide/water paste on microorganisms in carious dentin, *Br. Dent. J.*, 133, 19, 1972.
146. **Ostrom, C.,** Clinical cariology, in *The Biologic Basis of Dental Caries*, Menaker, L., Ed., Harper & Row, New York, 1980, 247.
147. **Darling, A. I.,** Response of the pulpodentinal complex to injury, in *Thomas' Oral Pathology*, Vol. 1, Gorlin, R. J. and Goldman, H. M., Eds., 1970, 308.
148. **Frank, R. M., Wolff, F., and Gutman, B.,** Microscopie Electronique de la Carie au Niveau de la dentin humaine, *Arch. Oral Biol.*, 9, 163, 1964.

149. **Avery, J. K.,** Mechanism of reparative dentin formation and pain, in *Mechanism And Control Of Pain,* Grossman, L. I., Ed., Masson, New York, 1979, 43.
150. **Bernick, S.,** Vascular and nerve changes associated with the healing of the human pulp, *Oral Surg.,* 33, 983, 1972.
151. **Johnsen, D. C. and Karlsson, U. L.,** Development of neural elements in apical portions of cat primary and permanent incisior pulp, *Anat. Rec.,* 189(1), 29, 1977.
152. **Gottlieb, B.,** The formation of secondary dentin and related problems, *J. Dent. Res.,* 25, 29, 1946.
153. **Avery, J. K., Cox, C. S., and Chiego, D. J.,** Presence and localization of adrenergic nerve endings in the dental pulps of mouse molars, *Anat. Rec.,* 198, 59, 1980.
154. **Matthews, B.,** Effect of sympathetic stimulation on the response of intradental nerves to chemical stimulation of dentin, in *Advances In Pain Research and Therapy,* Vol. 1, Bonica, J. J. and Albe-Frssard, D., Eds., Raven Press, New York, 1976, 195.
155. **Greenhill, J. D. and Pashley, D. H.,** The effects of desensitizing agents on the hydraulic conductance of human dentin, *in vitro, J. Dent. Res.,* 60(3), 686, 1981
156. **Gangarosa, L. P.,** Iontophoretic application of fluoride by tray techniques for desensitization ot multiple teeth, *J. Am. Dent. Assoc.,* 102(1), 50, 1981.
157. **Brännström, M.,** Dentinal and pulpal response. V. Application of pressure exposed to dentin, *J. Dent. Res.,* 40, 960, 1961.

Enamel

CHEMICAL COMPOSITION OF ENAMEL

M. E. J. Curzon and J. D. B. Featherstone

INTRODUCTION

Because dental caries affects such a large proportion of the human population, the emphasis in dental research has been on this disease and in particular the mechanism of caries in enamel. The composition of enamel must obviously affect the initiation and progression of caries, thus an early field of research was to identify the chemical components of enamel. Such studies have attempted to identify changes in the major and minor inorganic composition of enamel. The objective has been to find which components increase or decrease susceptibility to caries. Initial studies focused on the major components: calcium, phosphate, carbonate, and those elements easily analyzed, such as sodium, magnesium, and chloride. Lately, trace elements in whole and surface enamel have been studied in detail.

Early analytical techniques were not sensitive enough to detect many elements in tooth enamel so that most of these were recorded as present in "trace" quantities and hence the term trace elements. As analytical techniques have improved, it has been possible to characterize human enamel for all elements of the Periodic Table.

Research has concentrated on identifying the composition of enamel in relation to caries prevalence, tooth type, tooth surface, age, and geographic origin. In considering such categories a second differention needs to be made between whole and surface enamel. Much of the earlier work on enamel composition was based upon whole enamel because the large sample size made analysis much easier. Dental caries is however, initiated at the enamel surface within the outer 20 μm. Whether or not the initial lesion progresses to a cavity or becomes remineralized depends on a number of factors, but the chemical composition of the outer enamel layers is an important one. The trace and major elemental composition of this surface layer has therefore attracted considerable interest.

We shall therefore consider enamel composition first of all as whole enamel. Any variations in composition of whole enamel will reflect environmental variables operating during the period of tooth calcification and maturation before eruption. Secondly, the inorganic composition of surface enamel will be considered, for the reasons outlined above. We shall review the variations of age and geographic origin as well as the relationship of composition to dental caries.

ENAMEL COMPOSITION

Major Components

Our knowledge as to the considerations of the major components of enamel has not changed to any degree over the last 20 years. The data presented in two major reviews on this subject, by Brudevold and Söremark[1] and later by Jenkins[2] still holds true today.

From these two reviews and other published data, the calcium content of enamel is stated as approximately 36% (Table 1). The variations in concentrations that are reported are partly due to whether the calculations are based upon dry weight, ash weight, or on fresh enamel. There does not appear to be any variation in the calcium content of whole enamel due specifically to age or geographic origin. Similarly, the phosphorus content, present almost entirely as phosphate, is between 16 to 18%.

Depending on the authority quoted, the other major components of enamel include sodium, magnesium, chloride, and potassium. Some authors consider these elements as "trace elements", but the concentrations are found in terms of percentages and for the purposes of

Table 1
CONCENTRATIONS OF MAJOR COMPONENTS OF HUMAN DENTAL ENAMEL (EXCLUDING WATER AND ORGANIC MATERIAL)[1,2]

Component	Concentration range in percentage (dry weight)
Ca	33.6—39.4
P	16.1—18.0
CO_3	2.7— 5.0
Na	0.25— 0.90
Mg	0.25— 0.90
Cl	0.19— 0.30
K	0.05— 0.30

this review are considered as major components. Table 1 gives a summary of the concentrations of the major elements to be found in enamel. It will be noted that the variation in concentration of these four elements is much greater than for the calcium and phosphate. Part of this variation may be due to the different analytical methods used, but undoubtedly there are differences due to tooth type, age, and geographic origin. During tooth development a variation in dietary intake of sodium, chloride, and magnesium will be reflected in blood plasma levels and then presumably in the fluids in the tooth follicle. Further changes will occur during enamel maturation. Similar changes will occur for carbonate which is another major component of enamel. This will be discussed separately below.

These inorganic components are contained in an array of crystals closely related in structure to hydroxyapatite.[3,4] The geologically well-known apatite structure can accommodate many different ions with ease and enamel apatite is no exception. The crystals are approximately 30 to 40 nm in diameter and according to recent work by Orams et al.,[5] using argon ion beam-thinned samples, the crystal length may be 10 μm or more. The crystals are clustered together in so-called "rods" about 4 μm in diameter which are roughly perpendicular to the tooth surface. This physical arrangement has considerable bearing on the observed chemical properties of enamel since diffusion among the crystals is an important possible rate-controlling process.

Water and organic material are an intrinsic part of enamel and must also be considered as major components. The crystals of enamel are surrounded by a water-filled protein and lipid matrix which constitutes about 4% by weight. This translates to approximately 15% by volume. Some of the water is bound in the apatite-like crystals, but the majority is in the intercrystalline spaces and therefore provides an aqueous-organic-inorganic medium throughout the enamel. Thus, the enamel is a porous crystalline matrix with aqueous channels through the incorporated organic phase. The nature of the organic material is discussed briefly below.

Carbonate in Enamel

The carbonate content of dental enamel ranges from 1 to 5% w/w (as CO_3).[6,7] Results are mostly reported as CO_2 content but have been converted to carbonate as CO_3 for this review, as it is in this form in which it is present in enamel crystals. A wide variation in carbonate content of enamel occurs in each individual tooth and this has been neatly shown by the concentration contours reported by Weatherell et al.[7] The carbonate content rises in all regions of enamel from the surface to the interior. Carbonate in surface enamel is commonly 2 to 3% and rises sharply over the first 20 to 50 μm to about 5% and then plateaus off at this value across the enamel to the dentino-enamel junction. The gradient of increase is not always smooth and sometimes pockets of low carbonate concentration are found within the enamel. The variation of carbonate content in individual teeth appears to override any possible geographic variation,[8] and it is obvious that sampling and analytical procedures must be carefully considered in making comparisons.

At one time there was controversy as to whether the carbonate was incorporated into the apatite crystal lattice of enamel, whether it existed on crystal surfaces, or indeed occurred as a separate phase such as calcium carbonate. It is now well established that carbonate is an intrinsic part of the enamel apatite structure.[9-14] Most of the carbonate (about 95%) substitutes in or near phosphate positions,[13-16] and only a small amount substitutes for the hydroxyl ion. Charge balance is at least partly accomplished by calcium deficiency, hydroxyl deficiency, and sodium incorporation. A level of 4% CO_3 in apatite means that one in seven phosphate groups is replaced by a carbonate.

Carbonate disturbs the apatite structure and synthetic carbonated apatites have poorly resolved X-ray diffraction patterns due to these disturbances and small particle size. Dental enamel, however, has a remarkably sharp X-ray diffraction pattern for the amount of carbonate present.[17] Recent studies using synthetic carbonated-apatites synthesized in the presence of metal and/or fluoride[18] have shown that the disturbance in the apatite structure created by carbonate presence can be compensated for by the presence of some trace metal and fluoride combinations incorporated in the crystal. Thus, the apatite structure is very adaptive to ionic impurities and combinations of these. Enamel apatite is a well-crystallized material in which the carbonate content has been largely compensated for by other incorporated ions. The implications of this in terms of enamel reactivity and caries are discussed further below.

Protein and Lipid in Enamel

The organic content of mature dental enamel is approximately 1% by weight (3% by volume).[2] Approximately half of this is protein which has preferentially remained as enamel maturation occurred and the majority of the protein which was originally present during amelogenesis was excluded.[19,20] This remaining protein appears to be bound to the enamel apatite as evidenced by the high proportion of the amino acid residues; aspartic, glutamic, and phosphoserine, present[21-24] which would provide strong links to the apatite surface.

Lipid accounts for the balance of organic material in mature enamel. Odutuga and Prout[25] reported the presence of lipids from all the major classes, and their data suggest that 40 to 50% of the organic matter in enamel is lipid. Cholesterol, cholesterol esters, triglycerides, monoglycerides, diglycerides, free fatty acids, and phospholipids were all present. The structural interrelationships of the lipids, protein, and apatite is unknown.

The protein and lipid components briefly described above are an intrinsic and important part of enamel and must be clearly distinguished from the so-called pellicle on the outer enamel surface. This pellicle which is formed by the adsorption of salivary proteins to the outer enamel surface is described elsewhere.

Thus, dental enamel must be considered as an ordered array of inorganic apatite-like crystals intimately surrounded by a protein/lipid/water matrix. Diffusion of chemical species through the micropore structure of this matrix is one of the rate determining steps in the formation of subsurface carious lesions.[26-28] This is further discussed below in the section entitled "Composition of Enamel in Relation to Dental Caries".

Trace Elements

Most of the elements of the Periodic Table, the exceptions being largely the radioactive elements, have been identified in human enamel. A general summary of the elements to be found in enamel and at what concentration levels is given in Table 2. This data is derived from over 50 reports on trace elements in enamel, commencing with the first such study by Drea[29] and up to the present. For those elements listed as not detected (ND), it should be

Table 2
SUMMARY OF TRACE ELEMENTS IN DENTAL ENAMEL OF PERMANENT TEETH

Concentration range (μg/g or ppm)	Elements
1000	Na, Cl, Mg
100—1000	K, S, Zn, Si, Sr, F
10—100	Fe, Al, Pb, B, Ba
1—10	Cu, Rb, Br, Mo, Cd, I, Ti, Mn, Cr, Sn
0.1—0.9	Ni, Li, Ag, Nb, Se, Be, Zr, Co, W, Sb, Hg
0.1	As, Cs, V, Au, La, Ce, Pr, Nd, Sm, Tb, Y
Not detected	Sc, Ga, Ge, Ru, Pb, In, Te, Eu, Gd, Dy, Ho, Er, Tm, Lu, Hf, Ta, Re, Os, Ir, Pt, Tl, Bi, Rh

noted that in most cases these elements have been found in some enamel samples, but not all. Oftentimes these elements have been identified, but at concentrations below the quantitative detection limit of the analytical technique used. This table also includes sodium, chloride, magnesium, and potassium which have already been mentioned above, but they are included here for completeness and because many researchers have considered them as trace elements.

The analytical methods used have varied and include neutron activation, optical emission, X-ray spectrometry, atomic absorption spectrophotometry, volumetric analysis, flame photometry, and ion probe. It is not surprising that for many elements there has been a wide disparity of results. Different analytical methods are more suitable for certain elements. Thus, where only single trace elements have been analyzed, results are usually more accurate than when large numbers of elements are sought simultaneously. Other factors will also affect the results, such as the perfection of enamel separation, choice of teeth, whether teeth are analyzed singly or pooled, and how the enamel is prepared for analysis. Other factors mentioned before such as geographic origin may be very important as this will encompass the environmental factors of water, soil, and food composition, as well as racial and cultural variables which affect diet usage.

Because the presentation of data has varied between studies, it is not easy to calculate overall median and ranges of concentrations. Nevertheless, an attempt has been made and Table 3 gives such a summary. The trace elements tested have been ordered by median concentrations or frequency of occurrence by weight. It is quite clear from this table that sodium, chloride, and magnesium contribute more to the enamel composition than do all the other 66 trace elements put together. This is one reason why it is debatable whether they should be considered as trace elements.

The most commonly occurring trace elements in enamel, those above 100 μg/g (ppm), are all easily detected by most analytical methods. It is also of note that all reports on multielement analyses report these elements in all samples.

It is interesting that potassium, with a median concentration of 300 μg/g, is much lower in concentration than magnesium at 2800 μg/g. This emphasizes the difference in occurrence of trace elements when considered by weight. There appears to be several levels at which the elements can be grouped, as shown in Table 2. These are at concentrations greater than 1000 μg/g, between 100 and 1000 μg/g and below 1 μg/g. The relationship of trace elements to caries, to be discussed later, has largely been covered for these found in enamel above 1 μg/g with the exception of lithium.

Although it has been the custom to record trace elements by weight, Cutress[30] has noted that changes in chemical and crystallographic properties of enamel will be affected by the

Table 3
CONCENTRATIONS OF TRACE ELEMENTS IN HUMAN ENAMEL A SUMMARY FROM PUBLISHED ANALYTICAL REPORTS

	Concentration (μg/g or ppm)		
Element	Median	Range (3 × SE)	μmol/g
Na	7,200	4,500—23,600	313
Cl	3,200	2,500—15,500	90
Mg	2,800	1,310—5,700	115
K	300	60—1,000	7.7
S	214	0—200	6.7
Zn	210	9—1,500	3.2
Si	190	26—1,155	6.8
Sr	115	13—1,400	1.3
Fe	28	0—510	0.5
Al	23	0—510	0.9
B	15	0—190	1.4
Ba	15	0—510	0.1
Cu	7	0—208	0.1
Pb	5.5	0—1,000	0.03
Rb	5	0—30	0.06
Br	4	0.1—33.2	0.05
Mo	3	0.6—39	0.03
I	2	0.01—9.9	0.02
Ti	1.5	0.1—66	0.02
Mn	1	0.68—29	0.02
Cr	1	0.1—33	0.02
Sn	1	0.03—93	0.008
Cd	1	0.03—268	0.009
Ni	0.9	0.06—13	0.02
Li	0.9	0—13.2	0.1
Ag	0.7	0.01—37	0.006
Nb	0.6	0.1—1.0	0.006
Se	0.4	0.12—18.1	0.005
Be	0.3	0.01—15.9	0.03
Zr	0.3	0.02—12	0.003
Co	0.2	0.01—30	0.003
W	0.2	0.08—0.06	0.001
Sb	0.1	0.05—3.0	0.001
Hg	0.1	0.001—16	0.0005
As	0.06	0.001—0.4	0.0008
Cs	0.04	0.02—0.1	0.0003
V	0.02	0.06—0.2	0.0003
Au	0.02	0.02—0.36	0.0001

Note: Concentration by dry weight.

After Cutress, T. W., personal communication. With permission.

number of atoms present rather than purely by weight (concentration). Heavier atoms would appear to contribute more than lighter atoms when present in equal numbers. A ranking of trace elements in enamel by micromoles (Table 3) shows that lithium is more common by atom frequency than by weight while lead is less frequent. This may be important as both of these elements have been associated, in enamel, with caries prevalence. In the enamel crystallite lattice it would appear that the lithium could be more important than its weight contribution would suggest.

Table 4
SUMMARY OF TRACE ELEMENT CHANGES IN ENAMEL WITH AGE

No change	Decrease	Increase
Sr, Al, Si	Sr, Mn	Pb, Na
Cr, Hg, Sb	Zn, Cu	K, Co
Ag, Mn, Fe, Sn, I, Zn, Cu, Mg, Se	—	Se

Note: Derived from 14 reports in the literature. Where an element occurs in more than one category, it has been reported as such in several different reports.

Variations of Enamel Composition with Age and Geographic Origin

The literature is now extensive that shows changes of fluoride in enamel with age. During the lifetime of a tooth, with the wear associated with chewing, brushing, erosion, and abrasion the fluoride-rich surface enamel is lost. This may be accentuated by the routine use of abrasive prophylaxis pastes in the dentist's office. Depending on the oral hygiene practices, such as the use of fluoride toothpastes or mouth rinses, as well as foods, some fluoride will be taken up by the surface enamel. Nevertheless, there is still a net loss of fluoride with age. It is probable that the same occurs for a number of other trace elements.[31] Analysis of whole enamel from teeth derived from older donors will therefore show reduced concentrations of such elements as fluorine, tin, and lithium.

Changes with age for the composition of trace elements in enamel are not common for most elements. However, of note is that zinc levels are lower in unerupted teeth than in erupted teeth, but with only minor changes throughout life.[32] Modest changes have been noted for a few other trace elements such as a decrease in manganese[33] and tin[34]. Overall 23 trace elements, apart from fluoride, have been studied specifically for age changes (Table 4). Of these, 15 do not change. Four elements, copper, manganese, strontium, and zinc decrease and five elements, cobalt, lead, potassium, selenium, and sodium increase.

Many studies have recorded geographic variations in trace element composition of enamel. As would be expected, several elements do show considerable differences in enamel samples collected in different countries, over and above that due to analytical differences. Lead, manganese, strontium, selenium, and zinc all exhibit wide variations. Barium and cadmium may also vary, but the evidence is less substantial. The most extensive study on the question involves teeth collected from many areas of the U.S.[35] Enamel samples were chosen to reflect variables, such as soil type, latitude, longitude, distance from the sea, high and low fluoride drinking waters, and high and low caries prevalence areas. Substantial differences in concentrations were found for a number of elements as shown in Table 5. The results of this study were grouped by state as this tended to reflect the allowed variables.

The extensive studies on strontium, as a result of the fallout of ^{90}Sr, have all shown wide variations in concentrations. These reports have all indicated that large differences,[8,29,32,36-38] as much as several hundred parts per million, can be found between enamel from different geographic areas. This variation in Sr content of enamel as well as other calcified tissues, has been shown to vary with dietary intake, mainly via the water.[38]

In addition to the above, Brudevold et al.[40] found substantial differences in the zinc content of enamel between teeth from Maine (U.S.) at 2100 μg/g and teeth from Greenland at 430 μg/g. Curzon[35] noted lower zinc levels in New Zealand tooth enamel than in comparable samples from the U.S. Both of these studies used whole enamel, whereas in the most recent

Table 5
GEOGRAPHIC VARIATIONS IN TRACE ELEMENT COMPOSITION OF WHOLE HUMAN ENAMEL IN TOOTH SAMPLES FROM SEVERAL STATES IN THE U.S.A.

	Mean trace element concentration (μg/g)					
State	F	Mn	Cu	Pb	Cd	Sr
New England	72	0.73	1.80	6.0	1.85	136
New York	107	0.40	2.91	24.0	2.07	149
Oregon	51	0.46	0.29	35.5	1.13	83
Montana	201	0.40	0.40	55.0	7.33	252
California	55	0.48	0.25	36.4	0.72	137
South Carolina	123	0.69	1.13	2.7	0.80	151
Ohio	214	0.83	1.56	24.2	2.66	213
Florida	257	0.43	5.41	31.0	2.53	281

After Curzon, M. E. J., Trace Element Composition of Human Enamel and Dental Caries, Ph.D. thesis, University of London, London, 1977. With permission.

work, using surface enamel, Cutress[8] and Schamschula et al.[41] found much less variation. It would appear that the geographic differences during tooth development are quite influential in affecting whole enamel composition, but not so in posteruption changes in enamel which affect the surface composition.

COMPOSITION OF ENAMEL IN RELATION TO DENTAL CARIES

Major Components and Carbonate

Attempts to relate changes in calcium and phosphate in enamel to caries have not been successful. Neither does variation in these components appear to be related to caries. It is possible, however, that changes in localized areas of surface enamel will affect or even determine caries susceptibility. Analytical techniques are now becoming available which will enable this aspect to be studied. Enamel is a calcium-deficient apatite, and preferential acid dissolution of the central cores of enamel and synthetic apatite crystals has been demonstrated.[42,43] Studies using high-resolution lattice imaging suggested that calcium-deficient areas in enamel crystals were sites for acid attack during artificial carious lesion formation.[44] A recent study of synthetic carbonated apatites and enamel, using the same technique, indicated that these areas of the crystals were related to carbonate incorporation and the presence of dislocations.[45] The frequency of crystal defects and dislocations has also been related to carbonate content in a similar way.[46] The importance of these changes in relation to increased crystal reactivity and caries susceptibility has yet to be established, and so a specific relationship between enamel solubility and dental caries remains unproven.

Since carbonate is the third most abundant component of enamel and since it markedly affects apatite structure and increases apatite reactivity to acid, it would be expected to affect dental caries. Carbonate is preferentially lost from enamel crystals during caries formation.[17,47-49] Carbonate content and carbonate/fluoride ratios in teeth from different ethnic groups and geographical areas have been related to caries formation.[8] Patel and Brown[50] showed that a wide range of solubility exists in enamel. Cummulative weight percent dissolved when plotted against solubility product showed that about 20% is extremely soluble. This appears to be related to preferential loss of carbonate-containing apatite. Because the structural effect of carbonate in apatite can be modified by the incorporation of zinc, strontium and fluoride,[18] the interaction of trace elements, carbonate content, and structure requires further study in relation to caries susceptibility. Carbonate alone in its own right cannot be used as a caries susceptibility predictor.

The concentrations of sodium and magnesium increase from the enamel surface to the dentin and are loosely linked with carbonate content. However, epidemiologic studies have not related either of these elements to caries susceptibility. Magnesium has however, been shown to be preferentially lost in a similar way to carbonate during the early stages of caries.[49] The implications of this in caries susceptibility are unknown at this time.

The other major components of enamel form the water/protein/lipid matrix. This provides a micropore diffusion barrier to the ingress of organic acids during the caries process and the egress of calcium and phosphate to give mineral loss.[27] It is expected that this gel matrix will behave in a similar way as other body membranes and that diffusion rates will be determined by its physical and chemical composition. As crystal dissolution progresses, the pore structure will change permitting more rapid transport of chemical species. Adsorbed protein dramatically reduces apatite solubility,[51] and lipid markedly reduces diffusion rates of calcium and phosphate through a lipid/protein/apatite membrane.[52] Hence, although the organic matrix provides the route for acid entry to enamel during a caries challenge, it also provides a highly protective environment against demineralization.

Trace Elements

Most of the work on trace elements in enamel and dental caries has concerned fluoride. By and large the most useful work has considered surface fluoride concentrations because of early surface demineralization as discussed before.

The earliest study considering a relationship between individual surface enamel fluoride concentration and caries was reported by Keene et al. in 1975.[53] They demonstrated that caries-free sailors had higher mean enamel fluoride concentrations than those of a caries active group. No comparisons of individual enamel concentrations with individual caries scores were attempted. DePaola et al.[54] studied the relationship on an individual basis for over 1000 Boston area school children. After logarithmic transformation, fluoride concentrations (corrected for depth), and performing a square root transformation on DMFS scores, they found a correlation coefficient that showd $r = -0.38$. This indicated a 14% variance in caries scores explained by surface enamel fluoride variations. In a subsample of one community, where a wide range of fluoride exposure had occurred, the correlation coefficient was $r = -0.20$. In other communities, however, no significant correlation existed, probably due to the small range of fluoride concentrations present in the individual communities.

A similar difficulty was found by Englander and Mellberg,[55] who attempted to relate fluoride concentrations of extracted deciduous incisors to caries scores in children. Using a sample of 176 teeth, fluoride concentrations were grouped into 8 categories, and caries scores were classified as either 0 or greater than 0. Analysis by X^2 showed no association between the two factors, but the mean caries score was reported as 1.7. Use of such a restricted sample would probably not allow a relationship, even if present, to become apparent in the statistical analysis.

Poulsen and Larsen[56] had earlier tried to correlate caries experience with the fluoride content of the surface enamel of deciduous teeth. Caries examinations were carried out to determine the caries prevalence at age 3 years and again at age 6 years to calculate the caries incidence. At the time of the second examination, an enamel biopsy was carried out to determine the fluoride concentration in surface enamel, but no significant correlations were found. This lack of correlation was ascribed as due to errors in clinical examinations for dental caries and differences in fluoride exposure and wear of the enamel surface with time and therefore age.

Bischoff et al.[57] also attempted to relate individual fluoride concentrations to caries experience, using a population living in a high fluoride area. They reported no significant relationship, once again probably due to the small range of caries scores in the sample studied. Similarly, no significant relationship was discerned by van der Merwe et al.[58] in a sample drawn from both high and low fluoride areas.

Two previous studies have linked strontium in surface enamel to dental caries. Little and Barrett[59] reported that the strontium and fluoride concentrations of surface enamel were higher in a low caries (<3 DMFT) group than in a high caries (>7 DMFT) group. These relationships held for the entire sample and also when the sample was broken down into smaller geographical groups. No mention was made of attempted individual correlations. Later Spector and Curzon[60] found substantial differences in strontium in surface enamel in 439 teeth collected from Ohio and Wisconsin, U.S. Mean concentrations were 366.4 μg Sr per gram and 1169.6 μg F per gram. A significant, but weak correlation between fluoride concentration and caries as DMFT ($r = -0.16$) was found, and also between strontium and fluoride concentrations ($r = 0.48$). However, no correlation of strontium in surface enamel and caries was found, indicating that either the relationship was nonlinear or the mechanism of action of strontium in reducing caries, as seen epidemiologically,[61] did not depend upon the level of strontium incorporated individually in surface enamel.

In a study in Papua-New Guinea, Schamschula et al.[41] reported on a number of trace elements in surface enamel which they were able to relate to caries experience. In brief, their findings indicated significant negative associations for lithium and lead in enamel to caries prevalence and positive associations for fluoride, barium, potassium, and strontium. The association of greatest interest in this study was that of lithium as significant negative associations were also found for caries and lithium in saliva solids. These same authors have again recently reported an association of lithium to caries in another study.[62] The role of lithium in enamel to caries requires much further investigation.

Although the trace element content of surface enamel in relation to caries has recently been of great interest, the work on whole enamel should be mentioned. Previous studies have analyzed whole enamel in attempts to correlate trace elements with caries, but only considered simple relationships of individual elements to dental decay.[33,63-66] In one study however, all elements of the Periodic Table were looked for in 451 human enamel samples.[67] Using multiple regression statistical techniques, individual caries prevalence of each tooth donor was related to 31 trace elements identified in all samples. The results, in Table 6, not only considered the mean concentrations of the trace elements found to be related to caries, but also the estimated partial effect of DMFT per unit predictor. This latter was an attempt to determine how much of the caries could be attributed to the individual trace elements. Of interest here is the apparently large part of the DMFT per unit predictor explained by the elements manganese (0.90), copper (0.23), selenium (0.20), and cadmium (0.51), all of which, because of the positive sign, were associated with high caries. In contrast the most significant associations for trace elements and low caries were fluoride (t ratio -3.34) and strontium (t ratio -4.25), but the part of DMFT per unit predictor accounted for by these two elements was small at -0.10×10^{-1} and 0.11×10^{-1}, respectively. Thus, although highly significantly related to low caries, they explain only a very small part of the caries indexes.[67]

The question of by how much trace elements in enamel (either surface or whole) affect dental caries still remains to be adequately researched. It is a difficult subject because of the precision of analysis that is required on very small samples. The use of the newer analytical techniques, such as ion probe, may help to answer this question.

Enamel Composition and Remineralization

The outer few micrometers of enamel are regularly challenged by acid from plaque bacterial metabolism. The partial demineralization which occurs is followed by remineralization as

Table 6
SIGNIFICANT RELATIONSHIPS OF TRACE ELEMENTS IN WHOLE ENAMEL TO DENTAL CARIES

Element	n[a]	By analysis, mean µg/g	Caries relationship, partial effect[b]	*t* ratio
F	337	130.3	-0.10×10^{-1}[c]	−3.34
Al	335	22.9	-0.30×10^{-1}	−2.98
K	337	961.4	-0.62×10^{-2}	−2.93
V	239	0.02	0.62	2.02
Mn	336	0.6	0.90	2.28
Fe	337	27.9	-0.21×10^{-1}	−2.10
Ni	249	0.9	-0.33×10	−2.10
Cu	336	1.5	0.23	2.30
Zn	336	153.1	0.96×10^{-1}	2.41
Se	326	1.5	0.20	2.34
Rb	256	4.6	0.30×10^{-1}	2.32
Sr	337	157.1	-0.11×10^{-1}	−4.25
Cd	334	1.9	0.51	2.09
Pb	335	19.6	−0.96	−2.02

[a] n, Sample size is number of tooth donors represented in the regression model.
[b] Part of caries variable explained by trace element effect.
[c] Negative sign indicates that variable is associated with low caries.

After Curzon, M. E. J. and Crocker, D. C., *Arch. Oral Biol.*, 23, 647, 1978. With permission.

the partially dissolved crystals regrow with preferential exclusion of some ions and inclusion of others. The so-called intact surface layer of the carious lesion is formed by this remineralizing process and contains well-formed apatite crystals which are low in carbonate. Fluoride, when present, promotes remineralization and is preferentially incorporated into enamel during the remineralization process. It seems that strontium and zinc are also similarly incorporated during remineralization. This remineralization or maturation of surface enamel occurs even before a caries lesion is clinically evident. This process is part of the natural protective mechanism against caries. The enamel crystals which are formed are hence more resistant to subsequent acid challenge. Obviously calcium and phosphate must also be present at sufficient concentrations for crystal regrowth. Enhancement of this regrowth or remineralization is a promising means of caries prevention or reversal and is a process which is generally, and naturally, carried out by salivary components.

SUMMARY

Studies on the composition of enamel have shown that although the major structural components of calcium and phosphorus, and to a lesser extent sodium, magnesium, and chloride, do not vary much, there are substantial differences in carbonate and trace elements. Recent research has shown that, certainly for surface enamel, the carbonate and trace element content play major roles in the susceptibility of enamel to caries. More particularly, the subsequent history of the tooth as to further demineralization or remineralization is related to carbonate and the presence of certain trace elements such as fluoride, zinc and strontium.

REFERENCES

1. **Brudevold, F. and Söremark, R.,** Chemistry of the mineral phase of enamel, in *Structural and Chemical Organization of Teeth,* Miles, A. E. W., Ed., Academic Press, London, 1967, 247.
2. **Jenkins, G. N.,** *The Physiology and Biochemistry of the Mouth,* 4th ed., Blackwell Scientific, Oxford, 1978.
3. **Kay, M. I., Young, R. A., and Posner, A. S.,** Crystal structure of hydroxyapatite, *Nature (London),* 204, 1050, 1964.
4. **Young, R. A. and Spooner, S.,** Neutron diffraction of human tooth enamel, *Arch. Oral Biol.,* 15, 47, 1969.
5. **Orams, H. J., Zybert, J. J., Phakey, P. P., and Rachinger, W. A.,** Ultrastructural study of human dental enamel using selected-area argon-beam thinning, *Arch. Oral Biol.,* 21, 663, 1976.
6. **Little, M. F. and Brudevold, F.,** A study of inorganic carbon dioxide in intact human enamel, *J. Dent. Res.,* 37, 991, 1958.
7. **Weatherell, J. A., Robinson, C., and Hiller, C. R.,** Distribution of carbonate in thin sections of dental enamel, *Caries Res.,* 2, 1, 1968.
8. **Cutress, T. W.,** The inorganic composition and solubility of dental enamel from several specific population groups, *Arch. Oral Biol.,* 17, 93, 1972.
9. **McConnell, D.,** Recent advances in the investigation of the crystal chemistry of dental enamel, *Arch. Oral Biol.,* 3, 28, 1960.
10. **LeGeros, R. Z., Trautz, O. R., LeGeros, J. P., and Klein, E.,** Carbonate substitution in the apatite structure, *Bull. Soc. Chim. France,* Spec. Suppl., 1712, 1968.
11. **Elliott, J. C.,** Recent progress in the chemistry, crystal chemistry and structure of the apatites, *Calcif. Tissue Res.,* 3, 293, 1969.
12. **Ingram, G. S.,** The role of carbonate in dental mineral, *Caries Res.,* 7, 217, 1973.
13. **Voegel, J. C. and Garnier, P.,** Biological apatite crystal dissolution, *J. Dent. Res.,* 58(B), 852, 1979.
14. **Nelson, D. G. A. and Featherstone, J. D. B.,** Preparation, analysis and characterization of carbonated apatites, *Calcif. Tissue Int.,* 34, S69, 1982.
15. **Tochon-Danguy, H. J., Geoffrey, M., and Baud, C. A.,** E.S.R. study of the effects of carbonate substitution in synthetic apatites and apatites from human teeth, *Arch. Oral Biol.,* 25, 357, 1980.
16. **Elliott, J. C.,** The interpretation of the infra-red absorption spectra of some carbonate-containing apatites, in *Tooth Enamel,* Stack, M. V. and Fearnhead, R. W., Eds., Wright, Bristol, 1965, 20.
17. **Featherstone, J. D. B., Duncan, J. F., and Cutress, T. W.,** A mechanism for dental caries, *Arch. Oral Biol.,* 24, 101, 1979.
18. **Featherstone, J. D. B. and Nelson, D. G. A.,** The effect of F, Zn, Sr, Mg and Fe on the crystal-structural disorder in synthetic carbonated apatites, *Aust. J. Chem.,* 33, 2363, 1980.
19. **Robinson, C., Fuchs, P., and Weatherell, J. A.,** The fate of matrix proteins during the development of dental enamel, *Calcif. Tissue Res.,* 22(Suppl.), 185, 1977.
20. **Glimcher, M.J., Brickley-Parsons, D., and Levine, P. T.,** Studies of enamel proteins during maturation, *Calcif. Tissue Res.,* 24, 259, 1977.
21. **Eastoe, J. E.,** Enamel protein chemistry — past, present and future, *J. Dent. Res.,* 58b, 753, 1979.
22. **Termine, J. D., Torchis, D. A., and Conn, K. M.,** Enamel matrix: structural proteins, *J. Dent. Res.,* 58b, 773, 1979.
23. **Frank, R. M.,** Tooth enamel: current state of the art, *J. Dent. Res.,* 58(B), 684, 1979.
24. **Robinson, C., Lowe, N. R., and Weatherell, J. A.,** Amino acid composition, distribution and origin of tuft protein in human and bovine dental enamel, *Arch. Oral Biol.,* 20, 29, 1975.
25. **Odutuga, A. A., and Prout, R. E. S.,** Lipid analysis of human enamel and dentin, *Arch. Oral Biol.,* 19, 729, 1974.
26. **Moreno, E. C. and Zahradnik, R. T.,** Chemistry of enamel subsurface demineralization in vitro, *J. Dent. Res.,* 53, 226, 1974.
27. **Featherstone, J. D. B., Duncan, J. F., and Cutress, T. W.,** Crystallographic changes in human tooth enamel during in-vitro caries simulation, *Arch. Oral Biol.,* 23, 405, 1978.
28. **de Rooij, J. F., Kolar, Z., and Arends, J.,** Phosphate diffusion in whole bovine enamel at pH 7, *Caries Res.,* 14, 393, 1980.
29. **Drea, W. F.,** Spectrum analysis of dental tissues for "trace" elements, *J. Dent. Res.,* 15, 403, 1936.
30. **Cutress, T. W.,** Trace elements in human enamel, in *Trace Elements in Dental Disease,* Curzon, M. E. J. and Cutress, T. W., Eds., PSG-John Wright, New York, 1983.
31. **Little, M. F. and Steadman, L. T.,** Chemical and physical properties of altered and sound enamel, IV. Trace elements, *Arch. Oral Biol.,* 11, 273, 1966.
32. **Steadman, L. T., Brudevold, F., and Smith, F. A.,** Distribution of strontium in teeth from different geographic areas, *J. Am. Dent. Assoc.,* 57, 340, 1958.

33. **Derise, N. L. and Ritchey, S. J.,** Mineral composition of normal human enamel and dentin, and the relation of composition to dental caries. II. Microminerals, *J. Dent. Res.*, 53, 853, 1974.
34. **Battistone, G. C., Feldman, M. H., and Reba, R. C.,** The manganese content of human enamel and dentine, *Arch. Oral Biol.*, 12, 1115, 1967.
35. **Curzon, M. E. J.,** Trace Element Composition of Human Enamel and Dental Caries, Ph.D. thesis, University of London, London, 1977.
36. **Helsby, C.,** Determination of strontium in human tooth enamel by atomic absorption spectrometry, *Anal. Chim. Acta*, 69, 259, 1975.
37. **Nixon, G. S. and Helsby, C. A.,** The relationship between strontium in water supplies and human tooth enamel by atomic absorption spectrometry, *Anal. Chim. Acta*, 69, 259, 1975.
37. **Nixon, G. S. and Helsby, C. A.,** The relationship between strontium in water supplies and human tooth enamel, *Arch. Oral Biol.*, 22, 691, 1976.
38. **Wolf, N., Gedalia, I., Yariv, S., and Zuckermann, H.,** The strontium content of bones and teeth of human foetuses, *Arch. Oral Biol.*, 18, 233, 1973.
39. **Spector, P. C. and Curzon, M. E. J.,** Relationship of strontium in drinking water and surface enamel, *J. Dent. Res.*, 57, 55, 1978.
40. **Brudevold, F., Steadman, L. T., Spinelli, M. A., Amdur, B. H., and Gron, P.,** Distribution of strontium in teeth from different geographic areas, *Arch. Oral Biol.*, 8, 135, 1963.
41. **Schamschula, R. G., Adkins, B. L., Barmes, D. E.,** et al., WHO Study of Dental Caries Etiology in Papua-New Guinea, Publication No. 40, World Health Organization, Geneva, 1978.
42. **Scott, D. B., Simmelink, J. W., and Nygaard, V.,** Structural aspects of dental caries, *J. Dent. Res.*, 53, 165, 1974.
43. **Jongebloed, W. L., van den Berg, P. J., and Arends, J.,** The dissolution of single crystals of hydroxyapatite in citric and lactic acids, *Calcif. Tissue Res.*, 15, 1, 1974.
44. **Featherstone, J. D. B., Goodman, P., and McLean, J. D.,** Electron microscope study of defect zones in dental enamel, *J. Ultrastruct. Res.*, 67, 117, 1979.
45. **Featherstone, J. D. B., Nelson, D. G. A., and McLean, J. D.,** An electron microscope study of modifications to defect regions in dental enamel and synthetic apatites, *Caries Res.*, 15, 278, 1981.
46. **Nelson, D. G. A.,** The influence of carbonate on the atomic structure and reactivity of hydroxyapatite, *J. Dent. Res.*, 60(C), 1621, 1981.
47. **Hallsworth, A. S., Weatherell, J. A., and Robinson, C.,** Loss of carbonate during the first stages of enamel caries, *Caries Res.*, 7, 345, 1973.
48. **Cooledge, T. B. and Jacobs, M. H.,** Enamel carbonate in caries, *J. Dent. Res.*, 36, 765, 1957.
49. **Hallsworth, A. S., Robinson, C., and Weatherell, J. A.,** Mineral and magnesium distribution within the approximal carious lesion of dental enamel, *Caries Res.*, 6, 156, 1972.
50. **Patel, P. R. and Brown, W. E.,** Thermodynamic solubility product of human enamel powdered sample, *J. Dent. Res.*, 54, 728, 1975.
51. **Voegel, J. C., Belcourt, A., and Gillneth, S.,** Dissolution of hydroxyapatites treated with salivary glycoproteins and fluoride, *Caries Res.*, 15, 243, 1981.
52. **Oldershaw, M. D., Featherstone, J. D. B., and Rosenberg, H.,** Lipid effect on ionic transport in relation to enamel demineralization and remineralization, *J. Dent. Res.*, 60A (Abstr.), 819, 1981.
53. **Keene, H. J., Grossman, F. D., Pederson, E. D., Mellberg, J. R., and Nicholson, C. R.,** Caries experience and fluoride concentration of surface enamel in dental fluorosis patients, *Caries Res.*, 9, 247, 1975.
54. **DePaola, P. F., Brudevold, F., Aasendon, R., Moreno, E. C., Englander, H., Bakhos, Y., Bookstein, F., and Warram, J.,** A pilot study of the relationship between caries experience and surface enamel fluoride in man, *Arch. Oral Biol.*, 20, 859, 1975.
55. **Englander, H. R. and Mellberg, J. R.,** Failure to demonstrate an association between enamel F concentration and dental caries in the deciduous dentition, *J. Dent. Res.*, 55, 707, 1976.
56. **Poulsen, S. and Larsen, M. J.,** Dental caries in relation to fluoride content of enamel in the primary dentition, *Caries Res.*, 9, 59, 1975.
57. **Bischoff, J. I., van der Merwe, E. H. M., Retief, D. H., Barakow, F. H., and Cleaton-Jones, P. E.,** Relationship between fluoride concentrations in enamel, DMFT index, and degree of fluorosis in a community residing in an area with a high level of fluoride, *J. Dent. Res.*, 55, 37, 1976.
58. **van der Merwe, E. H. M., Bischoff, J. I., Fatti, L. P., Retief, D. H., Barkow, F. H., and Friedman, M.,** Relationships between fluoride in enamel, DMFT index, and fluorosis in high- and low-fluoride areas in South Africa, *Community Dent. Oral Epidemiol.*, 5, 61, 1977.
59. **Little, M. F. and Barrett, K.,** Sr and F content of surface and inner enamel versus caries prevalence in Atlantic coast of the USA, *Caries Res.*, 10, 297, 1976.
60. **Spector, P. C. and Curzon, M. E. J.,** Surface enamel fluoride and strontium in relation to caries prevalence in man, *Caries Res.*, 13, 227, 1979.

61. **Curzon, M. E. J., Spector, P. C., and Iker, H. P.,** An association between strontium in drinking water supplies and low caries prevalence in man, *Arch. Oral Biol.*, 23, 317, 1978.
62. **Schamschula, R. G., Cooper, M. H., Agus, H. M., and Un, P. S. H.,** Oral health in Australian children using surface and artisian water supplies, *Community Dent. Oral Epidemiol.*, 9, 27, 1981.
63. **Hoogland, P. L.,** Trace Elements and Dental Disease, Thesis, University of Amsterdam, Amsterdam, 1948.
64. **Healy, W. B., Ludwig, T. G., and Losee, F. L.,** Soils and dental caries in Hawkes Bay, New Zealand, *Soil Sci.*, 92, 359, 1963.
65. **Anderson, R. J.,** The relationship between dental conditions and the trace element molybdenum, *Caries Res.*, 3, 75, 1969.
66. **Vrbic, V. and Stupar, J.,** Dental caries and the concentration of aluminium and strontium in enamel, *Caries Res.*, 14, 141, 1980.
67. **Curzon, M. E. J. and Crocker, D. C.,** Relationships of trace elements in human tooth enamel to dental caries, *Arch. Oral Biol.*, 23, 647, 1978.

BIOSYNTHESIS OF ENAMEL PROTEINS

Harold C. Slavkin and Margarita Zeichner-David

INTRODUCTION

This review has several objectives: (1) to review the general problem area of enamel protein biosynthesis and (2) to discuss critical problems related to gene regulation, translation, and posttranslational processing of one or more unique enamel gene products. Since space limitations prevent an exhaustive review and citation of scientific publications, we shall indicate the most pertinent literature related to a biochemical approach of enamel biosynthesis. Our purpose is to provide the reader with a key to an expanding scientific literature. Our survey is complemented by several previous comprehensive reviews.[1-3]

The concept of enamel proteins as an heterogeneous family of proteins is rapidly changing. The notion of a multiaggregating assembly of numerous polypeptides, varying in molecular weight from 3,000 to 18,000 daltons, is being challenged. More recent evidence now suggests that secretory amelogenesis, the production of enamel proteins for a developing extracellular matrix, consists of the biosynthesis of two major enamel proteins, enamelins and amelogenins.[4] We now define, operationally two different enamel proteins, one consisting of a high molecular weight *enamelin* (~172,000 daltons) and the other a lower molecular weight *amelogenin* of approximately 30,000 daltons.[5] It is likely that the initial translation gene products for each of these different secretory enamel proteins will have still longer amino acid sequences and will also have additional "signal sequences". A number of studies indicate that the in vitro translation products of the mRNAs for various secretory proteins (e.g., collagens, insulin, ovalbumin) carry a 20- to 30-residue NH_2-terminal extension rich in hydrophobic amino acids.[6,7] This extension region is thought to provide a "signal sequence" which functions in the initiation of the polysome-membrane junction leading to segregation of nascent polypeptides.[6]

The archetype of secretory proteins found in developing extracellular matrices associated with mineralization is preprocollagen.[8,9] In view of the uncertainties concerning the actual number of structural genes for enamel, specific details regarding transcription, posttranscriptional processing, translation, and posttranslation processing within the ameloblast cells, as well as in the extracellular matrix, our discussion attempts to bring into focus biochemical studies and future opportunities of enamel biosynthesis.

BIOSYNTHESIS

Early investigations of developing enamel matrix proteins indicated a reversible aggregating system of heterogeneous and relatively low molecular weight polypeptides.[10] More recent investigations have been able to discriminate among enamel matrix constituents by considering hydrogen bonding, hydrophobic bonding, and electrostatic forces between diverse enamel polypeptides.[4,5,10-13] Advances in extraction methods using denaturing conditions in the presence of protease inhibitors, polyacrylamide gel electrophoresis in the presence of sodium dodecylsulfate (SDS), amino acid analysis including composition and partial sequence data, gel and ion exchange chromatography, and pulse/chase isotopic precursor labeling experiments have enabled additional penetration into this complex problem.

KINETICS AND LOCALIZATION OF ENAMEL PROTEIN BIOSYNTHESIS

Light microscopic and high resolution electron microscopic autoradiographic investigations of enamel protein biosynthesis in vivo and in vitro have provided kinetic and localization data describing the biosynthesis of enamel proteins in a number of different mammalian species.[14-19] Following the administration of various isotopic amino acid precursors *in situ*, label in the form of silver grains is first detected over the rough endoplasmic reticulum within 5 min.[16,19] Using isotopic monosaccharide precursors, it has been demonstrated that enamel proteins are glycosylated in the Golgi apparatus, packaged in condensing vacuoles derived from the inner saccules, transported in secretory vesicles to the secretory regions of the ameloblast, and then released into the forming enamel matrix within 30 min from the initiation of translation.[1,18,19] A number of obvious translational and posttranslational controls are apparent for the regulation of enamel biosynthesis.

ENAMEL GENE EXPRESSION

The number of structural genes for enamel is not as yet known. Current information suggests that most secretory proteins (e.g., globin, collagen, insulin, ovalbumin, immunoglobulins) are translated from unique poly[A]$^+$mRNAs. It is well-documented that in most eucaryotic cell types, formation of translatable mRNA is a complex process. Primary DNA transcripts, pre-mRNAs or hnRNAs, are synthesized in the nucleus and undergo posttranscriptional processing including cleavage of these extremely large RNA precursors, 3′-end polyadenylation, 5′-end capping and methylation, internal methylation, and finally splicing together of different segments of pre-mRNA into the final and functional mRNA molecule which is transported to the polysomes for subsequent translation into the unique gene products.[20] Transcription and posttranscriptional processing of enamel gene products is not as yet available. Data is available, however, which indicates that transcriptional inhibitors such as actinomycin D (30 μg/mℓ) or proflavine (60 μ*M*) reduce enamel protein synthesis by 60 to 72% in secretory ameloblast cells in vitro.[21]

TRANSLATION OF ENAMEL mRNAs

The isolation and characterization of nascent enamel polypeptides on polysomes has not been reported. Information is not as yet available regarding the details of nascent translation of enamel mRNAs; neither the number of mRNAs or the actual number of nascent polypeptides is as yet available.

Recently, our laboratory reported on the isolation and preliminary characterization of ameloblast-derived mRNAs and their respective gene products.[22-23] A poly[A]$^+$mRNA fraction of 16 to 26S was isolated from developing fetal New Zealand white rabbit molar ameloblast cells by ultracentrifugation on sucrose gradients.[23] When translated in a reticulocyte-lysate, cell-free system, the mRNA resulted in the synthesis of three proteins of mol wt 65,000, 58,000, and 43,000 daltons.[23] The higher molecular weight proteins were assumed to be enamel polypeptides. Further attempts to identify these translation gene products using immunoprecipitation assays indicate that they are antigenically cross-reactive with antisera which are produced against enamel matrix proteins. Attempts to produce a cDNA for enamel mRNAs are now in progress.

POSTTRANSLATIONAL PROCESSING OF ENAMEL PROTEINS

The available evidence now suggests that several major gene products, enamelin andamelogenin, characterize the ameloblast biochemical phenotype.[4,5,24] These proteins appear to have significant posttranslational modifications during the course of intracellular transport, secretion, extracellular matrix formation, and mineralization.[2,25-27] Our laboratory previously suggested the paradigm of a "preproenamel→→→proenamel→→→enamel" model[2,19,28] after the precedent discovered for a number of secretory proteins including the various types of collagen.[7-9,20,29] We further suggest that the kinetics and relative processing for enamelin will be found to significantly differ with those found for amelogenin.[4,5,24,30]

Evidence for a "precursor-product" relationship for enamel proteins has been suggested by a number of different laboratories.[2,12,19,21,28,30] Porcine, logamorph, rodent, and bovine enamel protein biosynthetic studies have all suggested proteolytic processing of nascent enamel proteins during secretion and during subsequent enamel matrix maturation.[25,27,28,30]

Available information suggests that amelogenins are processed from a 25,000- to 30,000-dalton to a 21,000-dalton polypeptide during secretion.[21,30,31] Of course, the nascent amelogenin polypeptide should prove to be of higher molecular weight. In contrast, enamelin polypeptides appear to be more stable, have different kinetic properties, and probably possess very different functions.

ENAMELIN BIOSYNTHESIS

A sequential dissociative extraction scheme has been reported in which fetal enamel matrix proteins are extracted first in 4 *M* guanidine HCl, pH 7.4, and subsequently in 4 *M* guanidine HCl, 0.5 *M* EDTA, pH 7.4, in the presence of protease inhibitors.[4,5] The latter procedure has been shown to be effective for demineralization and the extraction of those enamel proteins specifically associated with enamel calcium hydroxyapatite crystals.[5] The initial dissociative step extracts proline-rich amelogenins, whereas the subsequent extraction with guanidine HCl/EDTA removes enamelins.[5] The enamelins are approximately 72,000 daltons, stain positively for sialic acid, phosphate, and carbohydrates, and have an amino acid composition distinct from that reported for amelogenins.[5,24]

In addition to their amino acid composition differences, enamelins also differ from amelogenins in their very high affinity for apatite crystals. Biosynthetic data indicate that amelogenins are secreted before enamelins and that amelogenins are rapidly converted to increasingly smaller polypeptides by enamel proteases.[30] Enamelins appear to be highly stable during enamel maturation.[5] Immunoprecipitation data indicates that both enamelins and amelogenins are antigenically cross-reactive.[5] These features and many others now provide opportunities for future studies of enamelin biosynthesis.

AMELOGENIN BIOSYNTHESIS

The biosynthesis of amelogenins in ameloblast cells include translation on polysomes, glycosylation of the polypeptides in the Golgi, phosphorylation during condensation within secretor vesicles, and subsequent proteolytic posttranslational processing of secreted amelogenins into increasingly smaller polypeptides.[1,2,12,13,19,21,31-34] During fetal amelogenesis, newly secreted amelogenins are approximately 25,000 to 30,000 daltons, which is considered the archetype of amelogenins.[12,21,25-27,31] Although controversy had existed for many years regarding the actual number of different enamel gene products, current data now provides evidence to indicate that the larger molecular weight amelogenin is the precursor for the numerous small polypeptides.[35] Pulse/chase isotopic labeling experiments, polyacrylamide gel electrophoresis, and gel filtration chromatography have been used to establish the amelogenin precursor-product relationship; amelogenins of approximately 30,000 daltons are proteolytically cleaved to produce 21,000-dalton amelogenins which, in turn, produce 6,000-dalton amelogenin polypeptides.[4,5,12,13,21,25-28,31,35] Amino acid composition and partial amino acid sequence data further indicate that amelogenins of anomalous molecular weights were all derived from the same primary gene product.[35] Finally, partial characterization of enamel proteases, which mediate posttranslational processing of amelogenin, have a pH optimal activity of 6, are inhibited by phenylmethylsulfonyl fluoride (PMSF) and diisopropyl phosphofluoridate, and are not affected by EDTA.[25-27]

SUMMARY

A number of recent technical advances now enable biochemical investigations of enamel biosynthesis throughout vertebrate tooth development. Extraction techniques designed to minimize potential protein degradation and aggregation of enamel polypeptides during isolation and characterization, advances in two-dimensional SDS-slab gel electrophoresis, fluorography, the development of immunoprecipitation assays for enamel proteins,[36,37] and recent advances in DNA recombinant technology all provide promise for future elucidation of a molecular understanding of secretory amelogenesis.[38,39]

Of considerable evolutionary interest are the recent reports demonstrating that rabbit antimammalian enamel antibodies (i.e., antiamelogenins) are antigenically cross-reactive with a number of lower vertebrate enameloids (e.g., hagfish, shark, fish).[40,43] These studies suggest that enamel proteins throughout the 500 million years of vertebrate evolution have conserved amino acid sequence regions within the enamel protein molecules. Such immunological characteristics should be very useful in future studies employing the avidity of enamel antibodies for uses such as affinity chromatography.

Of considerable developmental interest is the recent report indicating that embryonic murine dental papilla mesenchyme can induce avian pharyngeal epithelial tissue to differentiate into secretory ameloblasts and produce enamel extracellular matrix.[44] Since *Aves* do not express a dentition, these xenoplastic heterologous tissue recombination experiments (mammalian mesenchyme plus avian epithelium) now suggest a number of unique experimental designs with which to understand the regulation of enamel gene transcription, translation, and posttranslational processing.

Finally, a number of biomedical problems can also be approached, including a further biochemical investigation of amelogenesis imperfecta, dental caries processes, and possibly improved methods for dental caries prevention.

ACKNOWLEDGMENTS

We are extremely grateful to Dr. John Termine of the National Institute of Dental Research for his valuable comments, suggestions, and continued collaboration. We would like to thank Ms. Gwen Airkens for her untiring assistance in the preparation of this manuscript. This work was supported by research grant DE-02848 and training grant DE-07006 from the National Institutes of Health, U.S. Public Health Service.

REFERENCES

1. **Weinstock, A.,** Matrix development in mineralizing tissues as shown by radioautography: formation of enamel and dentine, in *Developmental Aspects of Oral Biology*, Slavkin, H. C. and Bavetta, L. A., Eds., Academic Press, New York, 1972, 202.
2. **Slavkin, H., Trump, G., Schonfeld, S., Brownell, A., Sorgente, N., and Lee-Own, V.,** Epigenetic regulation of enamel protein synthesis during epithelial-mesenchymal interactions, in *Cell Interactions in Differentiation*, Saxen, L. and Weiss, L., Eds., Academic Press, New York, 1978, 209.
3. **Nylen, M. U. and Termine, J. D.,** Tooth enamel. III. Its development, structure, and composition, in Third International Symposium on Tooth Enamel, *J. Dent. Res.*, 58B, 1978.
4. **Termine, J. D., Torchia, D. A., and Conn, K. M.,** Enamel matrix: structural proteins, *J. Dent. Res.*, 58(B), 773, 1979.
5. **Termine, J. D., Belcourt, A. B., Christner, P. J., Conn, K. M., and Nylen, M. U.,** Properties of dissociatively extracted fetal tooth matrix proteins. I. Principal molecular species in developing bovine enamel, *J. Biol. Chem.*, 255, 9760, 1980.

6. **Blobel, G.,** Synthesis and segregation of secretory proteins: the signal hypothesis, in *International Cell Biology*, Brinkley, B. R. and Porter, K. R., Eds., The Rockefeller University Press, New York, 1977, 318.
7. **Steiner, D. F., Quinn, P. S., Chan, S. J., Marsh, J., and Tager, H. S.,** Processing mechanisms in the biosynthesis of proteins, in *Precursor Processing in the Biosynthesis of Proteins*, Zimmerman, M., Mumford, R. A., and Steiner, D. F., Eds., New York Academy of Sciences, New York, 1980, 1.
8. **Martin, G. R., Byer, P. H., and Piez, K. A.,** Procollagen, *Adv. Enzymol.*, 42, 167, 1975.
9. **Fessler, J. H. and Fessler, L. I.,** Biosynthesis of procollagen, *Annu. Rev. Biochem.*, 47, 129, 1978.
10. **Eastoe, J. E.,** Organic matrix of tooth enamel, *Nature (London)*, 187, 411, 1960.
11. **Eastoe, J. E.,** Enamel protein chemistry — past, present and future, *J. Dent. Res.*, 58(B), 753, 1979.
12. **Fukae, M. and Shimizu, M.,** Studies on the proteins of developing bovine enamel, *Arch. Oral Biol.*, 19, 381, 1974.
13. **Guenther, H., Croissant, R. D., Schonfeld, S. E., and Slavkin, H. C.,** Identification of four extracellular matrix enamel proteins during embryonic rabbit tooth organ development, *Biochem. J.*, 163, 591, 1977.
14. **Belanger, L. F.,** Autoradiographic visualization of the entry and transit of ^{35}S methionine and cystine in the soft and hard tissues of the growing rat, *Anat. Rec.*, 124, 555, 1956.
15. **Greulich, R. C. and Slavkin, H. C.,** Amino acid utilization in the synthesis of enamel and dentine matrices as visualized by autoradiography, in *The Use of Radioautography in Investigating Protein Synthesis*, LeBlond, C. P. and Warren, K. B., Eds., Academic Press, New York, 1965, 199.
16. **Warshawsky, H.,** Steps in secretion of enamel matrix protein, as shown by electron microscope radioautography of the ameloblasts of rat incisors following tyrosine-H^3 injection, *Anat. Rec.*, 154, 438, 1966.
17. **Frank, R. M.,** Autoradiographique Quantitative de L'amélogenése en Microcopie Électronique á L'aide de la Proline Tritiée Chez le Chat, *Arch. Oral Biol.*, 15, 569, 1970.
18. **Weinstock, A. and LeBlond, C. P.,** Elaboration of the matrix glycoprotein of enamel by the secretory ameloblasts of the rat incisor as revealed by radioautography after galatose-^{3}H-injection, *J. Cell Biol.*, 51, 26, 1971.
19. **Slavkin, H. C., Mino, W., and Bringas, P., Jr.,** The biosynthesis and secretion of precursor enamel protein by ameloblasts as visualized by autoradiography after tryptophan administration, *Anat. Rec.*, 185, 289, 1976.
20. **Revel, M. and Groner, Y.,** Post-transcriptional and translational controls of gene expression in eukaryotes, *Annu. Rev. Biochem.*, 47, 1079, 1978.
21. **Slavkin, H. C., Weliky, B., Stellar, W., Slavkin, M. D., Zeichner-Gancz, M., Bringas, P., Jr., Hyatt-Fischer, H., Shimizu, M., and Fukae, M.,** Ameloblast differentiation: protein synthesis and secretion in fetal New Zealand white rabbit molar tooth organs and isolated epithelia *in vitro*, *J. Biol. Buccale*, 6, 309, 1979.
22. **Lee-Own, V., Zeichner, M., Benveniste, K., Denny, P., Paglia, L., and Slavkin, H. C.,** Cell-free translation of messenger RNAs of embryonic tooth organs: synthesis of the major extracellular matrix proteins, *Biochem. Biophys. Res. Commun.*, 74, 849, 1977.
23. **Zeichner-David, M., Weliky, B. G., and Slavkin, H. C.,** Isolation and preliminary characterization of epithelial-specific messenger ribonucleic acids and their products during tooth development, *Biochem. J.*, 185, 489, 1979.
24. **Lyaruu, D. M., Belcourt, A., Fincham, A. G., and Termine, J. D.,** Neonatal hamster molar tooth development: extraction and characterization of amelogenins, enamelins and soluble dentin proteins, *Calcif. Tissue Int.*, 34, 86, 1982.
25. **Fukae, M., Tanabe, T., and Shimizu, M.,** Proteolytic enzyme activity in porcine immature enamel, *Tsurumi Univ. Dent. J.*, 3, 15, 1977.
26. **Shimizu, M., Tanabe, T., and Fukae, M.,** Proteolytic enzyme in porcine immature enamel, *J. Dent. Res.*, 58(B), 782, 1979.
27. **Moe, D. and Birkedal-Hansen, H.,** Proteolytic activity in developing bovine enamel, *J. Dent. Res.*, 58(B), 1012, 1979.
28. **Chrispens, J., Weliky, B., Bringas, P., and Slavkin, H.,** Proenamel — enamel — polypeptide: a concept, *J. Dent. Res.*, 58(B), 988, 1979.
29. **Patzelt, C., Labrecque, A. D., Duguid, J. R., Carroll, R. J., Keim, P. S., Heinrikson, R. L., and Steiner, D. F.,** Detection and kinetic behavior of preproinsulin in pancreatic islets, *Proc. Natl. Acad. Sci. U.S.A.*, 75(3), 1260, 1978.
30. **Zeichner-David, M., Slavkin, H. C., Lyaruu, D. M., and Termine, J. D.,** Biosynthesis and secretion of enamel proteins during hamster tooth development, *Calcif. Tissue Int.*, in press.
31. **Sasaki, S. and Shimokawa, H.,** Enamel proteins: biosynthesis and chemistry, *J. Dent. Res.*, 58(B), 765, 1979.
32. **LeBlond, C. P. and Warshawsky, H.,** Dynamics of enamel formation in the rat incisor tooth, *J. Dent. Res.*, 58(B), 950, 1979.

33. **Seyer, J. M. and Glimcher, M. J.,** Evidence for the presence of numerous protein components in immature bovine dental enamel, *Calcif. Tissue Res.*, 24, 254, 1977.
34. **Fincham, A. G.,** The amelogenin problem: a comparison of purified enamel matrix proteins, *Calcif. Tissue Int.*, 27, 65, 1979.
35. **Fukae, M., Tanabe, T., Ijiri, H., and Shimizu, M.,** Studies on porcine enamel proteins: a possible original enamel protein, *Tsurumi Univ. Dent. J.*, 6(2), 87, 1980.
36. **Slavkin, H.C., Zeichner–David, M., MacDougall, M., Bringas, P., Bessem, C., and Honig, L.S.,** Antibodies to murine amelogenins: localization of enamel proteins during tooth organ development in vitro, *Differentiation*, 23, 73, 1982.
37. **Slavkin, H.C., Zeichner–David, M., and Siddiqui, M.A.Q.,** Enamel extracellular matrix: differentiation specific gene products and the control of their synthesis and accumulation during development, in *Current Advances In Skeletogenesis*, Silbermann, M. and Slavkin, H.C., Eds. Excerpta Medica, Amsterdam, 1982, 24.
38. **Slavkin, H.C., Zeichner–David, M., MacDougall, M., Bessem, C., Bringas, P., Honig, L.S., Lussky, J., and Vides, J.,** Enamel gene products during murine amelogenesis in vivo and in vitro, *J. Dent. Res.*, 61 (Special Issue), 1497, 1982.
39. **Snead, M.L., Chandra, T., Robson, K.J.H., Zeichner–David, M., Woo, S.L.C., and Slavkin, H.C.,** Identification of a cDNA probe for investigating enamel gene expression during odontogenic epithelial–mesenchymal interactions, *J. Cell Biol.*, 95(2), 134, 1982.
40. **Herold, R.C., Graver, H.T., and Christner, P.,** Immunohistochemical localization of amelogenins in enameloid of lower vertebrate teeth,*Science*,207, 1357, 1980.
41. **Slavkin, H. C., Zeichner-David, M., Ferguson, M. W. J., Termine, J. D., Graham, E., MacDougall, M., Bringas, P., Bessem, C., and Grodin, M.,** Phylogenetic and immunogenetic aspects of enamel proteins, in *Oral Immunogenetics and Tissue Transplantation*, Riviere, G. R. and Hildemann, W. H., Eds., Elsevier/North-Holland, New York, 1982, 241.
42. **Slavkin, H., Graham, E. Zeichner–David, M., and Hildemann, W.,** Enamel–like antigens in hagfish: possible evolutionary significance, *Evolution*, in press.
43. **Slavkin, H.C., Samuel, N., Bringas, P., Jr., Nanci, A., and Santos, V.,** Selachian tooth development. II. Immunolocalization of amelogenin polypeptides in epithelium during secretory amelogenesis in *Squalus acanthias*, *J. Craniofacial Genet. Dev. Biol.*, in press.
44. **Kollar, E. J. and Fischer, C.,** Tooth induction in chick epithelium: expression of quiescent genes for enamel synthesis, *Science*, 207, 993, 1980.

EXPERIMENTAL APPROACHES TO THE STUDY OF ENAMEL MATRIX

A. G. Fincham, A. B. Belcourt, and J. D. Termine

INTRODUCTION

The extracellular matrix of developing dental enamel comprises a complex of proteins which account for some 20% of the dry weight of the structure.[1-3] As the enamel mineralizes with hydroxyapatite, this proportion of protein decreases to final levels of 0.1% in the fully mature enamel.[4] This change in the proportion of protein with increasing mineralization is accompanied by striking changes in the overall amino acid composition of the matrix. This change can even be seen within a single developing tooth by comparing the less mature enamel of the cervical region with more mineralized areas towards the incisal edge.[5-8]

The principal proteins of the young enamel matrix, the *amelogenins*, are characteristically rich in proline, leucine, histidine, and glutamyl residues, being distinct in composition and properties from other structural proteins such as the collagens or keratins.[3] The amelogenins consist of a heterogenous complex of proteins ranging in molecular size from approximately 5,000 to 30,000 daltons.[13,35,36] The smaller amelogenin components are richer in organic phosphorus content (~1 to 2%) than the larger proteins,[27,35] and may also be richer in conjugated carbohydrate, although the overall carbohydrate content of the amelogenin complex is quite small compared to conventional glycoproteins.[35,95] Concomitant with enamel maturation, the matrix amelogenin population becomes increasingly dominated by its smaller-sized constituents with progressive mineralization.[35,45,96]

The amelogenins are accompanied in the fetal matrix by a small proportion of acidic crystal-bound proteins, the *enamelins*,[9,35] which appear to be partially retained in the mature enamel while the amelogenins are completely lost from the structure. The enamelins consist of a heterogenous complex of proteins ranging in molecular size from approximately 7,000 to 70,000 daltons.[35,45] These proteins are slightly phosphorylated (~0.2% P), but extensively glycosylated (~3 to 4% hexosamine, ~4% sialic acid).[35] With progressive enamel mineralization, this population of matrix proteins also becomes dominated by its smaller-sized components. The detailed biochemistry and chemical interrelations of these enamelins and amelogenins are at present only poorly understood and their roles in the mineralization and ultrastructural organization of the enamel remain to be fully elucidated.

Studies directed to the separation and characterization of the several proteins of the fetal enamel matrix originate in the mid 1960s,[10-12] but despite some 15 years of subsequent work by several laboratories, definitive data on individual amelogenins or enamelins remains sparse.

This slow progress in our understanding of the enamel matrix proteins, coming during a period of striking technical advances in separative and analytical protein chemistry, reflects the unusual difficulties experienced by researchers in this field and has lead to a literature which is often difficult to interpret and conflicting in details. The reader is referred to several reviews for a wider discussion of this subject.[13-15]

In this section, we have attempted to outline the various techniques of preparation, purification, and analysis which have been applied to the study of the matrix of dental enamel, to provide the references to the original reports and to present in greater detail those methods, which in our experience, have proved useful and reproducible. No attempt has been made to be exhaustive in the description of methodologies and for a wider coverage of experimental detail the reader is referred to the papers cited in the text and in the reviews noted above.

THE ORGANIC MATRIX OF DEVELOPING MAMMALIAN ENAMEL

Species Studied

While there are data available on the overall amino acid compositions of the enamel matrices of some 15 mammalian species,[16-18] studies dealing with the fractionation of the matrix proteins or the investigation of their biosynthesis and structural transformations have been more restricted in range of subjects. For reasons of quantity, most protein separative studies have used either the fetal calf[10-12,20-36] or the immature (6-month-old) pig.[37-39] For studies of the developmental changes which accompany mineralization, the continuously erupting rodent (rat) incisor has proved of value,[40-44] and more recently, the neonatal hamster has been used in biochemical studies of overall changes in matrix protein proportions with increasing maturity.[45] Matrix protein biosynthesis has been investigated using the embryonic rabbit incisor,[46-48] but the marked difference in overall amino acid composition of the rabbit enamel matrix from that of most other species analyzed[16] has made interpretation difficult. Studies of the fetal human enamel matrix have been restricted to overall amino acid compositions[3,49] and more recently to changing amino acid proportions in areas of matrix of different degrees of mineralization.[8]

It appears probable that, for studies of matrix protein chemistry, the cow and the pig will remain the subjects of choice, although differences of detail between these and the human matrix should not be overlooked[8] (Figure 3). The determination of fetal age of specimens is frequently difficult, but is aided in the cow by the relationship of crown-to-rump measurement to jaw length and tooth size.[50]

Preparation of Tissues for Biochemical Studies

Most procedures involve, as first steps, dissection of unerupted tooth germs from the jaw as quickly as possible after death followed by manually depulping the tooth germ and stripping its mineralized matrix clear of adhering epithelial tissue. Usually a cold physiological salt solution, sometimes containing protease inhibitors,[35] is used to wash and/or hold the tissues during these operations. The ameloblast layer is then removed from the enamel surface by mechanical means or, more recently,[35,46] by sonication.

In the past, tissues were often stored frozen for convenience at various stages of the overall dissection process. However, to avoid possible post-mortem changes, it is recommended that the entire process be accomplished as quickly as possible without interim storage. Similarly, to avoid potential damage by leakage of intracellular enzymes during handling, the use of protease inhibitors in washing and/or holding solutions is also advisory. Once the mineralized matrices are rendered acellular, they may be safely stored frozen at $-20°$ or below until needed.

Tissues may, and indeed have been, studied at all stages of tooth development. However, at the very early stages, the next step, removal of the soft, less well-mineralized enamel from the tooth surface, is quite difficult. This is usually accomplished mechanically with the aid of traditional dissection tools.[11,29] Again, for best results, tissues should be kept cold during enamel matrix removal and soft scrapings stored frozen until needed.[11] Enamel can be removed from tooth matrices by direct guanidine extraction (see below) with vigorous stirring.[45] In this procedure, the enamel crystals are liberated as a free suspension from the underlying dentin substrates by action of the chaotropic agent. The amelogenin and enamelin fractions can then be prepared in the usual manner (see below).

Isolation of the Organic Matrix

Freshly isolated enamel matrix scrapings contain some 15 to 25% organic phase on a dry weight basis, depending on the stage of maturity of the enamel. Three principal procedures have been applied to the isolation of the matrix proteins from this material.

EDTA-dialysis demineralization — This procedure, involving the suspension of the enamel scrapings in EDTA solution (0.3 to 0.5 *M*, pH 7.0 to 7.4) within a dialysis membrane and dialysis against the EDTA solution in the cold for several days, has been widely applied in earlier studies.[18] The details are discussed by Stack.[51] Selective precipitation of amelogenin fractions in EDTA solutions has been applied.[11,28]

Acid demineralization — More recently, direct demineralization by dialysis in dilute acid has been routinely applied. The sample of enamel scrapings is suspended in dilute acetic acid (10% v/v) within a dialysis membrane ('Spectrapor'® #1) and exhaustively dialyzed against the acetic acid in the cold with frequent changes of the external solution. Finally, the acid concentration is reduced to low levels (0.1%) yielding a soluble preparation of the matrix proteins with some 1 to 2% of acid-insoluble material left in suspension. This small acid-insoluble fraction has been found to consist largely of enamelin material. The soluble matrix proteins are finally isolated from solution by lyophilization.

Guanidine/guanidine-EDTA extraction — This procedure provides a means for the isolation of the matrix proteins under dissociative conditions and in the presence of protease inhibitors. The amelogenin fraction of the matrix is first isolated by extraction of the enamel scrapings with guanidine hydrochloride solution (4.0 *M*, buffered to pH 7.4 with tris). The residual material, enamel crystals and enamelin proteins, is then treated with a guanidine-EDTA solution which permits the isolation of the enamelins. The guanidine is removed from the extracts by exhaustive dialysis (and/or ultrafiltration) and the proteins isolated by lyophilization.[35]

The three different extraction procedures outlined above have been found to yield similar preparations of amelogenins, although there are differences in detail (not as yet understood) which are apparent in electrophoretic or chromatographic analyses of the products.

However, in any isolation procedure used, the following points should be kept in mind. First, the matrix consists of a complex of proteins and polypeptides with a wide range of size, the lower limit probably being about 3000 daltons. Thus, dialysis and/or ultrafiltration conditions must be selected to permit retention of these smaller molecules and yet allow removal of the inorganic ions.

Second, several workers have reported the presence of proteases within the organic matrix.[52-53,78] While the precise nature and origins of these proteases are presently unclear, the conditions of matrix extraction should attempt to minimize the chance of any proteolysis, for example, through the use of inhibitors[35] and careful temperature control. It might be noted here, parenthetically, that the authors have been unable to demonstrate significant proteolytic action, at least towards the amelogenin components, during the preparation of bovine enamel matrix under conditions which deliberately omitted the above precautions.

Third, the principal amelogenin proteins, which comprise the bulk of the immature enamel matrix, exhibit striking reversible hydrophobic aggregation which is concentration, pH, and temperature dependent. These proteins will generally remain in solution at low pH and low temperature.

Fourth, the hydrophobic character of the amelogenins confers on them some unusual solubility properties, being both soluble in dilute inorganic buffer solutions (e.g., phosphate-buffered saline),[35,45] dilute acids, and also in certain organic solvents (e.g., chloroform to methanol, 2:1).[54]

Fifth, the enamelin proteins differ strikingly from the amelogenin proteins in their amino acid compositions (Table 1), being more acidic in nature and strongly bound to the mineral (hydroxyapatite) phase of the enamel. They tend to show low solubility in acid solutions.

In conclusion, the writers recommend the use of either direct acid extraction where rapid isolation of the matrix amelogenins is required or the use of the guanidine procedure which allows separation of the amelogenins and enamelins. As noted above, the electrophoretic

Table 1
AMINO ACID COMPOSITION OF BOVINE DENTAL ENAMEL MATRIX AND SOME SELECTED FRACTIONS (RESIDUES/1000)

	1	2	3	4	5	6
Aspartic acid	41	28	31	98	96	64
Threonine	27	23	24	55	51	30
Serine	55	30	67	81	72	114
Glutamic acid	166	220	94	146	130	135
Proline	221	274	197	139	64	94
Glycine	66	30	88	114	108	245
Alanine	26	25	0	50	79	
Cystine/2	0	0	0	2	3	—
Valine	35	42	25	39	60	30
Methionine	45	50	56	23	19	6
Isoleucine	32	38	40	28	43	21
Leucine	113	101	91	53	96	51
Tyrosine	44	26	145	21	17	21
Phenylalanine	31	17	28	49	37	23
Histidine	60	82	65	27	25	23
Lysine	17	7	24	32	53	31
Arginine	21	7	25	40	49	40

Note: 1, Total bovine fetal matrix; 2, principal amelogenin (30,000 daltons); 3, amelogenin polypeptide ("E4", 5,000 daltons); 4, enamelin preparation;[35] 5, total mature human enamel matrix;[14,92] 6, mature human EDTA-insoluble fraction.[72]

and chromatographic profiles of these preparations tend to differ slightly in details, the guanidine extracted matrix generally yielding a simpler product.

Matrix Protein Fractionation

Gel exclusion chromatography in dilute solutions — This procedure was among the first to be applied in attempts to obtain a fractionation of the complex of enamel matrix proteins. Typically, solutions of the proteins were chromatographed on columns of Sephadex® G50, G75, or G100 eluted with dilute formic acid,[11,22,28] acetic acid,[38] neutral tris-HCl buffers,[21] phosphate buffer,[20] tris-EDTA-borate,[55] or ammonium bicarbonate solution.[21,56] These procedures usually yielded profiles of some five to six overlapping peaks for samples of the fetal bovine matrix, although in some cases the patterns were more complex.[21,24] The fractions obtained in these separations were generally heterogeneous when analyzed electrophoretically and in some cases rechromatography of fractions was shown to produce a redistribution of proteins, which was attributed to reversible aggregations.[21] Exclusion chromatography in dilute acid solutions yielded fractions which showed discrete groups of components on electrophoretic analysis.[11,22] In more recent studies, experimentors have concentrated on the use of tandem columns of Biogel® P30 eluted with formic acid (0.1 *M*) at 10° (Figure 1). Under these conditions, it has been found that the principal chromatographic peaks represent discrete groups of electrophoretic components and that rechromatography yields fractions with the same elution volume as the parent fraction. (An exception to this statement has been found for the two trailing peaks of the bovine matrix chromatogram, each of which appears to contain the same electrophoretic component. The cause of this anomaly is not presently understood.) This P30 chromatographic procedure has proved reproducible and of value as a primary fractionation step, since the recovery of protein from the dilute solutions is readily accomplished by direct lyophilization. Further-

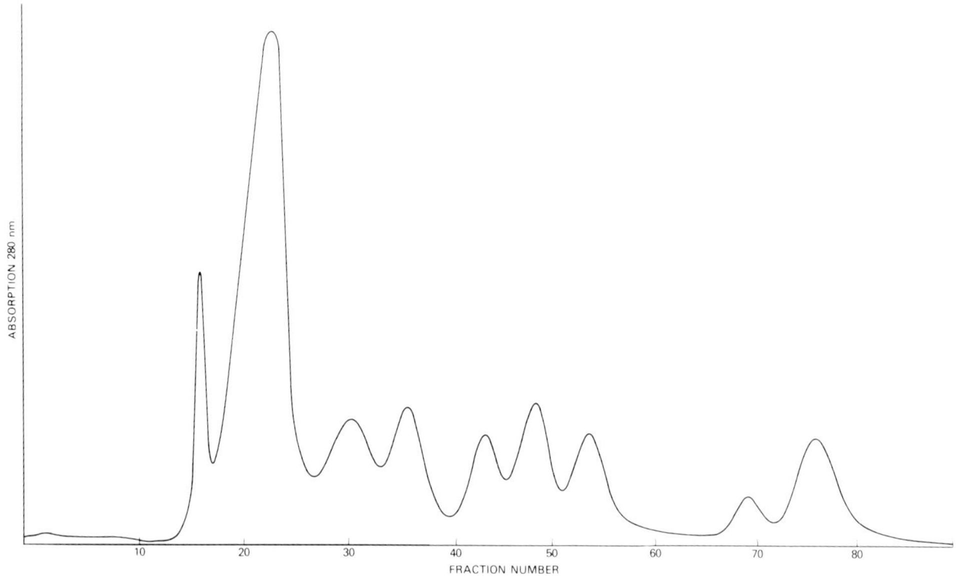

FIGURE 1. Typical elution profile of guanidine exracted total bovine fetal enamel matrix chromatographed on a tandem column of Biogel® P30 (2 × 95 × 2.5 cm) eluted with 0.1 *M* formic acid (6.7 mℓ/hr) at 10°C. Fraction volume was 4.7 mℓ.

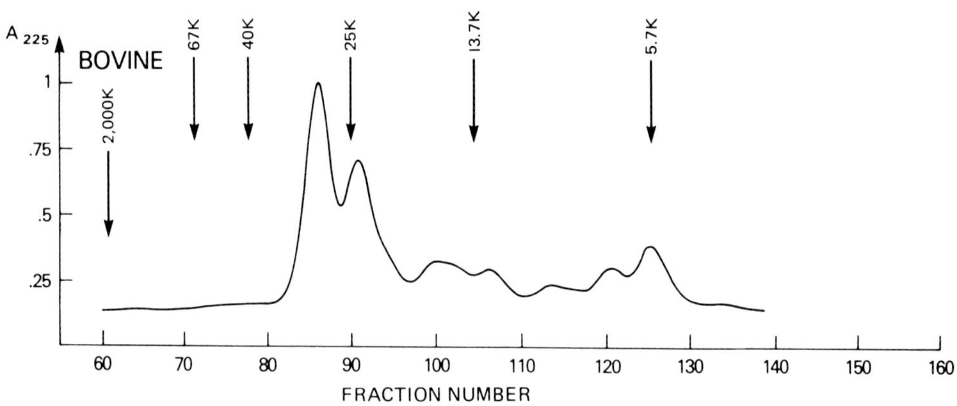

FIGURE 2. Typical elution profile of guanidine extracted total bovine fetal enamel matrix chromatographed on a tandem column of Sephacryl® S200 (2 × 95 × 2.6 cm) eluted with guanidine hydrochloride (4 *M*), tris (pH 7.4) (10 mℓ/hr), at 25°C. Fraction volume was 5 mℓ.

more, it has been found that the molecular weights of fractions estimated by calibration of the P30 column are in conformity with size data obtained by chromatography on Sephacryl® S200 in 4 *M* guanidine solution (see below).

Gel exclusion chromatography in disaggregating solvents — In attempts to overcome protein aggregations, several workers have employed chromatography of bovine or porcine enamel matrix in solutions of urea,[24] sodium deoxycholate,[54] or SDS-phosphate buffers.[36] These separations, as for the nondisaggregating system, have not generally yielded discrete fractions and in some instances the failure to obtain useful separations has been attributed to aggregation of the proteins in the disaggregating solvent.[24] Termine et al.[35] have used a system based on that of Belew et al.[57] in which the proteins are chromatographed in 4 *M* guanidine hydrochloride-tris (pH 7.4) on tandem columns of Sephacryl® S200 at 25° (Figure 2). The P30 and S200 chromatograms have proved highly reproducible and have been utilized for both preparative and molecular weight determination purposes. The fractionation range of the S200 system (separation from 3,000 to 75,000 daltons and linear below 30,000 daltons) has proved very suitable for the enamel matrix proteins.

Ion exchange chromatography — Partial fractionations of enamel matrix preparations or subfractions of the matrix, employing ion exchange column chromatography on a variety of media, have been widely reported. These have included DEAE-cellulose,[12,24,25,33] QAE-Sephadex®,[28] TEAE-cellulose,[38] CM-Sephadex®,[26] and SP-Sephadex®.[28] It is significant that none of these studies have provided routinely useful fractionation procedures, and it has been the experience of the writers that ion exchange fractionations of enamel matrix have proved irreproducible, the binding of the matrix amelogenins to the exchanger being extremely capricious. However, it has been found that the more acidic enamelins can be routinely separated from amelogenins by chromatography on a column (2.0 × 10 cm) of DEAE-Sephacel® equilibrated in 0.05 *M* tris-HCl (pH 7.4),7 *M* urea. In this system, the amelogenins are frontally eluted and the enamelin fraction may be subsequently recovered using a salt gradient to 1.0 *M* NaCl. More recently, it has been found that chromatography of the 27,000 dalton amelogenin complex on columns of QAE-Sephadex®, A-25 in 0.1 *M* tris-HCl (pH 9.4), 6 *M* urea buffer using a linear gradient from 0 to 1.5 *M* NaCl, is effective in the resolution of the several electrophoretically distinct components of this amelogenin fraction.[28,104]

Preparative electrophoresis — Seyer and Glimcher[27] employed free-flow horizontal and preparative flow-elution polyacrylamide gel electrophoresis (PAGE) in their isolation of the bovine matrix polypeptides "E1 to E4", although these techniques have not subsequently been successfully applied to the preparative fractionation of the enamel proteins. In later studies, Fincham[33] isolated small quantities of electrophoretically homogeneous amelogenins by extraction of slices cut from analytical scale PAGE. This gel-slicing study showed that the electrophoretically distinct components of the matrix behaved characteristically on reelectrophoresis and did not show properties of disaggregation as had previously been described for certain column fractions.[21] More recently, this gel slicing technique has been applied to larger PAGE gels (14 × 150 mm) prepared with the tris-EDTA-borate buffer system described below.

Preparative isoelectric focusing — There are no reports of the application of this technique to enamel matrix preparations, although the results of analytical PAGE and slab-gel isoelectric focusing would suggest that such preparative methods may be of value.[102]

Analytical Techniques

Electrophoresis — Early studies of the bovine matrix proteins employed starch gel slab methods.[11] The superior resolution of cylindrical gel polyacrylamide systems, applied to the enamel matrix proteins, has revealed a complex of some eight to ten major components and a further ten or more minor components.[54] Generally, there is good agreement between the data of different workers for the electrophoretic patterns of the matrix proteins[23,28,30,33,54] of the bovine enamel, and similarities are noted in the patterns found for different mammalian species.[23,57] The use of the tris-EDTA-borate-urea buffer system[58] has been adopted to allow the recovery of protein from gel slices without glycine contamination which occurs when the more conventional tris-glycine-urea buffer system is used. The incorporation of urea (6 *M*) into the gels provides enhanced resolution of bands and improves gel handling. The resolution of the complex of slower migrating bands (Figure 3) has been found to be improved by prolonging the electrophoresis after the elution of the tracker dye band.[33]

Sodium dodecylsulfate-PAGE (SDS-PAGE) — This conventional technique based on the procedures of Weber and Osborn[59] has been applied to enamel matrix by several groups of workers.[35-37,47] Termine et al.[35] have shown that the amelogenins behave in an anomalous

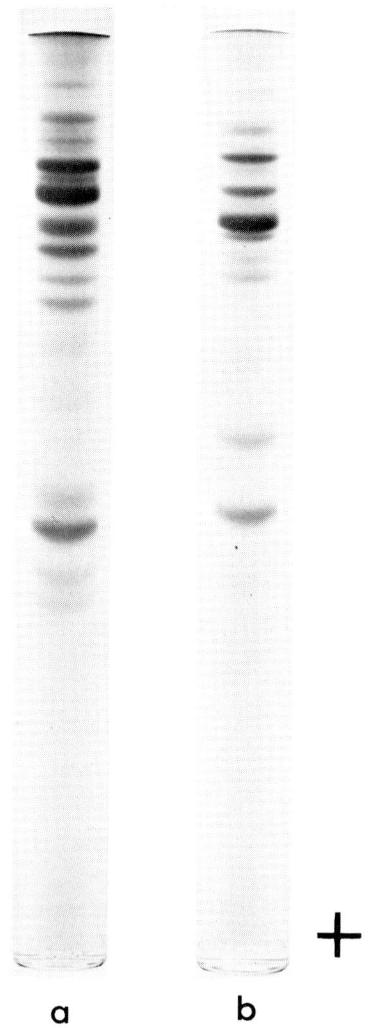

FIGURE 3A. Electrophoretic patterns of enamel matrix. Tris-EDTA-borate-urea PAGE, (pH 8.5), 7.5% acrylamide gel, stained with Amido black.[54] (a) Bovine; (b) human.

fashion with respect to molecular weights when electrophoresed in SDS-PAGE systems, whereas the enamelins show a normal Ferguson plot[60] when analyzed in gels of differing acrylamide concentrations both in the presence and absence of urea. Comparable anomalous behavior has been reported for tropomyosin when electrophoresed in SDS-PAGE and SDS-urea-PAGE systems,[98,99] where it was found that the tropomyosin subunit molecule (~35,000 daltons) electrophoresed in SDS-PAGE at some 30,000 daltons and in SDS-urea-PAGE with an apparent molecular weight of 45,000 daltons. This behavior is similar to that found for the principal bovine amelogenin.[35] The cause of this anomalous electrophoretic behavior is presently unknown, but may be related to the hydrophobic character of these proteins. In both PAGE and SDS-PAGE systems, the enamelins have been found to stain poorly with either Amido black or Coomassie® brilliant blue R250, but they can be effectively stained using the phosphate and/or sialic acid-sensitive "Stains-All" procedure of Green et al.[61]

Gel-isoelectric focusing — The application of this technique has not previously been reported in the literature of enamel matrix studies, although it was clear from earlier data

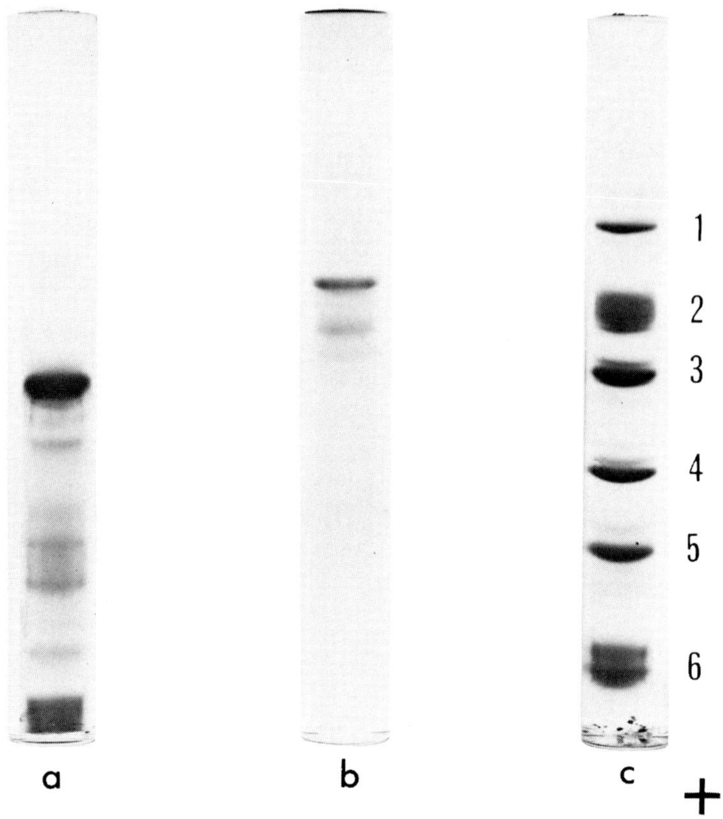

FIGURE 3B. Electrophoretic patterns of bovine matrix. SDS-urea PAGE, 7% acrylamide gel, stained with Coomassie® brilliant blue R250. (a) Amelogenin preparation; (b) enamelin preparation; (c) molecular weight markers. (1) Phosphorylase b (94,000); (2) bovine serum albumin (67,000); (3) ovalbumin (43,000); (4) carbonic anhydrase (30,000); (5) soy bean trypsin inhibitor (20,100); (6) α-lactalbumin (14,400).

that there were components within the bovine matrix complex of widely differing pI values. In the experience of the writers, horizontal isoelectric focusing in polyacrylamide gels (10%), using a pH 5 to 8 Ampholine® and staining by the procedure of Reisner et al.,[93] has proved to be a useful means of characterization of fractions from column separations of matrix proteins.

Molecular weight estimation — The aggregating properties of the enamel matrix proteins have lead, in certain cases, to the description of fractions with molecular weights of over 100,000 daltons,[21] resulting from such aggregations. Under dissociative conditions,[35] the matrix proteins appear to range in size from some 30,000 to 5,000 daltons in the case of the amelogenins and from 70,000 to some 8,000 daltons in the case of the enamelins, although higher molecular weights have been reported[62] and the existence of a "high molecular weight amelogenin" has been postulated.[46,62,101]

Of the classical procedures available, the principal techniques used to determine the sizes of purified enamel matrix proteins have been gel exclusion chromatography,[30,35,36] SDS-PAGE,[35,36,47] amino acid composition,[27,28,31-33] and in one case sedimentation equilibrium.[28] As has been noted above, the use of SDS-PAGE for molecular weight estimations of amelogenins is unsatisfactory, the results obtained varying with the acrylamide concentration and showing a lower size in SDS-gels as compared to SDS-urea-gels.[35]

Several studies have reported the molecular weights of the principal bovine amelogenins to be some 16,000 daltons (as estimated by amino acid composition), however, more recently, the use of the S200 guanidine columns and the P30 Biogel® have indicated a size of some 25,000 to 30,000 daltons for these components. Similarly, recent determinations on pig enamel suggest a size of 26,000 daltons for the principal amelogenin.[39] These reports appear to be in conflict with earlier data.[28,30]

Size data for the smaller amelogenin polypeptides have suggested molecular weights in the region of 5000 daltons as determined by amino acid composition and gel exclusion data. Recent data from P30 and S200 columns indicates sizes in the range of 5,000 to 10,000 daltons for these polypeptides.[100] Recent data for the complete sequences of two of the principal low molecular weight bovine amelogenins have confirmed sizes of some 5000 daltons.[103]

THE MATURE ENAMEL MATRIX

Preparation

Three principal difficulties are encountered in the preparation of mature enamel matrix: first, the organic matrix is present at only very low levels (~0.1% w/w);[13] second, it is desirable to obtain pure enamel free from contamination with the underlying dentin; and third, dissection procedures are extremely difficult due to the extreme hardness of the enamel.

The most commonly used preparative technique has been to remove the dentin from the pulp side of the tooth crown using a scalpel or dental burr, usually checking for dentinal contamination of the final enamel product by means of the Van Gieson stain.[3,49,63,64] In some studies,[65] enamel powder, obtained by grinding, was subjected to density fractionation purification by flotation in mixtures of bromoform-acetone[66] or, more recently, in a dense solution of cadmium tungstoborate.[67] This latter method is preferred as it prevents loss of constituents by organic solvent solution. Although these density fractionation procedures improve the degree of purification of the enamel preparation, the possibility of some dentinal contamination is usually still present.

Although less economical in time and materials, microdissection procedures[7,68] appear to be the most certain approach to the preparation of pure enamel. After removal of the superficial integuments (dental plaque and pellicle) by soft abrasion of the coronal surface with fine sandpaper, the crowns are serially cut into 750-μm-thick sections using a water-cooled diamond saw. Each section is removed from the sliced tooth, and a portion of the enamel at some distance from the dentin-enamel junction is separated by cleavage with a hard razor blade under the dissecting microscope. Each enamel chip is examined microscopically to ensure that it is free of any dentinal contamination, all doubtful specimens being rejected. In the case of mature bovine enamel, an additional step is necessary to remove the over-lying cementum layer.[7]

Isolation of the Organic Matrix

Various demineralizing solvents have been used to extract the organic matrix from enamel: hydrochloric acid,[69,71] acetic acid, trichloracetic acid,[49] lactic acid, and EDTA solutions. The most widely used procedure has been dissolution in cold EDTA solutions (0.5 *M*) at pH levels ranging from 7.2 to 8.3.[7,68,70,71] In each case, an insoluble residue was observed and described by Robinson et al.[49] as "tuft" protein.

Using a sequential extraction scheme with dissociative solvents,[9,72] it was demonstrated that the mature enamel matrix required prior solution of the apatite mineral phase before it could be completely extracted. An apatite-protein insoluble residue complex was obtained which was finally solubilized in lactic acid (60%).

Analytical Results

Apart from microquantities of citrate and lactate ions[73] and traces of carbohydrates,[74] the composition of the mature enamel matrix is dominated by protein[2] and lipid,[75] the latter accounting for almost half of the total organic content. Among the lipids, two classes have been distinguished: one directly extractable and the other soluble only after demineralization. The origin and function of these lipid constituents remains undetermined.[75,76,94]

For the protein fraction of mature enamel, the majority of studies have reported amino acid compositions only.[3,7,26,49,63,65,68-72,76-77,79] Alanine, glutamic acid, glycine, proline, and serine were the most abundant amino acid residues detected, but direct comparison of the results between different laboratories is difficult due to variations in enamel preparation and extraction procedures.

The question of the number and nature of the proteins present in the mature enamel is presently unclear. Only a few investigations showing PAGE patterns and gel-exclusion chromatographic profiles (Sephadex® G25 and G50) are available.[79-82]

More recently, the EDTA-soluble fraction of mature human enamel, analyzed by PAGE, was shown to contain a fast-migrating periodic acid-Schiff positive doublet.[68] This glycoprotein chromatographed on Sephacryl® S200 in 4 *M* guanidine as a single symmetrical peak of low molecular weight.[72] Following methanolysis and derivatization as their trimethylsilyl-ethers, the sugars were identified by gas chromatography as glucose, galactose, mannose, and fucose, Sialic acid, glucosamine, and galactosamine have also been detected along with low levels of organic phosphorus.[72] Although contamination with nucleic acids appeared to be absent, the UV spectrum of this material showed a greater absorption at 260 than at 280 nm,[68] as also observed by Spitzer and Ten Bosch[83] for thin slabs of bovine and human enamel. The EDTA-insoluble, lactic acid soluble residue was found to contain a glycoprotein of similar amino acid composition and molecular weight, but with a higher content of organic phosphorus.[72]

THE BIOLOGICAL ROLE OF THE ENAMEL MATRIX

A simple perusal of enamel ultrastructure presents an intriguing challenge to the dental research scientist. Nowhere else in mammalian biology are hydroxyapatite crystallites so large (both in cross-section and in length) and so highly co-orientated. Only in invertebrate systems are similar mineral structures observed and in those species such structures are composed of calcium carbonate not calcium phosphate. In fact, up to now it has been impossible to produce apatite crystals resembling those of enamel under synthetic conditions even remotely approximating those found in physiological systems.

Since the mineralization pattern of the invertebrates generally involves the deposition of discretely ordered inorganic mineral crystals on protein substrates, it is not unreasonable to assume that a similar process may also occur in developing dental enamel. In fact, even though bone hydroxyapatite crystals are very much smaller than those of enamel, the mineralization pattern of bone also involves the deposition of mineral within an underlying (collagen) protein fiber network. A major difference between the matrix-mineral relationship of enamel and those of other systems is that in enamel there is an absolute loss of matrix protein concomitant with increasing mineralization, while in other systems this is not usually the case. Further, the nature of the matrix in developing enamel changes from one rich in proline-dominated amelogenins early in enamel formation to one rich in the more acidic enamelins as mineralization approaches completion. Thus, the role of the enamel matrix in enamel crystal development is transitory and not intuitively obvious. Nevertheless, its role must be integral to the complexities of the unique mineral system upon which it operates and consequently essential to the structural integrity of the resulting erupted enamel.

Some old observations suggest a potential structural form for the fetal enamel matrix proteins (for a thorough discussion of these data see Reference 84). These observations were (1) that it is often possible to obtain a protein fiber X-ray diffraction pattern (cross β-pleated sheet conformation) from decalcified fetal bovine enamel matrix and (2) that this same matrix contains elements that appear fibrous when viewed in appropriately stained decalcified sections with the electron microscope. These intriguing data, unfortunately, induced diametrically opposite points of view about the natural form of the enamel matrix during active mineralization. One viewpoint suggested that these observations merely represented handling artifacts. That is, the diffraction and electron microscopic fiber patterns were a consequence of decalcification and/or drying conditions.[85,86] Thus, the visualized structural forms associated with the enamel matrix, e.g., helixes[87] and fibers,[86,88] are thought to be induced from enamel crystallites during preparation. This viewpoint argues then that most of the detailed potential matrix structures so far postulated (e.g., the so-called "prism sheath")[89] do not actually exist in vivo. The opposite viewpoint held that these structural observations were not artifacts and accurately represent reality in native tissues.[88] This conclusion extends the ideas of biological mineral deposition along fibrous protein structures (as found in bone and invertebrate calcified matrices) to developing enamel. Perhaps the resolution of this controversy will eventually accommodate both positions, at least partially. So far, neither viewpoint is wholly satisfactory in light of more modern evidence.

Recently, freshly dissected intact fetal bovine enamel tissue was examined by solid state nuclear magnetic resonance spectroscopy.[90] The measurements indicated that approximately 70% of the fetal bovine enamel protein contained peptide chains that were isotropically mobile with relaxation times on the order of 1 μsec or less. Thus, this portion of the enamel matrix consists of proteins which either exhibit a high degree of internal motion or which are freely tumbling very rapidly independent of their internal conformations. These extremely mobile developing enamel matrix constituents could not in any way be considered structural proteins in the same sense as collagen or the fibroins, for example. The remaining 30% of the fetal bovine enamel matrix exhibited restricted anisotropic chain motion consistent with that observed for structural proteins such as collagen. Therefore, this lesser fraction of the developing enamel matrix might represent structural protein in the usual sense of the term or protein chains which are rendered immobile by their attachment to the enamel mineral via chemical or physical bonding mechanisms.

Also recently, sequential dissociative extraction of fetal bovine[35] and neonatal hamster[45] enamel first with 4 *M* guanidine and then with 4 *M* guanidine-0.5 *M* EDTA showed that the amelogenin matrix components are completely solubilized without dissolution of the enamel apatite, while the enamelin components cannot be solubilized until the enamel apatite is also dissolved. Further, when total enamel matrix is resolubilized in guanidine and then synthetic apatite mineral is added, only the enamelin components adhere to the apatite.[35] These data suggest that the amelogenin components of the developing enamel are intercrystalline in location and easily solubilized, while the enamelin components are tightly bound to the enamel apatite crystal surfaces.[9] Additional support for this crystal-based partition of the amelogenins and enamelins comes from detailed electron microscopic experiments.[91] These investigators have found that in vivo, the enamelins appear to form a thin coat or sheath around individual enamel hydroxyapatite crystallites and that the amelogenins are located both abutting this enamelin coat and at positions between the growing crystallites. All these data are consistent with the preferential loss of amelogenin components and subsequent retention of enamelin-like species concomitant with progressive enamel maturation.[1,35,45]

The proteins of the mature enamel are acidic in character, low in molecular weight, and complex in composition.[68,72] These proteins are strikingly different from the fetal amelogenins and are better related compositionally to the embryonic enamelins which appear to undergo a dramatic reduction in size during enamel maturation.

The mature enamel TCA-insoluble "tuft" protein[49] and the EDTA-insoluble protein[72] have closely similar amino acid compositions and may be identical components, although there are some differences in the levels of glycine, serine, and hydroxyproline. Perhaps the presence in the mature enamel of such a protein fraction that resists solution by chelating agents results from a tight association between the phosphorylated organic material and the apatite crystals. The precise location within the enamel of this insoluble structure is of great interest to the understanding of the structural integrity of the normal mature tooth. Does this material occur in the tuft region near the enamel-dentin junction, at the surface, or throughout the enamel structure?

Even though we now have a better understanding of the overall organization of the major elements in enamel, we are extremely short of information on precisely what molecular mechanisms operate between those elements. Thus, we do not yet know how the process of initial enamel apatite nucleation occurs nor whether it is a cell- or matrix-mediated process. Further, we have no real comprehension of the mechanisms of ordered and coordinated crystal growth in developing enamel. However, we do now have enough information to form more intelligent and scientifically sound questions that might form the basis of future research efforts in this difficult area. For example, it would be useful to know how the properties of the hydrophobic amelogenins as an entropically driven coacervating system affect apatite solubility and/or crystal growth potentials. Further, it remains to be seen whether within the enamelin family, certain size- or charge-group specific members exist that affect directional growth along the different faces of growing enamel crystallites. Finally, it remains to be established just what specific submolecular configurations and/or amino acid milieu can operate as "active sites" in regulating enamel crystal maturation.

REFERENCES

1. **Deakins, M.,** Changes in ash, water and organic content of pig enamel during calcification, *J. Dent. Res.*, 21, 429, 1942.
2. **Stack, M. V.,** Organic constituents of enamel, *J. Am. Dent. Assoc.*, 48, 297, 1954.
3. **Eastoe, J. E.,** The amino acid composition of proteins from the oral tissues. II. The matrix proteins in dentine and enamel from developing deciduous human teeth, *Arch. Oral Biol.*, 8, 633, 1963.
4. **Stack, M. V.,** Mineral and protein levels in enamel from human, monkey and rat molars, *Odontol. Revy*, 8, 89, 1957.
5. **Robinson, C., Lowe, N. R., and Weatherell, J. A.,** Changes in amino-acid composition of developing rat incisor enamel, *Calcif. Tissue Res.*, 23, 19, 1977.
6. **Robinson, C., Fuchs, P., Deutsch, D., and Weatherell, J. A.,** Four chemically distinct stages in developing enamel from bovine incisor teeth, *Caries Res.*, 12, 1, 1978.
7. **Glimcher, M. J., Brickley-Parsons, D., and Levine, P. T.,** Studies of enamel proteins during maturation, *Calcif. Tissue Res.*, 24, 259, 1977.
8. **Fincham, A. G.,** Changing amino acid profiles of developing dental enamel in individual human teeth and the comparison of the protein matrix of developing human and bovine enamel, *Arch. Oral Biol.*, 25, 669, 1980.
9. **Termine, J. D., Torchia, D. A., and Conn, K. M.,** Enamel matrix: structural proteins, *(Tooth Enamel III), J. Dent. Res.*, 58(B), 773, 1979.
10. **Katz, E. P., Mechanic, G. L., and Glimcher, M. J.,** The ultracentrifugal and free zone electrophoretic characterization of the neutral soluble proteins of embryonic bovine enamel, *Biochim. Biophys. Acta*, 107, 271, 1965.
11. **Burgess, R. C. and MacLaren, C. M.,** Proteins in developing bovine enamel, in *Tooth Enamel I*, Stack, M. V. and Fearnhead, R. W., Eds., Wright Brothers, Bristol, 1965, 74.
12. **Nikiforuk, G. and Simmons, N. S.,** Purification and properties of enamel protein, *I.A.D.R., Toronto*, Abs. 441, 1965.
13. **Eastoe, J. E.,** Dental enamel, in *Comprehensive Biochemistry*, Vol. 26(C), Florkin, M. and Stotz, E., Eds., Elsevier, Amsterdam, 1971, 785.
14. **Eastoe, J. E.,** Enamel protein chemistry — past, present and future, *(Tooth Enamel III), J. Dent. Res.*, 58(B), 753, 1979.

15. **Fincham, A. G.,** The extracellular protein matrix of developing dental enamel, in *Perspectives in Differentiation and Hypertrophy,* Anderson, W., Ed., Elsevier, Amsterdam, in press.
16. **Levine, P. T., Seyer, J., Huddleston, J., and Glimcher, M. J.,** The comparative biochemistry of the organic matrix proteins of developing enamel. I., *Arch. Oral Biol.,* 12, 407, 1967.
17. **Eastoe, J. E.,** Organic matrix of tooth enamel, *Nature (London),* 187, 411, 1960.
18. **Glimcher, M. J., Mechanic, G. L., and Friberg, U. A.,** The amino acid composition of the matrix and neutral and acid soluble components of embryonic bovine enamel, *Biochem. J.,* 93, 198, 1964.
19. **Levine, P. T. and Glimcher, M. J.,** The isolation and amino acid composition of the organic matrix and neutral soluble proteins of developing rodent enamel, *Arch. Oral Biol.,* 10, 753, 1965.
20. **Kaneko, I.,** Protein in embryonic bovine enamel, *Bull. Tokyo Med. Dent. Univ.,* 14, 105, 1967.
21. **Mechanic, G. L., Katz, E. P., and Glimcher, M. J.,** The Sephadex gel filtration characteristics of the neutral soluble proteins of embryonic bovine enamel, *Biochim. Biophys. Acta,* 133, 97, 1967.
22. **Fincham, A. G.,** Electrophoretic and Sephadex gel filtration studies of bovine foetal enamel matrix at acid pH, *Calcif. Tissue Res.,* 2, 353, 1968.
23. **Katz, E. P., Seyer, J., Levine, P. T., and Glimcher, M. J.,** The comparative biochemistry of the organic matrix of developing enamel. II. Ultracentrifugal and electrophoretic characterization of proteins soluble at neutral pH, *Arch. Oral Biol.,* 14, 533, 1969.
24. **Mechanic, G. L.,** The multicomponent re-equilibriating protein system of bovine embryonic enamelin (dental enamel protein): chromatography in deaggregating solvents, in *Tooth Enamel II,* Stack, M. V. and Fearnhead, R. W., Eds., Wright Brothers, Bristol, 1971, 88.
25. **Fincham, A. G.,** Experiments on the DEAE-cellulose chromatography of bovine foetal enamel matrix and the isolation of a low molecular weight fraction, in *Tooth Enamel II,* Stack, M. V. and Fearnhead, R. W., Eds., Wright Brothers, Bristol, 1971, 79.
26. **Weidmann, S. M. and Eyre, D. R.,** The protein of mature and foetal enamel, in *Tooth Enamel II,* Vol. 2, Stack, M. V. and Fearnhead, R. W., Eds., Wright Brothers, Bristol, 1971, 72.
27. **Seyer, J. and Glimcher, M. J.,** The isolation of phosphorylated polypeptide components of the organic matrix of embryonic bovine enamel, *Biochim. Biophys. Acta,* 236, 279, 1971.
28. **Eggert, F. M., Allen, G. A., and Burgess, R. C.,** Amelogenins — purification and partial characterization of proteins from developing bovine dental enamel, *Biochem. J.,* 131, 471, 1973.
29. **Fukae, J. and Shimizu, M.,** Studies on the proteins of developing bovine enamel, *Arch. Oral Biol.,* 19, 381, 1974.
30. **Seyer, J. and Glimcher, M. J.,** Evidence for the presence of numerous protein components in immature bovine dental enamel, *Calcif. Tissue Res.,* 24, 253, 1977.
31. **Papas, A., Seyer, J., and Glimcher, M. J.,** Isolation from embryonic bovine dental enamel of a polypeptide (E_3) containing as its only phosphorylated sequence, Glu-O-phosphoserine-Leu, *FEBS Lett.,* 79, 276, 1977.
32. **Seyer, J. and Glimcher, M. J.,** Isolation, characterization and partial amino acid sequence of a phosphorylated polypeptide (E_4) from bovine embryonic dental enamel, *Biochim. Biophys. Acta,* 493, 441, 1977.
33. **Fincham, A. G.,** The amelogenin problem: a comparison of purified enamel matrix proteins, *Calcif. Tissue Int.,* 27, 65, 1979.
34. **Fincham, A. G.,** A simple procedure for the isolation of a major amelogenin polypeptide component, *Biochem. J.,* 181, 171, 1979.
35. **Termine, J. D., Belcourt, A. B., Christner, P. J., Conn, K. M., and Nylen, M. U.,** Properties of dissociatively extracted fetal tooth matrix proteins. I. Principal molecular species in developing bovine enamel, *J. Biol. Chem.,* 255, 9760, 1980.
36. **Sasaki, S. and Shimokawa, H.,** Enamel proteins: biosynthesis and chemistry, *(Tooth Enamel III), J. Dent. Res.,* 58(B), 765, 1979.
37. **Shimizu, M., Tanabe, T., and Fukae, M.,** Proteolytic enzyme in porcine immature enamel, *(Tooth Enamel III), J. Dent. Res.,* 58(B), 782, 1979.
38. **Fukae, M., Ijiri, H., Tanabe, T., Shimizu, M.,** Partial amino acid sequences of two proteins in developing porcine enamel, *(Tooth Enamel III), J. Dent. Res.,* 58(B), 1000, 1979.
39. **Fukae, M., Tanabe, T., Ijiri, H., and Shimizu, M.,** Studies of porcine enamel proteins: a possible original enamel protein, *Tsurumi Univ. Dent. J.,* 6, 87, 1980.
40. **Hiller, C. R., Robinson, C., and Weatherell, J. A.,** Variations in the composition of developing rat incisor enamel, *Calcif. Tissue Res.,* 18, 1, 1975.
41. **Elwood, W. K. and Apostolopoulos, A. X.,** Analysis of developing enamel of the rat. I. Fractionation: protein and calcium content, *Calcif. Tissue Res.,* 17, 317, 1975.
42. **Elwood, W. K. and Apostolopoulos, A. X.,** Analysis of developing enamel of the rat. II. Electrophoretic and amino acid studies, *Calcif. Tissue Res.,* 17, 327, 1975.
43. **Elwood, W. K. and Apostolopoulos, A. X.,** Analysis of developing enamel of the rat. III. Carbohydrate, DEAE-Sephadex, and immunological studies, *Calcif. Tissue Res.,* 17, 337, 1975.
44. **Robinson, C., Briggs, H. D., Atkinson, P. J., and Weatherell, J. A.,** Matrix and mineral changes in developing enamel, *(Tooth Enamel III), J. Dent. Res.,* 58(B), 871, 1979.

45. **Lyaruu, D. M., Termine, J. D., Belcourt, A. B., and Fincham, A. G.,** Neonatal hamster molar tooth development: extraction and characterization of amelogenins, enamelins and soluble dentin proteins, *Calcif. Tissue Int.*, 34, 86, 1982.
46. **Guenther, H. L., Croissant, R. D., Schonfeld, S. E., and Slavkin, H. C.,** Identification of four extracellular-matrix enamel proteins during embryonic-rabbit tooth-organ development, *Biochem. J.*, 163, 591, 1977.
47. **Slavkin, H. C., Weliky, B., Stellar, W., Slavkin, M. D., Zeichner-Gancz, M., Bringas, P., Hyatt-Fischer, H., Shimizu, M., and Fukae, M.,** Ameloblast differentiation: protein synthesis and secretion in fetal New Zealand white rabbit molar tooth organs and isolated epithelia "in vitro", *J. Biol. Buccale*, 6, 309, 1979.
48. **Zeichner-David, M., Weliky, B. G., and Slavkin, H. C.,** Isolation and preliminary characterization of epithelial-specific messenger ribonucleic acids and their products during embryonic tooth development, *Biochem. J.*, 185, 489, 1980.
49. **Robinson, C., Lowe, N. R., and Weatherell, J. A.,** Amino acid composition, distribution and origin of "tuft" protein in human and bovine dental enamel, *Arch. Oral Biol.*, 20, 29, 1975.
50. **Deutsch, D., El-Attar, I., Robinson, C., and Weatherell, J. A.,** Rate and timing of enamel development in the deciduous bovine incisor, *Arch. Oral Biol.*, 24, 407, 1979.
51. **Stack, M. V.,** Chemical organization of the organic matrix of enamel, in *Structural and Chemical Organization of Teeth*, Vol. 2, Miles, A. E. W., Ed., Academic Press, New York, 1967, 317.
52. **Suga, S.,** Histochemical observation of proteolytic enzyme activity in the developing dental hard tissues of the rat, *Arch. Oral Biol.*, 15, 555, 1970.
53. **Termine, J. D., Miyamoto, M. S., and Kuettner, K. E.,** Lysozyme, protease, and protease inhibitor proteins in fetal bovine enamel matrix extracts, *J. Dent. Res.*, 59, 1523, 1980.
54. **Fincham, A. G., Burkland, G., and Shapiro, I. M.,** Lipophilia of enamel matrix. A chemical investigation of the neutral lipids and lipophilic proteins of enamel, *Calcif. Tissue Res.*, 9, 247, 1972.
55. **Nikiforuk, G. and Gruca, M.,** Immunological and gel-filtration characteristics of bovine enamel protein, in *Tooth Enamel II*, Stack, M. V. and Fearnhead, R. W., Eds., Wright Brothers, Bristol, 1971, 95.
56. **Mechanic, G. L., Katz, E. P., and Glimcher, M. J.,** Gel filtration and chromatographic properties of neutral soluble bovine embryo proteins, in *Proc. 4th Eur. Symp. Calcif. Tissue*, Conference Series #120, Excerpta Medica, Amsterdam, 1966, 73.
57. **Belew, M., Fohlman, J., and Janson, J. C.,** Gel filtration of proteins on Sephacel S-200 superfine in 6 M guanidine HCl, *FEBS Lett.*, 91, 302, 1978.
58. **Ackroyd, P.,** Acrylamide gel slab electrophoresis in a simple glass cell for improved resolution and comparison of serum proteins, *Anal. Biochem.*, 19, 399, 1967.
59. **Weber, K. and Osborn, M.,** The reliability of molecular weight determinations by dodecyl sulphate polyacrylamide gel electrophoresis, *J. Biol. Chem.*, 244, 4406, 1969.
60. **Ferguson, K. A.,** Starch gel electrophoresis — application to the classification of pituitary proteins and polypeptides, *Metabolism*, 13, 985, 1964.
61. **Green, M. R., Pastewaka, J. V., and Peacock, A. C.,** Differential staining of phosphoproteins in polyacrylamide gels with a cationic carbocyanine dye, *Anal. Biochem.*, 56, 43, 1973.
62. **Chrispens, J., Weliky, B., Bringas, P., and Slavkin, H. C.,** "Proenamel — enamel — polypeptides:" a concept, *(Tooth Enamel III), J. Dent. Res.*, 58(B), 988, 1979.
63. **Weidmann, S. M. and Hamm, S. M.,** Studies of the enamel matrix of mature teeth, in *Tooth Enamel I*, Stack, M. V. and Fearnhead, R. W., Eds., Wright Brothers, Bristol, 1965, 83.
64. **LeGeros, R. Z.,** The unit-cell dimensions of human enamel apatite: effect of chloride incorporation, *Arch. Oral Biol.*, 20, 63, 1975.
65. **Belcourt, A., Frank, R. M., and Houver, G.,** Analyse des acides amines de la pellicule exogene acquise et des proteines de l'email superficiel chez l'homme, *J. Biol. Buccale*, 2, 161, 1974.
66. **Shapiro, I. M. and Hartles, R. L.,** An improved method with low material loss for the separation of teeth into fractions of differing density, *Arch. Oral Biol.*, 10, 155, 1965.
67. **Prout, R. E. S. and Shutt, E. R.,** Separation of enamel and dentine using cadmium tungstoborate solution, *Arch. Oral Biol.*, 15, 559, 1970.
68. **Belcourt, A. and Gillmeth, S.,** EDTA soluble protein of human mature normal enamel, *Calcif. Tissue Int.*, 28, 227, 1979.
69. **Weidmann, S. M. and Eyre, D. R.,** Amino acid composition of enamel protein in the fully developed human tooth, *Caries Res.*, 1, 349, 1967.
70. **Glimcher, M. J. and Levine, P. T.,** Studies of the proteins, peptides and free amino acids of mature bovine enamel, *Biochem. J.*, 98, 742, 1966.
71. **Weatherell, J. A., Weidmann, S. M., and Eyre, D. R.,** Histological appearance and chemical composition of enamel protein from mature human molars, *Caries Res.*, 2, 281, 1968.

72. **Belcourt, A. B., Fincham, A. G., and Termine, J. D.,** EDTA insoluble protein of adult human enamel, *Caries Res.,* 16, 72, 1982.
73. **Brudevold, F., Steadman, L. T., and Smith, F. A.,** Inorganic and organic components of tooth structure, *Ann. N.Y. Acad. Sci.,* 85, 110, 1960.
74. **Burgess, R. C., Nikiforuk, G., and MacLaren, C. M.,** Chromatographic studies of carbohydrate components in enamel, *Arch. Oral Biol.,* 3, 8, 1960.
75. **Odutuga, A. A., and Prout, R. E. S.,** Lipid analysis of human enamel and dentin, *Arch. Oral Biol.,* 19, 729, 1974.
76. **Hess, E. C., Lee, C. Y., and Neidig, B. A.,** The amino acid composition of enamel protein, *J. Dent. Res.,* 32, 585, 1953.
77. **Battistone, G. C. and Burnett, G. W.,** Studies of the composition of teeth. IV. The amino acid composition of human enamel protein, *J. Dent. Res.,* 35, 260, 1956.
78. **Moe, D. and Birkedal-Hansen, H.,** Proteolytic activity in developing bovine enamel, *(Tooth Enamel III), J. Dent. Res.,* 58(B), 1012, 1979.
79. **Burrows, L. R.,** An investigation of proteins of human enamel matrix with special reference to the imino-acid hydroxyproline, in *Tooth Enamel I,* Stack, M. V. and Fearnhead, R. W., Eds., Wright Brothers, Bristol, 1965, 59.
80. **Eyre, D. R.,** Studies on the Organic Marix of Dental Enamel, Ph.D. thesis, University of Leeds, 1969.
81. **Lezovic, J., Ziman, M., and Hocman, G.,** Gel chromatography of proteins from human teeth, *J. Chromatogr.,* 96, 258, 1974.
82. **Srinivasan, B. N. Brudevold, F.,** Soluble and total protein in human enamel, *J. Dent. Res.,* 54(A), 112, 1975.
83. **Spitzer, D. and Ten Bosch, J. J.,** The absorption and scattering of light in bovine and human dental enamel, *Calcif. Tissue Res.,* 17, 129, 1975.
84. **Nylen, M. U. and Termine, J. D., Eds.,** Tooth Enamel III, *J. Dent. Res.,* 58(B), 674, 1979.
85. **Fearnhead, R. W.,** Matrix — mineral relationships in enamel tissues, *(Tooth Enamel III), J. Dent. Res.,* 58(B), 909, 1979.
86. **Nylen, M. U.,** Matrix — mineral relationships — a morphologist's view point, *(Tooth Enamel III), J. Dent. Res.,* 58(B), 922, 1979.
87. **Smales, F. C.,** Structural subunit in prisms of immature rat enamel, *Nature (London),* 258, 772, 1975.
88. **Glimcher, M. J.,** Phosphopeptides of enamel matrix, *(Tooth Enamel III), J. Dent. Res.,* 58(B), 790, 1979.
89. **Glimcher, M. J., Travis, D. J., Friberg, U. A., and Mechanic, G. L.,** The electron microscope localization of the neutral soluble protein of developing bovine enamel, *J. Ultrastruct. Res.,* 10, 362, 1964.
90. **Termine, J. D. and Torchia, D. A.,** ^{13}C-^{1}H magnetic double-resonance study of fetal enamel matrix proteins, *Biopolymers,* 19, 741, 1980.
91. **Yanagisawa, T. and Nylen, M. U.,** The relationship between matrix and mineral in rat enamel — an electron microscopic study, *J. Dent. Res.,* 59(A), (Abstr. 303), 343, 1980.
92. **Eastoe, J. E.,** Composition of enamel protein from the unerupted human third molar, *J. Dent. Res.,* 55(D), 108, 1976.
93. **Reisner, A. H., Nemes, P., and Bucholtz, C.,** The use of Coomassie brilliant blue G.250 perchloric acid solution for staining in electrophoresis and isoelectric focusing on polyacrylamide gels, *Anal. Biochem.,* 64, 509, 1975.
94. **Prout, R. E. S. and Odutuga, A. A.,** Fatty acid composition of neutral lipids and phospholipids of enamel and dentine from human molars, *Arch. Oral Biol.,* 15, 293, 1974.
95. **Seyer, J. and Glimcher, M. J.,** The content and nature of the carbohydrate components of the organic matrix of embryonic bovine enamel, *Biochim. Biophys. Acta,* 184, 509, 1969.
96. **Fincham, A. G., Belcourt, A. B., and Termine, J. D.,** Changing patterns of enamel matrix proteins in the developing bovine tooth, *Caries Res.,* 16, 64, 1982.
97. **Fincham, A. G., Belcourt, A. B., Lyaruu, D. M., and Termine, J. D.,** Comparative protein biochemistry of developing dental enamel matrix from five mammalian species, *Calcif. Tissue Int.,* 34, 182, 1982.
98. **Ungewickell, E. and Branton, D.,** Assembly units of clathrin coats, *Nature (London),* 289, 420, 1981.
99. **Bretscher, A. and Weber, K.,** Tropomyosin from bovine brain contains two polypeptide chains of slightly different molecular weights, *FEBS Lett.,* 85, 145, 1978.
100. **Fincham, A. G., Belcourt, A. B., and Termine, J. D.,** The molecular composition of bovine fetal enamel matrix, in *Chemistry and Biology of Mineralized Connective Tissues,* Veis, A., Ed., Elsevier/North Holland, New York, 1981, 523.
101. **Belcourt, A. B., Fincham, A. G., and Termine, J. D.,** Bovine high molecular weight amelogenin proteins, *Calcif. Tissue Res.,* in press.
102. **Fincham, A. G., Belcourt, A. B., and Termine, J. D.,** Composition of the protein matrix of developing human enamel, *J. Dent. Res.,* in press.
103. **Fincham, A. G., Belcourt, A. B., Termine, J. D., Butler, W. T., and Cothran, W. C.,** Dental enamel matrix: sequences of two amelogenin polypeptides, *Bioscience Rep.,* 1, 771, 1981.
104. **Shimokawa, H. and Fincham, A. G.,** unpublished observations, 1982.

ULTRASTRUCTURAL PROPERTIES OF HUMAN ENAMEL APATITE

Racquel Zapanta LeGeros

INTRODUCTION

The human enamel, like other calcified or mineralized biological tissues (e.g., dentin, cementum, bone), is composed of an organic matrix and an inorganic or mineral phase. The inorganic phase of enamel is principally a calcium apatite, similar but not identical to pure calcium hydroxyapatite, $Ca_{10}(PO_4)_6(OH)_2$, in composition and structural properties.[1-15] In addition, the apatite crystallites of enamel significantly differ in size, shape, orientation, and concentrations of minor constituents from those of bone or dentin. Also, organic matrix of enamel uniquely differs in composition and concentration (organic/inorganic ratio) from those of either dentin or bone.[1-4,6-8,15-19] This chapter is concerned with the structural properties of enamel apatite and the factors (e.g., organic matrix, composition) affecting such properties.

MINERAL-MATRIX RELATIONSHIP IN HUMAN ENAMEL

The techniques of histology, microradiography, electron microscopy, and X-ray diffraction have been used to investigate the mineral-matrix interactions in human and other mammalian enamel tissues.[1-4,19-28] Some of the reported observations include

1. Remarkable fiber (matrix) and crystal (apatite) orientation.
2. Abrupt change in the orientation of crystals between prisms or between rod and interrod regions (Figure 1).
3. Initially formed crystals which appear as thin long ribbons with specific orientation with respect to the enamel prism or rod and with respect to each other (Figures 1 and 2).
4. Presence of elliptical tubules surrounding individual crystals in undemineralized sections, the tubules remaining unchanged and appearing as "crystal ghosts" after demineralization.
5. The nature of the organic matrix of human enamel is uniquely different from that of bone or dentin, decreasing in quantity and changing in composition with the maturation of enamel.

These observations have been interpreted in several and oftentimes divergent ways to explain the involvements of specific matrix components (e.g., structural proteins, phospholipids, etc.) in the mineralization processes, namely, nucleation of apatite crystals, control of crystal growth, and direction of crystal orientation.[1-4,19-36]

Nucleation

The process of nucleation starts with the clustering of single ions or molecules in solution (Ca^{2+}, PO_4^{3-}, OH^- ions, in the case of calcium hydroxyapatite) into some critical size, shape, and configuration, to form the nuclei of the solid phase, $Ca_{10}(PO_4)_6(OH)_2$. In the case of enamel apatite, the elegant model proposed by Travis and Glimcher[27,30] appears attractive. In this model (Figure 3), the organic matrix component, a highly structured protein, is phosphorylated at specific residues, thus providing a specific spatial arrangement for the binding of the Ca^{2+} ions from the solution (Figure 3A); the bound Ca^{2+} ions interact with other inorganic ions (e.g., PO_4^{3-}) from the solution to form the initial calcium-phosphate clusters or crystal embryos. The highly structured organic matrix in the form of tubules

Table 1
COMPARATIVE COMPOSITION OF HUMAN ENAMEL, DENTIN, BONE[1,6,15]

Constituents	Enamel	Dentin	Bone
Calcium, Ca^{2+}	36.00	27.00	24.50
Phosphorus, PO_4^{3-} as P	17.70	13.00	10.50
Ca/P weight	2.03	2.08	2.33
Ca/P molar	1.58	1.61	1.81
Sodium, Na^+	0.50	0.30	0.70
Potassium, K^+	0.08	0.05	0.03
Magnesium, Mg^{2+}	0.44	1.10	0.55
Carbonate, CO_3^{2-}	2.50	4.80	5.50
Fluoride, F^-	0.01	0.03	0.02
Chloride, Cl^-	0.30	0.01	0.10
Pyrophosphate, $P_2O_7^{4-}$	0.02	0.08	0.05
Trace elements: Sr^{2+}, Pb^{2+}, Ba^{2+}, Cu^{2+}, Zn^{2+}, Fe^{3+}, Li^+, etc.			
Ash (total inorganic)	97.80	70.00	65.00
Total organic	1.00	20.00	25.00
Adsorbed H_2O[a]	1.50	10.00	10.00

[a] Weight loss upon ignition at 100°C.

Arguments for and against the specific roles of enamel matrix components (phosphopeptides, phospholipids, etc.) in the nucleation, crystal growth, and orientation of the apatite crystallites, continue. In this paper, the organic matrix-mineral (apatite) relationship was briefly discussed only to serve as a background for the main focus which is the structural properties of enamel apatite.

STRUCTURAL PROPERTIES OF SYNTHETIC AND HUMAN ENAMEL APATITE CRYSTALLITES

As early as 1926[10,11] the inorganic phase of human enamel, dentin, and bone had been characterized as a basic calcium phosphate, with a structure similar to that of mineral hydroxyapatite, idealized as $Ca_{10}(PO_4)_6(OH)_2$. Other structural studies of synthetic (high-temperature) and mineral apatites compared enamel apatite to fluor-apatite, $Ca_{10}(PO_4)_6F_2$, or carbonate hydroxyapatites.[8,42,43] Most recent crystallographic studies combining X-ray and neutron diffraction data from powdered human enamel confirm the earlier observations of the similarity of enamel apatite to pure calcium hydroxyapatite in the positions of the principal atoms, Ca^{2+}, PO_4^{3-}, and OH^- in the apatite unit cell.[14] More important than the similarities observed in the structural properties between human enamel apatite crystallites and synthetic calcium hydroxyapatite (specifically those prepared from nonaqueous systems at high temperatures) are the striking dissimilarities which include "crystallinity", composition or stoichiometry, infrared absorption spectra, thermochemical properties, and presence of "impurity" ions in the case of the enamel apatite. (Some of these differences are summarized in Table 2). Such differences may play significant roles in properties of enamel apatite which are relevant to the understanding of the caries mechanism.

The simultaneous presence of several elements in enamel apatite (Table 1) makes extremely difficult the assessment of the individual and synergistic contributions of the "impurities" to the formation, stability and other structural properties of the enamel apatite. Results extrapolated from studies on synthetic, relatively pure, apatites have been valuable in this regard. However, it should be noted that some observations from synthetic apatites prepared

Table 2
COMPARABLE PROPERTIES OF SYNTHETIC CALCIUM HYDROXYAPATITE AND HUMAN ENAMEL APATITE

Properties	Syn A[a]	Syn B[b]	Enamel apatite
Composition	$Ca_{10}(PO_4)_6(OH)_2$	$Ca_{10}(PO_4)_6(OH)_2$ HPO_4^{2-} "H_2O"	$Ca_{10}(PO_4)_6(OH)_2$ Na^+ Co_3^{2-} F^- K^+, HPO_4^{2} Cl^- Mg^{2+} "H_2O" Trace elements
Stoichiometry (Ca/P)	1.67	1.60—1.64	1.55—1.64
Lattice parameters			
a-axis	9.422	9.433	9.441
c-axis	6.880	6.882	6.882
IR spectra			
Absorption bands	PO_4^{3-}, OH^-	PO_4^{3-}, OH^- HPO_4^{2-}	PO_4^{3-}, OH^- CO_3^{2-}
Ignition products (800°C)	Apatite	TCP + apatite	TCP[c] + apatite

[a] Syn A (from nonaqueous systems).
[b] Syn B (from aqueous systems, by precipitation or hydrolysis).
[c] Mg-substituted TCP.

from nonaqueous systems should be extrapolated with caution to biological apatites.[44] In the following section, the structural properties of synthetic calcium hydroxyapatite, $Ca_{10}(PO_4)_6(OH)_2$, and factors affecting its formation and stability will be discussed to better understand and interpret the structural properties of enamel apatite.

Synthetic Apatites

Crystallographic Properties of Calcium Hydroxyapatite, $Ca_{10}(PO_4)_6(OH)_2$

Apatite (from the Greek work, *apate*, meaning, to deceive) was originally a name given to calcium phosphate minerals which, because of their colorful and symmetric appearances, were often mistaken for valuable gem stones like topaz. The term apatite is now used to describe a crystal structure. Thus, several materials, although differing in composition, may have similar ''apatitic'' structures. The apatite structure was first established by Naray-Szabo and Mehmel, separately and simultaneously,[45,46] and subsequently refined by later investigators using mineral F-apatites and synthetic apatites prepared hydrothermally or prepared at high temperatures.[13,42,47,48] The decription below, is a summary of the work reported by Beevers and MacIntyre,[42] Trautz,[12] and Elliott[5] and from the extensive work of Young and co-workers.[13,14,47,48]

Crystallographically, apatite belongs to the hexagonal system, characterized by a sixfold *c*-axis perpendicular to three equivalent axes (a_1, a_2, a_3) at angles 120° to each other; is described by a space group symmetry, $P6_{3/m}$; and assumes a special crystal habit. The smallest building unit or unit-cell assumes a parallel-piped shape and contains 10Ca, $6PO_4$, and 20H atoms. The atomic arrangement in the unit cell shows spatial organization of the Ca, PO_4 tetrahedra, and OH groups with respect to each other. The OH ions located in the corners of the unit-cell are surrounded by two sets of Ca (II) atoms arranged in a triangle at levels $z = 1/4$ and $3/4$; by two sets of PO_4 tetrahedra also arranged in triangular positions at levels $z = 1/4$ and $3/4$; and by a hexagonal array of Ca atoms at a distance: four Ca(I) at $z = 0$ and two Ca(II) at $z = 1/2$. The ten Ca atoms are described Ca(I) and Ca(II) according to their environments; four Ca(I) atoms occupy positions at levels $z = 0$ and $z = 1/2$; six Ca(II) atoms occupy positions at levels $z = 1/4$ and $z = 3/4$ in sets of three. The OH groups shown as a single unit in this arrangement are in reality situated between the calcium triangles in a linear O-H-O-H arrangement parallel to the *c*-axis. The dimensions of the apatite unit cell is described by $a = b$ and *c*-axes (9.42 and 6.88 Å, respectively). The unit cell repeats itself in three dimensions to form the apatite crystal. Critical to the stability of the apatite structure is the network of PO_4 groups in a tightly packed arrangement.[5,42]

Substituents in the Apatite Structure

The apatite structure is capable of accommodating various substituents (e.g., Sr^{2+} for Ca^{2+}; F^- for OH^-; CO_3^{2-} for PO_4^{3-}) and still maintains its basic structure. Considering the properties of $Ca_{10}(PO_4)_6(OH)_2$ as the basic properties of unsubstituted OH-apatite, the properties affected by substituents are (1) Lattice parameters (*a*- and *c*-axis dimensions), (2) Crystal size and shape, (3) Crystal strain, (4) "Crystallinity" from X-ray diffraction patterns, (5) Infrared absorption spectral properties, (6) Dissolution properties, (7) Thermal stability. When the substituents are present simultaneously, they may have additive, synergistic, or opposite effects on the properties of the apatite crystallites.[5-9,43,48,49-52]

The effect on the lattice parameters depends on the size and amount of the substituent, in accordance with Vegard's law. Usually, if the substituent is larger than the ion substituted for (e.g., Sr^{2+} for Ca^{2+}; Cl^- for OH^-), the effect is to expand one or both of the lattice parameters. In some cases of substitutions, both *a*- and *c*-axes are affected either in the same or opposite directions ($a' > a$, $c' > c$; $a' < a$; $c' > c$; where a' and c' dimensions are those of substituted apatites); in other cases, only the *a*-axis dimensions are either increased or decreased, while the *c*-axis dimensions are not significantly different from the unsubstituted OH-apatite (Table 3). Some substitutions are coupled with others to maintain charge balance or neutrality in the apatite, e.g., CO_3^{2-} for PO_4^{3-} coupled with Na^+ for Ca^{2+}.[6,43]

The effect on "crystallinity" reflects the effect on crystallite size and/or strain and is deduced from the broadening characteristics of the X-ray diffraction patterns.[52] Broader diffraction peaks indicate smaller crystallites and/or greater strain; narrower peaks indicate larger crystallites and/or lesser strain. For example, an increase in "crystallinity" is observed in apatite as a consequence of F incorporation, while a decrease is observed as a consequence of carbonate incorporation.[6,7,43,53-59]

The infrared absorption spectra of calcium hydroxyapatite is characterized by the absorption bands reflecting the different vibration modes of the OH and PO_4 groups.[60] Since the PO_4^{3-} groups serve as the skeletal framework of the apatite structure, it is expected that any substitution will have the most pronounced effect on the environment of these groups, and this effect will be manifested in the frequencies and resolutions of their infrared absorption bands. The librational mode of the OH groups is specially sensitive to substitutions in the apatite structure.

The dissolution properties are also affected by the kind of substituent. For example, carbonate-substituted apatite crystals show preferential dissolution of the crystal core, while the carbonate-free synthetic apatites do not;[61] F-containing apatites dissolve to a lesser degree than F-free apatites, while the opposite is true of CO_3-containing apatites.

The nature and extent of substitution is greatly influenced by the manner of apatite preparation, i.e., whether from aqueous or nonaqueous systems[43,44] and in turn determines the type and magnitude of effects on the crystallite properties. For example, CO_3 for OH substitution is observed in apatites prepared from nonaqueous systems, 1000°C,[63-65] while CO_3 for PO_4 substitution is observed in those apatites from aqueous systems, 37 to 100°C.[6,43-65] These two types of carbonate substitution differ in their effects on the lattice parameters, on the infrared absorption spectra, and on thermal stabilities.[43,44] A consistent observation is that apatites prepared from aqueous systems, either by direct precipitation or hydrolysis at 80 to 100°C, always exhibit a longer *a*-axis dimension (longer by about 0.01 Å) than those prepared from nonaqueous systems or those prepared from aqueous systems and subsequently ignited at 800 to 1000°C.[43-66] The extent of Cl for OH substitution is less from

Table 3
QUALITATIVE EFFECTS OF SOME "SUBSTITUENTS" FOR Ca^{2+}, PO_4^{3-}, OR OH^- IN $\sim Ca_{10}(PO_4)_6(OH)_2$ ON THE LATTICE PARAMETERS AND "CRYSTALLINITY" OF APATITES

"Substituent"	Ionic rad.[a] (Å)	Lattice param. (*a*-axis)	(±0.003 Å) (*c*-axis)	"Crystallinity"
For calcium, Ca^{2+}	0.99	9.433	6.882	—
Strontium, Sr^{2+}	1.12	(+)[b]	(+)	(nc)
Barium, Ba^{2+}	1.34	(+)	(+)	(−)
Lead, Pb^{2+}	1.20	(+)	(+)	(−)
Potassium, K^+	1.33	(+)	(nc)[c]	(nc)
Sodium, Na^+	0.97	(nc)	(nc)	(NC)
Magnesium, Mg^{2+}	0.66	(nc)	(nc)	(−)
Cadmium, Cd^{2+}	0.97	(−)	(−)	(−)
Manganese, Mn^{2+}	0.80	(−)	(−)	(−)
Tin, Sn^{2+}	0.93	am	am[c]	am
For OH^- (O-H)				
Fluoride, F^-	1.36	(−)	(nc)	(+)
Chloride, Cl^-	1.81	(+)	(−)	(nc)
For PO_4^{3-}				
Carbonate, CO_3^{2-}	—	(−)	(+)	(−)
Citrate, $C_6H_5O_7^{3-}$	—	(+)#	(+)#	(−)
HPO_4^{2-}	—	(+)	(nc)	(nc)
Vanadate, VO_4^{3-}	—	(−)	(−)	(−)
Pyrophosphate, $P_2O_7^{4-}$	—	?	?	(−)

[a] Data from *Handbook of Chemistry and Physics*, 45th ed., Weast, R. C., Ed., CRC Press, Boca Raton, Fla., 1965, F89.

[b] (+), Increase; (−), decrease; (nc), no change; (am), amorphous.

[c] TCP phase formed in additional comparisons made with unsubstituted OH-apatite prepared from aqueous.

aqueous than from nonaqueous systems, 40 and 100 mol %, respectively.[48,62,66] These differences must be taken into account when extrapolating properties of synthetic apatites to biological apatites.

Substitutents for Ca^{2+}

The incorporation of *monovalent* cations, e.g., Na^+ and K^+, in synthetic apatite is limited and appears to be related to the incorporation of CO_3^{2-} or HPO_4^{2-}.[67] Since the atomic radius of Na^+ is similar to that of Ca^{2+}, a Na for Ca substitution is not expected to cause any significant change in lattice parameters. On the other hand, K^+ is larger than Ca^{2+}, and K for Ca substitution is expected to cause an increase in either or both of the lattice parameters. Such a difference has been reported for K^+ substitution in the apatite.[43,68] The substitution of the monovalent cations for divalent Ca^{2+} has to be coupled with another type of substitution to maintain charge neutrality in the apatite. The coupled substitution of Na for Ca with CO_3 for PO_4 is an example.[43] The incorporation of Na^+, K^+, and Li^+ have been studied in apatites prepared from aqueous systems.[67]

The incorporation of divalent ions, Sr^{2+}, Ba^{2+}, and Pb^{2+}, as substituents of Ca^{2+} in apatite is well known for both aqueous and nonaqueous systems.[6,43,69-74] These ions are capable of substituting for all the Ca^{2+} in the apatite to form $Sr_{10}(PO_4)_6(OH)_2$, $Ba_{10}(PO_4)_6(OH)_2$, and $Pb_{10}(PO_4)_6(OH)_2$ as end members of the isomorphous series. These ions being larger than Ca^{2+} cause expansion in both the *a*- and *c*-axis dimensions when they substitute in the apatite lattice. Their incorporation also causes changes in the infrared absorption spectra, notably on the PO_4^{3-} absorption bands and the disappearance or masking of the OH^- absorption bands ascribed to its librational vibration mode. Their effects on the ''crystallinity'' of apatites are not clear. Strontium-substituted apatites were reported to be more soluble than strontium-free apatites.[75]

The incorporation of Mg^{2+} in mineral and in synthetic apatites is very limited,[6,43] with no perceptible effect on the lattice parameters in spite of the smaller atomic radius of Mg^{2+} compared to Ca^{2+}. The presence of Mg^{2+} in the solution from which the apatite forms causes a reduction in apatite crystallite size, favors the formation of Mg-substituted tricalcium phosphate, TCP, and at 37°C promotes the formation and stabilization of amorphous calcium phosphate.[6,7,43,71,76-80] The effect of Mg^{2+} on the dissolution properties of apatite has not been investigated. Other cations which are incorporated in limited amounts are Mn^{2+}, Zn^{2+}, Fe^{2+}, Fe^{3+}, Al^{3+}, and Cr^{3+}. These ions cause a reduction in the crystallinity of apatites.[6,43,71] Qualitative effects of possible substituents for Ca^{2+} on the lattice parameters and on the crystallinity of apatites are shown in Table 3.

Substituents for PO_4^{3-}

Several anions are present in natural apatites as substituents for PO_4^{3-}, such as AsO_4^{3-}, SO_4^{2-}, SiO_4^{4-}, and CO_3^{2-}.[8,12] Of special interest in human enamel apatites are the anions CO_3^{2-} and HPO_4^{2-}.

HPO_4 for PO_4 — Substitution is implied when apatites obtained from solutions of higher HPO_4^{2-} concentrations demonstrate a larger *a*-axis dimension, as high as 9.47 Å.[138] The presence of HPO_4^{2-}, in synthetic apatites have been indirectly determined by the formation of $P_2O_7^{4-}$ upon ignition at 400°C and by the appearance of the B-TCP phase upon ignition at 700°C and above.[43,81-85] This indirect determination is even more inaccurate when CO_3 is also present since HPO_4^{2-} can react with CO_3^{2-} at these temperatures.[43,83,84] The effect of HPO_4 for PO_4 on the ''crystallinity'' and stability of apatite has not been determined. Preliminary results from the author's work indicates that this type of substitution occurs to a limited extent and causes a slight increase in the crystallinity of the apatite.

CO_3 for PO_4 — Substitution has been established in mineral OH- and F-apatites.[8] Such a type of substitution was demonstrated in apatites prepared by precipitation (37 to 100°C) or by hydrolysis of $CaHPO_4$, or $CaHPO_4 \cdot 2H_2O$ (DCPD), or of $Ca_8H_2(PO_4)_6 \cdot 5H_2O$ (OCP) in carbonate-containing solutions, or of $CaCO_3$ (calcite or aragonite) in phosphate-containing solutions.[6,43,54,86,87] The extent of carbonate incorporation is influenced by the presence of other ions in the solution from which the apatite forms. Up to a maximum of 3 mol CO_3 per apatite unit cell is obtained when CO_3 for PO_4 substitution is coupled with Na for Ca.[6,43,88] The presence of NH_4^+ or Sr^{2+} limits the amount of CO_3 incorporated into the apatite.[43,70,71] Compared to the CO_3-free OH-apatites, the CO_3 substituted apatites show the following properties:

1. Smaller *a*- and larger *c*-axis dimensions, with increasing carbonate content[6,43,44,53,54]
2. Smaller crystallite size[6,43,53-55]
3. Morphology changing from needle- or pencil-shaped crytals to rod-shaped to equiaxed crystals with increasing carbonate content[6,43,53,54]
4. Decreasing ''crystallinity'' of X-ray diffraction patterns with increasing carbonate contents[6,43,53-55]
5. Decreasing phosphate content of the apatite concomitant with increasing carbonate content[43,53]
6. Decreasing resolution of the PO_4^{3-} vibration bands with increasing intensity of the CO_3^{2-} absorption bands in the infrared absorption spectra[43,54]
7. Increasing extent of dissolution with increasing carbonate content

Substituents for OH^-

CO_3 for OH — Substitution has been observed in apatites from nonaqueous systems prepared at 1000°C.[63,64] The effect of such substitution on the lattice parameters is opposite to the CO_3 for PO_4 substitution, namely, the former causes an expansion of the *a*- and a contraction of the *c*-axes; the latter causes a contraction of the *a*- and expansion of the *c*-axes.[5,6,43,44,63,64] The frequencies of the CO_3^{2-} absorption bands are also different in these two types of carbonate substitution. The extent of CO_3 for OH substitution is 100 mol %, $Ca_{10}(PO_4)_6CO_3$.[63] The effect of this type of substitution on the morphology and stability of apatite crystallites has not been investigated. Although this type of substitution is suggested for enamel apatite, there is no direct evidence for such substitutions occurring in apatites prepared from aqueous systems.

Cl for OH — Substitution differs in extent for apatites obtained from aqueous and those from nonaqueous systems, 40 and 100 mol %, respectively.[6,44,62,66] This substitution is accompanied by:

1. Expansion in the *a*- with the contraction in the *c*-axis dimensions
2. Changes in the characteristics of the absorption bands of the OH groups in the case of partially substituted (Cl,OH)-apatites[62]
3. Decrease in thermal stability[44]

The effect of Cl substitution on the crystallite size, morphology, and dissolution properties of apatite is not known. However, it is predicted from the atomic arrangements of OH-, F- and Cl-apatites[48] that structural stability is in the decreasing order F-Ap > OH-Ap > Cl-Ap.

F for OH — Substitution predominates in mineral apatites.[8,12] In both aqueous and nonaqueous systems, F for OH substitution can be partial or full, $Ca_{10}(PO_4)_6(F,OH)_2$ or $Ca_{10}(PO_4)_6F_2$, respectively. The incorporation of F in the apatite is accompanied by:

1. Decrease in the *a*-axis without any significant change in the *c*-axis
2. Increase in crystallite size
3. Decrease in crystallite strain
4. Increase in "crystallinity" as deduced from their X-ray diffraction patterns
5. Increase in the resolution of their infrared absorption spectra
6. Decrease in dissolution rates and solubility in acid buffers
7. Increase in thermal stability[6,43,56-58,89,90]

Other Substituents

The presence of "lattice H_2O" possibly substituting for OH^- has been implied from the contraction of the *a*-axis dimensions of synthetic apatites (prepared from aqueous systems) pyrolyzed at 400°C.[82] Possible substitutions of monofluorophosphate, PO_3F for PO_4,[91] or of pyrophosphate, $P_2O_7^{4-}$ for PO_4^{3-}, have not been explored.

Factors Affecting Crystallite Size of Synthetic Apatites

In synthetic systems, the following factors have been shown to affect the "crystallinity" (reflecting crystallite size, strain) of apatites:

1. Presence of "impurity" ions in solution from which the apatite forms
2. Temperature of preparation
3. pH
4. Nature of the medium

The presence of other ions (referred to as "impurity" ions) besides Ca^{2+}, HPO_4^{2-}, PO_4^{3-}, and OH^- in the solutions or media from which the apatite forms causes these ions to be incorporated as a substituent(s) in the forming apatite or to be adsorbed on the surfaces and suppress further crystal growth. The composition of the apatite reflects the composition of the medium from which it forms. Changes in the lattice parameters (compared to unsubstituted apatites) indicate that the "impurity" ion(s) are incorporated as a substituent and are not just surface-bound.

Apatites prepared from nonaqueous systems are considerably larger in size than those prepared from aqueous systems.[6,44] Among those prepared in aqueous systems, apatite crystallinity increases with the temperature of preparation; crystals prepared at 100°C are much larger and more elongated than those prepared at 25 or 37°C.[6,43,54] The crystallinity of apatites prepared from lower pH are greater than those from higher pH ($5 > 7 > 10$). Calcium phosphates, e.g., DCPD, OCP, and apatite, prepared in gel systems are considerably larger than those prepared in solutions under similar conditions of composition, pH, and temperature.[95,96]

The simultaneous presence of some ions show synergistic effects on the crystallinity of apatite. The combined effects of Mg^{2+} and CO_3^{2-} on reducing crystallinity are greater than the addition of their individual effects.[43,55] The simultaneous presence of CO_3^{2-} and F^- gives an additive effect on the contraction of the *a*-axis dimensions.[43,88]

Compositional and Structural Factors Affecting Stability of Apatites

The extent of dissolution of synthetic apatites was found to decrease with increasing F^- content and to increase with increasing CO_3^- or increasing Sr^{2+} contents.[6,75,89,90,97-100] The simultaneous presence of CO_3^{2-} and F^- or Sr^{2+} and F^- causes a significant decrease in the extent of dissolution almost to the level of the CO_3^{2-}-free or Sr^{2+}-free apatites.

The simultaneous presence of CO_3^{2-} and HPO_4^{2-} causes the stoichiometry of the synthetic apatite to be closer to the pure calcium hydroxyapatite. The incorporation of F^- minimizes the incorporation of HPO_4^{2-}; the incorporation of Sr^{2+} minimizes the incorporation of CO_3^{2-}; and the incorporation of F^- suppresses almost completely, the incorporation of Cl^-. Such competitive incorporation is observed when two ions are simultaneously present in the solutions from which the apatite forms.[6,43,66,88,89,90] Thus, the stability of structure contributed by F is not only from its own incorporation, but from the exclusion of ions which would destabilize the apatite, e.g., CO_3^{2-} or HPO_4^{2-} or Cl^-.

Enamel Apatites

Ultrastructural Properties

Electron microscopic, electron microprobe, and chemical analyses of human enamel[1-4,15,16,22-24,28,35,36,39-41,101-107] have shown the following properties: enamel crystals as irregular hexagons, rod-shaped, flattened hexagons, and prismatic hexagons; closely packed in mature enamel and growing as thin-ribbons in immature enamel; and crystal growth progressing significantly and rapidly along the thickness and width of the crystals and to a least extent along its length. The crystal morphology (size and shape) differs as a function of age (deciduous > mature enamel crystals) and diseased states (fluorosed > normal enamel, carious > sound enamel crystals); composition of minor constituents or impurities changing from enamel surface to the dentin-enamel junction, between carious and sound enamel, between immature and mature enamel; enamel crystals being significantly larger than the crystals of bone or dentin. Some of these features are shown in Figures 1 and 2.

The X-ray diffraction patterns of human enamel are very similar, but not identical, to those of synthetic and mineral apatites. The "crystallinity" of enamel apatite is less than that of the mineral or synthetic (nonaqueous) apatites and more similar to that of apatites prepared from aqueous systems (60 to 80°C) or from gel systems (37°C). The "crystallinity" of enamel apatite is considerably greater than that of either dentin or bone as deduced from their X-ray diffraction patterns, indicating larger and/or less strained crystals in enamel compared to those in bone or dentin.

The infrared spectra of human enamel is almost identical to those of other biological apatites, in that they all exhibit the absorption bands of the CO_3^{2-} groups (although differing in intensities) and the characteristics of these bands are unique for CO_3^{2-} being a substituent in the apatite lattice rather than being surface-bound or present as a separate $CaCO_3$ phase.[43,108] Compared to the IR and resonance spectra of CO_3-substituted synthetic apatites, the spectra of human enamel is more akin to the spectra of apatites with CO_3 for PO_4 than those with CO_3 for OH substitution.[43,65,108,109]

The composition of human enamel mineral differs from pure calcium OH-apatite in stoichiometry (Ca/P ratio) and in the presence of several "impurity" ions which may or may not be substituents in the enamel apatite structure. The enamel mineral has been referred to as nonstoichiometric or calcium-deficient apatites. On the crystal level, the crystal core is speculated to have a different composition, e.g., carbonate-enriched compared to the rest of the apatite crystal.[110] This implies that the initial thin enamel apatite crystals grew under different conditions and solution composition than the mature crystals. The enamel mineral differs from the bone and dentin mineral in the concentrations of some of the major impurities, CO_3^{2-} and Mg^{2+} (Table 1).

Contributions of Some Impurities to the Crystal Properties of Enamel Apatite

Lattice Parameters

The average lattice parameters of human enamel apatite are a = 9.441 and c = 6.882; for mineral OH-apatite (Holly Springs) or nonaqueous synthetic OH-apatites, a = 9.422 and c = 6.880; for aqueous (precipitated or hydrolyzed at 100°C), a = 9.433 and c = 6.882 (±0.003 Å for all measurements). The difference in the a-axis dimensions between the two types of synthetic apatites is attributed to the presence of "lattice H_2O" and/or HPO_4 for PO_4 substitution in the apatites from aqueous systems.[6,43,44,66]

The observed expanded a-axis of human enamel apatite compared to pure calcium OH-apatite was explained in terms of a CO_3 for OH substitution.[63] Evidence against this explanation is

1. Similarity in spectral properties between human enamel and synthetic apatite from aqueous systems in which the manner of carbonate incorporation is by CO_3 for PO_4[6,43,65]
2. Lack of direct evidence for a CO_3 for OH substitution in apatites prepared from aqueous systems
3. The expansion (rather than a contraction) in the a-axis accompanying the loss of carbonate (as CO_2) in enamel pyrolyzed at 600 to 800°C[43,44]

Since a CO_3 for PO_4 substitution is known to cause a contraction in the a-axis rather than an expansion, the expanded a-axis of the enamel apatite cannot be attributed as a consequence of its carbonate content. Alternative and more reasonable explanations are the combined effects of Cl for OH, HPO_4 for PO_4, the concerted contributions of cations for Ca substitutions (in which the cations are larger than Ca^{2+} ion in ionic radii, e.g., K^+, Sr^{2+}, Ba^{2+}, and Pb^{2+}), and the possible presence of "lattice H_2O". The contraction in the a-axis of enamel and synthetic (aqueous) apatites pyrolyzed at 400°C supports the "lattice H_2O" and/or HPO_4 for PO_4 contributions to the expanded a-axis of the untreated human enamel.[6,66,82,111-113]

The incorporation of F^- in human enamel apatite as a consequence of ingested or topically applied F^- causes a contraction in the a-axis, the contraction being inversely proportional to the F^- content.[102] This observation is in agreement with observations made on naturally occurring F-apatite and synthetic (F,OH)- and F-apatites.[6,8,43,48,57,58] The simultaneous incorporation of Sr^{2+} and F^- in human enamel apatites causes opposite effects on the a-

axis.[6,114,115] The larger *a*-axis observed for carious or "white spot" enamel compared to normal enamel in the same tooth[105,116] may be partially due to the higher HPO_4^{2-} [116,117] and lower carbonate content[116,118] in these enamel apatites.

Crystal Size and Morphology

It was established from synthetic systems that the presence of F^- promotes the growth of larger and more elongated crystals while the presence of CO_3^{2-} favors the growth of smaller crystals, with morphology changing from needle-like or pencil-shaped to rod-like to equi-axed crystals. Other cations and anions present in the solutions from which the apatite formed were shown to cause reduction in the crystallite size of the apatite and in some cases facilitate the formation of other phases (e.g., TCP and ACP) at the expense of the apatite[6,43,53-59,71,89,90,92-94] The effect of F^- on promoting larger apatite crystallites is well documented.[6,16,57,90,99,100,102]

Considering the observations made in synthetic systems (Table 2), the much larger crystallite size of enamel apatite compared to that of either dentin or bone apatite may be explained in terms of the combined effects of the nature of their respective organic matrices and the concentrations of some of the "impurity" ions. For example, the concentrations of Mg^{2+} and CO_3^{2-} in enamel are much less than in dentin or in bone (Table 1). These ions are known to cause reduction in crystallite size of synthetic apatites when individually present, and the effects are even more pronounced when both ions are present simultaneously.[6,43,55] The presence of slightly higher concentrations of $P_2O_7^{4-}$ in dentin and bone compared to that in enamel (Table 1) could also contribute to the smaller crystallites of bone and dentin apatite since the presence of pyrophosphate has been shown to suppress the crystal growth of apatite.[6,43,92-94] In synthetic systems, under similar conditions of pH, temperature, and calcium/phosphate ratio in the media, calcium phosphates prepared from gel systems grow larger than those from spontaneous precipitation,[95,96] indicating that conditions that favor slow growth promote the growth of larger crystals. The gel properties of the human enamel matrix compared to the more rigid properties of the dentin or bone organic matrices would facilitate the growth of larger enamel apatite crystallites.

Thermochemical Properties

Analyses of pyrolyzed human enamel (100 to 950°C) using thermogravimetry, nuclear magnetic resonance, X-ray diffraction, infrared absorption, and scanning electron microscopy have helped elucidate some of the structural properties of enamel apatite and the contribution of some of the minor constituents or "impurities" to the stability of enamel apatite.[6,43,82,113-115,119,120] It was observed that pyrolysis of human enamel is accompanied by weight losses, changes in lattice parameters, changes in the resolutions of the IR absorption spectra, changes in crystal morphology, and appearance of β-TCP phase.

Considering that the organic content is only a minor amount (less than 2 wt %), the changes observed are reasonably attributed to the enamel mineral (apatite). These changes have been attributed to

1. Presence of adsorbed and "lattice H_2O"
2. Presence of HPO_4^{2-}
3. Reaction of Mg^{2+} with HPO_4^{2-} to form the Mg-substituted β-TCP phase
4. Thermal instability due to the CO_3^{2-} constituent

Dissolution Properties

Enamel crystallites subjected to acid dissolution in natural or artificial caries show a preferential dissolution of the central core extending anisotropically along its *c*-axis.[39,41,105] The schematic representation of such dissolution is shown in Figure 4. This dissolution

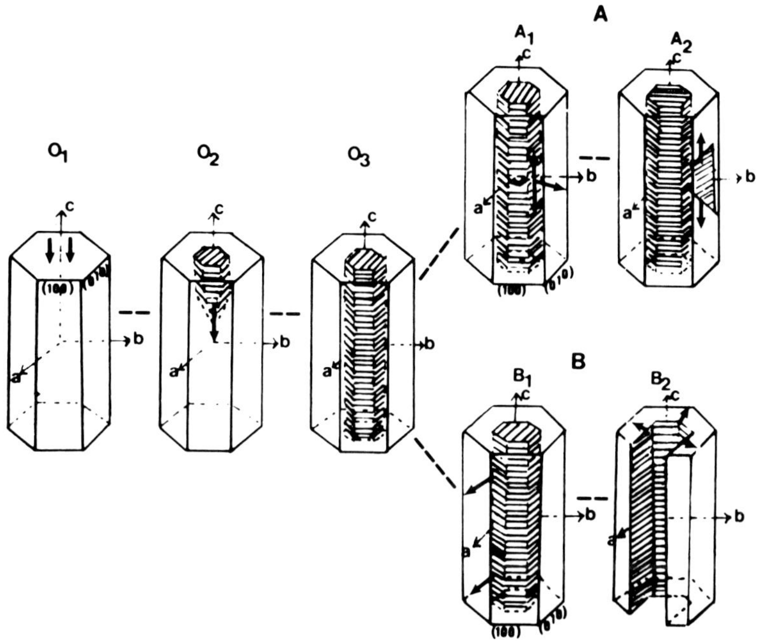

FIGURE 4. Schematic representation of the pattern of dissolution of enamel apatite crystals showing preferential dissolution along the *c*-axis. (From Voegel, J. C. and Frank, R. M., *Calcif. Tissue Res.*, 24, 19, 1977. With permission.)

behavior has been related to presence of dislocations in the crystals caused by growth mechanisms and/or presence of high concentrations of impurities such as CO_3^{2-} and Mg^{2+}.[110]

One of the mechanisms for the well-documented cariostatic effect of fluoride[121] is believed to be due to the reduction in apatite solubility as a consequence of fluoride incorporation in the enamel apatite. Referring to the opposite effects of carbonate and fluoride on the structural properties of synthetic apatites, their opposite effects on the stability (e.g., susceptibility to acid dissolution) of enamel apatite are predictable — the incorporation of CO_3^{2-} contributing to the growth of smaller, more strained, and therefore weaker and more easily dissolved crystallites contrasted with the incorporation of F^- contributing to the growth of larger, less strained, and therefore stronger and less easily dissolved apatite crystallites.

The nebulous associations of some elements with caries resistance or caries susceptibility[122-125] could be better defined by determining their effects on the formation and stability of apatites when such elements act as substituents in the apatite or as crystal growth suppressors or both.

Relationship of Enamel Apatite with Other Calcium Phosphates

The morphology (ribbon-like or plate-like) of the initially formed human enamel apatite crystallites, the nonstoichiometry of enamel apatite, the structural similarities with some calcium phosphates (notably with octacalcium phosphate, OCP) and the formation of pyrophosphate or tricalcium phosphates upon ignition of the enamel mineral, suggesting the presence of HPO_4^{2-}, were among the bases for the speculations that enamel apatite, idealized as $Ca_{10}(PO_4)_6(OH)_2$, forms indirectly by way of other calcium phosphate precursors.[126-130] Among the calcium phosphates implicated as precursors are dicalcium phosphate dihydrate, DCPD; octacalcium phosphate, OCP; tricalcium phosphate, TCP; and amorphous calcium phosphate, ACP.

In pathological tissue calcifications, some of these calcium phosphates co-exist with apatite, e.g., in dental calculi and urinary stones[6,7,78,92,96,131,132] In synthetic systems, these

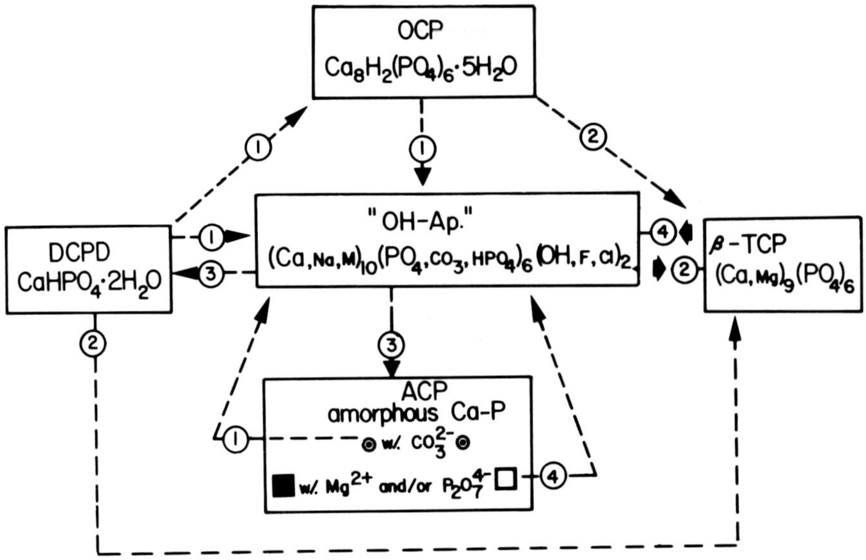

FIGURE 5. Interactions among apatite (substituted OH-Ap) and related calcium phosphates, dicalcium phosphate dihydrate (DCPD), octacalcium phosphate (OCP), tricalcium phosphate (TCP), and amorphous calcium phosphate (ACP). The processes involved are (1) hydrolysis; (2) hydrolysis in the presence of Mg^{2+}; (3) dissolution and reprecipitation; (4) no reaction.[7]

calcium phosphates form under specific conditions of pH, temperature, and solution composition, but can transform to one another through the processes of dissolution, reprecipitation or hydrolysis at higher pH, and/or upon the introduction of other ions such as F^-, Mg^{2+}, and CO_3^{2-}.[7,43,92,95,96,126] Such interrelationship among several calcium phosphates related to apatite is shown in Figure 5. Several calcium phosphates can co-exist in synthetic systems under conditions of fluctuating pH (i.e., unbuffered) and/or varying solution composition,[7,43,92,95,96] e.g., co-existence of OCP and DCPD in gel systems.

DCPD, $CaHPO_4$•$2H_2O$

Suspension of human teeth in acidic phosphate solution resulted in the growth of DCPD crystals on the tooth surfaces,[95] the size and abundance of the crystals depending on the pH: the lower the pH, the *larger* the crystals. Formation of DCPD crystals were also observed in powdered human enamel treated with phosphoric acid.[133] The partial dissolution of the tooth mineral (apatite) at pH 4.5 caused the release of Ca^{2+}, HPO_4^{2-}, and PO_4^{3-} ions in the solution which subsequently cause the crystallization of DCPD, which is the more stable calcium phosphate form at this low pH. An increase in pH can cause the hydrolysis of DCPD to OCP or to apatite; introduction of F^- ions will cause the hydrolysis of DCPD to F-containing apatites even at this low pH. The morphology of the DCPD crystals is influenced by the media (solution or gel), pH, and presence of other ions.[7,92,95,96] For example, the presence of pyrophosphate ions causes the growth of DCPD crystals into spherules, similar in morphology to OCP, instead of the usual tabular or plate-like morphology.[7,92,95] DCPD is also known as brushite.

OCP, $Ca_8H_2(PO_4)_6$•$5H_2O$

The similarity in structure and in composition between octacalcium phosphate, OCP, and OH-apatite makes OCP the most likely candidate for apatite precursor.[86,126,128] under conditions where apatite must form by way of a calcium phosphate precursor. The OCP precursor model or Brown's model appears to account for the morphology (ribbon-like) of the initially formed human apatite crystallites and the presence of "lattice H_2O" and/or HPO_4^{2-} and assumes that the incorporation of "impurity" ions (e.g., F^-, CO_3^{2-}, and Mg^{2+}) in the apatite occurs during the hydrolysis of OCP to apatite.[86]

TCP, $Ca_3(PO_4)_2$

Biological whitlockites or TCP is always Mg-substituted.[6,7,43,78,79] In synthetic aqueous systems, pure TCP does not form and can only be obtained by solid state reactions at high (700 to 1000°C) temperatures. The presence of Mg^{2+} facilitates the formation of TCP from aqueous systems at low temperatures, 25 to 100°C.[6,7,43,55,78,79,134] Pure TCP can hydrolyze to OH-apatite, however, Mg-substituted TCP resists hydrolysis to apatite.[43,77]

ACP, Amorphous Calcium Phosphates

Under conditions of high supersaturation, the formation of apatite in synthetic systems were shown to proceed by an amorphous calcium phosphate precursor.[135-137] Carbonate-containing ACP were shown to hydrolyze to carbonate-containing apatite,[43,55,84] while Mg-containing ACP were shown to be indefinitely stable.[6,7,43,55,79] Because of the unknown contributions of several "impurities" on the formation and stability of ACP, the role of ACP in the formation of enamel apatite cannot be definitely established.

REFERENCES

1. **Miles, A. E. W., Ed.,** *Structural and Chemical Organization of Teeth,* Vol. 1 and 2, Academic Press, New York, 1967.
2. **Stack, M. V. and Fearnhead, R. W. Eds.,** *Tooth Enamel I,* John Wright and Sons, Bristol, 1965.
3. **Fearnhead, R. W. and Stack, M. V. Eds.,** *Tooth Enamel II,* John Wright and Sons, Bristol, 1971.
4. **Nylen, M. U., Termine, J. D. Eds.,** Tooth Enamel III, *J. Dent. Res.,* 58B, 684, 1979.
5. **Elliott, J. C.,** The problems of the composition and structure of the mineral components of hard tissues, *Clin. Orthoped.,* 93, 313, 1973.
6. **LeGeros, R. Z.,** Apatites in biological systems, in *Progress in Crystal Growth and Characterization,* Vol. 4, Pamplin, B., Ed., Pergamon Press, Elmsford, N.Y., 1981, 1.
7. **LeGeros, R. Z.,** Phosphate minerals in human tissues, in *Phosphate Minerals,* Nriagu, J. R. and Moore, R., Eds., University of Chicago Press, in press.
8. **McConnell, D.,** *Apatite. Its Crystal Chemistry, Mineralogy, Utilization and Biologic Occurrences,* Springer-Verlag, New York, 1973.
9. **Eanes, E. D.,** Enamel apatite: chemistry, structure and properties, *J. Dent. Res.,* 58B, 829, 1979.
10. **Gross, R.,** Die kristalline struktur von Dentine und Zahnschmelz, *Fetschr. Zahn. Inst. Univ. Greifswald,* 59, 1926.
11. **de Jong, W. F.,** La substance minerale dans les os, *Rec. Trav. Chim.,* 45, 445, 1926.
12. **Trautz, O. R.,** The crystalline organization of dental mineral, in *Structural and Chemical Organization of Teeth,* Miles, A. E. W., Ed., Academic Press, New York, 1967, 165.
13. **Young, R. A.,** Biological apatites vs. hydroxyapatite at the atomic level, *Clin. Orthop.,* 113, 249, 1975.
14. **Young, R. A. and Mackie, P. E.,** Crystallography of human tooth enamel: initial structure refinement, *Mater. Res. Bull.,* 15, 17, 1980.
15. **Zipkin, I.,** The inorganic composition of bones and teeth, in *Biological Calcification,* Schraer, H., Ed., Appleton-Century-Crofts, New York, 1970, 69.
16. **Kerebel, B., Daculsi, G., and Vervaere, A.,** High resolution electron microscopy and crystallography study of some biological apatites, *J. Ultrastruct. Res.,* 57, 266, 1976.
17. **Deutsch, D. and Pe'er, E.,** Development of enamel in human fetal teeth, *J. Dent. Res.,* 61, 1543, 1982.
18. **Kerebel, B., Daculsi, G., and Kerebel, L. M.,** Ultrastructural studies of enamel crystallites, *J. Dent. Res.,* 58B, 844, 1979.
19. **Angmar-Mansson, B.,** A quantitative microradiographic study on the organic matrix of developing human enamel in relation to the mineral content, *Arch. Oral Biol.,* 16, 83, 1971.
20. **Eastoe, J. E.,** Enamel protein chemistry — past, present, future, *J. Dent. Res.,* 58B, 753, 1979.
21. **Decker, J. D.,** Fixation effects on the fine structure of enamel crystal-matrix relationship, *J. Ultrastruct. Res.,* 44, 58, 1973.

22. **Orams, H. J., Zybert, J. J., Phakey, P. P., and Rachinger, W. A.,** Ultrastructural study of human dental enamel using selected-area argon-ion-beam thinning, *Arch. Oral Biol.*, 21, 663, 1976.
23. **Bhaskar, S. N., Ed.,** *Orban's Oral Histology and Embryology*, C. V. Mosby, St. Louis, 1976, 45.
24. **Osborn, J. W.,** Variations in structure and development of enamel, *Oral Sci. Rev.*, 3, 3, 1973.
25. **Scott, D. B. and Nylen, M. U.,** Organic-inorganic interrelationships in enamel and dentine. A possible key to the mechanisms of caries, *Int. Dent. J.*, 12, 432, 1962.
26. **Skobe, Z.,** The secretory stage of amelogenesis in rat mandibular incisor teeth observed by SEM, *Calcif. Tissue Res.*, 21, 83, 1976.
27. **Travis, D. F. and Glimcher, M. J.,** The structure and organization of, and the relationship between the organic matrix and inorganic crystals of embryonic bovine enamel, *J. Cell Biol.*, 23, 447, 1964.
28. **Boyde, A.,** The structure of developing mammalian dental enamel, in *Tooth Enamel I*, John Wright and Sons, Bristol, 1965, 163.
29. **Fearnhead, R. W.,** Matrix-mineral relationships in enamel tissues, *J. Dent. Res.*, 58B, 909, 1979.
30. **Glimcher, M. J.,** Phosphopeptides of enamel matrix, *J. Dent. Res.*, 58, 753, 1979.
31. **Irving, J. T. and Wuthier, R. E.,** Histochemistry and biochemistry of calcification with special reference to the role of lipids, *Clin. Orthop.*, 56, 237, 1968.
32. **Bonucci, E.,** The organic-inorganic relationships in calcified organic matrices, in *Physico-Chimie et Cristallographie des Apatites d'interet Biologique*, CNRS No. 230, 1975, 231.
33. **Butler, W. J., Hallt, W. T. L., Boles, J. M.,** The phospho-proteins of teeth and bones, *J. Dent. Res.*, 52A, 133, 1973.
34. **Nylen, M. U.,** Matrix-mineral relationships — a morphologist's viewpoints, *J. Dent. Res.*, 58B, 922, 1979.
35. **Nylen, M. U., Eanes, E. D., and Omnell, K. A.,** Crystal growth in rat enamel, *J. Cell. Biol.*, 18, 109, 1963.
36. **Daculsi, G. and Kerebel, B.,** High resolution electron microscope study of human enamel crystallites: size, shape and growth, *J. Ultrastruct. Res.*, 65, 163, 1978.
37. **Marsh, M. G., Hamilton, G., and Sass, R.,** The crystal sheaths from bivalve hinge ligaments, *Calcif. Tissue Res.*, 25, 45, 1978.
38. **Griebstein, W. J.,** in, *Tooth Enamel I*, Fearnhead, R. W. and Stack, M. V., Eds., John Wright and Sons, Bristol, 1965, 190.
39. **Jongebloed, W. L., Molenaar, I., and Arends, J.,** Morphology and size-distribution of sound and acid-treated enamel crystallites, *Calcif. Tissue Res.*, 19, 109, 1975.
40. **Johnson, N. W., Poole, D. F. G., and Tyler, J. E.,** Factors affecting the differential dissolution of human enamel in acid and EDTA. A scanning electron microscope study, *Arch. Oral Biol.*, 16, 385, 1971.
41. **Silverstone, L. M.,** Structure of carious enamel, including the early lesion, *Oral Sci. Rev.*, 3, 100, 1973.
42. **Beevers, C. A. and MacIntyre, D. B.,** The atomic structure of fluor-apatite and its relation to that of tooth and bone minerals, *Mineral. Mag.*, 27, 254, 1947.
43. **LeGeros, R. Z.,** Crystallographic Studies of the Carbonate Substitution in the Apatite Structure, Ph.D. thesis, New York University, 1967.
44. **LeGeros, R. Z.,** Apatites from aqueous and non-aqueous systems; relation to biological apatites, in *Proc. 1st Int. Congr. Phosphorus Compounds*, IMPHOS, Paris, 1977, 347.
45. **Mehmel, M.,** Uber die Struktur des Apatit, *Z. Kristallogr.*, 75, 323, 1930.
46. **Naray-Szabo, S.,** The structure of apatite $(CaF)Ca_4(PO_4)_3$, *Z. Kristallogr.*, 75, 387, 1930.
47. **Kay, M. T., Young, R. A., and Posner, A. S.,** Crystal structure of hydroxyapatite, *Nature (London)*, 204, 1051, 1964.
48. **Young, R. A. and Elliott, J. C.,** Atomic scale bases for several properties of apatites, *Arch. Oral Biol.*, 11, 699, 1966.
49. **Elliott, J. C. and Mackie, P. E.,** Monoclinic hydroxyapatite in, *Physico-Chimie et Cristallographie des Apatites d'interet Biologique*, CNRS No. 230, Centre National de la Recherche Scientifique, Paris, 1975, 69.
50. **Young, R. A.,** Some aspects of crystal structural modeling of biological apatites, in *Physico-chimie et Cristallographie des Apatites d'interet Biologique*, CNRS No. 230, Centre National de la Recherche Scientifique, Paris, 1975, 21.
51. **Montel, G.,** Conceptions actuelles sur la structure et la constitution des apatites synthetiques comparables aux apatites biologiques, in *Physico-chimie et Cristallographie des Apatites d'interet Biologiques*, CNRS No. 230, Centre National de la Recherche Scientifique, Paris, 1975, 13.
52. **Klug, H. P. and Alexander, L. E.,** *X-ray Diffraction Procedures for Polycrystalline and Amorphous Materials*, 2nd ed., John Wiley & Sons, New York, 1974.
53. **LeGeros, R. Z., Trautz, O. R., LeGeros, J. P., and Shirra, W. P.,** Apatite crystallites: effects of carbonate on morphology, *Science*, 155, 1409, 1967.

54. **LeGeros, R. Z., LeGeros, J. P., Trautz, O. R., and Shirra, W. P.,** Conversion of monetite, CaHPO4, to apatites: effect of carbonate on the crystallinity and morphology of apatite crystallites, in *Advances in X-ray Analysis*, Barrett, C. S., Newkirk, J. B., and Rudd, C. O., Eds., 1971, 57.
55. **LeGeros, R. Z., Shirra, W. P., Miravite, M., LeGeros, J. P.,** Amorphous calcium phosphates: synthetic and biological, *Physico-chimie et Cristallographie des Apatites d'interet Biologique*, CNRS No. 230, 1975, 105.
56. **LeGeros, R. Z., Shirra, W. P., and LeGeros, J. P.,** Apatite crystallites: effect of fluoride on morphology, *J. Dent. Res.*, 52, 170, 1973.
57. **LeGeros, R. Z. and Suga, S.,** The crystallographic nature of fluoride in fish enameloids, *Calcif. Tissue Int.*, 32, 169, 1980.
58. **LeGeros, R. Z. and Tung, M.,** Factors affecting fluoride incorporation in apatites, *J. Dent. Res.*, 59B, 973, 1980.
59. **Blumenthal, N. C., Betts, F., and Posner, A. S.,** Effect of carbonate and biological macromolecules on the formation and properties of hydroxyapatite, *Calcif. Tissue Res.*, 18, 81, 1975.
60. **Herzberg, G.,** *Infrared and Raman Spectra of Polyatomic Molecules*, von Nostrand, New York, 1945.
61. **Arends, J. and Jongebloed, W. L.,** Ultrastructural studies of synthetic apatite crystals, *J. Dent. Res.*, 58B, 837, 1979.
62. **Dykes, E. and Elliott, J. C.,** The occurrence of chloride ions in the apatite lattice of Holly Springs hydroxyapatite and dental enamel, *Calcif. Tissue Res.*, 7, 241, 1971.
63. **Elliott, J. C.,** The Crystallographic Structure of Dental Enamel and Related Apatites, Ph.D. thesis, University of London, London, 1964.
64. **Bonel, G. and Montel, G.,** Sur une nouvelle apatite carbonatee synthetique, *C. R. Acad. Sci.*, 258, 923, 1964.
65. **LeGeros, R. Z., LeGeros, J. P., and Trautz, O. R.,** Two types of carbonate substitution in the apatite structure, *Experientia*, 24, 5, 1969.
66. **LeGeros, R. Z.,** The unit-cell dimensions of human enamel apatite: effect of chloride incorporation, *Arch. Oral Biol.*, 20, 63, 1974.
67. **LeGeros, R. Z., Quirolgico, G., and LeGeros, J. P.,** Trace elements: incorporation of monovalent cations, Na^+,K^+,Li^+, *J. Dent. Res.*, 60B, 452, 1981.
68. **Simpson, D.,** The nature of alkali carbonate apatites, *Am. Mineral.*, 49, 363, 1964.
69. **Collins, R. L.,** Strontium-calcium hydroxyapatite solid solutions: preparations and lattice constant measurements, *J. Am. Chem. Soc.*, 82, 5067, 1960.
70. **LeGeros, R. Z., Quirogico, G., and LeGeros, J. P.,** Incorporation of strontium in apatite: effect of pH and temperature, *J. Cent. Res.*, 58A, 168, 1979.
71. **LeGeros, R. Z., Taheri, M. H., and LeGeros, J. P.,** Formation and stability of apatites: effects of some cationic substituents, in *Proc. 2nd Int. Congr. Phophorus Compounds*, IMPHOS, Paris, 1980, 89.
72. **Hiejligers, H. J. M., Driessens, F. C. M., and Verbeeck, R. M. H.,** Lattice parameters and cation distribution of solid solutions of calcium and strontium hydroxyapatites, *Calcif. Tissue Int.*, 29, 127, 1979.
73. **Mohseni-Koutchesfehani, S.,** Contribution a l'etude des apatites barytiques, Thesis, Universite de Paris, 1961.
74. **Muller, M.,** Die Fallund und die Rongernographische Untersuchung des Mischkristallsystemen, $Ca_{10}(PO_4)_6(OH)_2$ — $Pb_{10}(PO_4)_6(OH)_2$, *Helv. Chim. Acta*, 38, 2065, 1947.
75. **LeGeros, R. Z., Quirolgico, G., and LeGeros, J. P.,** Consequences of strontium incorporation in apatite, *J. Dent. Res.*, 58, 169, 1979.
76. **Trautz, O. R. and LeGeros, R. Z.,** Effect of magnesium on various calcium phosphates, *Proc. Gen. Meet. Int. Assoc. for Dental Research*, Abstr. #101, 1962.
77. **LeGeros, R. Z., Miravite, M. A., and Klein, E.,** Biological whitelockites, its formation and occurrences, *Proc. Gen. Meet. Int. Assoc. for Dental Research, J. Dent. Res.*, 53 (Abstr. 580), 199, 1974.
78. **LeGeros, R. Z.,** Variations in the crystalline components of human dental calculus. I. Crystallographic and spectroscopic analysis, *J. Dent. Res.*, 53, 45, 1974.
79. **LeGeros, R. Z., Contiguglia, S. R., Alfrey, A. C.,** Pathological calcifications associated with uremia: two types of calcium phosphate deposits, *Calcif. Tissue Res.*, 13, 173, 1973.
80. **Boskey, A. L. and Posner, A. S.,** Mg stabilization of amorphous calcium phosphates: a kinetic study, *Mater. Res. Bull.*, 9, 907, 1974.
81. **Gee, A. and Deitz, F. R.,** Pyrophosphate formation upon ignition of precipitated basic calcium phosphates, *J. Am. Chem. Soc.*, 77, 2692, 1965.
82. **LeGeros, R. Z., LeGeros, R., and Bonel, G.,** Types of "H_2O" in human enamel and in precipitated apatites, *Calcif. Tissue Res.*, 26, 111, 1978.
83. **Heughbaert, J. C.,** Contribution a l'Etude de l'Evolution des Orthophosphates de Calcium Precipites en Orthophosphates Apatitiques, Ph.D. These, L'Institute National Polytechnique de Toulouse, 1977.

84. **Greenfield, D. J., Termine, J. D., and Eanes, E. D.,** A chemical study of apatites prepared by hydrolysis of amorphous calcium phosphate in carbonate containing aqueous solutions, *Calcif. Tissue Res.*, 14, 131, 1974.
85. **Berend, G. and Hageus, A. J.,** Thermo-analytische untersuchung von sekundarem Calciumphosphate-dihydrat, *Thermochim. Acta,* 11, 367, 1975.
86. **Chickerur, N. S., Tung, M. S., and Browne, W. E.,** A mechanism for incorporation of carbonate into apatite, *Calcif. Tissue Int.*, 32, 55, 1980.
87. **LeGeros, R. Z., Vandemaele, K. H., Go, P., Quirolgico, G. B., and Legeros, D. J.,** Transformation of calcium carbonates and calcium phosphates to carbonate apatites: possible mechanisms for phosphorite formation, in *Proc. 2nd Int. Congr. Phosphorus Compounds,* IMPHOS, Paris, 1980, 41.
88. **LeGeros, R. Z., Miravite, M. A., and Bonel, G.,** CO_3 in apatites: effect of other ions on CO_3 incorporation, *J. Dent. Res.*, 55B(Abstr. 955), 300, 1976.
89. **LeGeros, R. Z.,** Consequences of fluoride incorporation on some properties of apatites, Int. Assoc. for Dental Research (IADR) Symposium, Osaka, Japan, 1980.
90. **LeGeros, R. Z.,** Contribution of fluoride to the stability and stoichiometry of apatites, *J. Dent. Res.*, 60A(Abstr. 279), 380, 1981.
91. **Ingram, G. S.,** The reaction of monofluorophosphate with apatite, *Caries Res.*, 7, 217, 1972.
92. **LeGeros, R. Z., Miravite, M. A., Quirolgico, G., and Morales, P.,** Factors influencing crystal growth and morphology of urinary stone crystallites, in *Colloquium on Renal Lithiasis,* Finlayson, B., Thomas, W. C., Jr., Eds., University of Florida Press, Gainesville, 1976, 187.
93. **Fleisch, H. and Russell, R. G. G.,** A review of the physiological and pharmacological effects of pyrophosphates on bones and teeth, *J. Dent. Res.*, 51, 324, 1972.
94. **Fleisch, H.,** Physico-chemical and biological effects of pyrophosphate and diphosphonates, *Physico-chimie et Cristallographie des Apatites d'interet Biologique,* CNRS No. 230, Centre National de la Recherche Scientifique, Paris, 1975, 263.
95. **LeGeros, R. Z. and LeGeros, J. P.,** Brushite crystals grown by diffusion, *J. Cryst. Growth,* 13, 476, 1972.
96. **LeGeros, R. Z. and Morales, P.,** Renal stone crystals grown in gel systems, *J. Invest. Urol.*, 11, 12, 1973.
97. **Moreno, E. C., Kresak, M., and Zahradnik, R. T.,** Physico-chemical aspects of fluoride-apatite systems relevant to the study of dental caries, *Caries Res.*, 11, 142, 1977.
98. **LeGeros, R. Z. and Tung, M. S.,** Dissolution properties of CO_3-containing apatites, *J. Dent. Res.*, 61, 232, 1982 (Abstr.); *Caries Res.*, in press.
99. **LeGeros, R. Z., Kerebel, L. M. and Silverstone, L.,** Comparative properties of fluoridated and non-fluoridated biological and synthetic apatites, *J. Dent. Res.*, 59A, 523, 1980.
100. **LeGeros, R. Z., Singer, L., Ophaug, R. H., Thein, A., and LeGeros, J. P.,** The effect of fluoride on the stability of synthetic and biological (bone mineral) apatites, in *Proc. Int. Symp. Osteoporosis,* Jerusalem, in press.
101. **Arends, J. and Jongbloed, W. L.,** Crystallite dimensions of enamel, *J. Biol. Bucc.*, 6, 161, 1976.
102. **Frazier, P. D.,** Adult human enamel: an electron microscopic study of crystallite size and morphology, *J. Ultrastruct. Res.*, 22, 1, 1968.
103. **Scott, D. B., Simmelink, J. W., and Nygaard, V.,** Structural aspect of dental caries, *J. Dent. Res.*, 53, 165, 1974.
104. **Selvig, K. A.,** Periodic lattice images of hydroxyapatite crystals in human bone and dental hard tissues, *Calcif. Tissue Res.*, 6, 227, 1970.
105. **Voegel, J. C. and Frank, R. M.,** Stages in the dissolution of human enamel crystals in dental caries, *Calcif. Tissue Res.*, 24, 19, 1977.
106. **Glick, P. L.,** Patterns of enamel maturation, *J. Dent. Res.*, 58B, 883, 1979.
107. **Weatherell, J. A., Robinson, C., and Hallsworth, A. S.,** Variations in the chemical composition of human enamel, *J. Dent. Res.*, 53, 180, 1974.
108. **LeGeros, R. Z., LeGeros, J. P., Trautz, O. R., and Klein, E.,** Spectral properties of carbonate in carbonate-containing apatites, *Dev. Appl. Spec.*, 7, 3, 1970.
109. **Doi, Y., Aoba, Y., Okazaki, M., Takahashi, J., and Moriwaki, Y.,** Analysis of paramagnetic centers in x-ray irradiated enamel, bone, and carbonate-containing hydroxyapatite by electron spin resonance spectroscopy, *Calcif. Tissue Int.*, 28, 107, 1979.
110. **Boyde, A.,** Carbonate concentration, crystal center, core dissolution, caries cross striations, circadian rhythms and compositional contrast in SEM, *J. Dent. Res.*, 58B, 981, 1979.
111. **LeGeros, R. Z., Trautz, O. R., LeGeros, J. P., and Klein, E.,** Pyrolysis of biological apatites: x-ray diffraction and infrared studies, *Proc. 48th General Assembly of the Int. Assoc. for Dental Research,* Abstr. 177, 1970.
112. **LeGeros, R. Z., Piliero, J. A., Go, P., and Pentel, L.,** Thermochemical properties of enamel, dentine and bone, *J. Dent. Res.*, 60A, Abstr. 720, 4, 1981.

113. **Holcomb, D. W. and Young, R. A.**, Thermal decomposition of human tooth enamel, *Calcif. Tissue Int.*, 31, 189, 1980.
114. **Baud, C. A. and Very, J. M.**, Ionic substitutions *in vivo* in bone and tooth apatite crystals, in *Physico-chimie et Cristallographie des Apatites d'interet Biologique*, CNRS No. 230, Montel, B., Ed., Centre de la Recherche Scientifique, Paris, 1975, 405.
115. **LeGeros, R. Z., Miravite, M. A., Quirolgico, G. N., and Curzon, M. E. J.**, The effect of some trace elements on the lattice parameters of human enamel and synthetic apatites, *Calcif. Tissue Res.*, Suppl. 22, 111, 1977.
116. **LeGeros, R. Z., Kazimiroff, J., Go, P.**, Characterization of human enamel "white spots", *J. Dent. Res.*, 58A, 137, 1979.
117. **Arends, J. and Davidson, C. L.**, HPO_4^{2-} content in enamel and artificial caries lesions, *Calcif. Tissue Res.*, 18, 65, 1975.
118. **Hallsworth, A. S., Weatherell, J. A., and Robinson, C.**, Loss of carbonate during the first stages of enamel caries, *Caries Res.*, 7, 345, 1973.
119. **LeGeros, R. Z., Shirra, W. P., and Pentel, L.**, Thermal stability of human enamel, *Caries Res.*, 13, 96, 1979.
120. **Davidson, C. L. and Arends, J.**, Thermal analysis studies on sound and artificially decalcified tooth enamel, *Caries. Res.*, 11, 313, 1977.
121. **Brown, W. E. and Konig, K. G., Eds.**, Cariostatic mechanisms of fluoride *Caries Res.*, 11 (Suppl. 1), 1, 1977.
122. **Curzon, M. E. J. and Spector, P. C.**, An association between strontium in drinking water supplies and low caries prevalence in man, *Arch. Oral Biol.*, 23, 317, 1978.
123. **Curzon, M. E. J., Spectro, P. C., Losee, F. L., and Crocker, D. C.**, Dental caries related to Cd and Pb in whole human dental enamel, in *Trace Substances in Environmental Health*, Hemphill, D. D., Ed., University of Missouri Press, Columbia, 1977, 23.
124. **Cuzress, T. W.**, The inorganic composition and solubility of dental enamel from several specified population groups, *Arch. Oral Biol.*, 17, 93, 1977.
125. **Lappalainen, R. and Knuutila, M.**, The distribution and accummulation of Cd, Zn, Pb, Cu, Co, Ni, Mn and K in human teeth from five different geological areas of Finland, *Arch. Oral Biol.*, 24, 363, 1979.
126. **Brown, W. E., Smith, J. P., Lehr, J. R., and Frazier, A. W.**, Crystallographic and chemical relations between octacalcium phosphate and hydroxyapatite, *Nature (London)*, 196, 1048, 1962.
127. **Fowler, B. O., Moreno, E. C., and Brown, W. E.**, Infrared spectra of hydroxyapatite, OCP and pyrolyzed OCP, *Arch. Oral Biol.*, 11, 477, 1966.
128. **Brown, W. E.**, in Nylen, M. U., and Termine, J. D. Eds., Tooth Enamel III, *J. Dent. Res.*, 58B, 857, 1979.
129. **Francis, M. D. and Webb, N. C.**, Hydroxyapatite formation from a hydrated calcium monohydrogen phosphate precursor, *Calcif. Tissue Res.*, 6, 335, 1971.
130. **Termine, J. D.**, Mineral chemistry and skeletal biology, *Clin. Orthop.*, 85, 207, 1972.
131. **Schroeder, H.**, *Formation and Inhibition of Dental Calculus*, Han Hubert, Vienna, 1969.
132. **Wetsbury, E. J.**, Some observations on the quantitative analysis of over 1000 urinary calculi, *Br. J. Urol.*, 46, 215, 1974.
133. **Chow, L. C. and Brown, W. E.**, Formation of $CaHPO_4 \cdot 2H_2O$ in tooth enamel as an intermediate product in topical fluoride treatments, *J. Dent. Res.*, 54, 65, 1975.
134. **Rowles, S. L.**, The precipitation of whitlockite from aqueous solutions, in *Colloque International sur les Phosphates Mineraux Solides, Bull Soc. Chim. Fr.*, 151, 1968.
135. **Eanes, E. D.**, Amorphous intermediates in the formation of biological apatites, *Physico-chimie et Cristallographie des Apatites d'interet Biologique*, CNRS No. 230, Centre de la Recherche Scientifique, Paris, 1975, 295.
136. **Furedi-Milhofer, H., Bilinski, H., Brecevic, L. J., and Despotovic, D.**, The formation of calcium phosphate precipitates metastable equilibria and kinetics, in *Physico-Chemie et Cristallographie des Apatites d'interet Biologique*, Centre de la Recherche Scientifique, 1975, 303.
137. **Posner, A. S., Blumenthal, N. C., and Betts, F.**, Formation and structure of chemically precipitated hydroxyapatites, in *Proc. 2nd Int. Congr. Phosphorus Compounds*, IMPHOS, Paris, 1980, 25.
138. **LeGeros, R. Z.**, unpublished, 1972.

Periodontium

PERIODONTAL LIGAMENT: METABOLISM

J. Sodek

INTRODUCTION

In discussing metabolic activity of a tissue it is convenient to subdivide the topic into metabolism of the cells contained within the tissue, the kinetics of the cellular components, and the metabolism of the extracellular matrix components. Information in each of the areas is limited for the periodontal ligament primarily because of the difficulties in studying such small amounts of tissue. However, as technologies are being refined and analytical techniques become more sensitive, knowledge in these areas is beginning to accumulate.

CELLULAR METABOLISM

Information on any aspect of cell metabolism of periodontal ligament cells is severely limited. Kawase et al.[1] have followed the metabolism of radiolabeled succinate in excised periodontal ligament and have shown its transfer into a number of amino acids that were incorporated into matrix macromolecules. Based on their results they anticipate active gluconeogenesis in ligament. Although these workers found differences with pulp and liver tissues, there is no reason to suspect that the metabolism of periodontal ligament fibroblasts is very different from other connective tissue fibroblasts, although characteristic differences related to the specialized function of these cells might be expected. Although biochemical studies on these cells in vivo are extremely difficult, in vitro systems have been described which are potentially useful for metabolic studies.

ORGAN CULTURE

Since a functional periodontal ligament exists only in fully erupted teeth that are normally in occlusion, only mature tissues can be studied. Melcher and Turnbull[2-4] have described an in vitro system for maintaining a single mouse molar and its associated periodontium in organ culture for several days using a modified New circulator. Although, this provides a viable system for in vitro studies of mature tissues, there are difficulties in regulating the oxygen tension and since the medium is continuously recycled, metabolites, some of which may be toxic, are accumulated. To avoid these problems the culture system has been modified to a continuous-flow culture system (CFCS)[5,6] to permit studies on the effect of oxygen tension[7] and ascorbic acid[8] on protein metabolism in the periodontal tissues. In both studies on protein metabolism, the cultures were found to be comparable with tissue metabolism in vivo.

A particular advantage of the CFCS for biochemical studies is that the levels of nutrients and hormones can be maintained accurately. Further, metabolites can be analyzed in samples of the medium effluent, although it should be appreciated that the metabolites are generated from other dental tissues including the alveolar bone as well as the periodontal ligament.

A particular problem in the organ culture system is the difficulty in providing a uniform distribution of nutrients throughout the tissues. The problem stems from the limited diffusion through large explants and is unlikely to be resolved without disrupting the periodontal ligament. It should also be recognized that since the ligament is not subject to occlusal forces in culture, some aspects of ligament metabolism may be impaired. Nevertheless, the culture system would seem to be valuable in obtaining basic biochemical data on metabolic processes in a ligament which is intact, if not fully functional.

TISSUE CULTURE

A number of attempts have been made to culture periodontal ligament in vitro.[9,10] In each case it was apparent that the tissue undergoes rapid degeneration. Although survival of some cells can occur, it is doubtful if these cells are behaving with any semblance of cells in vivo since the pattern of protein synthesis is dramatically altered.[10] This change may reflect the rapid proliferation of localized cells in the explanted tissue. To achieve success with such a tissue culture system seems highly unlikely because of the difficulty in avoiding trauma associated with the physical operations necessary to collect periodontal ligament. Further, the removal of the tissue from its functioning environment undoubtedly will have deleterious effects.

CELL CULTURE

Perhaps the most profitable approach to studying cellular metabolism in periodontal ligament is to isolate the cells from the tissue and grow the matrix-free cells in tissue culture. Cells can be obtained from periodontal ligament using either cell suspension or explant techniques which have been described by Brunette et al.[11] Briefly, periodontal ligament is scraped from the middle third of extracted teeth using a sharp scalpel and placed in a minimal essential medium (MEM) with antibiotics. Pieces of explant are placed in culture dishes and covered by a cover glass which is held in place by sterile silicon grease along two opposing edges. Culture medium containing 15% fetal bovine system (FBS) and antibiotics are then added. After 4 to 6 days, fibroblast cells can be seen around the explants. These cells form a monolayer after 20 days, at which point the cells can be subcultured.

In the cell suspension technique, an enzyme mixture containing collagenase, elastase, and DNAse in Kreb's buffer is used to release the cells from excised tissue. Following incubation at 37°C for 20 min, the cells are separated from tissue fragments by filtration through a 200-mesh stainless steel seive. An equal volume of FBS is added to inhibit the protease activity and the cells sedimented at 180 $\times$ g. The cells are resuspended in growth medium and filtered through a glass capillary array (35- to 50-μm pore size). These cells can be plated at approx 1×10^5 cells in 60-mm dishes and grown in α-MEM with 15% FBS.

Using these procedures, periodontal ligament cells have been successfully cultured from porcine molars,[11] monkey incisors,[12] and dog premolars.[13] Limited success has been achieved with rat ligament cells.[14]

Several factors have been found to be important for success in these culture systems. Tissue must be collected quickly from freshly extracted teeth, and, if teeth are collected post-mortem, extractions should be carried out as soon as possible. Trauma should be minimized during extraction; removal of attached gingiva will facilitate extraction. The incubation medium is also important. Generally minimal essential media are used and these are supplemented with 10 to 15% FBS. However, the detailed composition of the media may vary with the species from which fibroblasts are obtained.[15]

Studies on monkey incisor periodontal ligament fibroblasts have shown that these cells differ from skin fibroblasts from the same animal in some cultural characteristics, in their c-AMP content and response to prostaglandin E_1.[12] These observations are important because they justify further studies on ligament-derived cells. Had the ligament cells not shown these characteristic features it would have been questionable as to why ligament cells, with the inherent problems of obtaining these cells, should be studied at all.

In the original studies of Marmary et al.[12] it was observed that two distinct cell types generally grew out from ligament explants. The predominant cells were fibroblastic in appearance. However, epithelial cells were also recognized which were subsequently characterized and shown to be derived from the epithelial "rests of Malassez".[11] The "rest" cells are generally located in groups on the tooth side of the ligament and are believed to be remnants of Hertwig's epithelial root sheath. Their function in vivo is however, unknown. In mixed cultures, multilayers of epithelial and fibroblasts form. The sandwiching of epithelial cells between fibroblast layers mimics the in vivo relationship of these cells.[16]

Methods for separating the fibroblast and epithelial cells are available. Kanoza et al.[17] have grown mixed cells on collagen gels and have utilized the ability of the epithelial cells to protect the gel from digestion by bacterial collagenase. Under these conditions collagen is degraded under the fibroblasts and these cells are released, whereas the epithelial cells remain attached. The two cell types have also been separated by their differential susceptibility to trypsinization from plastic culture dishes[12] and the selective loss of attachment of fibroblasts following exposure to hormones.[18]

The heterogeneity of fibroblasts in connective tissues has been reported.[19-22] It is perhaps not surprising, therefore, to expect heterogeneity of cells in the periodontal ligament since cells in this tissue are responsible for maintaining the cementum and bone as well as the soft connective tissue of the ligament itself.[23] In addition, it might be expected that cells obtained from the ligament may be isolated at various stages of differentiation from less differentiated forms to fully differentiated cells. Cultures of fibroblasts from the periodontal ligament should, therefore, be considered to comprise a variety of cell populations and not as a homogeneous population of fibroblasts.

Evidence for ligament fibroblast heterogeneity has been observed by Birek[24] who has used a reverse hemolytic plaque assay (RHPA) to demonstrate variation in the synthesis of collagen by individual fibroblasts. Besides being able to analyze the synthetic capabilities of individual cells, this assay is not destructive so that following analysis the cells remain viable and available for subsequent studies. The RHPA can be applied to the analysis of any antigenic component secreted by the cells or on the cell surface provided that a specific antibody is available for the component.

Study of individual fibroblast populations can also be achieved by cloning techniques. Although, ideally, plating individual cells into separate microwell culture dishes is the best approach, practically it is laborious and tedious. A more useful approach is to prepare suspensions of cells at a known cell concentration and to plate microwells with 0.5 to 10 cells per well according to the plating efficiency of the cells. On a statistical basis less than one cell per well should attach and grow into a clone colony. Should more than one cell attach in an individual well, multiple colonies will be apparent and such cultures can be discarded. The success of these methods depends upon the culture conditions, conditioned culture media often being advantageous in early growth stages, and the proliferative capacity of the attached cells. In cells from mature tissues, the number of potential divisions of the isolated cells is much lower than for cells from embryonic tissue.[25] Since periodontal ligament is a mature tissue, cloning for large numbers of cells might be difficult. However, cloned populations of porcine and monkey ligament fibroblasts have been obtained, albeit in low numbers, in our laboratories. Heterogeneity has been observed between these cloned cells with respect to their growth characteristics, detailed morphology, and their synthesis of collagens.[14]

Cloned fibroblasts from periodontal ligament are particularly useful for studies of cellular metabolism. Large numbers of cells would not be required for metabolic studies in which radiolabeled substrates can be used to trace metabolic pathways. Further, any differences between cloned populations may be linked to the function or differentiated state of the cell.

Although in vitro methods provide a distinct advantage over in vivo experiments, the major disadvantages are that a functioning tissue is not being studied in vitro and in cell culture studies the natural environment of the cells, the matrix, has been stripped away. It is important, therefore, to interpret results obtained in vitro with caution. Wherever possible, studies should be compared in different in vitro systems and also, if possible, in vivo before attempting to extrapolate in vitro data to the in vivo situation. Nevertheless, valuable information can be obtained from cultured tissue and advantage can be taken of the relatively simple system that culture methods provide.

THE KINETICS OF CELLS IN THE PERIODONTAL LIGAMENT

The periodontal ligament is an interesting connective tissue in that it maintains, within narrow limits, a constant width between the calcified tissues of alveolar bone and cementum besides functioning in tooth support. Consequently, it is not surprising to find both formative and resorptive cells for the maintenance of all three tissues within the ligament itself.[26] Whereas the fibroblasts are prevalent throughout the ligament, osteoblasts and cementoblasts form pallisades along the bone and cementum surfaces, respectively. In addition, resorptive cells such as osteoclasts and cementoclasts for the calcified tissues, and fibroblasts and macrophages in the soft tissue are also present.

To determine the turnover of cells in tissues, measurements of proliferation rates have largely been made because determination of cell loss is difficult. Since mitotic figures are rare in connective tissues, cellular proliferation has generally been analyzed from the incorporation of tritiated thymidine (^{3}H-TdR) into dividing cells. From the studies reviewed by Cleaver[27], it is apparent that the necessary assumptions regarding the specificity of labeling and the stability of nuclear DNA are valid and that ^{3}H-TdR labeling combined with radioautography is sufficiently reliable to measure cellular proliferation.

Leblond et al.[28] have devised a system for the classification of cell populations. In the periodontal ligament, it appears that the cell number remains constant[75] and there is significant proliferation which would imply a situation of renewing cell populations compared to an expanding cell population where there is a continual increase in cell number.

Studies by Cameron,[29] in which a continuous labeling procedure was used, indicate that connective tissues comprise populations of slowly renewing cells. In this situation, turnover can be measured from analyses of the labeling index and the time required for DNA synthesis. The rate of cell formation is given by the expression:*

$$K_b = \frac{\text{L.I.}}{T_s}$$

where K_b is the cell birth rate, L.I. is the labeling index, and T_s is the time for DNA synthesis. In combination with the L.I., this suggests a K_b of 0.0001 to 0.003 cells per hour. Values of 1 to 3% have been obtained for the labeling index.[32-38] Variations are due to the inverse relationship of L.I. with age,[32,37] to regional variations,[38] to circadian periodicity,[39] and possibly to functional considerations.[34] Roberts et al.[35] have measured the time for DNA synthesis in orthodontically stimulated teeth to be 8.5 to 9.5 hr. If the ligament was composed of a homogeneous population of cells derived from a stem compartment, the turnover time (which is equivalent to the reciprocal of K_b) would, therefore, be in the range of 12 to 40 days. This contrasts with data that were obtained in mouse molar ligament by Cameron,[29] using continuous administration of ^{3}H-TdR. Extrapolating from Cameron's data, a value closer to 60 days can be calculated. Clearly the ligament is not composed of a single cell population and further detailed studies are necessary to obtain accurate data on cellular kinetics in the periodontal ligament.

Although a low proliferative rate is evident for periodontal ligament cells, a rapid increase in ^{3}H-TdR labeling is observed when orthodontic forces are applied to the tooth.[34] This has been observed in regions of the ligament under tension[36] and under pressure,[40] although in

* From Bertalanffy[30] and Lala et al.[31]

the latter instance it should be emphasized that differentiation between resident cells and monocyte-derived cells is difficult. Increased ^{3}H-TdR labeling has also been observed in ligament following surgical wounding.[41,42] In these studies, dividing cells were localized paravascularly and subsequently migrated towards the wound site with some of the migrating cells undergoing further division. At least two and possibly three populations of dividing cells were identified, one adjacent to bone, one in the body of the ligament, and the third adjacent to the cementum. If the dividing cells represent progenitor cells and the wound situation considered to be an exacerbation of normal cell turnover, these results would suggest three identifiable pools of progenitor cells which may serve to repopulate the osteoblast, fibroblast, and cementoblast populations.

Studies on ^{3}H-TdR labeled cells in the ligament of continuously erupting mouse incisors by Beertsen[43,44] have revealed two distinct groups of fibroblasts. Those in the bone-related half of the ligament are similar to the cells in ligaments associated with teeth of limited eruption in that the labeling index is low and labeling occurs throughout the length of the ligament. In the tooth-related half of the ligament, high labeling is observed at the formative end of the tooth and labeled cells appear to migrate with the eruption of the tooth thereby demonstrating a form of "life span" turnover.

It is evident that there are a number of considerations that must be satisfied before analyses on cellular turnover in a tissue are initiated. Some of these considerations that are pertinent to the periodontal ligament are listed below.

1. It is important that resident cells are distinguished from nonresident cells such as monocytes, which as precursors of macrophages and osteoclasts can strongly influence the labeling index in areas of matrix resorption. The use of thin (3-μm) sections can overcome this problem by allowing accurate identification of cells.[45] Gould et al.[42] have attempted to circumvent the problem of labeled monocytes by using radiation to lower the number of circulating monocytes prior to thymidine labeling.
2. In connective tissues, the low number of proliferating cells can lead to inaccurate values for labeling indexes. This can be remedied by continuous administration of ^{3}H-TdR by adding the isotopic precursor to the drinking water.[29] In this way, increasing numbers of cells will be labeled vs. time, thereby, increasing the accuracy of measurement.
3. The nature of cell turnover experiments promotes problems in statistical analysis since the most economical method is to examine large numbers of tissue sections and cells rather than large numbers of animals to obtain statistically acceptable data. Often this results in few animals being used in the studies. However, Chambers et al.[46] have suggested that experimental designs with more animals are preferable to increasing the number of sections analyzed per animal.
4. Cellular heterogeneity in the tissues is a problem that is difficult to circumvent. More work is needed to develop means of distinguishing between different subpopulations of fibroblast in vivo.
5. Variations of cellular turnover relating to location, function, species age, and diurnal rhythm can all contribute to variance in results obtained.

MATRIX METABOLISM

The metabolism of the matrix components in the periodontal ligament is an extension of the metabolism of those cells responsible for the synthesis and degradation of the matrix components. The properties of a connective tissue are a reflection of the nature and amounts of the matrix components. As with most connective tissues, the most abundant component in periodontal ligament is collagen which exists almost entirely in fiber forms. Other fibrous components may exist as elastin or oxytalan fibers. These fiber systems are embedded in a matrix of ground substance comprising glycoproteins of which the proteoglycans are prominent.

The observation that teeth became loose and are readily lost in conditions of avitaminosis C have suggested that collagen metabolism is an important aspect of ligament function.[47,48] The suggestion is based on the knowledge that vitamin C (ascorbic acid) is a necessary cofactor in the hydroxylation of collagen and is required for polysome formation in fibroblasts. Underhydroxylated collagen, which is formed when vitamin C levels are low, is unstable and poorly secreted by the fibroblasts so that an imbalance between synthesis and degradation can occur. Subsequently, numerous radioautographic studies have indicated that collagen turnover and remodeling is particularly high in the periodontal ligament.[49-56] Whereas the radioautographic studies do suggest a high metabolism in periodontal tissues, they cannot discriminate between the metabolism of the individual components. Even when radiolabeled proline is used to preferentially label collagen, it must be recognized that this precursor is not sufficiently specific. In some studies, bacterial collagenase has been used to selectively remove the collagenous components from the tissue sections. Although there is some merit in this approach, it is imperative that the enzyme preparation be highly specific and that noncollagenous proteins are not solubilized as a result of a close association to the degraded collagen.

Specific analysis of the metabolism of a particular matrix component requires the use of radiolabeled precursor molecules that are exclusive to that component. If the precursor is not sufficiently selective, the component must be purified before analysis. With respect to collagen, advantage can be taken of the essentially unique presence of hydroxyproline which is formed as a postribosomal modification of proline residues. By using radiolabeled proline as a precursor, specific radioactivities of hydroxyproline can be used as a measure of collagen metabolism without purification of the protein. Alternatively, other radiolabeled amino acids could be used and the collagen radioactivity measured by selectively degrading this protein with highly purified bacterial collagenase.[57] In this approach, more accurate results will be obtained if radiolabeled amino acids that are incorporated in high amounts into collagen, such as glycine, are used.

The major problem in measuring the turnover of proteins generally is the recycling and reutilization of the precursor when radiolabeled amino acids are used. The problem associated with recycling is discussed in detail in reviews by Waterlow et al.[58] and Zak et al.[59] Recycling is basically due to the slow excretion of amino acids by the kidney and the continual incorporation and release of amino acids from proteins being continually synthesized and degraded. This problem applies to both radioautographic and biochemical studies, but has been largely ignored, leading to large errors in the calculation of protein turnover values.

Most of the methodologies developed for circumventing the errors of precursor recycling cannot be applied to small tissues such as the periodontal ligament. However, the combination of careful dissection and microanalytical methods have provided valuable information on collagen turnover in the periodontal ligament.

By separating the newly synthesized collagen from the mature cross-linked collagen with extractions in 0.45 *M* sodium chloride, Sodek[60,61] has followed the incorporation of ^{3}H-proline into both of these pools. Samples are hydrolyzed in 6 *N* HCl under standard conditions to degrade the proteins to amino acids and a Beckman 121 *M* amino acid analyzer is used to separate the amino acids for colorimetric and radioactivity analyses. From these analyses, the specific radioactivity of hydroxyproline was obtained. The specific radioactivity of hydroxyproline was shown to rise rapidly in the newly synthesized pool, which represents only a few percent of the total collagen, peak between 1 to 2 hr following i.p. injection of precursor, and rapidly fall thereafter. By measuring the rate of fall of isotope from this pool and comparing the rate of incorporation into the insoluble collagen over the same time period, it was found that collagen maturation in the molar ligament,[60] but not in the rodent incisor ligament,[62] is highly efficient. The half-life of the salt-extractable collagen in the molar ligament was calculated to be 40 min. From this value, the half-life of the insoluble pool was calculated to be approximately 1 day.[61] Subsequently, the half-life of rat incisor ligament was calculated to be 3 days. These half-lives are considerably shorter than values obtained radioautographically and also biochemically[63,64] using experimental systems where recycling of precursor is effective and was not compensated for.

In more recent studies,[74] half-lives for the molar and incisor ligaments of 5.5 and 9 days, respectively, were obtained by following the rate of loss of radiolabeled collagen from the mature collagen pool. However, after correcting for the precursor specific radioactivity as described by Poole,[65] these values reduce to half-lives approaching those values obtained from the earlier experimental system.[61,62]

The reason for the rapid turnover of collagen in the periodontal ligament is unknown. In the ligament of continuously erupting incisors, a rapid remodeling of collagen in the region of the intermediate plexus is predicted. This is supported by the observation of apparently higher collagen phagocytic activity in this region[66] in mouse incisors. The rapid remodeling of collagen fibers in the intermediate plexus is believed to permit the movement of the tooth and tooth-related ligament proximally, whereas the bone-related ligament remains stationary. It is conceivable that collagen in the tooth-related and bone-related ligament is relatively stable providing a structural support for the attachment of tooth to bone. However, since the tooth related ligament migrates with the tooth, it is likely to be subject to a ''life span'' turnover, being synthesized at the formative end of the tooth and degraded at the gingival margin. Extensive phagocytosis of collagen has been observed at the gingival margin by Beertsen and Everts.[66]

Since the rat molar is a tooth of limited eruption, the even higher turnover of collagen in this tissue is surprising. The high turnover has been suggested to be related to functional forces.[60] However, in hypofunctional teeth, protein[54,55] and collagen[67] turnover is increased; the half-life of collagen being estimated at 0.5 days compared to 1 day in ligament of control teeth. It is probable that the increased turnover is due to reactivated eruption[68] which appears to be accompanied by increased protein turnover across the width of the ligament.[54,55] Since there is no evidence of an intermediate plexus in the molar ligament, it is likely that the increased collagen turnover occurs throughout the ligament which would explain the reduced tooth support observed by Kanoza et al.[67]

Since collagen metabolism appears to be high several weeks following reactivated eruption in rats,[74] the rapid turnover of collagen would not appear to be related to function. If the rapid turnover is an inherent property of the cells of the periodontal ligament, it is not apparent in cell culture experiments since collagen synthesis by monkey ligament cells was not significantly different from that in skin fibroblasts.[69] However, the cell density and higher vascularity of the ligament may in part account for increased collagen remodeling in this tissue. Another consideration in ligament remodeling is the physiological drift that some teeth undergo throughout life. The importance of drift with respect to the rate of matrix turnover remains to be evaluated.

In contrast to collagen, the metabolism of the other matrix components has received little attention. Although there are difficulties with small amounts of material, technology is already available for studying proteoglycans. The glycosaminoglycan side chains can be labeled with ^{3}H-glucosamine (or ^{3}H-acetate). The metabolism of the individual glycosaminoglycan types can be assessed either by digestion with chondroitinase A B C or hyaluronidase.[70] Alternatively, the GAG chains can be separated by electrophoresis on acetate strips[71] and the appropriate bands can be cut out and the radioactivity measured. Although recycling problems would be anticipated with glucosamine and acetate, this should not be a problem with radiolabeled sulfate ($^{35}SO_4$) which is rapidly cleared from the tissue fluids.

This precursor is useful for measuring the metabolism of sulfated GAGs. From the study of proline incorporation into noncollagenous proteins of the rat molar ligament, it would appear that the turnover of the structural noncollagenous proteins generally is similar to collagen turnover.[61]

The rapid remodeling of collagen in the periodontal ligament allows more detailed analysis of collagen synthesis and maturation. Using ^{14}C-glycine as a radiolabeled precursor, newly synthesized collagen and mature collagens have been analyzed over a 6-hr period using a combination of sodium dodecyl sulfate-polyacrylamide gel electrophoresis (SDS-PAGE) and fluorography to separate and quantitate type I and III procollagens and collagens.[72,73] It has been shown that conversion of type I procollagen to collagen α chains is rapid and proceeds largely by the successive removal of the amino-terminal propeptides followed by the carboxyl-terminal propeptides. However, it was apparent that type III procollagen was converted into a procollagen intermediate form (p α-III) which was relatively stable. Similar observations were made in rat molar and incisor ligaments. Following pepsin digestion, the salt-extractable collagens are converted to α-chains from which the relative amounts of type I and III collagen can be determined. A significantly higher proportion of type III collagen was found in incisor ligament (16 to 20%) compared to molar ligament (14 to 15%), however, the proportion was constant for each over the study period, indicating a similar rate of synthesis for the two collagen types. A lower level of type V collagen synthesis was also observed.

The salt-insoluble collagen, solubilized by pepsin digestion, was also analyzed on SDS-PAGE using Coomassie® blue to locate and quantitate the collagen α chains and fluorography to measure the incorporation of labeled collagen. A rapid increase in specific radioactivity of both type I and III collagen α chains was observed, demonstrating a similar turnover rate for the two collagen types. Further, the rapid increase in type III collagen radioactivity indicates that the p α-III intermediate must be incorporated into the mature collagen fraction. A similar conversion pattern for the different procollagens was observed in vitro for monkey periodontal ligament fibroblast.

In summary, knowledge of the metabolism of periodontal ligament is limited. The use of culture techniques provides a valuable system to study metabolic processes within the cells and to some degree secreted proteins. Cell kinetics, which is a particularly interesting problem in the ligament, is difficult to study and necessitates in vivo analysis. The high rate of matrix protein turnover is a unique feature of the periodontal ligament and provides a unique opportunity to study the detailed metabolism of these proteins in vivo.

ACKNOWLEDGMENT

I would like to thank Dr. D. M. Brunette for his advice in the preparation of this review on ligament metabolism.

REFERENCES

1. **Kawase, T., Enomoto, Y., Yokomizo, S., and Saito, S.,** In vitro study on succinate metabolism of bovine periodontal ligament: proteoglycan biosynthesis, *J. Dent. Res. Special Issue A*, 60, 343, 1981.
2. **Melcher, A. H. and Turnbull, R. S.,** Organ culture in studies on the periodontium, in *Organ Culture in Biomedical Research*, Balls, M. and Monnickendam, M. A., Eds., Cambridge University Press, Cambridge, 1976, 149.
3. **Turnbull, R. S. and Melcher, A. H.,** Method for maintaining large explants of mature mixed tissue in organ culture: molar tooth and supporting structure from the mouse, *In Vitro*, 10, 6, 1974.

4. **Melcher, A. H., Turnbull, R. S., and Hiddleston, W. J.,** In-vitro culture of periodontium of young adult mouse, *Arch. Oral Biol.*, 18, 933, 1973.
5. **Yen, E. H. K. and Melcher, A. H.,** A continuous flow culture system for organ culture of large explants of adult tissue: effect of oxygen tension on mouse molar periodontium, *In Vitro*, 14, 811, 1978.
6. **Feng, J., Melcher, A. H., Brunette, D. M., and Moe, H. K.,** Determination of L-ascorbic acid levels in culture medium: concentrations in commercial media and maintenance of levels under conditions of organ culture, *In Vitro*, 13, 91, 1977.
7. **Yen, E. H. K., Sodek, J., and Melcher, A. H.,** The effect of oxygen partial pressure on protein synthesis and collagen hydroxylation by mature periodontal tissues maintained in organ culture, *Biochem. J.*, 178, 605, 1979.
8. **Sodek, J., Feng, J., Yen, E. H. K., and Melcher, A. H.,** The effect of ascorbic acid on protein synthesis and collagen hydroxylation in continuous flow organ cultures of adult mouse periodontal tissues, *Calcif. Tissue Int.*, 34, 408, 1982.
9. **Rossman, L. E., Rosenbloom, J., and Robinson, P.,** Biosynthesis of collagen in isolated periodontal ligament, *J. Dent. Res.*, 54, 1115, 1975.
10. **Sodek, J., Brunette, D. M., Feng, J., Heersche, J. N. M., Limeback, H. F., Melcher, A. H., and Ng, B.,** Collagen synthesis is a major component of protein synthesis in the periodontal ligament in various species, *Arch. Oral Biol.*, 22, 647, 1977.
11. **Brunette, D. M., Melcher, A. H., and Moe, H. K.,** Culture and origin of epithelial-like and fibroblast-like cells from porcine periodontal ligament explants and cell suspensions, *Arch. Oral Biol.*, 21, 393, 1976.
12. **Marmary, Y., Brunette, D. M., and Heersche, J. N. M.,** Differences *in vitro* between cells derived from periodontal ligament and skin of *Macaca irus*, *Arch. Oral Biol.*, 21, 709, 1976.
13. **Boyko, G. A., Melcher, A. H., and Brunette, D. M.,** Formation of a new periodontal ligament by periodontal ligament cells implanted *in vivo* after culture *in vitro*, *J. Periodontal Res.*, 16, 73, 1981.
14. **Limeback, H. F.,** Ph.D. thesis, University of Toronto, 1979.
15. **Barnes, D. and Sato, A.,** Methods for growth of cultured cells in serum-free medium, *Anal. Biochem.*, 102, 255, 1980.
16. **Brunette, D. M., Kanoza, R. J., Marmary, Y., Chan, J., and Melcher, A. H.,** Interactions between epithelial and fibroblast-like cells in cultures derived from monkey periodontal ligament, *J. Cell Sci.*, 27, 127, 1977.
17. **Kanoza, R. J. J., Brunette, D. M., Purdon, A. D., and Sodek, J.,** Isolation and identification of epithelial-like cells in culture by a collagenase-separation technique, *In Vitro*, 14, 746, 1978.
18. **Rao, L. G., Moe, H. K., and Heersche, J. N. M.,** In vitro culture of porcine periodontal ligament cells: response of fibroblast-like and epithelial-like cells to prostaglandin E_1, parathyroid hormone and calcitonin and separation of pure population of fibroblast-like cells, *Arch. Oral Biol.*, 23, 957, 1978.
19. **Martin, G. M., Sprague, C. A., Norwood, T. H., and Pendergrass, W. R.,** Clonal selection, attenuation, and differentiation in an *in vitro* model of hyperplasia, *Am. J. Pathol.*, 74, 137, 1974.
20. **Kaufman, M., Pinskey, L., Straisfeld, C., and Zilahi, B.,** Qualitative differences in testosterone metabolism as an indication of cellular heterogeneity in fibroblast monolayers derived from human preputial skin, *Exp. Cell Res.*, 96, 31, 1975.
21. **Ko, S. D., Page, R. C., and Narayanan, A. S.,** Fibroblast heterogeneity and prostaglandin regulation of subpopulations, *Proc. Natl. Acad. Sci. U.S.A.*, 74, 3429, 1977.
22. **Hurum, S., Sodek, J., and Aubin, J. E.,** Synthesis of collagen, collagenase and collagenase inhibitors by cloned human gingival fibroblasts, *Biochem. Biophys. Res. Commun.*, 107, 356, 1982.
23. **Melcher, A. H. and Eastoe, J. E.,** The connective tissue of the periodontium, in *Biology of the Periodontium*, Melcher, A. H. and Bowen, W. H., Eds., Academic Press, New York, 1969, 167.
24. **Birek, C.,** Ph.D. Thesis, University of Toronto, 1981.
25. **Hayflick, L.,** The limited *in vitro* lifetime of human diploid cell strains, *Exp. Cell Res.*, 37, 614, 1965.
26. **Melcher, A. H.,** On the repair potential of periodontal tissues, *J. Periodontol.*, 47, 256, 1976.
27. **Cleaver, J. E.,** *Thymidine Metabolism and Cell Kinetics*, North-Holland, Amsterdam, 1967.
28. **Leblond, C. P., Messier, B., and Kopriwa, B.,** Thymidine H^3 as a tool for the investigation of the renewal of cell populations, *Lab. Invest.*, 8, 296, 1959.
29. **Cameron, I. L.,** Cell renewal in the organs and tissues of non-growing adult mouse, *Tex. Rep. Biol. Med.*, 28, 203, 1970.
30. **Bertalanffy, F. D.,** Tritiated thymidine versus colchicine technique in the study of cell population cytodynamics, *Lab. Invest.*, 13, 871, 1964.
31. **Lala, P. K., Maloney, M. A., and Patt, H. M.,** Measurement of DNA-synthesis time in myeloid-erythroid precursors, *Exp. Cell Res.*, 38, 626, 1965.
32. **Jensen, J. J. and Toto, P. D.,** Radioactive labelling index of the periodontal ligament in aging rats, *J. Dent. Res.*, 47, 149, 1968.

33. **Baumrind, S. and Buck, D. L.,** Rate changes in cell replication and protein synthesis in the periodontal ligament incident to tooth movement, *Am. J. Orthod.*, 53, 109, 1970.
34. **Roberts, W. E. and Jee, W. S. S.,** Cell kinetics of orthodontically stimulated and non-stimulated periodontal ligament in the rat, *Arch. Oral Biol.*, 19, 17, 1974.
35. **Roberts, W. E., Chase, D. L., and Jee, W. S. S.,** Counts of labelled mitoses in the orthodontically-stimulated periodontal ligament in the rat, *Arch. Oral Biol.*, 19, 665, 1974.
36. **Yee, J. A., Kimmel, D. B., and Jee, W. S. S.,** Periodontal ligament cell kinetics following orthodontic tooth movement, *Cell Tissue Kinet.*, 9, 293, 1976.
37. **Tonna, E. A., Weiss, R., and Stahl, S. S.,** The cell proliferative activity of paradontal tissues in aging mice, *Arch. Oral Biol.*, 17, 969, 1972.
38. **Weiss, R., Stahl, S. S., and Tonna, E. A.,** Functional demands on the cell proliferative activity of the rat periodontium studied radioautographically, *J. Dent. Res.*, 47, 1153, 1968.
39. **Roberts, W. E., Auber, M. M., Sparaga, J. M., and Smith, R. K.,** Circadian periodicity of the rat molar periodontal ligament, *Am. J. Orthod.*, 76, 316, 1979.
40. **Kvam, E.,** Cellular dynamics on the pressure side of the rat periodontium following experimental tooth movement, *Scand. J. Dent. Res.*, 80, 364, 1972.
41. **Gould, T. R. L., Melcher, A. H., and Brunette, D. M.,** Location of progenitor cells in periodontal ligament of mouse molar stimulated by wounding, *Anat. Rec.*, 188, 133, 1977.
42. **Gould, T. R. L., Brunette, D. M., and Melcher, A. H.,** Migration and division of progenitor cell populations in periodontal ligament after wounding, *J. Periodontal Res.*, 15, 20, 1980.
43. **Beertsen, W.,** Tissue dynamics in the periodontal ligament of the mandibular incisor of the mouse; a preliminary report, *Arch. Oral Biol.*, 18, 61, 1973.
44. **Beertsen, W.,** Migration of fiboblasts in the periodontal ligament of the mouse incisor as revealed by autoradiography, *Arch. Oral Biol.*, 20, 659, 1975.
45. **Kimmel, D. and Jee, W. S. S.,** in *Proc. 1st Workshop on Bone Morphometry*, Jaworski, Z. F. G., Ed., University of Ottawa Press, 1976.
46. **Chambers, D. W., Roberts, W. E., Perko, J., and Grigas, E. J.,** Variance analysis of experimental design in a study of osteoclast histogenesis, *J. Dent. Res. Special Issue A*, 59, 534, 1980.
47. **Hunt, A. M. and Paynter, K. J.,** The effects of ascorbic acid deficiency in the teeth and periodontal tissues of guinea pigs, *J. Dent. Res.*, 38, 232, 1959.
48. **Ostergaard, E. and Loe, H.,** The collagen content of skin and gingival tissues in ascorbic acid-deficient monkeys, *J. Periodontal Res.*, 10, 103, 1975.
49. **Carneiro, J. and Leblond, C. P.,** Suitability of collagenase treatment for the radioautographic identification of newly synthesized collagen labelled with [^{3}H]-glycine or [^{3}H]-proline, *J. Histochem. Cytochem.*, 14, 334, 1966.
50. **Skougaard, M., Frandsen, A., and Baker, D. G.,** Collagen metabolism of skin and periodontal membrane in the squirrel monkey, *Scand. J. Dent. Res.*, 78, 374, 1970.
51. **Carneiro, J. and de Moraes, F. F.,** Radioautographic visualization of collagen metabolism in periodontal tissues of the mouse, *Arch. Oral Biol.*, 10, 833, 1965.
52. **Crumley, P. J.,** Collagen formation in the normal and stressed periodontium, *Periodontics*, 2, 53, 1964.
53. **Stallard, R. E.,** The utilization of [^{3}H]-proline by the connective tissue elements of the periodontium, *Periodontics*, 1, 185, 1963.
54. **Rippin, J. W.,** Collagen turnover in the periodontal ligament under normal and altered functional forces. I. Young rat molars, *J. Periodontal Res.*, 11, 101, 1976.
55. **Rippin, J. W.,** Collagen turnover in the periodontal ligament under normal and altered functional forces. II. Adult rat molars, *J. Periodontal Res.*, 13, 149, 1978.
56. **Skougaard, M. R. and Carsten, A. L.,** Collagen metabolism in skin and periodontal ligament of rats. Effect of aging and irradiation, *J. Dent. Res.*, 57, 334, 1978.
57. **Peterkofsky, B. and Diegelmann, R.,** Use of a mixture of proteinase-free collagenases for the specific assay of radioactive collagen in the presence of other proteins, *Biochemistry*, 10, 988, 1971.
58. **Waterlow, J. C., Garlick, P. J., and Millward, D. J.,** in *Protein Turnover in Mammalian Tissues and in the Whole Body*, North-Holland, Amsterdam, 1978.
59. **Zak, R., Martin, A. F., and Blough, R.,** Assessment of protein turnover by use of radioisotopic tracers, *Physiol. Rev.*, 59, 407, 1979.
60. **Sodek, J.,** A new approach to assessing collagen turnover using microassay. A highly efficient and rapid turnover of collagen in rat periodontal tissues, *Biochem. J.*, 160, 243, 1976.
61. **Sodek, J.,** A comparison of the rates of synthesis and turnover of collagen and non-collagen proteins in adult rat periodontal tissues and skin using a microassay, *Arch. Oral Biol.*, 22, 655, 1977.
62. **Sodek, J.,** A comparison of collagen and non-collagenous protein metabolism in rat molar and incisor periodontal ligaments, *Arch. Oral Biol.*, 23, 977, 1978.
63. **Orlowski, W. A.,** Biochemical studies of collagen turnover in rat incisor periodontal ligament, *Arch. Oral Biol.*, 23, 1163, 1978.

64. **Tonino, G. J. M.,** Discussion of collagen as the generator of tooth eruption, in *The Eruption and Occlusion of the Teeth,* Colston Papers No. 27, Poole, D. F. G. and Slack, M. V., Eds., Butterworths, London, 1976, 307.
65. **Poole, B.,** The kinetics of disappearance of labeled leucine from the free leucine pool of rat liver and its effects on the apparent turnover of catalase and other hepatic proteins, *J. Biol. Chem.,* 246, 6587, 1971.
66. **Beertsen, W. and Everts, V.,** The site of remodelling of collagen in the periodontal ligament of the mouse incisor, *Anat. Rec.,* 189, 479, 1977.
67. **Kanoza, R. J. J., Kelleher, L., Sodek, J., and Melcher, A. H.,** A biochemical analysis of the effect of hypofunction on collagen metabolism in the rat molar periodontal ligament, *Arch. Oral Biol.,* 25, 663, 1980.
68. **Melcher, A. H. and Correia, M. A.,** Remodelling of periodontal ligament in erupting molars of mature rats, *J. Periodontal Res.,* 6, 118, 1971.
69. **Limeback, H. F., Sodek, J., and Brunette, D. M.,** Nature of collagens synthesized by monkey periodontal-ligament fibroblasts *in vitro, Biochem. J.,* 170, 63, 1978.
70. **Saito, H., Yamagata, T., and Susuki, S.,** Enzymatic methods for the determination of small quantities of isomeric chondroitin sulfates, *J. Biol. Chem.,* 243, 1536, 1968.
71. **Niebes, P. and Schifflers, M. H.,** Micromethod for the determination of glycosaminoglycans in the serum. Results from the serum of healthy and varicose subjects, *Clin. Chim. Acta,* 62, 195, 1975.
72. **Sodek, J. and Limeback, H. F.,** Comparison of the rates of synthesis, conversion, and maturation of Type I and Type III collagens in rat periodontal tissues, *J. Biol. Chem.,* 254, 10496, 1979.
73. **Limeback, H. F. and Sodek, J.,** Procollagen synthesis and processing in periodontal ligament *in vivo* and *in vitro.* A comparative study using slab-gel fluorography, *Eur. J. Biochem.,* 100, 541, 1979.
74. **Sodek, J.,** unpublished.
75. **McCulloch, C. A. G. and Melcher, A. H.,** Cell density and cell generation in the periodontal ligament of mice, *Am. J. Anat.,* in press.

DISEASES OF THE PERIODONTIUM

Paul B. Robertson

INTRODUCTION

The normal periodontium consists of four interrelated structures, gingiva, cementum, periodontal ligament, and alveolar bone, which serve to support and protect the teeth (Figure 1). The *gingiva* has both epithelial and connective tissue components which begin apically at a junction with alveolar mucosa, slope coronally to end at a thin edge at the gingival margin, and then plunge apically for 1 to 3 mm to form the gingival sulcus. The depth of the sulcus, measured with a probe calibrated in millimeters, is a major clinical indicator of periodontal disease. Recent evidence, however, suggests that histological sulcus depths rarely correspond to clinically determined depths. In general, the probe tip does not reach to the base of the sulcus in health and passes well into gingival connective tissue in disease. The interior wall of the sulcus is lined coronally by sulcular epithelium and apically by junctional epithelium. The junctional epithelium joins the gingival connective tissue to the tooth surface by a system of hemidesmosomes and basal lamina analagous to structures which mediate intercellular attachment. *Cementum* has many of the chemical and physical properties of bone and functions to anchor fibers of the periodontal ligament to the tooth. This mineralized layer averages 16 to 200 μm in thickness and is continually deposited throughout life, although the cellular proliferative activity of cementum is exceptionally low. The *periodontal ligament* is composed primarily of collagen fibrils arranged in bundles, greater than 85% of which are $[\alpha_1(I)]_2\alpha_2$. The *alveolar bone* encases the ends of periodontal ligament fibers and displays a high metabolic rate, allowing for eruption and movement of the tooth. The ligament is highly vascular by virtue of vessels which arise from adjacent bone marrow and to a lessor extent from periapical vessels. In addition lymph vessels, nerve bundles, a variety of connective tissue cells, and epithelial rests populate the healthy ligament space.

The term "periodontal disease" includes essentially all pathologic processes affecting the gingiva, alveolar bone, cementum, and periodontal ligament. The predominant periodontal diseases are characterized by inflammation in response to products of microorganisms which colonize tooth and adjacent tissue surfaces. These inflammatory diseases of the periodontium are usually classified clinically into two broad groups, *gingivitis* and *periodontitis*, based primarily on whether the inflammatory condition is limited to the gingiva or involves cementum, periodontal ligament, and alveolar bone, in addition to the gingiva.

PERIODONTAL DISEASES

Several acute and chronic forms of gingivitis and at least three subclasses of periodontitis have been described. *Chronic gingivitis* and *acute necrotizing ulcerative gingivitis* (ANUG, Vincent's infection) are the most common forms of gingivitis. In general, chronic gingivitis is a painless, generalized condition which results in gingival erythemia, changes in gingival contour due to edema, and hemorrhage when the affected area is disturbed. ANUG is clinically distinct from chronic gingivitis in that ANUG is usually painful and exhibits frank ulceration and necrosis primarily in interdental areas. A number of subclasses of periodontitis have also been reported. For present purposes, these will be categorized into three groups, *chronic adult periodontitis, aggressive adult periodontitis,* and *juvenile periodontitis*. All subclasses of periodontitis result in loss of periodontal ligament attachment to cementum, alveolar bone resorption, and formation of epithelial lined pockets between gingiva and tooth surfaces as depicted in Figure 2. Clinically, suppuration and bleeding may be observed

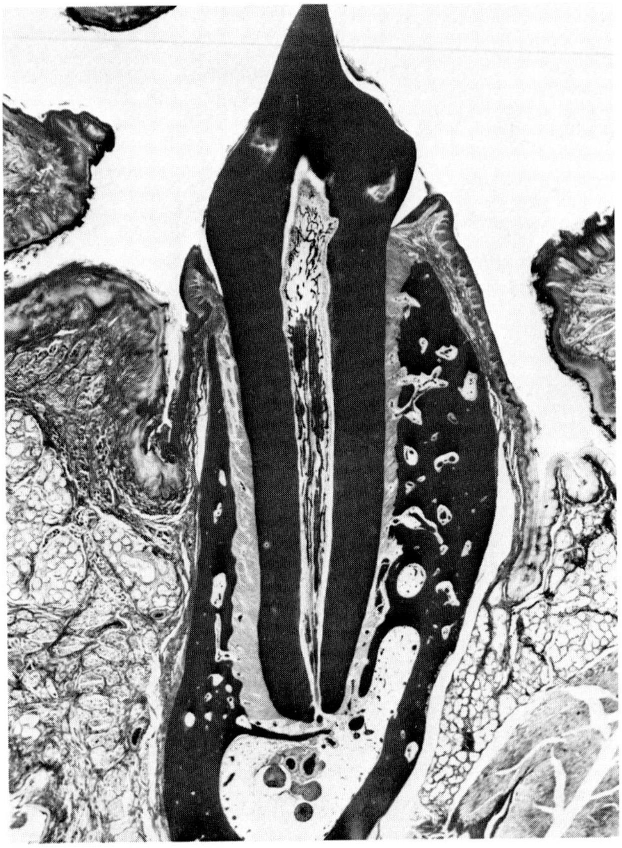

FIGURE 1. The healthy periodontium. The periodontal attachment apparatus consists of alveolar bone, cementum, and intervening periodontal ligament which serve to support the tooth in the jaw. These attachment structures are covered by gingiva which provides protection and a junctional epithelial seal between the gingival connective tissue and the tooth surface.

from the pocket, and the entire process results in tooth mobility and eventual tooth exfoliation. In chronic adult periodontitis, these events occur slowly and are observed affecting most teeth. The cumulative loss of periodontal support rarely exceeds 0.3 mm/year, and these lesions are usually associated with visible calcified and uncalcified microbial deposits. Aggressive adult periodontitis and juvenile periodontitis are associated with rapid loss of periodontal support in young adults and adolescents, respectively. The former usually involves only a few teeth, but no distributional pattern has been established. Juvenile periodontitis is primarily localized to permanent molars and incisors, although more generalized patterns have been observed. In both rapidly progressing diseases, the severity of periodontal destruction is not consistent with visual accumulations of microorganisms nor are calcified deposits usually encountered.

Chronic gingivitis and chronic adult periodontitis are the most common periodontal diseases, constituting by some estimates over 90% of periodontal pathology. Hence, it is not uncommon to see the terms "chronic gingivitis-periodontitis" and "periodontal disease" used interchangeably. Collectively, the several forms of gingivitis and periodontitis are among the most common afflictions of mankind. Bone resorption consistent with periodontitis has been observed in the fossil remains of Neanderthal man, and detailed descriptions of various periodontal diseases are given in Chinese, Egyptian, and Hebrew writings dating

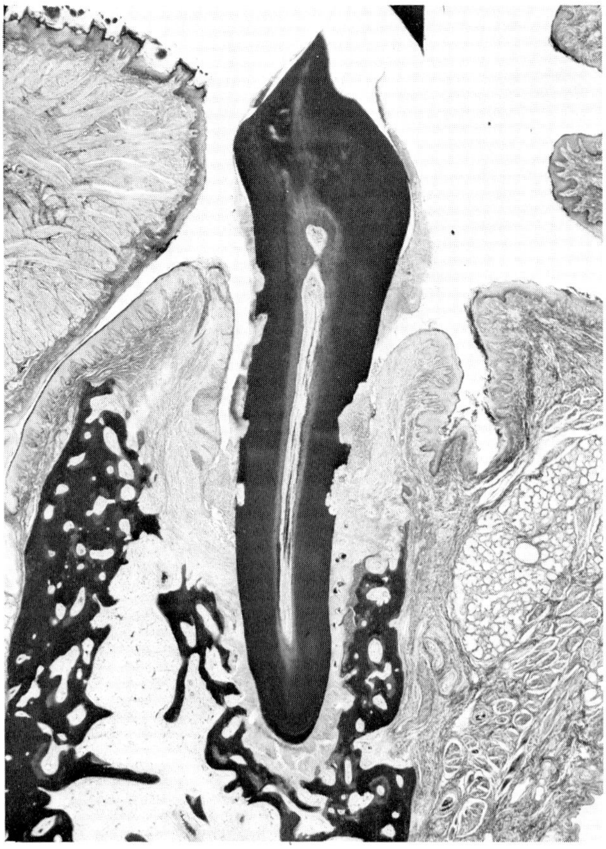

FIGURE 2. Periodontitis. This inflammation, which results in loss of periodontal supporting structures, is mediated by products of oral microorganisms and is responsible for the majority of tooth loss in the world. Several forms of periodontitis exist, all of which cause loss of alveolar bone, formation of epithelial lined pockets adjacent to the root surface, mobility, and eventual tooth exfoliation.

well before the fall of the Roman Empire. Epidemiological surveys conducted during this century suggest that essentially all of the world population have experienced some form of periodontal disease. Gingivitis is frequently observed during the first decade of life, and periodontitis is responsible for the majority of tooth loss in adults.

It is well established that dental plaque, a general term for tooth adherent material composed principally of microorganisms and microbial products, is the primary etiological agent in gingivitis and periodontitis. Emerging evidence suggests that different clinical forms of periodontal disease, including chronic gingivitis and periodontitis, may also have different microbial etiologies. The subgingival microflora associated with clinical health is composed mainly of Gram-positive organisms, primarily species of *Streptococcus*. Gingivitis and periodontitis are characterized by an increasingly greater dominance of the subgingival microflora by Gram-negative anaerobes and spirochetes. For example, Gram-negative facultative and anaerobic rods have been reported to comprise approximately 40% of the cultivable subgingival flora in chronic gingivitis and greater than 65% of the subgingival flora in chronic periodontitis. In addition, recent studies have begun to associate specific groups of organisms with specific forms of periodontal disease.

Since microorganisms are rarely found within periodontal connective tissues, the destruction of supporting structures associated with inflammatory periodontal disease is generally assumed to result from the activity of microbial products. The direct effects of microbial enzymes, organic acids, and various cytotoxic materials may contribute to changes in epithelium and adjacent connective tissue. The chronicity of the disease, extent of involvement, and typical lymphocyte and plasma cell infiltrate suggest that the host response to products of oral microorganisms may play a major role in the destruction of collagenous and osseous connective tissue characteristic of periodontitis. Agents elaborated by microorganisms appear to traverse the junctional epithelium and may then interact with a variety of host cells which in turn mediate destruction of supporting periodontal structures. Indeed, immune and nonimmune host mechanisms have been postulated for essentially all events in periodontitis based primarily on in vitro observations. The ensuing discussion will consider some host-microbial flora interactions which may be important in the etiology, pathogenesis, and treatment of gingivitis and periodontitis.

ETIOLOGY OF PERIODONTAL DISEASE

It seems clear that the initiation and progression of gingivitis is dependent on the accumulation of microorganisms proximate to the gingival margin. In addition, a consistently strong correlation has been observed between the presence and severity of periodontal disease and measures of oral hygiene, regardless of whether the disease process is expressed in terms of gingivitis, periodontitis, or loss of alveolar bone. Despite a variety of experimental approaches using markedly differing disease indexes and populations with divergent cultural, socioeconomic, and geographic backgrounds, the results of epidemiological surveys have been remarkably uniform with respect to the universality of periodontal disease and the strong positive correlation between inflammatory periodontal disease and the presence of microbial plaque. Comprehensive reviews of these epidemiological studies are provided by Waerhaug,[1] Ramfjord et al.,[2] Chilton and Miller,[3] and Sheiham.[4] Experimental evidence supporting the role of microorganisms as the primary etiological agent in gingivitis and periodontitis is based on induction of the disease process solely by allowing the accumulation of bacteria, reversal of the periodontal lesion by mechanical or chemotherapeutic plaque removal, and association of differing clinical types of periodontal disease with particular groups of oral microorganisms. Detailed reviews of the role of microorganisms in gingivitis and periodontitis are given by Socransky,[5] Kelstrup and Theilade,[6] Socransky,[7] and Slots.[8]

ROLE OF MICROORGANISMS

Three general theories have been proposed to explain the origin of microbial products injurious to the periodontium. First, such products may originate from exogenous organisms not normally present in the oral flora. Second, an increased toxic load may result from a nonspecific increase in total numbers of the indigenous flora, or third, from shifts in the indigenous flora to dominance by specific bacterial species. Distinct patterns of flora associated with periodontal health and disease have not been clearly demonstrated, but the available evidence favors the role of indigenous microflora rather than a more classical infection by particular exogenous organisms. In longitudinal studies of gingivitis induced by allowing the unrestricted accumulation of plaque over a 21-day period,[9] both quantitative and qualitative changes in the microflora were noted. Withdrawal of all active oral hygiene procedures in subjects with initially healthy gingiva consistently resulted in clinical gingivitis, the development of which was highly correlated to the amount of plaque accumulation at the gingival margin. Reinstitution of oral hygiene measures resulted in prompt remission of gingivitis. In addition, major differences were observed in the distribution of organisms by microscopic counts of smears obtained from early and late plaques. At the beginning of the experimental period, the microflora consisted almost entirely of Gram-positive cocci and short rods. By 2 to 3 weeks, the total flora comprised 50% Gram-positive cocci and short rods, 30% Gram-negative cocci and small rods, 8% fusobacteria, and 4% spirilla and spirochetes.

Longitudinal human studies of the microbiological changes associated with periodontitis are difficult to conduct for both practical and ethical reasons. These studies have been performed in nonhuman primates. One such study monitored the cultivable subgingival microflora in the cynomolgus monkey, *Macaca fascicularis*, during the ligature-induced progression of naturally occurring gingivitis to periodontitis.[10] A silk ligature was placed at the cementoenamel junction to facilitate plaque development and the clinical and microbiological progression of the lesion was followed for 17 weeks as shown in Table 1. At the time of ligature placement all sites manifested chronic gingivitis. The subgingival flora was characterized by Gram-positive cocci and rods with *Bacteroides melaninogenicus* ssp. *intermedius* the dominant Gram-negative organism. By 3 weeks following ligation, an increase in gingival inflammation was observed coincident with a significant increase in motile and surface translocating Gram-negative rods, primarily *Capnocytophaga* sp. and *Campylobacter sputorum*. Major increases in pocket depths and radiographic evidence of alveolar bone loss was observed 3 to 7 weeks after ligation. This stage was dominated by Gram-negative anaerobic rods which represented almost 62% of the cultivable flora. *B. gingivalis* increased significantly, accounting for over 34% of the total cultivable microflora. Gingival inflammation decreased and no further evidence of alveolar bone loss was observed during the remainder of the experimental period. This quiescent stage corresponded to decreasing levels of *B. gingivalis*. The microflora of the cynomolgus monkey is comparable to that reported for human chronic gingivitis and adult periodontitis and may be valuable in determining host-parasite interactions in periodontal disease.

A summary of cross-sectional studies of percent cultivable subgingival microflora from humans with experimental gingival health,[11] clinical gingival health,[12] chronic gingivitis,[13] chronic adult periodontitis,[14,15] juvenile periodontitis,[8,15,16] and adult aggressive periodontitis[17] is shown in Table 2. Sites exhibiting chronic gingivitis show up to 20 times the number of organisms than sites which are healthy but all species present in health are also present in gingivitis. A major shift to Gram-negative anaerobic rods is characteristic of chronic adult periodontitis. The most common isolate is *B. gingivalis*, formerly classified as *B. melaninogenicus* ssp. *asaccharolyticus*, and *B. asaccharolyticus*. Two organisms, *Capnocytophaga* sp. and *Actinobacillus actinomycetemcomitans* have been associated with juvenile periodontitis. While *Capnocytophaga* is also found in sites from individuals with chronic adult periodontitis, *Actinobacillus* appears to be primarily associated with rapidly progressing bone loss in young patients. A Gram-negative obligately anaerobic short rod, *B. capillus*, has recently been reported to comprise greater than 50% of the cultivable flora in three cases of aggressive adult periodontitis.

It seems clear that microbial plaque is responsible for inflammatory periodontal disease and, in addition, that some microbial specificity may exist in different periodontal diseases. However, neither cause and effect relationships by Koch's postulates nor clear microbial specificity have been established. The microbiological contents of human plaque varies widely among individuals and among sites within the same individual. Supragingival microflora is different from that found subgingivally. In periodontitis, the constitution of subgingival plaque appears to vary both with the sampling method and the depth at which the sample is obtained. Finally, large numbers of organisms in cultural studies remain unidentified, particularly in samples obtained from juvenile periodontitis and adult aggressive periodontitis. The identification of responsible pathogens remains one of the major goals of periodontal research.

Table 1
PERCENT CULTIVABLE SUBGINGIVAL MICROFLORA FROM HUMAN SITES

Periodontal condition	Number of sites	G+ cocci	G+ rods		G− cocci	G− rods		Dominant isolates
		FAC & AN	FAC	AN	FAC & AN	FAC	AN	
Experimental gingival health	20	61.0	16.0	3.2	13.6	—	2.0	*Streptococcus sanguis*, *S. mitior*, *Veillonella* sp.
Clinical gingival health	7	40.4	35.1	9.5	2.3	—	12.7	*S. sanguis*, *S. mitior*, *Veillonella* sp. *Actinomyces israellii*
Chronic gingivitis	9	29.8	16.9	9.2	4.3	14.8	25.0	*S. sanguis*, *S. mitior*, *A. naeslundii*, *Haemophilus* sp. *Bacteriodes* sp.
Chronic adult periodontitis	17	6.2—10.8	3.9—9.8	4.1—15.1	0.6—1.6	0.0—7.6	66.9—74.3	*A. israellii*, *Bacteroides gingivalis*, other *Bacteroides* sp. *Fusobacterium nucleatum*
Juvenile periodontitis	24	14.0—16.3	3.1—10.6	9.9—15.3	1.8—8.4	1.5—9.0	55.1—59.2	*B. melaninogenicus* ssp. *intermedius*, other *Bacteroides* sp., *Actinobacillus actinomycetemcomitans*, *Capnocytophaga* sp.
Aggressive adult periodontitis	3	5.5	4.0		2.0	2.0	82.7	*B. capillus*

Table 2
PREDOMINANT CULTIVABLE SUBGINGIVAL MICROFLORA DURING LIGATURE-INDUCED PERIODONTITIS

	Stage I (Time = 0)	Stage II (1 to 3 weeks)	Stage III (4 to 7 weeks)	Stage IV (8 to 17 weeks)
N[a]	7	4	7	5
Total count ($\times 10^6$)[b]	7.78	11.05	5.48	44.70
	(2.6 − 73.6)	(0.5 − 27.6)	(0.8 − 11.4)	(23.3 − 73.6
G+ cocci	25.9 ± 7.7	14.9 ± 11.8	7.1 ± 7.1	12.5 ± 8.5
Anaerobic	12.3 ± 12.7	6.9 ± 7.4	3.0 ± 5.1	12.1 ± 8.6
Facultative	13.6 ± 10.9	8.0 ± 7.1	4.1 ± 3.2	0.4 ± 0.9
G+ rods	17.8 ± 12.7	18.9 ± 5.8	5.3 ± 6.3	16.5 ± 11.7
Anaerobic	4.3 ± 10.5	7.2 ± 12.5	3.9 ± 6.4	4.9 ± 3.1
Facultative	13.5 ± 10.1	11.7 ± 11.4	1.4 ± 3.4	11.6 ± 11.8
G− cocci	2.2 ± 3.8	3.2 ± 3.6	4.1 ± 10.0	3.0 ± 4.9
G− rods	49.0 ± 14.6	56.8 ± 30.7	77.1 ± 14.2	59.7 ± 8.2
Anaerobic	31.4 ± 7.5	28.8 ± 9.1	61.9 ± 15.3	41.7 ± 12.8
Bacteroides melaninogenicus ssp. *intermedius*	17.2 ± 9.5	6.8 ± 10.1	0.9 ± 1.3	1.5 ± 3.3
B. asaccharolyticus	1.3 ± 3.5	5.3 ± 9.2	[34.2 ± 15.7]	[17.8 ± 4.5]
Fusobacterium sp.	8.2 ± 6.4	8.3 ± 3.5	14.8 ± 5.0	10.4 ± 3.8
Facultative	17.6 ± 8.2	28.1 ± 5.2	15.2 ± 14.6	18.0 ± 8.2
Motile and surface translocating G− rods	6.6 ± 2.9	[20.0 ± 3.5[c]]	14.3 ± 20.9	14.7 ± 7.6

[a] Number of sample sites.
[b] Mean total colony forming units on ETSA with range of counts in parentheses.
[c] Mean percentage of total cultivable flora ± SD.
☐ Values within box are significantly different (p <0.01, student t test) from Stage I values in row.

ROLE OF LOCAL NONMICROBIAL FACTORS

Both experimental and epidemiological studies have failed to demonstrate any local factors, aside from microorganisms, which cause gingivitis or periodontitis. Several of these, including dental appliances, occlusion, and anatomic oral abnormalities, may modify the progress of the disease, once established. Modifying local factors, such as defective restorations, prosthetic appliances, ligatures, and calculus, may interfere with the patient's ability to remove plaque or may provide an environment that facilitates the establishment of a more pathogenic flora. The potential role of mechanical local factors has been reviewed by Pennell and Keagle.[18]

Occlusal trauma refers to alterations in the periodontium either caused by excessive occlusal forces or associated with a decreased capacity of inadequate periodontal connective tissue to withstand normal forces. Such alterations include mobility, widening of the periodontal ligament space, and changes in alveolar bone which appear to represent an adaptation by the periodontium to excessive force. The relationship between occlusion and periodontal disease has recently been reviewed by Polson.[19] In the absence of plaque, excessive occlusal forces will not initiate gingivitis, induce periodontitis, or result in loss of connective tissue attachment. The effect of occlusion superimposed on existing disease is less well defined, but studies in animal models suggest that under such conditions alveolar bone loss may be increased.

A number of clinical reports suggest that an inadequate width of keratinized gingiva may predispose to periodontal disease. An overview of research in this area is provided by Caffesse.[20] The problem has been addressed in several studies involving bilateral areas of insufficient gingiva by surgically increasing the zone of keratinized gingiva in one site and leaving the other site untreated. In general, no postoperative differences in periodontal health have been observed between surgically treated and contralateral untreated areas. It would appear that periodontal health can be maintained in the absence of keratinized gingiva, but the protective role of broad zones of gingiva under conditions of existing disease has not been established.

HOST RESPONSE IN PERIODONTAL DISEASE

The histopathological events which occur during the progression from periodontal health to chronic gingivitis and adult periodontitis have been divided into four general overlapping stages by Page and Schroeder.[21] The *initial stage* consists of vasculitis concomitant with increased migration of neutrophilic leukocytes subjacent to the junctional epithelium. Alterations in the junctional epithelium and some loss of perivascular collagen may be observed. The *early stage* begins after approximately 4 to 7 days of plaque accumulation and is characterized by the appearance of an inflammatory exudate consisting primarily of lymphocytes, continuing collagen destruction, and proliferation of basal junctional epithelial cells. Over ensuing weeks, plasma cells become more numerous and in the *established stage*, dominate the inflammatory infiltrate. This established lesion is compatible with a clinical diagnosis of chronic gingivitis and may persist without further involvement of deeper supporting structures. Under conditions not well understood, the established lesion may progress to the *advanced stage*, or periodontitis, with extension of the epithelium apically along the root surface, subsequent pocket formation, destruction of alveolar bone and periodontal ligament, and eventual tooth loss. In addition to frank loss of connective tissue, several alterations in the distribution of collagen have been described, including an increase in A and B collagens, the appearance of a Type-I trimer, ($[\alpha_1(I)]_3$), and relative increases in poorly cross-linked, soluble collagen. Both direct effects of bacterial products and indirect effects, primarily immune system-mediated responses, of the host to bacterial products have been postulated to account for changes which occur in periodontal disease. Both mechanisms have recently been discussed in detail by Page and Schroeder,[22] Ranney,[23] Rizzo and Mergenhagen[24], and Nisengard.[25]

DIRECT MICROBIAL EFFECTS

Oral bacteria produce a variety of toxic substances which may induce direct tissue damage. Nonspecific substances demonstrated both in mixed cultures and in periodontal pockets include ammonia, hydrogen sulfide, indole, and a variety of toxic amines and organic acids. Proteolytic enzymes detected in plaque include hyaluronidase, chondrosulfatase, neuraminidase, and β-glucuronidase. One enzyme which may be of particular importance in periodontal disease is microbial collagenase. The collagenase from *Clostridium histolyticum* has been extensively characterized and is generally considered representative of microbial collagenolytic enzymes. Multiple forms of the *Clostridium* enzyme have been described in crude culture fluid extracts which may reflect subunit structure, isozymes, and/or proteolysis following initial secretion from the cell. *Clostridium* collagenase A, a 105,000-dalton fraction, has a high degree of substrate specificity and cleaves the bond between X-Gly in the amino acid sequence Pro-X-Gly-Pro-Y to release a Gly-NH_2 terminus.[26] Elaboration of multiple scission enzymes has been described for a number of other organisms, including *Mycobacterium tuberculosis*, *Streptomyces madurae*, *Pseudomonas aeruginosa*, *Achromobacter iophagus*, *Vibro* B-30, and *Bacillus cereus*. Degradation of collagen by *Bacteroides melaninogenicus* was first reported by Gibbons and MacDonald.[27] The *Bacteroides* collagenase was heat labile, had a pH optimum from 6.8 to 7.3, and appeared to be partially activated by L-cyteine.

A survey of over 35 human oral isolates grown anaerobically showed elaboration of collagenolytic activity by all tested species of *Bacteroides*, including *B. melaninogenicus* ssp. *melaninogenicus* and *intermedius*, *B. fragilis*, *B. gingivalis*, and *B. capillus*, as well as *Actinobacillus actinomycetemcomitans*.[28] Enzyme activity was not detected in species of *Actinomyces*, *Fusobacterium*, *Capnocytophaga*, *Eikenella*, or *Selenomonas*. Characteristics of these collagenases were similar to the enzyme from *Clostridium* except that *Bacteroides* collagenase is located between the cell membranes and is not released into the media during log phases of growth.

HOST RESPONSES TO MICROBIAL PRODUCTS

A variety of pathways have been described by which microbial products may induce inflammation and mediate connective tissue degradation in periodontal disease. Some interactions of microbial products with polymorphonuclear leukocytes (neutrophils), macrophages, lymphoid cells, and mast cells are shown in Table 3.

Page and Schroeder[22] have characterized neutrophils and macrophages as the ''shock troops'' and ''long standing army'', respectively, in the battle between host and challenging microorganisms. The migration of neutrophils through affected periodontal connective tissue and into the periodontal pocket has long been recognized as a characteristic feature of gingivitis and periodontitis. Indeed, the neutrophil response is so constant that several diagnostic indexes have been proposed based on the emigration of neutrophils out of the pocket and into the oral cavity. A destructive role of neutrophils in periodontal disease is consistent with tissue changes associated with the Arthus reaction and with the potential degradative effects of neutrophilic lysosomal enzymes. The association of increased susceptibility to gingivitis and periodontitis in humans and animal model systems with known neutrophil disorders suggest that a failure of neutrophil defense against plaque microorganisms may also be important in the pathogenesis of periodontal disease. This protective role of neutrophils is supported by reports of unusually severe periodontal disease in patients with Chediak-Higashi syndrome,[29] cyclic neutropenia, and ''lazy leukocyte syndrome''. Several reports suggest that patients manifesting juvenile diabetes often have severe periodontitis. Juvenile diabetes is frequently associated with impaired neutrophil function which may be the link between the two diseases. Recently, the majority of patients with juvenile periodontitis[30,30a] and many patients with aggressive adult periodontitis[31] have been shown to have a defective function of peripheral blood neutrophils. In addition, there is evidence that several organisms, including *Capnocytophaga* species and *Actinobacillus actinomycetemcomitans*, produce substances which seriously impair neutrophil function. Systemic defects or impairment by substances elaborated locally by certain bacteria can cause the loss of the protective ability of neutrophils. This may allow the establishment of a flora with greater pathogenic potential, leading to a more rapidly progressing disease process typical of juvenile and aggressive periodontitis.

Macrophages are similar to neutrophils in that these mononuclear phagocytes are chemotactically responsive, carry a host of hydrolytic enzymes, and have the ability to engulf and kill microorganisms. Unlike neutrophils, however, macrophages are capable of replication, extensive protein synthesis, and participation in a broad range of biological functions beyond phagocytosis. Macrophages may be activated by bacterial substances, immune complexes, and lymphokines. Once activated the macrophage synthesizes and secretes lysosomal hydrolases, prostaglandins, cytotoxic factors, and a variety of neutral proteases. In addition, the macrophage appears to play a major role in the regulation of B- and T-lymphocyte function as well as antigen presentation to lymphoid cells.

One common feature of both cells is the elaboration of a tissue collagenase capable of causing hydrolytic scission of peptide bonds in the helical region of undenatured collagen.

Table 3
ORAL MICROBIAL PRODUCTS

Neutrophils	Macrophages	Lymphoid cells	Mast cells
Chemotaxis	Synthesis and secretion	Lymphokine production	FC-IgE (acute inflammation)
Inhibition of phagocytosis and killing	Pyrogen	Macrophage control factors (MIF, MAF, MCF, MSF)	Degranulation
Release of acid hydrolases, neutral proteases, and collagenase	PMN chemotaxis	Lymphotoxin	Histamine
	Prostaglandins	Chemotactic factor	Slow reacting substance A (SRS-A)
	Cytotoxins	Osteoclast activating factor	Heparin
	Acid hydrolases	Antibody production	Proteases
	Collagenase and elastase	IgA (regulation of microbial colonization)	
	Antigen processing and presentation to lymphoid cells	IgE (mast cells)	
	Increased phagocytosis and killing	IgM and IgG (immune complexes and complement activation)	

The enzymes from both cells have a pH optima from 7 to 9, molecular weights ranging from 40,000 to 65,000 daltons, absolute requirements for calcium, and are inhibited by serum α2 macroglobulin. Neutrophil and macrophage collagenases are typical of other tissue collagenases in that they cleave undenatured collagen at a single peptide bond between residues 772 and 773 in the α_1 chain of the Type I molecule. This single cleavage characteristic applies to Types I, II, and III collagen, although some differences in relative susceptibility among collagen types has been observed and is markedly different from the multiple cleavage affected by microbial collagenases.

LYMPHOID CELLS AND THE IMMUNE SYSTEM

The immune system, orchestrated by thymic-dependent (T) and bone marrow-derived (B) lymphocytes, has long been held responsible for many of the destructive effects of periodontal disease. Several investigations have shown that specific antibody production can be stimulated by host cells by applying antigenic substances to the gingiva. Antibodies in human serum have been detected to species of *Leptotrichia*, *Fusobacterium*, *Treponemia*, *Actinomyces*, *Veillonella*, and *Bacteroides*, although a relationship between antibody titer and severity of disease for these organisms remains unclear. Patients with chronic adult periodontitis do demonstrate significantly higher antibody titers reactive with *Bacteroides gingivalis* and recent studies have shown high titers to *Actinobacillus actinomycetemcomitans* in subjects with *juvenile periodontitis*. Immunofluorescent studies have demonstrated the deposition of antigen-antibody complexes in inflamed gingiva and such complexes could activate compliment. Much of the evidence on the role of complement in the disease process stems from studies of gingival fluid, a serum transudate which flows out of the gingival sulcus. The fluid flow rate increases with inflammation, and the protein constituents of the fluid represent a 15 to 30% dilution of serum. In severe periodontitis, native C3 is reduced to less than 25% of serum levels, and the presence of terminal cleavage products, C3c and C3d, strongly support the probability of complement activation. Cleavage of Factor B to the active enzyme Factor Bb (alternative complement pathway) has been demonstrated in gingival fluid from lesions of both chronic adult and juvenile periodontitis. Activation of C4 (classical complement pathway) appears common in juvenile periodontitis, but is relatively uncommon in chronic adult periodontitis.

Cell-mediated immunity, initiated by competent T-lymphocytes, has also been implicated in periodontal destructon based on early studies which suggested that antigens from dental plaque stimulated lymphocyte blastogenesis in proportion to the clinical severity of periodontal disease manifested by the individual from whom the cultured lymphocytes were obtained. Later studies[32] found quite different patterns of blastogenesis, depending on the microbial agent employed. Antigens derived from *Actinomyces*, for example, were potent stimulators of lymphocytes in essentially all patients tested. However, strong blastogenic responses to *Treponema denticola* and *B. gingivalis* were limited to lymphocytes from patients with severe periodontitis. The lymphokines, lymphotoxin, and osteoclast activating factor have been demonstrated in cultures of plaque-stimulated lymphocytes from healthy patients or from unstimulated lymphocytes from patients with periodontitis. Recent evidence suggests that most implicated oral microorganisms are mitogenic and that blastogenesis induced in peripheral lymphocytes by exposure to bacterial substances may be polyclonal rather than specifically antigenic.

In summary, many pathways can be postulated to explain the acute inflammation, vascular permeability, accumulation of inflammatory cells, collagen degradation, and bone resorption characteristic of periodontal disease. It is possible that the pathway varies with the nature of the etiologic organism and the severity of the disease process.

TREATMENT OF PERIODONTAL DISEASE

Mechanical removal of dental plaque continues to be the principle means of maintaining oral health, and, in general, toothbrushing supplemented by some form of interproximal cleaning appears to provide the most effective means of plaque removal. While the superiority of a particular toothbrushing technique has not been established, most investigations support the use of floss for interproximal cleansing, especially in normal interdental spaces. In healthy subjects, oral hygiene performed every 24 to 48 hr will maintain gingival health. It may be important to note, however, that toothbrushes, toothpicks, and other oral hygiene aids and the rationale for their use have been available for over 40 centuries. Apparently, individual patient commitment to thoroughly cleaning all surfaces of the teeth is considerably more important than the choice of a particular oral hygiene device.

The difficulty in manipulating presently available oral hygiene aids and the time commitment required to maintain oral cleanliness by mechanical methods has prompted a number of investigations directed to the use of antimicrobial agents. Systemic antibiotics do not appear to offer a practical means of life-long plaque control for the general population. Antibiotics have been successfully used to treat periodontitis which is ''advanced'' or resistant to other forms of therapy. In chronic adult periodontitis, however, available studies indicate that the clinical effects of adjunctive antibiotic administration are similar with conventional therapy alone.

A number of topical antimicrobial agents have been reported for oral use. The most extensively tested is chlorhexidine gluconate.[33] Chlorhexidine mouth rinses appear to be effective in preventing development of gingivitis in initially healthy individuals, both short-term and over a 2-year period. In patients manifesting periodontitis with associated pocket formation, the effect of the agent on subgingival plaque has not been established. Other agents which may reduce plaque formation, and in some cases improve existing gingivitis, but have not been shown effective in treatment of periodontitis, include cetylpyridinium chloride and related compounds, hexylresorcinal, alexidine, urea peroxide and other oxygenating agents, benethamine chloride, and a variety of proteolytic enzymes.[34-36]

Keyes et al.[37] have recently renewed interest in the use of saturated salt and hydrogen peroxide solutions in periodontal therapy. While clinical results are positive in a limited number of case reports, the value of such solutions has not been demonstrated in controlled studies. Furthermore, it is unclear whether the improvement in periodontal health reported by this group was the result of antimicrobial activity on the part of the agent or diligence in mechanical oral hygiene procedures by the patient and therapist.

Scaling is a general term for removal of bacteria, bacterial products, and calculus from the crown and root surface, while the term root planing is limited to procedures used to smooth the root surface. Since the process of smoothing root surfaces will result in removal of bacterial debris and calculus, a considerable operational component, as well as clinical benefit, can be ascribed to scaling. Thus, when the root surface is involved, the two terms are commonly used interchangeably, and in clinical practice the procedures are often combined. Scaling and root planing, combined with effective oral hygiene, will reverse gingivitis and, in many cases, control chronic adult periodontitis. In the case of the latter, scaling and root planing appears to be limited by accessibility of the root surface. In initially moderate pockets in areas accessible to instrumentation, scaling and root planing is as effective as other forms of therapy. Deep pockets, tortuous defects, or furcations may require surgical exposure of the root surface to facilitate instrumentation. Under these conditions, most surgical periodontal procedures are equally effective in restoring and maintaining periodontal health in patients willing to maintain relatively plaque-free conditions thereafter. Unfortunately, present techniques do not routinely result in regeneration of cementum, periodontal ligament, and alveolar bone lost to the disease process. Available evidence suggests that osseous grafting may be clinically useful in certain types of osseous defects, but biologic tissue relationships adjacent to affected root surfaces following either graft or repair techniques remain unclear.

Progress in therapy for all diseases of the periodontium requires new information on the microflora and host responses associated with periodontal health and disease. The microflora and associated pathogenic pathways for each of the periodontal diseases have not been clearly established, nor are sensitive methods available for the assessment of present disease activity. Such information is necessary to the development of methods which prevent colonization of the teeth or foster the establishment of a microflora compatible with periodontal health or interfere with host mechanisms responsible for disease progression. In addition, biological relationships at the connective tissue-root surface interface require clarification to allow progress in regeneration of lost periodontal structures.

REFERENCES

1. **Waerhaug, J.,** Epidemiology of periodontal disease, in *World Workshop in Periodontics,* Ramfjord, S. P., Kerr, D. A., and Ash, M. M., Eds., W. B. Saunders, Philadelphia, 1966, 751.
2. **Ramfjord, S. P., Emslie, R. D., Green, J. C., Held, A. F., and Waerhaug, J.,** Epidemiological studies of periodontal disease, *Am. J. Publ. Health,* 58, 1713, 1968.
3. **Chilton, N. W. and Miller, M. I.,** Epidemiology of periodontal disease, in *Int. Conf. Research in the Biology of Periodontal Disease,* The College of Dentistry, University of Illinois, Urbana, 1977.
4. **Sheiham, A.,** The epidemiology of dental caries and periodontal disease, *J. Clin. Periodontol.,* 6, 7, 1979.
5. **Socransky, S. S.,** Relationship of bacteria to the etiology of periodontal disease, *J. Dent. Res.,* 49, 203, 1970.
6. **Kelstrup, J. and Theilade, E.,** Microbes and periodontal disease, *J. Clin. Periodontol.,* 1, 15, 1974.
7. **Socransky, S. S.,** Microbiology of periodontal disease — present status and future considerations, *J. Periodontol.,* 48, 497, 1977.
8. **Slots, J.,** Subgingival microflora and periodontal disease, *J. Clinical Periodontol.,* 6, 351, 1979.
9. **Löe, H.,** Human research model for the production and prevention of gingivitis, *J. Dent. Res.,* 50, 256, 1971.
10. **Kornman, K. S., Holt, S. C., and Robertson, P. B.,** The microbiology of ligature-induced periodontitis in the cynomolgus monkey, *J. Periodont. Res.,* 16, 363, 1981.
11. **Loesche, W. J. and Syed, S. A.,** Bacteriology of human experimental gingivitis: effect of plaque and gingivitis score, *Infect. Immun.,* 21, 830, 1978.
12. **Slots, J.,** Microflora in the healthy gingival sulcus in man, *Scand. J. Dent. Res.,* 85, 247, 1977.
13. **Slots, J., Moembo, D., Langeback, J., and Frandsen, A.,** Microbiota of gingivitis in man, *Scand. J. Dent. Res.,* 86, 174, 1978.
14. **Slots, J.,** The predominant cultivable microflora of advanced periodontitis, *Scand. J. Dent. Res.,* 85, 114, 1977.
15. **Tanner, A. C. R., Haffer, C., Bratthall, G. T., Visconti, R. A., and Socransky, S. S.,** A study of the bacteria associated with advancing periodontitis in man, *J. Clin. Periodontol.,* 6, 278, 1979.
16. **Newman, M. G. and Socransky, S. S.,** The microbiota of periodontosis, *J. Periodont. Res.,* 12, 120, 1977.
17. **Kornman, K. S. and Holt, S. C.,** Physiological and ultrastructural characterization of a new *Bacteroides* species (*Bacteroides capillus*) isolated from severe localized periodontitis, *J. Periodont. Res.,* 16, 467, 1981.
18. **Pennel, B. M. and Keagle, J. C.,** Predisposing factors in the etiology of chronic inflammatory periodontal disease, *J. Periodontol.,* 48, 517, 1977.
19. **Polson, A. M.,** Interrelationship of inflammation and tooth mobility (trauma) in pathogenesis of periodontal disease, *J. Clin. Periodontol.,* 7, 351, 1980.
20. **Caffesse, R. G.,** Longitudinal evaluation of periodontal surgery, *Dent. Clin. North Am.,* 24(4), 751, 1980.
21. **Page, R. C. and Schroeder, H. E.,** Pathogenesis of inflammatory periodontal disease. A summary of current work, *Lab. Invest.,* 33, 235, 1976.
22. **Page, R. C. and Schroeder, H. E.,** Current status of the host response in chronic marginal periodontitis, *J. Periodontol.,* 52, 477, 1981.

23. **Ranney, R.,** Pathogenesis of periodontal disease, in *Int. Conf. Research in the Biology of Periodontal Disease,* The College of Dentistry, University of Illinois, Urbana, 1977.
24. **Rizzo, A. A. and Mergenhagen, S. E.,** Host responses in periodontal disease, in *Int. Conf. Research in the Biology of Periodontal Disease,* The College of Dentistry, University of Illinois, Urbana, 1977.
25. **Nisengard, R. J.,** The role of immunology in periodontal disease, *J. Periodontol.,* 48, 505, 1977.
26. **Lwebuga-Mukasa, J. E., Harper, E., and Taylor, P.,** Collagenase enzymes from *Clostridium:* characterization of individual enzymes, *Biochemistry,* 15, 4736, 1976.
27. **Gibbons, R. J. and MacDonald, J. B.,** Degradation of collagenous substrates by *Bacteroides melaninogenicus, J. Bacteriol.,* 81, 614, 1961.
28. **Robertson, P. B., Lantz, M., Marucha, P. T., Kornman, K. L., Trummel, C. L., and Holt, S. C.,** Collagenolytic activity associated with *Bacteroides* species, *J. Periodont. Res.,* 17, 275, 1982.
29. **Clark, R. A. and Kimball, H. R.,** Defective chemotaxis in Chediak-Higashi syndrome, *J. Clin. Invest.,* 50, 2645, 1971.
30. **Cianciola, L. G., Genco, R. J., Patters, M. P., McKenna, J., and VanOss, C. J.,** Defective polymorphonuclear leukocyte function in human periodontal disease, *Nature (London),* 265, 445, 1977.

30a. **Clark, H., Page, R. C., and Wilde, G.,** Defective neutrophil chemotaxis in juvenile periodontitis, *Infect. Immun.,* 18, 694, 1977.

31. **Lavine, W. S., Maderazo, E. G., Stolman, J., Ward, P. A., Cogen, R. B., Greenblatt, I., and Robertson, P. B.,** Impaired neutrophil chemotaxis in patients with juvenile and rapidly progressing periodontitis, *J. Periodont. Res.,* 14, 10, 1979.
32. **Patters, M. R., Genco, R. J., Reed, M. J., and Mashimo, P. A.,** Blastogenic response of human lymphocytes to oral bacterial antigens: comparison of individuals with periodontal disease to normal and edentulous subjects, *Infect. Immun.,* 14, 1213, 1976.
33. **Löe, H.,** Chlorhexidine in the prophylaxis of dental diseases, *J. Periodont. Res.,* 8(Suppl. 12), 5, 1973.
34. **Loesche, W. J.,** Chemotherapy of dental plaque infections, *Oral Sci. Rev.,* 9, 65, 1976.
35. **Gjermo, P., Baastad, K. L., and Rolla, G.,** The plaque inhibiting capacity of 11 antibacterial compounds, *J. Periodont. Res.,* 5, 102, 1970.
36. **Robinson, P. J.,** Evaluation of therapy, in *International Conference on Research in the Biology of Periodontal Disease,* The College of Dentistry, University of Illinois, Urbana, 1977.
37. **Keyes, P. H., Wright, W. E., and Howard, S. A.,** The use of phase contrast microscopy and chemotherapy in the diagnosis and treatment of periodontal lesions — an initial report (I) and (II), *Quintessence Int.,* 1, 1, 1978; 2, 7, 1978.

Gingiva

COMPOSITION OF GINGIVA

Walter A. Orlowski

INTRODUCTION

The gingiva is composed of variable amounts of connective tissue and epithelium and covers the teeth and the surface of the alveolar bone.[1,2] The epithelium covers a rather dense lamina propria which is characterized by the presence of high and slender papilla. The bulk of gingiva is formed by dense collagen fibers which may insert either into teeth, the lamina propria, or blend with the periosteum.

Since the last reviews of the biochemistry of oral tissues by Gerson and Meyer[3] in 1970 and by Page[4] in 1972, major advances have been made in our knowledge of the biochemistry of gingiva. These advances reflect the general growth of scientific knowledge of the last decade and the increased utilization of biochemical techniques by dental researchers.

ANALYSIS OF GINGIVA

Methods of Sampling for Biochemical Analysis

For biochemical analysis, adequate samples of gingiva can be readily dissected grossly, at the most, with the aid of a magnifying glass. When working with small animals, lyophilization or just deep freezing of tissues will aid in dissection. This author found it useful even when working with larger animals.

When separation of epithelium and connective tissue is desired, microdissection, enzymes, and chemicals have been utilized. Microdissection techniques have been extensively utilized by dental researchers, especially for enzyme studies.[5-8] These methods are largely adaptations of the method of Lowry.[9] Enzymes such as trypsin and collagenase, and chemicals such as EDTA or thiocyanate, are effective in separating epithelium from subjacent connective tissue and have been extensively utilized for separating cells for tissue culture studies. Felton et al.[10] and Rabinowitz et al.[11] utilized clostridial collagenase, whereas Dale et al.[12] in a recent study used trypsin and EDTA. Thiocyanate has also been utilized,[13] but to a lesser extent than other agents. Eastoe[14] separated the epithelium by autoclaving the tissue, thus destroying the subjacent connective tissue. This method was utilized for amino acid analysis of the epithelium.

For histochemical analysis, the method of fixation of tissue varies depending on the histochemical procedure. For glycosaminoglycans, 10% formalin with 2% calcium acetate[15] is quite effective. For immunohistochemical techniques, unfixed tissues are required, although the transport medium of Michel et al.[16] probably can also be utilized. Wang et al.[17] did use formalin fixed tissue successfully to study the distribution of Type I and Type III collagens in oral tissues using the immunoperoxidase technique.[18]

The relevant analytical techniques will be discussed in appropriate sections on the individual components of the gingiva.

Sources of Error in Biochemical Analysis

There are several important considerations in evaluating biochemical data on gingiva. Sampling is probably the major source of error. The portion of the gingiva sampled may markedly alter the results, especially when working with small animals such as rats. Based on histologic observations of gingival tissue, there is every reason to suppose that attached gingiva would have a different chemical composition than free marginal gingiva and different from movable oral mucosa. Also, there is considerable difference in composition and morphology of gingiva between different species.

Table 1
COMPOSITION OF WHOLE GINGIVA[a]

Component	Human[c]	Human[20,d]	Porcine[22,d]
Collagen	45.0[4,7]	37.7	45.0
NCP[b] (total)	—	47.7	28.6
NCP-structural glycoproteins	20.0[51,59]	—	—
Lipids	7.2[20]	7.2	7.2
Glycosaminoglycans (GAG)	1.4[102]	—	—
DNA	0.85[7]	1.42	0.72
RNA	2.12[7]	1.52	2.2
Plasma proteins	—	—	3.8
RBC	—	—	3.0
Uncharacterized	24.43	—	—

[a] Expressed as percent of dry weight of tissue.
[b] NCP-noncollagenous proteins; total measured by biuret reaction.
[c] Values selected from the literature.
[d] Water content estimated as 74%.[4]

When working with human tissues, one major source of error is the difficulty of obtaining normal tissue for analysis. Even clinically normal tissue of adults usually has some inflammation. This fact probably accounts for most of the discrepancies in the literature.

Another consideration is the handling of tissue prior to analysis or extraction. Many authors routinely lyophilize tissue in order to preserve them and facilitate dissection. Lyophilization probably does not alter the results of the chemical analysis for most constituents, but would affect extraction of certain substances such as enzymes. The same reservation applies even more so to storing tissue under refrigeration.

Another source of difficulty in interpreting biochemical data is the many different dimensions used in reporting experimental values. The results are variably reported in terms of dry weight, wet weight, or even of stored wet weight of tissue. Sometimes they are expressed as a percent of total extract or the weight of the whole sample. In this review, an attempt will be made to convert all data to micrograms per milligram (or milligrams per gram) of dry weight of tissue to enable us to compare different studies.

CHEMICAL COMPOSITION OF WHOLE GINGIVA

Biochemically, gingiva is characterized by high water content, estimated to be 74% by Page[4] in human gingiva. Golub et al.,[19] working with normal tissue overlying unerupted third molars, estimated the water content to be 81 to 82%. The bulk of gingiva is formed by the connective tissue matrix composed of collagen, proteoglycans, glycoproteins, and lipids. In addition, the epithelium will contribute to the content of noncollagenous proteins, lipids, nucleic acids, and basement membrane components.

A representative composition of human and porcine gingiva is given in Table 1. For the purpose of comparison, all data were transposed into micrograms per milligram of dry weight, and where water content of the tissue was not given, an estimate of 74% was used.[4] The values in the first column were arbitrarily selected by the author and are representative of the overall composition of human gingiva. Collagen accounts for the bulk of gingiva, with a relatively large uncharacterized protein fraction, in addition to structural insoluble noncollagenous glycoproteins. The DNA accounted for less than 1% and RNA accounted for approximately 2.0% of the total gingiva. The GAGs, although present in small amounts relative to collagen, are in fact relatively more abundant in gingiva than in other soft

Table 2
ANALYSIS OF SUGARS IN THE GINGIVA (μg/mg OF DRY WEIGHT)[a]

	Hexosamine	Neutral sugars	Sialic acid	Fucose	Total sugar
Bovine gingivae	5.14 ± 0.29	14.0 ± 2.3	2.7 ± 0.16	1.55 ± 0.22	23.4
Porcine gingivae	6.45 ± 0.58	14.5 ± 1.9	2.3 ± 0.51	1.76 ± 0.16	25.00

[a] Mean of four to eight determinations ± SD.

From Mickalites, C. and Orlowski, W. A., *J. Dent. Res.*, 56, 1023, 1977. With permission.

connective tissues. Structural noncollagenous glycoproteins, proteoglycans, and lipids will be further discussed in other sections in this chapter.

Noncollagenous Proteins (NCP)

Different studies estimated the relative proportions of NCP as being either higher, equal to, or lower than collagen. For human gingiva, it is probably lower since collagen constitutes 50% of the total (see Table 1, Column 1). In inflamed tissues, the relative proportion of NCP would be higher. Rabinowitz and co-workers,[20] using a biuret method,[21] estimated the NCP fraction in human gingiva to be 47%, much higher than the amount of collagen (37%) (see Table 1, Column 2). This may reflect the presence of inflammation with the concomitant increase in serum proteins and a decrease in collagen.[7] In pig gingiva, the amount of NCP was 28.6%[22] (see Table 1, Column 3). Flieder et al.[7] estimated that 25% of dry weight of tissue of normal gingiva was noncollagenous protein. Golub et al.[19] found that soluble NCP accounted for 17% of the total protein and is probably the major portion of the uncharacterized protein fraction in Table 1, Column 1.

Little work has been done on characterization of the soluble pool of NCP, presumably soluble cellular components, and serum proteins. Most of the work to date has been on the insoluble matrix noncollagenous glycoproteins which will be discussed under a separate section.

Serum Proteins

Various attempts were made to estimate the amount of serum proteins in gingiva. Flieder et al.[7] assumed that most of the 25% of NCP identified in his study was derived from serum. This is probably not the case. Rabinowitz et al.,[22] using an isotope dilution method, estimated plasma content to be approximately 1% of wet weight or 3.8% of dry weight of tissue. They also estimated that 3.0% of dry weight were red blood cells. Both of these values appear somewhat high, but are probably close to the actual value.

Carbohydrate Analysis

Another estimate of noncollagenous components is carbohydrate analysis, since both Type I and Type III collagens contain very little bound carbohydrate (Table 2). Mickalites and Orlowski[23] found a relatively high sugar content in porcine and bovine gingiva and periodontal ligament, accounting for approximately 2.5 and 3.6% of dry weight, respectively. In the gingiva, neutral sugars comprised 60% of the total, hexosamines 25%, and fucose and sialic acids 15%. A total of 50% of hexosamine was recovered as glycosaminoglycans, the other 50% presumably were in glycoproteins. No effort was made to follow the distribution of sugars or acid mucopolysaccharides in the insoluble portion of the gingiva. Pearson et al.[24] also found that the periodontal ligament was rich in neutral sugars, bound to noncollagenous proteins.

Table 3
COLLAGEN ANALYSIS IN GINGIVA

Tissue	OH-Proline[a]	Collagen[b]	Ref.
Human	60.0	448	4
Human	91.0	678[c]	7
Human	42.3	317	35
Marmoset	42.0	313	4
Bovine	68.0	507	23
Porcine	45.0	335	23
Rat	27.3	204	37
Monkey	50.0	370	38

[a] Hydroxyproline (in μg/mg) of dry weight of gingiva.
[b] Calculated by multiplying hydroxyproline concentration by 7.46 (in μg/mg of dry weight).
[c] Connective tissue was separated by microdissection from the epithelium. Values are for connective tissue only.

COLLAGEN

Methods of Analysis

Collagen content of tissues is best estimated by hydroxyproline analysis. The method of Woessner[25] is most commonly used. In view of the fact that Type III collagen contains approximately 13 more residues of hydroxyproline per chain, this analysis should be viewed as a close estimate rather than an exact measurement. The calculated collagen concentration based on hydroxyproline content of Type I collagen would yield slightly inflated values. This consideration is much more important when evaluating metabolic studies or when comparing different tissues.

For more precise measurements and for the study of collagen polymorphism, collagen has to be isolated and the individual chains separated and characterized. For initial extraction of collagen from tissues, neutral salt solutions and weak acid solvents are utilized. Type I, Type III, Type II, and basement membrane collagens are also released from the insoluble residue by partial pepsin digestion[26] and then can be further separated by differential salt precipitation.[27,28] The different chains can be separated by liquid chromatography on CM-cellulose columns[29,27] or on SDS-polyacrylamide gel electrophoresis.[30] The α-chains separated by liquid chromatography are further characterized by cyanogen bromide (CNBr) cleavage and by mapping of CNBr peptides either by liquid chromatography[31-33] or by SDS-polyacrylamide gel electrophoresis.[30] The collagens are then identified either by the migration of individual α-chains on SDS gel electrophoresis, or by the pattern of CNBr peptides.

Recently developed immunohistochemical methods[18,34] further allow localization of different collagens in tissues. Both immunofluorescent[34] and immunoperoxidase[18] methods have been successfully utilized in the study of oral tissues.

The above is an overview of methods used in the isolation and characterization of collagen and is by no means complete. For more specific application and adaptation of the methods and additional methodology, the reader is referred to actual studies which will be discussed in the following section.

Collagen Content

Representative values for total collagen are given for different species in Table 3. Again all values were converted to micrograms per milligram of dry weight of tissue using the value of 74% for water content where appropriate. The marked differences in collagen content, 20 to 50%, in different species reflected the different amounts of epithelium and connective tissue and possible differences in sampling (attached vs. free gingiva). In earlier human studies, Schultz-Haudt and Aas[36] and From and Schultz-Haudt[35] obtained values that were considerably lower than those reported by more recent studies.[4,7] However, they did use a method which yielded lower results and was more difficult to control. A more accurate estimate was made by Flieder et al.,[7] who isolated noninflamed connective tissue by microdissection and obtained much higher values for human gingiva (Table 2). For comparison with other results, this value should be corrected for the absence of epithelium. A correction factor of 1.5 was estimated by Ostegaard and Löe,[39] and 1.57 was estimated by From and Schultz-Haudt[35] based on projected histometric measurement of the relative surface area of epithelium and connective tissue. By dividing the value obtained by Flieder et al.[7] by 1.5, a corrected value of 452 μg/mg is obtained and is approximately equal to value obtained by Page.[4] Golub et al.[19] did obtain a higher value of 664 μg/mg for mucosa overlying unerupted third molars. This value decreased markedly in inflamed tissues.

Hydroxyproline analysis has limited value when studying adult human gingiva because of the presence of inflammation which markedly alters the results. From and Schultz-Haudt,[35] in fact, often had difficulty correlating the histologic appearance of gingiva with hydroxyproline values. Hara et al.[40] clearly demonstrated that there was little correlation between the clinical appearance of gingiva and collagen content. Values for patients with the same clinical index of inflammation varied by 100%. On the other hand, there was a much better correlation with the histopathologic index of inflammation and collagen content. Unfortunately, the method of expressing results makes a comparison of this study with others difficult, nevertheless, their conclusions are justified on the basis of their results. Most individuals who attempted to obtain precise analysis of human tissue probably have had a similar experience.

Collagen Polymorphism

Our knowledge of the composition of gingiva is a reflection of the major advances in science in the last decade. Some of these will be briefly reviewed in order to put the recent advances on composition of gingiva in perspective.[41-49]

The discovery of Type II collagen by Miller and Matukas[41] in the chick embryo cartilage opened the field of the heterogeneity of collagen. To date, four matrix collagens have been discovered with the following chain composition: Type I-$(\alpha 1(I))_2\alpha 2$, Type II-$(\alpha 1(II))_3$, Type III-$(\alpha 1(III))_3$, and Type I-trimer-$(\alpha 1(I))_3$.[43,44]

Type I collagen is widely distributed and forms the bulk of larger collagen fibers in soft connective tissues, bone, and dentin. Type III collagen is associated with reticulin fibers and is usually located around blood vessels or in the smaller fibers that surround the larger Type I collagen fibers. Type II collagen is found only in hyaline cartilage, whereas Type I trimer has been isolated only from diseased tissues or cultured cells from a variety of tissues.[42,43]

In addition, several basement membrane collagens have also been identified. Kefalides[49] isolated Type IV basement membrane collagen composed of three $\alpha 1(IV)$-chains. More recent studies suggest that basement membrane collagens are a heterogenous group of proteins. To date at least five basement membrane collagen chains have been identified: chains A,B,C, D and αC.[44,50] There is a possibility that other membrane collagens exist in other tissues. Although there is still considerable controversy as to the nature of the basement membrane collagens, it is currently believed that Type IV collagen is composed of either three C chains, $(\alpha 1(IV))_3$ or three D chains, $(\alpha 2(IV))_3$.[50] Chains A,B, and αC are believed to be part of a different collagen, Type V, and are designated either as $\alpha 2(V)$, $\alpha 1(V)$, and $\alpha 3(V)$, respectively. Three Type V collagens have been identified with the following chain composition: $(\alpha 1(V))_2\alpha 2$, $(\alpha 2(V))_3$, and $(\alpha 3(V))_3$. Type IV collagen is largely distributed at the dermo-epithelial junction, glomerular basement membrane, and lens capsule. Type V collagen is usually located perivascularly and around certain cells. For more complete discussion of the types of collagen and their distribution, a recent review of Bornstein[44] and a monograph by Gay and Miller[43] are recommended.

Polymorphism of Gingival Collagen

The collagens of gingiva have been studied extensively and their polymorphism well demonstrated. To date, the following collagens have been identified either in normal or inflammed gingiva: Type I, Type I-trimer, Type III, Chains A and B (Type V), and Chain C (Type IV). The bulk of collagen is Type I forming over 80% of total collagen; however, precise proportions have been difficult to ascertain.[51-59]

Ballard and Butler[51] were first to isolate and characterize Type III collagen in gingiva. Schneir et al.[53] found that approximately 20% of gingival collagen was Type III collagen and that there was no difference in the relative proportion of Type I and Type III collagen in patients treated with Dilantin® (diphenylhydantoin), when compared to untreated patients. Dabbous et al.,[54] working with bovine gingiva, found that the insoluble gingival collagen contained 17% of Type III and 83% Type I collagen. Butler et al.[55] found similar amounts of Type III collagen, approximately 20%, in periodontal ligament. Narayanan et al.,[56] in comparing gingiva and edentulous ridge, estimated the proportion of Type III collagen to be 5 to 25% of the total. Finally, Wang et al.[17] estimated Type III collagen content in porcine tissues to be 7.5% in gingiva, 12% in periodontal ligament, 6% in alveolar bone, and 2.5% in cementum. Birkedal-Hansen et al.[57] found approximately 5% of Type III collagen in cementum. When working with inflamed tissues, lower estimates of Type III collagen are usually obtained: 3.8% by Narayanan et al.[58] and 10% by Hammouda et al.[59]

Metabolic studies further demonstrated active synthesis of Type III collagen by gingival tissues in vivo[60] and in vitro.[52,61] Sodek and Limeback[60] found active synthesis of Type III collagen in rat periodontal ligament and gingiva accounting for 17 and 10% of the total newly synthesized collagen, respectively. Narayanan and Page[52] found decreased synthesis of Type III collagen in cultured fibroblasts from human inflamed gingiva. Engel et al.[61] demonstrated that most fibroblasts from human gingiva synthesized both Type I and Type III collagens, although a small population of fibroblasts only synthesized Type I.

In addition, several basement membrane collagens have been identified both biochemically and immunohistochemically. Narayanan et al.,[58] using biochemical techniques, found high levels of A and B chains (Type V collagen) and a lesser amount of C chains in inflamed human gingiva. The concentration of Chains A and B in normal gingiva was 0.23 and 0.67%, respectively.[62] Sauk,[63] using immunohistochemical techniques, identified C chains (Type IV) as the only basement membrane collagen present in the dento-epithelial attachment and between lamina propria and the surface epithelium. Chains A and B (Type V) were absent in these sites which is consistent with the other observations that they mainly form the cytoskeleton of certain cells, notably smooth muscle cells.

Immunohistochemical methods have proved useful in localizing different types of collagens in gingiva.[18,64] Wang et al.[17] found that the bulk of larger collagen fibers in periodontal ligament such as Sharpey's fibers was Type I collagen and that Type III collagen is distributed at the periphery of larger fibers and around blood vessels. The distribution of Type III collagen is the same as that of reticulin fibers, and it is generally believed that it forms the bulk of reticulin fibers. Sauk,[63] using specific antibodies against Chains A,B, and C, found only Chains C in the dento-epithelial junction as discussed previously.

The immunohistochemical techniques have also been used to study certain diseases of connective tissue other than periodontitis. Sauk et al.[65] found that in dentinogenesis imperfecta, there was an accumulation of Type III collagen in bone and dentin, which may reflect the inability of collagen to mature. Other diseases have not been studied as yet, but hopefully the use of these techniques will greatly elucidate their pathogenesis.

Another important finding has been the isolation of Type I trimer $(\alpha 1(I))_3$ by Narayanan and Page[52] in fibroblast cultures from inflamed human gingiva. These cultures also secreted less Type III collagen than do normal fibroblast lines. Narayanan et al.[58] conclusively demonstrated that Type I trimer is also present in inflamed gingiva in vivo. The presence of this type of collagen may markedly affect the properties of adult human gingiva and its resistance to disease, since most clinically normal human gingiva would exhibit some degree of inflammation histopathologically. In addition to human inflamed gingiva, Type I trimer was identified in both the salt and the guanidine chloride extracts of dentin collagen from lathyritic rats[66,67] in normal predentin, and in cultures of odontoblasts in vitro.[68]

It has been difficult to ascertain the exact proportion of different collagens in adult human gingiva because the relative concentrations of each type varies with inflammation. Narayanan et al.[58] found that inflamed human gingiva contained 3.8% of Type III, 86.8% of Type I, 8.0% of Type A and B chains (Type V), and 1.5% of Type I trimer. Hammouda et al.[59] also found a lower proportion of Type III collagen in inflamed tissues, being approximately 10% of the total. The high levels A and B chains (Chains B were three to four times more abundant) appear to be directly related to inflammation, and this finding was further supported by subsequent studies of Narayanan et al.[62]

Collagen Cross-Linking

The main intramolecular cross-link has been identified as an aldol condensation product between two allylsyl residues in the amino terminal nonhelical portion of the tropocollagen molecule. There are also many reducible intermolecular cross-links which are derived from adjacent lysine and/or hydroxylysine residues.[69]

The cross-links in gingiva have been characterized by Furuto et al.[70] and by Kuboki et al.[71] Furuto et al.[70] found that in the soluble collagen of rat gingiva, the main cross-links comprising 60 to 70% of the total were dihydroxylysinonorleucine (DHLNL) and hydroxylysinonorleucine (HLNL) in a ratio of 1.2:1.0. In addition, they did find an appreciable amount of histidinohydroxymerodesmosine (H_{is}-HMD). Kuboki et al.,[71] on the other hand, reported that the major cross-link in the insoluble collagen of bovine gingiva was H_{is}-HMD, 57.4% of the total. Also the ratio of DHLNL to HLNL was 0.18, the lowest of any·tissue tested. It is somewhat difficult to reconcile some of these data. It should be noted that different collagen fractions, soluble vs. insoluble and different species were examined. Pearson et al.[24] found that the main cross-link in the periodontal ligament was DHLNL. The role of these cross-links is not clear. They appear to decrease in number with age and may give rise to new more stable covalent cross-links.[69]

Solubility of Gingival Collagen

One of the important properties of gingival collagen is its marked insolubility. This is particularly true of rat collagen.[72,73] Schneir et al.[72] also found that human gingival collagen was much less soluble in acid solutions. Sodek,[73] in studying the metabolism of periodontium, found that collagen in oral tissues was much less soluble. The amount of salt soluble collagen in periodontal ligament, gingiva, and skin was 2.8, 3.0, and 6.3%, respectively. This insolubility was unrelated to cross-linking[74,75] and is probably a reflection of interaction with other components of connective tissue such as glycoproteins.

In our study on the effects of cross-linking on the turnover of gingival collagen,[76] we found that gingival collagen was at least three times less soluble than skin collagen and that lathyrism had less of an effect on collagen solubility in gingiva than in skin. This decreased effect of lathyrism on oral collagen was previously reported by Schneir et al.[75]

NONCOLLAGENOUS MATRIX GLYCOPROTEINS (STRUCTURAL GLYCOPROTEINS) (NCP)

The separation and identification of structural glycoproteins by Robert et al.[77] have opened the field of the structural noncollagenous proteins of connective tissue matrix. These glycoproteins are rich in amino acids, serine, threonine, and glycine. They also contain a variety of sugars including glucose, galactose, mannose, fucose, and sialic acid. They are highly insoluble because of close binding to collagen and are released only by degrading collagen either by gelation with heat[77] or by collagenase digestion.[78] Recently, Ballard and Butler[51] used cyanogen bromide cleavage to solubilize collagen and release noncollagenous proteins.

The noncollagenous proteins (NCP) associated with insoluble collagen were extracted by Ballard and Butler[51] from gingival tissue of stillborn infants and patients treated by diphenylhydantoin (Dilantin®). They extracted the gingival tissues exhaustively with neutral salt and acetic acid solution and then degraded the insoluble collagen with cyanogen bromide. The NCP comprised 7% of dry weight of gingiva in stillborn infants and 20% in patients treated with Dilantin.® Dabbous and Brinkley[79] found that 13 to 15% of bovine gingiva was NCP and that it had an amino acid composition typical of that reported for structural glycoproteins. In gingiva from patients with chronic periodontitis, Hammouda et al.[59] found that 17 to 20% of the insoluble matrix was NCP. Whether this represents an increase in NCP due to inflammation is not clear. The marked insolubility of these glycoproteins and their close association with insoluble collagen has led to the hypothesis that they play an important role in aggregating small collagen fibrils into larger fibrils.[80]

Whether any of these structural matrix proteins are tissue specific is uncertain. However, Scott and Veis[81,82] found that they were tissue specific for skin and bone. Many of these proteins are apparently bound to α1-CNB-6 of Type I collagen since there was a 30% decrease of this peptide in collagen preparations. This observation was confirmed by Dabbous and co-workers.[54,79] This somewhat difficult area of research offers great possibilities for the understanding of the biologic function of different connective tissue matrices.

Fibronectin

Fibronectin (synonyms: large external transformation sensitive protein, LETS; cold insoluble globulin; cell surface protein) is a newly discovered glycoprotein found in serum, which forms a coating over the surface of the cells.[83-85] It is the means of attachment of cells to the matrix collagen. Using indirect immunofluorescence and radiolabeling, Baum and Wright[83] demonstrated that gingival fibroblasts synthesized appreciable amounts of fibronectin in tissue culture. A recent review of the chemistry and biology of fibronectin by Ruoslahti[85] is recommended for additional background information.

The full significance of this protein in both physiologic and pathologic states is still being explored. Only one study explored fibronectin in oral diseases. Zetter et al.[84] found that this protein (LETS) was altered in several pathologic conditions affecting oral mucosa, including lichen planus and lupus erythematosus. Another important finding was that fibronectin synthesis is markedly increased in cultures of cancer cells, raising a question as to whether lack of fibronectin is responsible for the invasiveness of cancer. A few preliminary reports on the significance of fibronectin and oral tissues have also appeared.[85a,b]

PROTEOGLYCANS*

Analysis

The proteoglycans can be extracted from soft connective tissue by the methods essentially developed for cartilage,[86] using either 4*M* guanidine or 2*M* calcium acetate. Agarose gel columns have been utilized to further purify and separate the proteoglycans.

For analysis of the individual GAGs, the proteins are first degraded by enzymes, either

* The following abbreviations will be used in this section: GAG — glycosaminoglycans; HA — hyaluronic acid; C-4-S — chondroitin-4-sulfate; C-6-S — chondroitin-6-sulfate; DS — dermatan sulfate; HS — heparan sulfate; Hep — heparin; KS — keratosulfate.

papain, pepsin, or ficin.[87] Because of high negative charge they are readily precipitated with cationic detergents. They can be further separated on a cetylpyridinium chloride (CPC) impregnated microcellulose column originally developed by Antonopoulos et al.[88] The modification of Svejcar and Robertson[89] is generally used today. This method uses differential solubility of CPC and GAG complexes in $MgCl_2$ salt solutions for separation. However, there is considerable overlap of the different isomers of CS as well as CS and HS in different fractions and further characterization of these fractions is essential for accurate results.[90]

The discovery of specific chondroitinases by Saito et al.[91,92] and further separation of digestion products by paper chromatography[93] has made possible precise estimation of different isomers of CS, HS, and other GAGs even when present in small quantities. There is a wide variety of adaptation of these methods, and the reader is referred to specific articles to be discussed in the succeeding section for application of the methodology.

Histochemical Analysis

Earlier studies were largely confined to histochemical techniques. Because of their high negative charge, GAGs lend themselves readily to histochemical study. Fullmer,[94,95] in the course of studying the development of oxytalan fibers using azure A and Reinhardt's stains, observed that these fibers appeared to arise from abundant GAGs (acid mucopolysaccharides) which were present between collagenous fibers. Quintarelli,[96] using toluidine blue and azure A metachromasia, found that the gingiva gave more pronounced metachromasia than skin, but less than cartilage. From this study and Fullmer's,[95] it appears that there may be more acid mucopolysaccharides in the gingival tissues than in skin. Fullmer[97] and Melcher[98] also made extensive studies on changes in GAGs in disease.

More recent histochemical studies have proved useful in following the distribution of GAGs in the tissues. Today Alcian® blue has replaced other stains used in the past for characterization and identification of GAG in tissues. The method generally adopted is the critical salt concentration at pH 5.8 of Scott and Dorling[99] or modifications of it. Using that method and also selective digestion with specific chondroitinases and hyaluronidase, Pedlar[100] demonstrated that GAGs are confined to the connective tissue and there there was considerable difference in their distribution and much more intense staining was observed in gingiva than in the tongue. Most sulfated GAGs were closely associated with collagen fibers, whereas in the lamina propia more hyaluronic acid was present.

Plecash[101] used ruthenium red binding to observe the distribution of proteoglycans under the electron microscope. He found three structures which bound ruthenium red:

1. Fine collagen fibrils of 30 to 50 Å which appeared to extend from one collagen fiber to another
2. Dense coating around collagen fibers of about 100 Å in thickness
3. Dense deposits of 150 to 200 Å distributed randomly along dense collagen fibers

Although ruthenium red does bind to proteoglycans, it will also bind to other acidic substances such as acidic glycoproteins.

Glycosaminoglycan (GAG) Analysis (Biochemical Studies)

Biochemical analysis revealed considerable variation in the total amount of GAGs in different species (Table 4). For most species, the content is approximately 0.5% of dry weight of tissue. An estimate of GAG content can be obtained by multiplying uronic acid content in Table 4 by 2.5. The relatively high value of 1.4% obtained by Embery et al.[102] for human gingiva differs from that of Ciancio and Mather,[103] who found only 0.5%, and is much higher than any results reported for soft connective tissues. Embery's[102] observation is, however, supported by a study of Chiappa and Tussing,[104] who found that HA and CS together accounted for 1.11% of dry weight of tissue.

Table 4
GLYCOSAMINOGLYCANS IN GINGIVA (μg/mg OF DRY WEIGHT)

Tissue	Uronic acid	Ref.
Human	4.8	102
Human	2.2	103
Bovine	2.02	23
Bovine	2.13	108
Porcine	2.52	23
Porcine	2.30	90
Rat	1.08	111
Rat	1.84	110

Table 5
COMPONENTS OF GINGIVAL GLYCOSAMINOGLYCANS[a]

Component	Human[102]	Bovine[108,b]	Porcine[90]
HA	43.0	24.2	15.7
C-4-S	23.0	37.3	31.1
C-6-S	0.0	0.0	8.0
DS	33.0	36.0	29.9
HS	0.0	2.6	15.3
Total uronic acid[c]	0.48	0.21	0.23

Note: Abbreviations: hyaluronic acid (HA); chondroitin-4-sulfate (C-4-S); chondroitin-6-sulfate (C-6-S); dermatan sulfate (DS); heparan sulfate (HS); heparin (Hep); keratosulfate (KS); glycosaminoglycans (GAG).

[a] Expressed as percent of total uronic acid.
[b] Values obtained for 1-year-old animals were used for the purpose of comparison.
[c] Expressed as μg/mg of dry weight.

Schultz-Haudt[105] was first to use biochemical techniques for the study of GAGs in gingiva. He identified both hyaluronic acid and chondroitin sulfate in human gingiva and was able to identify a testicular hyaluronidase resistant component, presumably DS. Thonard and Blustein[106] in gingiva and Paunio[107] in periodontal ligament found similar results. In our study,[23] we found that 50% of hexosamine in the tissue was actually found in GAG, with the remainder presumably being bound glycoproteins.

In the last few years, major progress has been made in characterization of individual GAGs in the gingiva of different species including man (Table 5). The methods described in the previous section were used in these studies. There is a general agreement that C-4-S, DS, and HA account for the bulk of GAGs. Sakamoto et al.[108] also isolated smaller amounts of HS, approximately 7.5% of the total. Hiramatsu et al.[90] found higher levels of HS (15.3%) and in addition isolated a lesser amount of C-6-S(8.0%). Embery et al.[102] did not find either C-6-S or HS in their preparations from human gingiva.

Judging from the methods employed by these authors, one would have to conclude that it is unlikely the differences were due to the methods and that instead they may reflect species differences or, more likely, different sampling techniques. It is possible that the relative proportion of C-6-S and HS will be dependent on the amount of periodontal ligament and periosteum that is incorporated into the sample. Munemoto et al.[109] found appreciable amounts of both GAGs in the preparations from periodontal ligament. Embery et al.[102] used human gingiva obtained from gingivectomies and thus probably only dealt with free marginal gingiva. This may explain the absence of HS and C-6-S in their preparations.

Kofoed and Tocci[110] and Kofoed[111] reported findings of other GAGs in the gingiva of a rat, including Hep. They used a microcellulose column for isolation of their GAGs. The fractions obtained from such a column usually contain some contamination from related GAGs, as was demonstrated by Hiramatsu in his study of gingival GAGs. It is not known whether the Hep fraction obtained by Kofoed[111] contained any CS.

Sakamoto et al.[108] also studied the age changes in the pattern of GAGs in bovine gingiva. With age there was an overall decrease in the total amount of GAGs with a relatively greater decrease in the amount of C-4-S and DS. As a result, the relative amount of hyaluronate increases with age, a somewhat paradoxical finding initially reported by Paunio.[107] This finding differs from those reported in other tissues such as skin.[53]

In summary, it appears that the gingiva of different species is rich in C-4-S, DS, and HA and contains lesser amounts of HS. C-6-S may or may not be present. In addition, age changes are rather striking and more pronounced than those of skin. This aspect has not been studied in man.

Isolation of Proteoglycans from Gingiva

The new advances in the isolation and separation of proteoglycans and glycosaminoglycans (GAGs) increased our understanding of their structure and their aggregation, and has enabled characterization of a large number of them including those of the gingiva. Most of the work was done on cartilage proteoglycans. A typical cartilage proteoglycan is composed of a protein core to which are attached GAGs, namely, C-6-S and KS. These proteoglycans are in turn aggregated into large molecules of 50,000,000 mol wt on the hyaluronic acid chain through a specific link-peptide.[113,114] It is not known whether hyaluronate plays an aggregating role in the soft tissues.

The proteoglycans of the soft tissues appear to be a diverse group of substances.[161] They usually contain more proteins and fewer chains of GAGs.[102,115] They are much more difficult to isolate and characterize since they are present in small amounts, often in close association with collagen and glycoproteins.

Recently it was also discovered that the HS proteoglycan is a major component of basement membranes, along with Type IV collagen and laminin, and its contents are markedly reduced in diabetes.[115a,b] This finding has opened a new area of interest in proteoglycan research.

There are two recent studies on the isolation of proteoglycans from gingiva. The human gingival proteoglycan isolated by Embery et al.[102] contained two GAGs, C-4-S, and DS, with a total protein content of 46% and an equimolar ratio of uronic acid and hexosamine. It was rich in aspartate, glutamate, serine, threonine, glycine, alanine, and leucine. It thus resembled the proteoglycan of the skin.[115] The proteoglycan isolated from bovine gingiva by Dziewiatkowski et al.[116] contained 8% protein, 30% uronic acid (glucuronic and iduronic acid), 32% hexosamine (*N*-acetylglucosamine and *N*-acetylgalactosamine), 15% sulfate, 3 to 5% of neutral sugars, and 3 to 5% fucose. HA acid was associated with the proteoglycan in some manner. Whether DS was the only chondroitin sulfate was not ascertained.

Analysis of Glycosaminoglycans in the Epithelium

The proteoglycans and their GAGs are basically the products of fibroblasts and an integral part of the ground substance of connective tissue matrix. However, several papers reported findings of acid mucopolysaccharides in the gingival epithelium.[117-119] Most of these were histochemical studies using toluidine blue or Alcian® blue at pH 2.5. Neither of these methods are specific for GAGs. Thonard and Wiebken[119] also demonstrated histochemically that

Table 6
LIPIDS IN GINGIVA[a]

	Human[20]	Human[122]	Porcine[22]	Bovine[124]
Total[a]	5.65	5.65	7.12	7.67
Nonpolar	70.0	67.0	69.8	70.7
Free fatty acids	17.0	19.2	21.0	17.0
Triglycerides	14.34	17.6	17.2	16.9
Monodiglycerides	4.14	10.3	8.2	7.0
Cholesterol esters	11.6	12.9	16.3	22.8
Cholesterol	10.6	7.0	8.2	7.0
Phospholipids (total)	30.0	33.0	30.2	29.3
Phospho-choline	15.9	13.4	11.3	10.0
Phospho-ethanolamine	8.2	6.3	10.4[a]	9.4
Shingomyelin	1.3	4.7	3.8	8.8
Other	4.6	8.6	5.1	Trace

[a] Total lipids expressed as percent of dry weight of tissue. The remaining values are expressed as percent of total lipids.

amniotic epithelial-like cells can produce GAG. Recently, Bartold et al.[120] separated epithelium from the underlying connective tissue and biochemically isolated and identified the GAGs. They found that about 10% of GAGs were recovered in the epithelial fraction. Whether these are bound to basement membranes or possibly represent GAGs that diffused into the epithelium is still to be determined. Using a more specific histochemical technique, Pedlar[100] was unable to demonstrate any GAGs in the gingival epithelium. It is possible that GAG identified histochemically in the epithelium may have diffused out of the connective tissue following degradation during inflammation.

LIPIDS

Lipid Analysis in Gingiva

The use of silica gel column, gas chromatography and paper, and thin layer chromatography has greatly facilitated analysis and quantitation of lipids of gingiva. Rabinowitz and co-workers,[20,22,121] Pellat et al.,[122] and Das et al.[123,124] have isolated and characterized lipids in porcine, human, and bovine gingiva. The lipid content of human gingiva is approximately 1.5% on a wet weight basis or 5.8% on a dry weight basis (Table 6). Human and porcine gingiva were remarkably similar in their lipid composition. Approximately 70% of the total were neutral lipids and free fatty acids, whereas phospholipids accounted for 30.0%. Characteristically, gingiva contained large amounts of free fatty acids and cholesterol esters, which accounted for two thirds of the total cholesterol in pig[22] and bovine gingiva.[124] Triglycerides and diglycerides accounted for over 20% of the total lipids. The most common phospholipids were phosphotidyl choline and ethanolamine which together accounted for 70% of total phospholipids or about 20% of total lipids.[20,22,122] The human gingiva differed from porcine in that it did contain more phosphotidyl choline and methyl esters but considerably less sphingomyelin. Das et al.[123] also found a somewhat similar composition in bovine gingiva (Table 6). In addition to the common occurring lipids, Wood et al.[125] also identified ceramides and cerebrosides in monkey gingiva. These are important components of cell membranes and are difficult to extract.

Prostaglandins

Another important finding has been the identification of considerable amounts of arachidonic acid in gingiva. It is an important precursor of prostaglandins and thus gingiva appears to possess a high capacity for prostaglandin synthesis (PGE_2, PGF).[126-130] Lower levels of thromboxane and prostacycline[130] were also identified in human gingiva. These substances are important in both physiologic and pathologic bone resorption and in modulating the inflammatory process.

Distribution of Lipids (Histochemical and Biochemical Studies)

The distribution of lipids in different parts of gingiva was also studied histochemically[131] and biochemically.[121] Anneroth and Ivemark[131] found that keratinizing epithelium stained most intensely and identified phospholipids, free fatty acids, and cholesterol in high amounts in epithelium. They were heavily concentrated in keratinizing portions of the epithelium. Cholesterol was localized in large amounts in the outermost keratinized layer. The intensity of staining appeared to be directly proportional to loss of water from tissues and degree of keratinization, however, keratohyalin granules stained only weakly for lipid.

Rabinowitz et al.[121] separated pig gingiva into epithelial and connective tissue portions. They found that most of the fatty acids were found in the epithelium, whereas the connective tissue portion contained larger amounts of triglycerides and cholesterol. Almost all of the arachidonic acid was present in the connective tissue. Phospholipids appeared to be equally distributed when expressed in terms of total lipids. However, when the values were expressed in terms of the weight of tissue, the concentration of both neutral lipids and phospholipids were approximately 50% higher in the epithelial portion of the gingiva, thus confirming the histochemical observations. Most of the arachidonic acid is located in connective tissue which suggest that fibroblasts may be the main endogenous source of prostaglandins.

Role of Lipids

The exact role of lipids in the gingiva is not known. In addition to being important constituents of cell membranes and a source of energy, lipids do appear to play a protective role in the epithelium and seem to be intimately related to keratinization. Also Nikkari and Heikkinen[132] found that about 10% of collagenous matrix is composed of lipid and that only a small amount, or approximately 10%, is extracted by the usual methods employed. Surprisingly, few studies exist on the interaction of lipids and collagen. Recently, Sauk and Roszkowski[133] found that bacterial lipids have an inhibiting capacity on collagen aggregation. More studies are needed on the lipids of connective tissue and their interaction with collagen and other matrix components.

ENZYMES

Gerson and Meyer[3] exhaustively reviewed the literature on this subject prior to 1970. These included the studies on both the connective tissue and the epithelium. It is not the purpose of this section to go into details of enzyme action. Nevertheless, the more important studies on the gingiva will be reviewed and in particular the isolation and characterization of collagenase because of its importance in connective tissue metabolism and because so much of the original research was done on human gingiva.[134-137]

Isolation of Collagenase from Gingiva

Collagenase is the only gingival enzyme that has been isolated and characterized.[135,137] Initially Gibson and Fullmer[138] identified the enzyme in inflamed human gingival tissue. In subsequent studies, Fullmer et al.[134] extracted and purified the enzyme from cultures of human gingival tissue. It is a protein of 40,000 mol wt which acts at neutral pH and requires calcium for its activity.

In a series of papers from Fullmer's laboratory, Birkedal-Hansen et al.[134,139] describe the isolation, purification, and distribution of bovine gingival collagenase. It is localized in fibroblasts subjacent to the epithelium.[140] It is a protein of 60,000 mol wt and is usually present in the tissues in an inactive form which can be readily activated either by trypsin or by thiocyanate. The ability of thiocyanate to activate the enzyme suggests that this enzyme is also bound to an inhibitor. Such inhibitors have been identified in serum: α-2 macroglobulin and β-anticollagenase.[141] Uitto et al.[137] isolated collagenase from inflamed human gingiva and found that most of it was strongly bound to collagen. Along with collagenase, α-2 macroglobulin was also present. Thus, collagenase may exist in tissues either intracellularly or extracellularly and either as a precursor or bound to an inhibitor.

Other enzymes involved in collagen metabolism have not been studied to the same extent in gingiva. However, major advances have been made in isolation and characterization of prolyl hydroxylase,[47] lysyl oxidase,[142] and glucosyltransferase.[143] The reader is referred to respective reviews for further information. The only studies on these enzymes in the dental literature are Ramamurthy et al.[144] and Ramamurthy and Golub.[145,146] They used the prolyl hydroxylase and lysyl hydroxylase activity as a measure of collagen synthesis and found that levels were depressed in inflammation,[144] but were unaffected by diabetes.[145] They also found that lysyl oxidase activity[146] was decreased in lathyrism, but relatively increased in diabetes.

Lysosomal Enzymes

Many lysosomal enzymes have been studied histologically and biochemically, and their activity is readily detected in inflamed gingiva.[147-150] A large proportion of these exist in precursor or latent form and are only released during inflammation.[138,147] The following lysosomal enzymes were studied in gingival tissues: hyaluronidase,[149] β-glucuronidase,[8,147] alkaline phosphatase,[147] acid phosphatase, and cathepsin B.[149,150] These enzymes are important in the normal metabolism of cells. When released during inflammation, they are responsible for the breakdown of connective tissues. Gerson[149] found a much higher level of acid phosphatase in keratinizing when compared to nonkeratinizing epithelium, suggesting a role for lysosomal enzymes in keratinization.

Enzymes of Energy Metabolism

The enzymes of the glycolytic and the hexose monophosphate shunt have also been studied extensively.[3,151-154] Gerson and Meyer[3] reviewed most of the studies prior to 1970. More recently Berquist and Nuki[152-154] in a series of papers reported on distribution of hexokinase, fructoaldolase, and glucose-phosphate dehydrogenase in the epithelium. They utilized microdissection of frozen sections and microanalytical techniques. The activity was invariably higher in the epithelium than in the connective tissue. Hexokinase activity was highest in the surface layers of the epithelium and lowest in the basal cell layer. The distribution of fructoaldolase, on the other hand, was the opposite of hexokinase. This suggests that hexokinase is more important for the synthetic activity of the epithelium.

EPITHELIAL PROTEINS

There is only a handful of studies dealing with biochemical analyses of keratin and related proteins. The first attempt to identify major epithelial proteins was by Eastoe,[155] who extracted human oral epithelium by destroying the connective tissue by autoclaving. Using amino acid analysis, he found that the major protein of oral epithelium was keratin, but that it was quite different from skin keratin. Dale et al.[156] isolated three proteins from palatal epithelium of a rat and found that they were similar to fibrous proteins of the epidermis. They did find one component in the epidermis that was absent in the oral epithelium.

Recently there were successful attempts at culturing keratinocytes[157,158] in vitro, both from the gingiva and from the buccal mucosa. Taichman et al.[157] showed that they readily proliferate and resemble stratified squamous epithelium, and the electron microscope reveals most of the structures associated with squamous epithelium. In a subsequent study, Taichman and Prokop[159] demonstrated that approximately 50% of the cell mass of keratinocytes may be keratin-like proteins.

Another important product of the epithelial cell is laminin, a protein responsible for the binding of cells in the basement membrane.[160] Terranova and Martin[85a] recently reported on its role in periodontal reattachment, but its full significance in oral tissues awaits further exploration.

REFERENCES

1. **Stern, I. R.,** Oral mucous membranes, in *Orban's Oral Histology and Embryology,* 9th ed., Bashkar, S. N., Ed., C. V. Mosby, St. Louis, 1980, 288.
2. **Melcher, A. H. and Eastoe, J. E.,** The connective tissues of the periodontium, in *Biology of the Periodontium,* Melcher, A. H. and Bowen, W. H., Eds., Academic Press, New York, 1969, 167.
3. **Gerson, S. and Meyer, J.,** Biochemical assay of heterogenous soft tissues of the oral cavity, in *Advances in Oral Biology,* Vol. 4, Staple, P. H., Ed., Academic Press, New York, 1970, 289.
4. **Page, R. C.,** Macromolecular interactions in the connective tissues of the periodontium, in *Developmental Aspects of Oral Biology,* Slavkin, H. C. and Bavetta, L. A., Eds., Academic Press, New York, 1972, 291.
5. **Nuki, K. and Bonting, S.,** Quantitative histochemistry of the developing hamster tooth: alkaline phosphatase and lactic dehydrogenase, *J. Histochem. Cytochem.,* 9, 117, 1961.
6. **Gerson, S.,** Quantitative histochemistry of rat oral mucosa: glucose-6-phosphate dehydrogenase, *Arch. Oral Biol.,* 12, 891, 1967.
7. **Flieder, D. E., Sun, C. N., and Schneider, B. C.,** Chemistry of normal and inflamed gingival tissue, *Periodontics,* 4, 320, 1966.
8. **Goggins, J. F. and Billups, L. C.,** Cytochemical measurement of beta-glucuronidase activity in normal and inflamed human gingiva, *J. Histochem. Cytochem.,* 19, 416, 1971.
9. **Lowry, O.,** The quantitative histochemistry of the brain, *J. Histochem. Cytochem.,* 1, 420, 1953.
10. **Felton, J. H., Person, P., and Stahl, S. S.,** Biochemical and histochemical studies of aerobic oxidative metabolism of oral tissues. II. Enzymatic dissection of gingival and tongue epithelia from connective tissue, *J. Dent. Res.,* 44, 392, 1965.
11. **Rabinowitz, J. L., Taveres, C., and Marsh, J. B.,** Polar and neutral lipids of epithelium and connective tissue of pig gingiva, *J. Periodontal Res.,* 7, 230, 1972.
12. **Dale, B. A., Stern, I. B., and Clagett, J. A.,** Initial characterization of the proteins of keratinized epithelium of rat oral mucosa, *Arch. Oral Biol.,* 22, 75, 1977.
13. **Araya, K., Imagawa, Y., and Shizya, H.,** Separation of oral epithelium from connective tissue, *J. Dental Res.,* 41 (Abstr.), 732, 1962.
14. **Eastoe, J. E.,** The amino acid composition of proteins from the oral tissue. I. A comparison of human oral epithelium, epidermis and nail proteins, *Arch. Oral Biol.,* 8, 449, 1963.
15. **Pedlar, J.,** Histochemistry of glycosaminoglycans in the skin and oral mucosa of the rat, *Arch. Oral Biol.,* 24, 777, 1979.
16. **Michel, B., Milner, Y., and David, K.,** Preservation of tissue-fixed immunoglobulins in skin biopsies of patients with lupus erythematosus and bullous diseases — preliminary report, *J. Invest. Dermatol.,* 59, 449, 1973.
17. **Wang, H-M., Nanda, V., Rao, L. G., Helcher, A. H., Heersche, J. N., and Sodek, J.,** Specific immunohistochemical localization of Type III collagen in porcine periodontal tissues using the peroxidase-antiperoxidase method, *J. Histochem. Cytochem.,* 28, 1215, 1980.
18. **Sternberger, L. A., Hardy, P., Jr., Cuculi, J. J., and Meyer, H. G.,** The unlabelled antibody-enzyme method of immunohistochemistry. Preparation and properties of soluble antigen antibody complex (horseradish peroxidase-antihorseradish peroxidase) and its use in identification of spirochetes, *J. Histochem. Cytochem.,* 18, 315, 1970.
19. **Golub, L. M., Curran, J. B., and Kennett, S.,** Collagen metabolism in pericoronitis, *J. Oral Pathol.,* 2, 222, 1973.
20. **Rabinowitz, J. L., Rutberg, M., Cohen, D. W., and Marsh, J. B.,** Human gingival lipids, *J. Periodontal Res.,* 8, 381, 1973.

21. **Gornwall, A. T., Bardawill, C. J., and David, M. M.,** Determination of serum proteins by means of the biuret reaction, *J. Biol. Chem.*, 177, 751, 1949.
22. **Rabinowitz, J. L., Bailey, T. A., and Marsh, J. B.,** Polar and neutral lipids of pig gingiva, *Arch. Oral Biol.*, 16, 1195, 1971.
23. **Mickalites, C. and Orlowski, W. A.,** Study of the non-collagenous components of the periodontium, *J. Dent. Res.*, 56, 1023, 1977.
24. **Pearson, C. H., Wohllebe, M., Carmichael, D. J., and Chovelon, A.,** Bovine periodontal ligament: an investigation of the collagen, glycosaminoglycans and insoluble glycoprotein components at different stages of tissue development, *Connect. Tissue Res.*, 3, 195, 1975.
25. **Woessner, J. G., Jr.,** The determination of hydroxyproline in tissues and proteins. Samples containing small proportions of this amino acid, *Arch. Biochem. Biophys.*, 93, 440, 1961.
26. **Miller, E. J.,** Structural studies on cartilage collagen employing limited cleavage and solubilization with pepsin, *Biochemistry*, 11, 4903, 1972.
27. **Chung, E. and Miller, E. J.,** Collagen polymorphism: characterization of molecules with the chain composition $(\alpha\ 1(III))_3$ in human tissues, *Science*, 183, 1200, 1974.
28. **Epstein, E. H.,** $(\alpha\ 1(III))_3$ Human skin collagen. Release by pepsin digestion and preponderance in fetal life, *J. Biol. Chem.*, 249, 3225, 1974.
29. **Piez, K., Eigner, E., and Lewis, M.,** The chromatographic separation and amino acid composition of the subunits of several collagens, *Biochemistry*, 2, 58, 1963.
30. **Furthmayr, H. and Timpl, R.,** Characterization of collagen peptides by sodium dodecylsulfate-polyacrylamide electrophoresis, *Anal. Biochem.*, 41, 510, 1971.
31. **Bornstein, P. and Piez, K. A.,** The nature of the intramolecular cross-links in collagen. The separation and characterization of peptides from the cross-link region of rat skin collagen, *Biochemistry*, 5, 3460, 1966.
32. **Butler, W. T., Piez, K. A., and Bornstein, P.,** Isolation and characterization of the cyanogen bromide peptides from the α 1-chain of rat skin, *Biochemistry*, 6, 3771, 1967.
33. **Lichtenstein, J. R., Byers, P. H., Smith, B. D., and Martin, G. R.,** Identification of the collagenous proteins synthesized by cultured cells from human skin, *Biochemistry*, 14, 1589, 1975.
34. **Timpl, R.,** Immunological studies on collagen, in *Biochemistry of Collagen*, Ramachandran, G. N. and Reddi, A. H., Eds., Plenum Press, New York, 1976, 319.
35. **From, S. HJ. and Schultz-Haudt, S. D.,** Comparative histological and microchemical evaluations of the collagen content of human gingiva, *J. Periodontol.*, 34, 216, 1963.
36. **Schultz-Haudt, S. D. and AAS, E.,** Observations on the status of collagen in human gingiva, *Arch. Oral Biol.*, 2, 17, 1960.
37. **Orlowski, W. A.,** The incorporation of H^3-Proline into the collagen of the periodontium of a rat, *J. Periodontal Res.*, 11, 96, 1976.
38. **Østegaard, E. and Löe, H.,** The collagen content of skin and gingival tissues in ascorbic acid-deficient monkeys, *J. Periodontal Res.*, 10, 103, 1975.
39. **Østegaard, E. and Löe, H.,** The correction coefficient for the ratio of total tissue to connective tissue in the gingiva, *J. Periodontal Res.*, 2, 240, 1967.
40. **Hara, K., Takahashi, T., Nohara, D. M., and Kobayashi, S.,** A correlation between microscopic numerical evaluation, clinical scoring and total collagen content in inflamed gingivae, *J. Periodontol.*, 47, 459, 1975.
41. **Miller, E. J. and Matukas, V. J.,** Chick cartilage collagen: a new type of α-chain not present in bone or skin of the species, *Proc. Natl. Acad. Sci. U.S.A.*, 64, 1264, 1969.
42. **Miller, E. J.,** Biochemical characteristics and biological significance of the genetically distinct collagens, *Mol. Cell. Biochem.*, 13, 165, 1976.
43. **Gay, S. and Miller, E. J.,** *Collagen in the Physiology and Pathology of Connective Tissue*, Gustav Fischer Verlag, New York, 1978.
44. **Bornstein, P. and Sage, H.,** Structurally distinct collagen types, *Annu. Rev. Biochem.*, 49, 957, 1980.
45. **Piez, K.,** Primary structure, in *Biochemistry of Collagen*, Ramachandran, G. N. and Reddi, A. H., Eds., Plenum Press, New York, 1976, 1.
46. **Martin, G. R., Byers, P. H., and Piez, K.,** Procollagen, *Adv. Enzymol.*, 42, 167, 1975.
47. **Prockop, D. J., Kivirikko, K. I., Tuderman, L., and Guzman, N. A.,** The biosynthesis of collagen and its disorders. Part I and II *N. Engl. J. Med.*, 301, 13, 1979; 301, 77, 1979.
48. **Prockop, D. J., Berg, R. A., Kivirikko, K. I., and Uitto, J.,** Intracellular steps in the biosynthesis of collagen, in *Biochemistry of Collagen*, Ramachandran, G. N. and Reddi, A. H., Eds., Plenum Press, New York, 1976, 163.
49. **Kefalides, N. A.,** Structure and biosynthesis of basement membranes, *Int. Rev. Connect. Tissue Res.*, 6, 63, 1973.

50. **Gay, S. and Miller, E. J.,** Characterization of lens capsule collagen: evidence for the presence of two unique chains in molecules derived from major basement membrane structures, *Arch. Biochem. Biophys.*, 198, 370, 1979.
51. **Ballard, J. B. and Butler, W. T.,** Proteins of the periodontium. Non-collagenous proteins of human gingivae, *J. Oral Pathol.*, 3, 176, 1974.
52. **Narayanan, A. S. and Page, R. C.,** Biochemical characterization of collagen synthesized by fibroblasts derived from normal and diseased human gingiva, *J. Biol. Chem.*, 251, 5464, 1976.
53. **Schneir, M., Ogata, S., and Fine, A.,** Confirmation that neither phenotype nor hydroxylation of collagen is altered in overgrown gingiva from diphenyl-hydantoin-treated patients, *J. Dent. Res.*, 57, 506, 1978.
54. **Dabbous, M. Kh., Seif, M., Brinkley, B., Butler, W. T., and Braswell, C. W.,** Gingival matrix proteins: the nature of insoluble bovine gingival collagen, *J. Periodontal Res.*, 4, 204, 1979.
55. **Butler, W. T., Birkedal-Hansen, H., Beegle, W. F., and Taylor, R.,** Proteins of the periodontium: identification of collagens with the $(\alpha 1\ (I))_2\alpha 2$ and $(\alpha 1\ (III))_3$ structures in bovine periodontal ligament, *J. Biol. Chem.*, 250, 8907, 1975.
56. **Narayanan, A. S., Hassell, T. M., Page, R. C., Hoke, J. A., and Meyers, D. F.,** Human edentulous ridge collagens. Characterization and comparison with gingival collagens, *J. Periodontal Res.*, 16, 504, 1981.
57. **Birkedal-Hansen, H., Butler, W. T., and Taylor, R. E.,** Proteins of the periodontium: characterization of the insoluble collagen of bovine dental cementum, *Calcif. Tissue Res.*, 23, 39, 1977.
58. **Narayanan, A. S., Page, R. C., and Meyers, D. F.,** Characterization of collagens of diseased human gingiva, *Biochemistry*, 19, 5037, 1980.
59. **Hammouda, O., Seif, M., Brinkley, B., Dabbous, M. Kh., and Jurand, J.,** Gingival matrix collagen in chronic periodontitis, *J. Dent. Res.*, 59, 17, 1980.
60. **Sodek, J. and Limeback, H. F.,** Comparison of the rates of synthesis. Conversion and maturation of Type I and Type III collagen in rat periodontal tissues, *J. Biol. Chem.*, 254, 10496, 1979.
61. **Engel, D., Schroeder, H. E., Gay, R., and Cladett, J.,** Fine structure of cultured human gingival fibroblasts and demonstration of simultaneous synthesis of Type I and III collagen, *Arch. Oral Biol.*, 25, 283, 1980.
62. **Narayanan, A. S., Engel, L. D., and Page, R. C.,** Collagen αA, αB content is elevated in inflamed gingiva, *J. Dent. Res.*, 60, A337, 1981.
63. **Sauk, J. J.,** Immunohistochemical localization of C-chain collagen peptides in gingiva and the dental-epithelial attachment, *J. Dent. Res.*, 59, 1997, 1980.
64. **Rao, L. G., Wang, H-M., Kalliechararan, R., Heersche, JN. M., and Sodek, J.,** Specific immunohistochemical localization of Type I collagen in porcine periodontal tissue using the peroxidase-labelled antibody technique, *Histochem. J.*, 11, 73, 1979.
65. **Sauk, J. J., Gay, R., Miller, E. J., and Gay, S.,** Immunohistochemical localization of Type III collagen in the dentin of patients with osteogenesis imperfecta and hereditary opalescent dentin, *J. Oral Pathol.*, 4, 210, 1980.
66. **Wohllebe, M. and Carmichael, D. J.,** Type I trimer and Type I collagen in neutral salt soluble lathyritic rat dentin, *Eur. J. Biochem.*, 92, 183, 1978.
67. **Wohllebe, M. and Carmichael, D. J.,** Biochemical characterization of guanidinium chloride soluble dentine collagen from lathyritic-rat incisors, *Biochem. J.*, 18, 667, 1979.
68. **Munksgaard, E. C. and Moe, D.,** Types of collagen in an extract of odontoblasts and dentine from developing bovine teeth, *Arch. Oral Biol.*, 25, 483, 1980.
69. **Tanzer, M. L.,** Cross-linking, in *Biochemistry of Collagen*, Ramachandran, G. N. and Reddi, A. H., Eds., Plenum Press, New York, 1976, 137.
70. **Furuto, D., Mechanic, G. L., and Schneir, M.,** Identification of reducible crosslinks in rat gingiva and palatal mucosa collagen, *J. Dent. Res.*, 56, 916, 1977.
71. **Kubuki, T., Takagi, S., Sasaki, S., Saito, S., and Mechanic, G. L.,** Comparative collagen biochemistry of bovine perodontium, gingiva and dental pulp, *J. Dent. Res.*, 60, 159, 1981.
72. **Schneir, M., Vogan, I., Yu, H., Yavelow, J., Liu-Montcalm, A., and Furuto, D.,** Ability of salt and acetic acid to extract human and animal gingival collagen, *J. Dent. Res.*, 54, 1095, 1975.
73. **Sodek, J.,** A highly efficient and rapid turnover of collagen in rat periodontal tissues: a new approach to assessing collagen turnover by using a micro-assay, *Biochem. J.*, 160, 243, 1976.
74. **Yavelow, J., Furuto, D., Ogata, S., and Schneir, M.,** Possible alternative mechanisms for collagen maturation in rat oral soft tissues, *J. Dent. Res.*, 56, B125, 1977.
75. **Schneir, M., Yavelow, J., Ogata, S., and Furuto, D.,** Probing the mechanism(s) of rapid collagen maturation in rat gingiva, *J. Dent. Res.*, 56 (Abstr.), A123, 1977.
76. **Carey, M. A. and Orlowski, W. A.,** The effects of crosslinking on collagen in rat gingiva, *J. Dent. Res.*, 56 (Abstr.), A123, 1977.

77. **Robert, A. M., Robert, B., and Robert, L.,** Chemical and physical properties of structural glycoproteins, in *Chemistry and Molecular Biology of the Intercellular Matrices*, Vol. 2, Balazs, E. A., Ed., Academic Press, New York, 1971, 237.
78. **Wolf, I., Fuchswans, W., Weiser, M., Furthmayr, H., and Timpl, R.,** Acidic structural proteins of connective tissue: characterization of their heterogenous nature, *Eur. J. Biochem.*, 20, 426, 1971.
79. **Dabbous, M. and Brinkley, B.,** Bovine gingiva. II. Proteins of the insoluble matrix, *J. Periodontol.*, 48, 630, 177.
80. **Jackson, D. S. and Bentley, P.,** Collagen-glycosaminoglycan interaction, in *Treatise of Collagen*, Vol. 2A, Gould, G. S., Ed., Academic Press, New York, 1968, 189.
81. **Scott, P. G. and Veis, A.,** The cyanogen bromide peptides of bovine soluble and insoluble collagen. I. Characterization of peptides from soluble Type I collagen by sodium dodecylsulfate polyacrylamide gel electrophoresis, *Connect. Tissue Res.*, 4, 107, 1976.
82. **Scott, P. G. and Veis, A.,** The cyanogen bromide peptides of bovine soluble and insoluble collagens. II. Tissue specific cross-linked peptides of insoluble skin and dentine collagen, *Connect. Tissue Res.*, 4, 117, 1976.
83. **Baum, J. B. and Wright, W. E.,** Demonstration of fibronectin as a major extracellular protein of human gingival fibroblasts, *J. Dent. Res.*, 59, 631, 1980.
84. **Zetter, B. R., Daniels, T. E., Guillen, C., and Greenspan, J. S.,** LETS protein in normal and pathological human oral epithelium, *J. Dent. Res.*, 58, 484, 1979.
85. **Ruoslahti, E.,** Fibronectin, *J. Oral Pathol.*, 10, 3, 1981.
85a. **Terranova, V. P. and Martin, G. R.,** Possible role for attachment proteins in a periodontal reattachment, *J. Dent. Res.*, 60(Abstr.), A392, 1981.
85b. **Slavkin, H. V., Melnick, M., Brownell, A., Bessen, C., and Bringas, P., Jr.,** Anti-fibronectin antibody mediates dentine and enamel remineralization *in vitro*, *J. Dent. Res.*, 60 (Abstr.), A402, 1981.
86. **Gregory, J. D., Sadjera, S. W., Hascall, V. C., and Dziewiatkowski, D. D.,** The proteoglycans of bovine nasal cartilage: dissociative extraction, in *The Chemistry and Molecular Biology of the Intercellular Matrix*, Vol. 2, Balazs, E. A., Ed., Academic Press, London, 1970, 843.
87. **Schiller, S., Slover, G., and Dorfman, A.,** A method for the separation of acid mucopolysaccharides: its application to the isolation of heparin from the skin of rats, *J. Biol. Chem.*, 236, 983, 1961.
88. **Antonopoulos, C. A., Gardell, S., Szirmai, J. A., and DeTyssonsk, E. ER.,** Determination of glycosaminoglycans (mucopolysaccharides) from tissues on microgram scale, *Biochim. Biophys. Acta*, 83, 1, 1964.
89. **Svejcar, J. and Robertson, W. van B.,** Microseparation and determination of mammalian acidic glycosaminoglycans (mucopolysaccharides), *Anal. Biochem.*, 18, 333, 1967.
90. **Hiramatsu, M., Abe, I., and Minani, N.,** Acid mucopolysaccharides in porcine gingiva, *J. Periodontal Res.*, 13, 224, 1978.
91. **Saito, H., Yamagata, T., and Suzuki, S.,** Enzymatic methods for the determination of small quantities of isomeric chondroitin sulfates, *J. Biol. Chem.*, 243, 1536, 1968.
92. **Yamagata, T., Saito, S., Osami, H., and Suzuki, S.,** Purification and properties of bacterial chondroitinases and chondrosulfatases, *J. Biol. Chem.*, 243, 1523, 1968.
93. **Suzuki, S., Saito, H., Yamagata, T., Anno, K., Seno, N., Kawai, Y., and Furuhashi, T.,** Formation of three types of disulfated disaccharides from chondroitin sulfates by chondroitinase digestion, *J. Biol. Chem.*, 243, 1543, 1968.
94. **Fullmer, H. M.,** The critique of normal connective tissues of the periodontium and some alterations with periodontal disease, *J. Dent. Res.*, 41, 225, 1962.
95. **Fullmer, H. M.,** A comparative histochemical study of elastic, pre-elastic and oxytalan connective tissue fibers, *J. Histochem. Cytochem.*, 8, 290, 1960.
96. **Quintarelli, G.,** Histochemistry of the gingiva. IV. Preliminary investigations on the muco-polysaccharides of connective tissue, *Arch. Oral Biol.*, 2, 277, 1960.
97. **Fullmer, H. M.,** A histochemical study of periodontal disease in the maxillary alveolar process of 135 autopsies, *J. Periodontol.*, 32, 206, 1960.
98. **Melcher, A. H.,** Some histological and histochemical observation on the connective tissue of chronically inflamed gingiva, *J. Periodontal Res.*, 2, 127, 1967.
99. **Scott, J. E. and Dorling, J.,** Differential staining of acid glycosaminoglycans (mucopolysaccharides) by Alcian blue in salt solutions, *Histochemie*, 5, 221, 1965.
100. **Pedlar, J.,** Histochemistry of glycosaminoglycans in the skin and oral mucosa of the rat, *Arch. Oral Biol.*, 24, 777, 1979.
101. **Plecash, J. M.,** Associations of acid mucopolysaccharides with normal gingival collagen fibrils in the rat molar periodontium, *J. Can. Dent. Assoc.*, 7, 246, 1972.
102. **Emery, G., Oliver, W. M., and Stanbury, F. B.,** The metabolism of proteoglycans and glycosaminoglycans in inflamed human gingiva, *J. Periodontal Res.*, 14, 512, 1979.

103. **Ciancio, S. G. and Mather, M. L.,** Acid mucopolysaccharides in gingivitis and periodontitis, *J. Periodontal Res.*, 6, 188, 1971.

104. **Chiappa, J. and Tussing, G.,** Quantitation of acid mucopolysaccharides in gingival specimens of human diabetic and non-diabetics, *J. Dent. Res. IADR Abstr.* #169 95, 1971.

105. **Schultz-Haudt, S. D.,** Observations on the acid mucopolysaccharides of human gingiva, *Odontol. Tidskr.*, 66, 3, 1958.

106. **Thonard, J. C. and Blustein, R.,** Sialic acid in human gingiva, *J. Dent. Res.*, 44, 379, 1965.

107. **Paunio, K.,** Periodontal connective tissue: biochemical studies of disease in man, *Soum. Hammaslääk. Toim*, 65, 249, 1960.

108. **Sakamoto, N., Okuda, K., and Okamoto, H.,** Qualitative and quantitative analyses of bovine gingival glycosaminoglycans, *Arch. Oral Biol.*, 23, 983, 1978.

109. **Munemoto, K., Iwayama, G., Yoshida, M., Sera, M., Aono, M., and Yokomizo, I.,** Isolation and characterization of acid mucopolysaccharides of bovine periodontal membrane, *Arch. Oral Biol.*, 15, 369, 1970.

110. **Kofoed, J. A. and Tocci, A.,** The acidic glycosaminoglycans of gingiva of diabetic rats, *J. Periodontal Res.*, 5, 323, 1973.

111. **Kofoed, J. A.,** Effects of hypophysectomy on gingival glycosaminoglycans, *J. Dent. Res.*, 50, 423, 1971.

112. **Loewi, B.,** The acid mucopolysaccharides of human skin, *Biochem. Biophys. Acta*, 52, 435, 1961.

113. **Roden, L.,** Structure and metabolism of connective tissue proteoglycans, in *The Biochemistry of Glycoproteins and Proteoglycans*, Lennarz, J. W., Ed., Plenum Press, New York, 1980, 267.

114. **Hascall, V. C. and Heinegard, D.,** The structure of cartilage proteoglycans, in *Extracellular Matrix Influences on Gene Expression*, Slavkin, H. C. and Greulich, R. C., Ed., Academic Press, New York, 1975, 423.

115. **Fransson, L. A.,** Structure and metabolism of the proteoglycans of dermatan sulphate, in *The Chemistry and Molecular Biology of the Intercellular Matrix*, Vol. 2, Balazs, E., Ed., Academic Press, New York, 1970, 823.

115a. **Hassel, J. R., Robey, P. G., Barrach, H. H., Wilczek, J., Rennard, S. T., and Martin, G. R.,** Isolation of heparan sulfate-containing proteoglycan from basement membrane, *Proc. Natl. Acad. Sci. U.S.A.*, 77, 4494, 1980.

115b. **Rohrback, D. H., Hassel, J. R., Kleinman, H. K., and Martin, G. R.,** Alterations in the basement membranes (heparan sulfate) proteoglycans in diabetic mice, *J. Am. Diabetic Assoc.*, 31, 185, 1982.

116. **Dziewiatkowski, D. D., LaValley, S. J., and LaValley, J. A.,** Characterization of a proteoglycan from bovine gingivae, *J. Dent. Res.*, 56, A327, 1977.

117. **Cimasoni, G. and Held, A. J.,** Intercellular mucopolysaccharides in human gingival epithelium, *Arch. Oral Biol.*, 9, 751, 1964.

118. **Thonard, J. C. and Sherp, H. W.,** Histochemical demonstration of acid mucopolysaccharides in human gingival epithelial intercellular spaces, *Arch. Oral Biol.*, 7, 125, 1962.

119. **Thonard, J. C. and Wiebkin, O. W.,** Localization of mucopolysaccharides in epithelial-like cells cultured *in vitro*, *J. Periodontal Res.*, 8, 101, 1973.

120. **Bartold, P. M., Wiebkin, O. W., and Thonard, J. C.,** Glycosaminoglycans of human gingival epithelium, *Connect. Tissue Res.*, 9, 99, 1981.

121. **Rabinowitz, J. L., Tavares, C., and Marsh, J. B.,** Polar and neutral lipids of epithelium and connective tissue of pig gingiva, *J. Periodontal Res.*, 7, 230, 1972.

122. **Pellat, B., Dargent, P., and Di Costanzo, G.,** Les Lipides de la Gencive Humaine, *J. Biol. Buccale*, 3, 247, 1975.

123. **Das, S. K., Adhikary, P. K., and Bhattacharyya, K. K.,** Surface and lamina propria lipids of bovine gingiva, *J. Dent. Res.*, 55, 602, 1976.

124. **Das, S. K., Adhikary, P. K., and Bhattacharyya, D. K.,** Study on the composition of bovine gingiva lipids, *J. Dent. Res.*, 55, 182, 1976.

125. **Wood, N., Rabinowitz, J. L., Bremer, E. C., and Fillios, L. C.,** Evidence for hydroxyacyl-ceramides and hydroxacyl-cerebrosides in female rhesus monkey gingiva, *J. Periodontol.*, 51, 716, 1980.

126. **Das, S. K., Adhikary, P. K., and Bhattacharyya, D. K.,** Detection of prostaglandins in bovine gingival tissue, *J. Dent. Res.*, 55, 185, 1976.

127. **Albers, H. K., Loning, T., and Lisboa, B. P.,** Biochemical and morphologic studies on prostaglandins E and F in the normal and inflamed gingiva (in German), *Dtsch. Zahnaerztl. Z.*, 34, 440, 1979.

128. **El Attar, T. M. and Lin, H. S.,** Cyclic AMP and prostaglandins in periodontal disease, *Adv. Prostaglandin Thromboxane Res.*, 8, 1739, 1980.

129. **Goodson, J. M., Dewhirst, F., and Brunetti, A.,** Prostaglandin E_2 levels in human periodontal disease, *Prostaglandins*, 6, 81, 1974.

130. **Dewhirst, F. E., Offenbacher, S., Moss, D., and Goodson, J. M.,** Relative levels of prostaglandin E, thromboxane and prostacycline in periodontal tissues, *J. Dent. Res.*, 60, A606, 1981.

131. **Anneroth, G. and Ivemark, B. I.,** Histochemical localization of lipids in the human gingival epithelium and their relation to the dry mass concentration, *Odontol. Revy*, 29, 9, 1971.
132. **Nikkari, T. and Heikkinen, E.,** The lipids of collagen preparations, *Acta Chem. Scand.*, 22, 3047, 1968.
133. **Sauk, J. J. and Roszkowski, M.,** Interaction of lipopolysaccharides and lipid A with collagen, *J. Dent. Res.*, 60, A643, 1981.
134. **Fullmer, H. M., Taylor, R. E., and Guthrie, R. W.,** Human gingival collagenase: purification, molecular weight, and inhibitor studies, *J. Dent. Res.*, 51, 349, 1972.
135. **Birkedal-Hansen, H., Cobb, C. M., Taylor, R. E., and Fullmer, H. M.,** Bovine gingival collagenase: demonstration and initial characterization, *J. Oral Pathol.*, 3, 232, 1974.
136. **Woolley, D. E.,** Characterization and serum inhibition of neutral collagenase from cultured dog gingival tissues, *Biochem. Biophys. Acta*, 522, 205, 1978.
137. **Uitto, V., Turoto, H., and Saxen, L.,** Extraction of collagenase from human gingiva, *J. Periodontal Res.*, 13, 207, 1978.
138. **Gibson, W. and Fullmer, H. M.,** Collagenolytic activity in gingival tissues *in vitro, J. Dent. Res.*, 45, 1225, 1966.
139. **Birkedal-Hansen, H., Cobb, C. M., Fullmer, H. M., and Taylor, R. E.,** Activation of latent bovine gingival collagenase, *Arch. Oral Biol.*, 20, 681, 1975.
140. **Birkedal-Hansen, H., Cobb, C. M., Taylor, R. E., and Fullmer, H. M.,** Fibroblastic origin of bovine gingival collagenase, *Arch. Oral Biol.*, 21, 297, 1976.
141. **Woolley, E. D. and Evanson, J. M.,** Collagenase and its natural inhibitors in relation to the rheumatoid joint, *Connect. Tissue Res.*, 5, 31, 1977.
142. **Siegel, R. C.,** Lysyl oxidase, in *International Review of Connective Tissue Research*, Vol. 8, Hall, D. A. and Jackson, D. S., Eds., Academic Press, New York, 1979, 73.
143. **Kivirikko, K. I. and Myllyla, R.,** Collagen glycosyltransferases, in *International Review of Connective Tissue Research*, Vol. 8, Hall, D. A. and Jackson, D. S., Eds., Academic Press, New York, 1979, 23.
144. **Ramamurthy, N. S., Garant, P. R., and Golub, L. M.,** Prolyl and Lysyl hydroxylase activities in rat gingiva: effects of aging and inflammation, *J. Dent. Res.*, 56, B57, 1977.
145. **Ramamurthy, N. S. and Golub, L. M.,** Collagen biosynthetic enzyme activity in diabetic rat gingiva, *J. Dent. Res.*, 57, A311, 1978.
146. **Ramamurthy, N. S. and Golub, L. M.,** Effect of diabetes on lysyl oxidase activity in rat gingiva and skin, *J. Dent. Res.*, 59, A535, 1980.
147. **Hasegawa, K. and Cimasoni, G.,** The relative amounts of free and latent acid hydrolases in homogenates of human gingiva, *Arch. Oral Biol.*, 20, 521, 1975.
148. **Goggins, J. F. and Fullmer, H. M.,** Enzymes of human gingiva involved with mucopolysaccharide catabolism, *J. Periodontal Res.*, 4 (Suppl. 4), 31, 32, 1969.
149. **Gerson, S.,** Activity of glucose-6-phosphate dehydrogenase and acid phosphatase in non-keratinized and keratinized oral epithelium and epidermis in the rabbit, *J. Periodontal Res.*, 8, 151, 1973.
150. **Goggins, J. F., Fulimer, H. M., and Steffek, A.,** Hyaluronidase activity of human gingiva, *Arch. Pathol.*, 85, 272, 1968.
151. **Nicollau, J., Tamer, A. M., and Bergamasxchi, O.,** Activities of hexokinase, phosphofructokinase and pyruvate kinase in the gingival tissues of the rat, hamster, guinea pig and human, *J. Periodontal Res.*, 12, 279, 1977.
152. **Berquist, J. J. and Nuki, K.,** Pentose shunt enzyme activities in the non-inflamed attached gingiva of the dogs, *J. Periodontal Res.*, 8, 11, 1973.
153. **Berquist, J. J. and Nuki, K.,** Hexokinase activity in the non-inflammed attached gingiva of the rhesus monkey, *J. Periodontal Res.*, 8, 215, 1973.
154. **Berquist, J. J. and Nuki, K.,** Fructoaldolase activity in non-inflammed attached gingiva of rhesus monkeys, *J. Periodontal Res.*, 8, 222, 1973.
155. **Eastoe, J. E.,** The amino acid composition of proteins from the oral tissues. I. A comparison of human oral epithelium, epidermis and nail proteins, *Arch. Oral Biol.*, 8, 449, 1963.
156. **Dale, B. A., Stern, I. B., and Clagett, J. A.,** Initial characterization of the proteins of keratinized epithelium of rat oral mucosa, *Arch. Oral Biol.*, 22, 75, 1977.
157. **Taichman, L., Reilly, S., and Garant, P. R.,** *In vitro* cultivation of human oral keratinocytes, *Arch. Oral Biol.*, 24, 335, 1979.
158. **Birkedal-Hansen, H., Moe, K., Dabelsteen, E., and Rheinholdt, J.,** Propagation of keratinizing human gingival epithelial cell line *in vitro, J. Dent. Res.*, 57, A98, 1978.
159. **Taichman, L. B. and Prokop, C. A.,** The synthesis of keratin proteins in cultured oral keratinocytes, *J. Dent. Res.*, 60, A404, 1981.
160. **Timpl, R., Rohde, H., Robuy, P. G., Rennard, S. I., Foidart, J. M., and Martin, G. R.,** Laminin-glycoprotein from basement membranes, *J. Biol. Chem.*, 254, 9933, 1979.
161. **Gregory, J. D.,** personal communication.

GINGIVAL METABOLISM

John W. Simpson

INTRODUCTION

Gingiva in conjunction with an attachment apparatus (cementum, periodontal ligament, and alveolar and supporting bone) forms a tissue complex known as the periodontium. Gingival and periodontal tissues invest and give structural support to the tooth by epithelial and connective tissue fiber attachment.

Physiologically, the gingiva and other periodontal unmineralized connective tissue function to (1) attach tooth to bone, (2) provide a barrier to penetration by oral bacteria and other potentially harmful substances, and (3) assist in absorbing the forces exerted by speech, swallowing, and mastication.

Histologically, the gingiva consists of keratinized stratified squamous epithelium that surfaces the underlying connective tissue (lamina propria). The vascular connective tissue layer projects into the nonvascular epithelium providing a pathway for nutrient flow into the epithelium component. The connective tissue of the gingiva, like other connective tissues, consists of fibers, cells, and extrafibrillar, extracellular, and extravascular ground substance. Gingival fibers are collagenous, reticular (young collagen), elastin, and oxytalan in composition. Fullmer[1] and Melcher and Eastoe[2] have comprehensively reviewed the structure of these fibers. The cells of the lamina propria include fibroblasts, macrophages, melanophages, mast cells, lymphocytes, and plasma cells. These are distributed within the gingiva to meet requirements unique to this tissue. The connective tissue fibers and cells are dispersed in ground substance comprised of a mixture of proteoglycans and glycosaminoglycans. Although the ground substances are distributed in a heterogenous manner specific for gingiva, the specifics and subtleties of the distribution have not been elucidated.

COMPOSITION

Chemically, the composition of gingiva has been partly determined. The percent composition of the major substances in normal human gingiva is given in Table 1. The fibrous nature of gingiva is reflected in the high concentration of protein. Collagen with the associated ground substance and noncollagenous protein is present in abundant amounts. The collagen of healthy gingiva is predominately type I. Smaller amounts of type III are also present.[8]

Glycosaminoglycans of bovine gingiva have been analyzed by two-dimensional electrophoresis on cellulose acetate strips (Table 2). The major components of bovine gingival glycosaminoglycans are chondroitin-4-sulfate, 32 to 40%; dermatan sulfate, 33 to 37%; hyaluronic acid, 12 to 27%; and heparan sulfate in limited amounts. Neither chondroitin-6-sulfate, heparin, nor keratan sulfate were found. The total uronic acid content of bovine gingiva decreases with age; the decrease is striking during the first 3 years and diminishes after the 3rd year.

Lipid constitutes 7.35% of the dry weight of human gingiva, with the percent concentration shown in Table 3. On a comparative basis, the lipid composition of human gingiva is similar to that of porcine gingiva. The one striking difference is that human gingival phospholipids contain 4% sphingomyelin, while in the pig this lipid amounts to 12% of the total. In both human and pig gingiva, there is a high content of free fatty acids similar to that in skin.[11,12]

Another chemical characterization of gingiva is the pool of free amino acids. Amino acid pools of rat gingiva have been determined (Table 4) and the range of concentration of these substances is within the range occurring in other mammalian tissues. There are only small

Table 1
PERCENT COMPOSITION OF HUMAN GINGIVA

	Dry weight (%)	Ref.
Collagen	49.05[a]	3
	31.00	4
	27.60	5
Protein (noncollagenous)	62.00[a]	3
Glycosaminoglycans	0.32	6
Hexosamine	0.37	5
	0.20	7
Glucuronic Acid	0.20	7
Total lipid	7.35[a]	3
RNA	1.80[a]	3
DNA	1.90[a]	3

[a] Calculated on the basis that gingiva contains 80% water.

Table 2
GAG[a] CONTENTS IN BOVINE GINGIVA AT VARIOUS AGES

	Percent				
Age	Chondroitin-4-sulfate	Dermatan sulfate	Hyaluronic acid	Heparan sulfate	Uronic acid
2 weeks	40.1	36.8	17.4	5.8	30.6
1 year	37.3	36.0	24.2	2.6	43.1
3 years	37.1	33.4	26.7	2.9	37.8
4 years	34.7	36.3	25.8	3.2	40.3
7 years	32.3	35.3	24.1	8.3	37.6
8 years	33.3	35.7	24.6	6.3	38.1
10 years	23.3	36.3	24.2	7.3	37.1

Note: These values were obtained from pooled samples of three animals.

[a] GAG, glycosaminoglycan.

From Sakamoto, N., Okamoto, H., and Okudo, K., *Arch. Oral Biol.*, 23, 983, 1978. With permission.

differences in free amino acids between gingiva and palate. Tongue has a decidedly higher free amino acid content than gingiva or palate, probably because of the large muscle mass. Except for cysteic acid, free amino acids in female rat gingiva are higher than in male rat gingiva.

METABOLISM

Periodontal metabolism has undoubtedly evolved as a means of maintaining the architecture of the periodontium. Periodontal structure is continually adjusting to wear and aging through remodeling and regeneration. Thus, the biochemistry of gingiva is a dynamic system aimed at preserving structural integrity.

Two experimental approaches have been used to evaluate the overall biochemical activity

Table 3
PERCENT LIPID CONCENTRATION OF HUMAN GINGIVA

Mono and diglycerides	4.14 ± 0.97[a]	28[b]
Cholesterol	10.56 ± 0.45	27
Free fatty acids	16.98 ± 1.05	24
Triglycerides	14.34 ± 0.55	28
Methyl esters	9.34 ± 0.65	25
Cholesterol esters	14.58 ± 2.18	28
Total phospholipids	30.03 ± 1.18	28

[a] SE of the mean.
[b] Number of individual analyses.

From Rabinowitz, J. L., Rutberg, M., Cohen, D. W., and Marsh, J. B., *J. Periodontal Res.*, 8, 381, 1973. © 1973 Munksgaard International Publishers, Ltd., Copenhagen, Denmark. With permission.

Table 4
FREE AMINO ACIDS FROM RAT GINGIVA, PALATE, AND TONGUE

Amino acid	Gingiva (female)	Gingiva (male)	Palate (male)	Tongue (male)
Alanine	3.08	0.95	1.65	17.25
Arginine	<0.50	—[a]	0.97	0.57
Aspartic acid	2.05	0.26	0.86	6.03
Cysteic acid	0.68	0.85	0.20	1.18
Glutamic acid	6.27	1.04	1.84	24.28
Glutamine	0.76	—[a]	0.51	8.82
Glycine	1.71	0.67	0.73	7.16
Histidine	0.75	—[a]	0.67	1.85
Isoleucine	0.57	0.17	0.27	0.39
Leucine	1.37	0.45	0.85	0.85
Lysine	0.59	—[a]	0.70	0.83
Methionine	0.80	0.47	0.52	0.38
Phenylalanine	1.41	0.29	0.95	ND[b]
Serine	2.06	0.78	0.96	8.98
Taurine	11.61	2.35	3.90	51.17
Threonine	1.63	—[a]	0.58	5.25
Tyrosine	1.73	0.35	1.23	ND[b]
Valine	1.19	0.29	0.82	2.01

Note: The values are averages from two measurements; values are expressed as milligrams per 100 g wet weight.

[a] Not determined.
[b] ND, not detected.

From Simpson, J. W., *J. Dent. Res.*, 50, 1361, 1971. With permission.

for gingiva. The first measures the respiratory rate for gingiva, while the second estimates the turnover rate of the major constituents of gingival connective tissue.

The respiratory rate for normal human gingiva measured as Qo_2 is 1.49 (Table 5). This value was obtained from 6 different experiments using 236 gingival samples. The Qo_2 for

Table 5
ENDOGENOUS Qo_2 VALUES FOR NORMAL HUMAN GINGIVA

Number of samples	State of tissue	Endogenous Qo_2	Ref.
8	Normal	1.60	14
5	Normal	1.90	15
10	Normal	1.36	16
36	Normal	1.39	17
75	Normal	1.43	18
	Age Group		
21	0—20	1.64	
36	21—40	1.39	
18	41—60	1.26	
27	Normal	1.47	19
Average		1.49	

Adapted from Eichel, B. and Shahrik, H. A., *Advances in Oral Biology*, Vol. 1, Staple, P. H., Ed., Academic Press, New York, 1964, 131. With permission.

human gingiva is within the range of Qo_2 values for other human tissues.[21] When compared with tissues elsewhere in the body, human gingiva has a comparatively low respiratory rate. The gingival Qo_2 is only slightly lower than that of skin and higher than uterine tissue and gastric smooth muscle.[21]

Gingival Qo_2 values are affected by variations in the experimental parameters, namely, tissue slice thickness and volume, reaction mixture contents, and several factors unique to gingival tissue. The ratio of epithelial to connective tissue in the specimen may influence Qo_2 values since epithelium has a higher respiratory rate than connective tissue. In general, samples with a light to moderate degree of inflammation or those in a proliferative state tend to have a higher Qo_2 value, while gingiva with moderate to heavy inflammation or degeneration have lower Qo_2 values. The average endogenous Qo_2 of human gingiva decreases with increasing age (Table 5), while sex and race (Negroid vs. Caucasian) have comparatively little impact.

A second index of biochemical activity of gingiva is derived from the turnover rates of tissue constituents. Fritz[22] evaluated the incorporation and release of ^{15}N-glycine by proteins of rat gingival mucosa, palatal mucosa, tongue membrane, and incisor pulp (Table 6). All oral tissues studied had a relatively high initial uptake of ^{15}N. The rates of release of labeled nitrogen from the proteins of gingival mucosa, palatal mucosa, and tongue membrane were relatively rapid. Slower activity in the incisor pulp proteins approximately paralleled that of liver proteins. The ^{15}N concentration of liver protein declined less rapidly than those of the mucosal tissue proteins. Thus, the protein turnover rate in the gingiva is faster than the turnover rate of liver proteins. Experimentally, it has been calculated that the half-life for turnover of mature collagen in rat gingiva is 5 days, in periodontal ligament 1 day, in alveolar bone 6 days, and in skin corium 15 days.[23] Thus, gingival and periodontal ligament have a relatively rapid protein turnover rate when compared with other connective tissue structures.

Half-life times for the glycosaminoglycans in rat gingiva, skin, and cartilage are displayed in Table 7. The half-life times were estimated graphically from a line drawn through the experimental data points at the site where half the maximal radioactivity was obtained. In general, the glycosaminoglycan fraction of gingiva is metabolized more quickly than in skin

Table 6

Ratio of $\frac{\text{Atom Percent Excess } ^{15}\text{N in Liver Proteins}}{\text{Atom Percent Excess } ^{15}\text{N in Tissue Proteins}}$

	Fasting Time in Hours Since Last ^{15}N-Glycine Feeding			
Tissue protein	10	18	90	161
Gingival mucosa		$\frac{0.256}{0.219} = 1.17$	$\frac{0.226}{1.151} = 1.50$	$\frac{0.182}{0.113} = 1.61$
Palate mucosa	$\frac{0.311}{0.390} = 0.8$	$\frac{0.256}{0.304} = 0.84$	$\frac{0.226}{0.215} = 1.05$	$\frac{0.182}{0.140} = 1.30$
Tongue membrane		$\frac{0.256}{0.286} = 0.90$	$\frac{0.226}{0.220} = 1.03$	$\frac{0.182}{0.129} = 1.41$
Incisor pulp	$\frac{0.311}{0.263} = 1.18$	$\frac{0.256}{0.229} = 1.12$	$\frac{0.226}{0.200} = 1.13$	$\frac{0.182}{0.166} = 1.10$

From Fritz, I., *J. Dent. Res.*, 34, 446, 1955. With permission.

Table 7
HALF-LIFE TIMES OF RAT GINGIVA, SKIN, AND TRACHEAL CARTILAGE GLYCOSAMINOGLYCANS

	T1/2 days		
	Gingiva	**Skin**	**Cartilage**
Hyaluronic acid	2.67	3.09	14.10
Heparitin sulfate	7.70	10.34	16.70
Chondroitin-4-sulfate	7.20	8.42	7.70
Chondroitin-6-sulfate	6.70	7.80	13.70
Dermatan sulfate	7.00	5.76	12.00
Heparin	9.80	8.00	—

From Kofoed, J. A., Bozzini, C. E., and Tocci, A. A., *J. Dent. Res.*, 50, 171, 1971. With permission.

or cartilage. Except for heparin, all gingival glycosaminoglycan fractions had a half-life of 7 days or less, suggesting that collagen and the supporting ground substance may have similar turnover rates. On the basis of respiratory rate and turnover rates it is evident that although the Qo_2 of human gingiva is relatively low, the turnover rates for protein and glycosaminoglycans in rat gingiva reflect a relatively high degree of biochemical activity in this structure.

INTERMEDIARY METABOLISM

There is a paucity of data on intermediary metabolism in the gingiva. Gingiva has considerable synthetic capacity as evidenced by biosynthesis of protein,[25] especially collagen[26-29] and the associated proteoglycans.[30,31] There is, however, little information concerning the enzymatic steps of the biosynthetic processes in gingiva. The one area that has received considerable experimental attention is mechanisms of gingival glucose metabolism.

Table 8
ACTIVITIES OF SOME ENZYMES OF GLUCOSE METABOLISM IN RAT GINGIVA

Enzyme	Activity[a]
Phosphofructokinase	4.2
Glucose-6-phosphate dehydrogenase	2.1
Pyruvate oxidase complex	0.4
Lactic dehydrogenase	11.9

[a] Units are defined as the amount of enzyme forming 1 μmol of product per minute at 25°C. Values are given as units per gram tissue (wet weight).

From Simpson, J. W., *J. Dent. Res.*, 53, 938, 1974. With permission.

Animal studies have demonstrated that enzymes representative of the Embden-Meyerhof, pentose phosphate, and citric acid cycle pathways are present in the gingiva of rabbit,[20] rat,[32-34] guinea pig,[34-39] dog,[40] monkey,[41] and cow.[42] The evidence is that the three pathways are operative in animal gingiva and that the enzyme levels for these pathways are within the range found in other animal tissues.

Enzyme levels at pivotal points of metabolic pathways have been used to estimate the relative participation of these pathways in tissue metabolism. From the data in Table 8 it can be seen that the ratio of phosphofructokinase to glucose-6-phosphate dehydrogenase is 2:1. This ratio connotes that although the highest capacity for glucose metabolism in rat gingiva is via the Embden-Meyerhof sequence, it is only slightly greater than that for the pentose-phosphate pathway. Flow of carbon through the various metabolic paths may be monitored by specifically labeled glucose. The ratio of $^{14}CO_2$ derived from glucose-1-^{14}C and glucose-6-^{14}C of 2·1 to 2·5 from rat gingiva[33] is consistent with the interpretation that both the Embden-Meyerhof and pentose phosphate sequences are operative in rat gingiva. Rat gingiva also has a high potential for converting pyruvate to lactate compared to the ability to oxidize pyruvate through the pyruvate oxidase complex (Table 8). When the in vitro fate of glucose-U.L.-^{14}C in rat gingiva was determined however, it was obvious that the radioactive content of carbon dioxide, glutamate, and aspartate was an order of magnitude greater than the radioactive content in lactic acid.[43] It would seem therefore, that the disposition of pyruvate by rat gingiva is predominately by the citric acid cycle.

It seems plausible that glucose is metabolized in human gingiva by pathways similar to that in animal gingiva. The relative rates of operation for the pathways in human gingiva may, however, be different from animal gingiva. Eichel and Shahrik[20] reviewed in detail the oxidative enzyme systems in human gingiva. From their data (Table 9) they concluded that human gingiva possesses a relatively high glycolytic metabolism and a very low aerobic or respiratory metabolism. Human gingiva is thus similar to other animal gingiva in this respect. Since potentials do not necessarily indicate the flow of substrate through the various pathways, further experiments are required to establish the relative participation of the various pathways for glucose metabolism in human gingiva.

GINGIVAL DISEASE

Several diseases affect the gingiva and other components of the periodontium. Diseases unique to the periodontium are listed in Table 10. There are other systemic conditions that

Table 9
ACTIVITIES OF SOME OXIDATIVE ENZYMES OF HUMAN GINGIVA

Enzyme	Activity[a]
Lactic dehydrogenase	161.7
Malic dehydrogenase[b]	33.6
Glucose-6-PO_4 dehydrogenase	14.4
$DPNH_2$ cytochrome C reductase	10.2
Cytochrome C oxidase	3.6
$TPNH_2$ cytochrome C reductase	2.7
Succinic dehydrogenase cytochrome C reductase	1.5

[a] Data expressed as 10^{-1} μmol cytochrome C, TPN, or DPN reduced or oxidized per 3 mℓ of reaction mixture per minute per milligram wet weight.
[b] Assayed at pH 9.5 — other enzymes assayed at pH 7.4.

From Eichel, B. and Shahrik, H. A., *Advances in Oral Biology*, Vol. 1, Staple, P. H., Ed., Academic Press, New York, 1964, 131. With permission.

lead to gingival pathology. The best known is scurvy in which an early symptom is gingival bleeding. Scurvy results from a deficiency in the activity of prolyl hydroxylase due to the diminished quantity of the normal cofactor, ascorbic acid. The end result is aberration in the collagen molecule.

One of the most prevalent of all human diseases is chronic destructive periodontitis. Accumulation of microorganisms in dental plaque in close proximity to the gingiva is thought to initiate chronic destructive periodontitis. Much of the effort in periodontal disease research has been directed toward understanding the mechanisms of tissue loss during inflammatory periodontal disease.

There is ample evidence that periodontal collagen is destroyed during periodontal disease[45-49] and that there are changes in the proteoglycan and glycosaminoglycan content of gingiva during periodontitis.[5,50,51] It seems certain that periodontal tissue destruction results from alterations in the metabolism of the macromolecules. Whether the defect is one of anabolism, catabolism, or both remains controversial.

Three different processes of collagen synthesis have been considered as possible sites where defects might lead to collagen loss during periodontal disease.

1. Alterations in synthetic rates might result from inflammatory disease. This concept stems from the high turnover rate of periodontal protein, including collagen, when compared with connective tissue elsewhere in the body.[22,24] Reduction in the activities of lysyl and prolyl hydroxylase from inflamed rat gingiva when compared to normal gingiva has been interpreted as evidence that collagen synthesis is reduced in the inflamed tissue.[52]
2. It has been suggested that diseased gingiva contains collagen types not found in healthy gingiva. This is supported by observations that fibroblasts derived from healthy gingiva produce type III collagen (5 to 30%), while fibroblasts from diseased tissue produce collagen with composition $[\alpha 1(I)]_3$.[53] The α1 trimer has been found in diseased gingiva, also suggesting that there may be a change of collagen in diseased tissue.[54]
3. It has been of interest to determine whether there are interferences in the normal process of collagen maturation in diseased tissue. Initiative for considering a possible maturational defect comes from the observation that tissue collagen exists in three

Table 10
CLASSIFICATION OF PERIODONTAL DISEASES

Acute gingivitis
- Acute gingivitis
- Excludes:
  - Acute necrotizing ulcerative gingivitis (syn. fusospirochaetal gingivitis, Vincent's infection)
- Herpetic gingivostomatitis
- Pericoronitis
- Streptococcal gingivostomatitis

Chronic gingivitis
- Simple marginal
- Hyperplastic
- Ulcerative
- Desquamative
- Other
- Unspecified

Gingival recession
- Localized
- Generalized
- Unspecified

Acute periodontitis
- Acute periodontitis
  - Periodontal abscess (syn. parodontal abscess)
- Excludes:
  - Acute apical periodontitis
  - Acute periapical abscess

Chronic periodontitis
- Complex
- Simplex
- Pericoronitis
- Unspecified

Periodontosis
- Periodontosis

Other and unspecified
- Gingival fibromatosis and localized enlargement; e.g., tuberosity periodontal lesions, due to traumatic occlusion
- Includes:
  - Stillman's cleft, McCall's festoons
  - Gingival cyst
  - Unspecified

From Williams, R. C. and Howell, T. H., *Textbook of Oral Biology*, Shaw, J. H., Sweeney, E. Z., Coppuccino, C. C., and Miller, S. M., Eds., W. B. Saunders, Philadelphia, 1978, 1020.

distinct pools depending on the maturational state. When collagen solubilities of healthy gingiva are compared with those of inflamed gingiva, there is generally an increase in soluble collagen in the inflamed tissue;[55,56] the increased collagen solubility is an observation compatible with an increase in amount of immature collagen in diseased tissue.

Alternatively, collagen destruction during inflammatory gingival disease may occur by accelerated collagen catabolism. A generally accepted pathway for in vivo collagen degradation is where the process is initiated by a specific collagenase with possible synergistic action by neutral proteinases. The collagenous fragments formed by collagenase activity are denatured at body temperature and are susceptible to the sequential action of gelatinase and peptidases. Since collagenase is a pivotal enzyme in collagen metabolism, collagen catabolism research has centered around this enzyme.

Collagenase is a metalloproteinase that catalyzes cleavage of peptides in the helical region of undenatured collagen at physiological pH and temperature. Collagenase derived from human gingival explants has properties similar to other mammalian collagenases in that:

1. It catalyzes cleavage of peptide bonds across the triple helix with the production of the one quarter, three quarter collagen fragments.
2. It is operative over a broad pH range from pH 7 to 9.5.
3. It has a molecular weight of approximately 40,000 daltons.
4. It is inhibited by EDTA, serum, and cysteine.
5. It has an absolute requirement for calcium.[57]

There is evidence that gingival collagenase activity correlates with the clinical state of the tissue;[58,59] collagenolytic activity is least in normal and mildly inflamed tissue and greatest in severely inflamed hyperplastic tissue.

Two hypotheses have been advanced to explain the elevation of collagenase in inflamed tissue:

1. It has been suggested that the collagenase increase results from the influx of cells in the inflammatory infiltrate. All cells of the inflammatory infiltrate have a collagenase producing capacity.[60]
2. It has been postulated that the collagenase increase may result from perturbation of normal regulatory mechanisms for the enzyme. The most likely modes for alterations in regulatory mechanisms are either proteolytic activation of procollagenase[61] and a collagenase-inhibitor complex[62,63] or induction of increased synthesis of the enzyme.[64]

Accelerated biodegradation of connective tissue macromolecules during inflammatory disease may also affect the concentration of proteoglycan and glycosaminoglycan in the tissues. Gingival hyaluronidase levels that may participate in the degradation of hyaluronic acid and chondroitin sulfate are increased during inflammation.[65-67] Gingival cathepsin D, a lysosomal enzyme that may function to degrade proteoglycan core protein, is also increased during gingivitis.[68] It may be that the same mechanisms operative in collagenase increase during disease may cause an increase in the enzymes that catalyze the degradation of proteoglycans and glycosaminoglycans.

The processes that influence the severity of periodontal disease are multifactorial and may include a hormonal influence. During inflammation of the human gingiva, there is accelerated conversion of adrostenedione to testosterone[69] and estrone to estradiol-17β.[70] Both of the products are biologically more active than the substrates. Further, estradiol incubated with inflamed gingival tissue causes an increase in the inflammatory mediating substance prostaglandin E_2;[71] cyclic AMP which may control prostaglandin levels is also increased. Factors such as aging, nutritional status, and immunological responses may also contribute to the pathological changes in the soft connective tissue during chronic destructive periodontitis and thus modify the inflammatory components.

Research on the destruction of gingival connective tissue during inflammatory disease has centered around mechanisms at the substrate and enzyme level. From the accumulated data, there are suggestions that both biosynthetic and biodegradative defects may lead to a loss of gingival constituents during disease. Future research needs to be directed toward elucidating the multifactors that influence the severity of chronic destructive periodontal disease.

REFERENCES

1. **Fullmer, H. M.,** Connective tissue components of the periodontium, in *Structural and Chemical Origination of Teeth,* Vol. 2, Miles, A. E. W., Ed., Academic Press, New York, 1967, 349.
2. **Melcher, A. H. and Eastoe, J. E.,** The connective tissue of the periodontium, in *Biology of the Periodontium,* Melcher, A. H. and Bowen, W. H., Eds., Academic Press, New York, 1969, 167.
3. **Rabinowitz, J. L., Rutberg, M., Cohen, D. W., and Marsh, J. B.,** Human gingival lipids, *J. Periodontal Res.,* 8, 381, 1973.
4. **Schultz-Haudt, S. D. and Aas, E.,** Observations on the status of collagen in human gingiva, *Arch. Oral Biol.,* 2, 131, 1960.
5. **Aas, E.,** Hyperplasia gingivae diphenylhydantoinea, *Acta Odontol. Scand.,* 21(Suppl. 34), 11, 1963.
6. **Schultz-Haudt, S. D.,** Separation of acid mucopolysaccharides of human gingiva, *Acta Chem. Scand.,* 11, 1070, 1957.
7. **Ciancio, S. G. and Mather, M. L.,** Acid mucopolysaccharides in gingivitis and periodontitis, *J. Periodontal Res.,* 6, 188, 1971.
8. **Ballard, J. B. and Butler, W. T.,** Proteins of the periodontium. Biochemical studies on the collagen and noncollagenous proteins of human gingivae, *J. Oral Pathol.,* 3, 176, 1974.
9. **Sakamoto, N., Okamoto, H., and Okudo, K.,** Qualitative and quantitative analyses of bovine gingival glycosaminoglycans, *Arch. Oral Biol.,* 23, 983, 1978.
10. **Rabinowitz, J. L., Bailey, T. A., and Marsh, J. B.,** Polar and neutral lipids of pig gingiva, *Arch. Oral Biol.,* 16, 1195, 1971.
11. **Koplon, B. S., Luddy, F. E., and Rabinowitz, J. L.,** Studies on a case of sclerederma adultorum, *Acta Derm. Venereol.,* 48, 516, 1968.
12. **Wilkinson, D. I.,** Variability in composition of surface lipids. The problem of the epidermal contribution, *J. Invest. Dermatol.,* 52, 339, 1969.
13. **Simpson, J. W.,** Free amino acids in rat gingival tissue, *J. Dent. Res.,* 50, 1361, 1971.
14. **Glickman, I., Turesky, S., and Hill, R.,** Determination of oxygen consumption in normal and inflamed human gingiva using the Warburg manometric technique, *J. Dent. Res.,* 28, 83, 1949.
15. **Schroder, H. K. and Schroder, R.,** Oxygen uptake by normal and inflamed gingiva and saliva, *Helv. Odontol. Acta,* 1, 13, 1957.
16. **Senter, A. D., Eiler, J. J., and Lee, K. H.,** Endogenous respiration of human gingival tissue, *Proc. Soc. Exp. Biol. Med.,* 100, 323, 1958.
17. **Manhold, J. H., Bolden, T. E., and Katz, S.,** Microrespirometer technique for study of human gingival tissue, *J. Dent. Res.,* 39, 746, 1960.
18. **Volpe, A. R., Manhold, J. H., and Manhold, B. S.,** Effect of age and other factors upon normal gingival tissue respiration, *J. Dent. Res.,* 41, 1060, 1962.
19. **Manhold, J. H. and Volpe, A. R.,** Effect of inflammation in the absence of proliferation on the oxygen consumption of gingival tissue, *J. Dent. Res.,* 42, 103, 1963.
20. **Eichel, B. and Shahrik, H. A.,** Cytochemical aspects of oxidative enzyme metabolism in gingiva, in *Advances in Oral Biology,* Vol. 1, Staple, P. H., Ed., Academic Press, New York, 1964, 131.
21. **Spector, W. A.,** *Handbook of Biological Data,* W. B. Saunders, Philadelphia, 1956, 260.
22. **Fritz, I.,** Studies on the incorporation and release of N^{15} by tissue proteins of rats fed N^{15}-glycine. II. The activities of muscle, oral mucosa and incisor pulp proteins, *J. Dent. Res.,* 34, 446, 1955.
23. **Sodek, J.,** A comparison of the rates of synthesis and turnover of collagen and non-collagen proteins in adult rat periodontal tissues and skin using a microassay, *Arch. Oral Biol.,* 22, 655, 1977.
24. **Kofoed, J. A., Bozzini, C. E., and Tocci, A. A.,** Half-life times of the glycosaminoglycan fractions in gingiva, skin, and cartilage of rats, *J. Dent. Res.,* 50, 171, 1971.
25. **Steinberg, A. D. and Jeffay, H.,** Incorporation of radioactive amino acids into oral tissue proteins, *J. Dent. Res.,* 45, 812, 1966.
26. **Claycomb, C. K., Summers, G. W., and Dvorak, E. M.,** Oral collagen biosynthesis in the guinea pig, *J. Periodontal Res.,* 2, 115, 1967.
27. **Skougaard, M. R., Levy, B. M., and Simpson, J.,** Collagen metabolism in skin and periodontal membrane of the marmoset, *Scand. J. Dent. Res.,* 78, 256, 1970.
28. **Page, R. C. and Ammons, W. F.,** Collagen turnover in the gingiva and other mature connective tissues of the marmoset *Sanguinus oedipus, Arch. Oral Biol.,* 19, 651, 1974.

29. **Sodek, J. and Limeback, H. F.,** Comparison of the rates of synthesis, conversion, and maturation of type I and type III collagens in rat periodontal tissue, *J. Biol. Chem.*, 254, 10496, 1979.
30. **Goggins, J. F. and Billups, L. C.,** In vitro synthesis of acid mucopolysaccharides by human gingival fibroblasts, *J. Dent. Res.*, 51(Suppl. 2), 303, 1972.
31. **Orlowski, W. A.,** The incorporation of S^{35}-sulfate and C^{14}-glucosamine into the gingivae and periodontal ligament of a rat, *J. Dent. Res.*, 55, B67, 1976.
32. **Simpson, J. W.,** Pathways for glucose metabolism in the rat gingiva. I. Patterns of enzymes of glucose metabolism, *J. Dent. Res.*, 53, 938, 1974.
33. **Simpson, J. W.,** Studies of the pentose phosphate pathway of glucose metabolism in rat gingival tissue, *Arch. Oral Biol.*, 15, 689, 1970.
34. **Nicolau, J., Tamer, A. N., and Bergamaschi, O.,** Activities of hexokinase, phosphofructokinase and pyruvate kinase in the gingival tissue of the rat, hamster, guinea pig and human, *J. Periodontal Res.*, 12, 279, 1977.
35. **Honjo, K., Tsukamoto, Y., Nakamura, R., Yamazaki, T., Tsunemitsu, A., and Matsumura, T.,** Malate dehydrogenase in the gingiva and alveolar bone in scurvy, *J. Osaka Univ. Dent. School*, 5, 11, 1965.
36. **Charreau, E. H., Kofoed, J. A., and Houssay, A. B.,** Enzymes of the tricarboxylic cycle in periodontal tissues of the guinea pig, *Arch. Oral Biol.*, 10, 875, 1965.
37. **Charreau, E. H., Kofoed, J. A., and Houssay, A. B.,** Enzymes of the glycolytic cycle in periodontal tissues of the guinea pig, *Arch. Oral Biol.*, 11, 709, 1966.
38. **Charreau, E. H., Kofoed, J. A., and Houssay, A. B.,** Enzymatic activities in the incisor and molar regions of the guinea pig, *J. Dent. Res.*, 46, 241, 1967.
39. **Charreau, E. H., Kofoed, J. A., and Houssay, A. B.,** Oxidoreductases in periodontal tissues in the guinea pig, *J. Dent. Res.*, 46, 268, 1967.
40. **Bergquist, J. J.,** Lactic dehydrogenase in non-inflamed attached gingiva of dogs, *J. Periodontal Res.*, 7, 8, 1972.
41. **Bergquist, J. J., Organ, R., and Nuki, K.,** Isocitric dehydrogenase activity in non-inflamed attached gingiva of rhesus monkeys, *J. Periodontal Res.*, 9, 193, 1974.
42. **Rosett, T., Garner, P. S., and Gangarosa, L. P.,** Studies in the biochemistry of oral tissues. III. Pathways of carbohydrate metabolism in bovine attached gingiva, *J. Dent. Res.*, 51, 867, 1972.
43. **Simpson, J. W.,** Pathways for glucose metabolism in the rat gingiva. II. In vitro studies of the fate of glucose-UL-^{14}C, *J. Dent. Res.*, 54, 190, 1975.
44. **Williams, R. C. and Howell, T. H.,** Periodontology, in *Textbook of Oral Biology,* Shaw, J. H., Sweeney, E. Z., Coppuccino, C. C., and Miller, S. M., Eds., W. B. Saunders, Philadelphia, 1978, 1020.
45. **Fullmer, H. M.,** A histochemical study of periodontal disease in the maxillary alveolar processes of 135 autopsies, *J. Periodontol.*, 32, 206, 1961.
46. **Melcher, A. H.,** Some histological and histochemical observations on the connective tissue of chronically inflamed human gingiva, *J. Periodontal Res.*, 2, 127, 1967.
47. **Page, R. C. and Schroeder, H. E.,** Pathogenesis of inflammatory periodontal disease, *Lab. Invest.*, 33, 235, 1976.
48. **Schroeder, H. E., Munzel-Pedrazzoli, S., and Page, R.,** Correlated morphometric and biochemical analysis of gingival tissue in early chronic gingivitis in man, *Arch. Oral Biol.*, 18, 899, 1973.
49. **Flieder, D. E., Sun, C. N., and Schneider, B. C.,** Chemistry of normal and inflamed human gingival tissues, *Periodontics*, 4, 302, 1966.
50. **Quintarelli, G.,** Histochemistry of the gingiva. IV. Preliminary investigations on the mucopolysaccharides of connective tissue, *Arch. Oral Biol.*, 2, 227, 1960.
51. **Schultz-Haudt, S. D.,** Observations in the acid mucopolysaccharides of human gingiva, *Odontol. Tidskr.*, 66, 3, 1958.
52. **Ramamurthy, N. S., Garant, P. R., and Golub, L. M.,** Prolyl and lysyl hydroxylase activities in rat gingiva: effects of aging and inflammation, *J. Dent. Res.*, 56, B57, 1977.
53. **Narayanan, A. S. and Page, R. C.,** Biochemical characterization of collagens synthesized by fibroblasts derived from normal and diseased human gingiva, *J. Biol. Chem.*, 18, 5464, 1976.
54. **Narayanan, A. S., Page, R. C., and Meyers, D. F.,** Characterization of collagens of diseased human gingiva, *Biochemistry,* 19, 5037, 1980.
55. **Stern, B. D.,** Collagen solubility of normal and inflamed human gingiva, *Periodontics*, 5, 167, 1967.
56. **Golub, L. M., Curran, J. B., and Kennett, S.,** Collagen metabolism in pericoronitis, *J. Oral Pathol.*, 2, 222, 1973.
57. **Fullmer, H. M., Taylor, R. E., and Guthrie, R. W.,** Human gingival collagenase: purification, molecular weight and inhibitor studies, *J. Dent. Res.*, 51, 349, 1972.
58. **Beutner, E. H., Thrifthauser, C., and Hazen, S. P.,** Collagenase activity of gingival tissue from patients with periodontal disease, *Proc. Soc. Exp. Biol. Med.*, 121, 1082, 1966.

59. **Fullmer, H. M., Baer, P. N., Driscoll, E. J., and Lightbody, P. M.,** Collagenase and gingival disease: a quantitative preliminary report, in *Proc. 1st Pan-Pacific Congress of Dental Research,* Tokyo, 1969, 167.
60. **Simpson, J. W., Mackler, B. F., O'Neill, P., and Mailman, M. L.,** Identification of collagenase in cultured blood mononuclear cells, *J. Dent. Res.,* 59, 2, 1980.
61. **Birkedal-Hansen, H., Cobb, C. M., Taylor, R. E., and Fullmer, H. M.,** Procollagenase from bovine gingiva, *Biochim. Biophys. Acta,* 429, 229, 1976.
62. **Pettigrew, D. W., Sodek, J., Brunette, D. M., and Wang, H. M.,** Collagenolytic enzyme and inhibitor production by periodontal cells *in vitro, J. Dent. Res.,* 58, A154, 1979.
63. **Simpson, J. W. and Mailman, M. L.,** Synthesis of a collagenase inhibitor by gingival fibroblasts in culture, *Biochim. Biophys. Acta,* 673, 279, 1981.
64. **Pettigrew, D. W., Ho, G. H., Sodek, J., Brunette, D. M., and Wang, H. M.,** Effect of oxygen tension and indomethacin on production of collagenase and neutral proteinase enzymes and their latent forms by porcine gingival explants in culture, *Arch. Oral Biol.,* 23, 767, 1978.
65. **D'Alliard, F. J., Ciancio, S. G., and Hausmann, E.,** Hyaluronidase activity of the gingival environment, *J. Periodont. Periodontics,* 40, 17, 1969.
66. **Goggins, J. F., Fullmer, H. M., and Steffek, A. J.,** Hyaluronidase activity of human gingiva, *Arch. Pathol.,* 85, 272, 1968.
67. **Tynelius-Bratthal, G.,** Hyaluronidase activity in gingival crevicular material and in periotoneal exudate leukocytes in dogs, *J. Periodontal Res.,* 7, 3, 1972.
68. **Tzmouranis, A., Matthys, J., Ishikawa, I., and Cimasoni, G.,** Increase of extracellular cathepsin D activity in gingival washings during experimental gingivitis in man, *Arch. Oral Biol.,* 22, 375, 1977.
69. **ElAttar, T. M. A.,** The *in vitro* conversion of male sex steroid (1,2-^{3}H)-androstenedione in normal and inflamed human gingiva, *Arch. Oral Biol.,* 19, 1185, 1974.
70. **ElAttar, T. M. A. and Hugoson, A.,** The *in vitro* conversion of female sex steroid, oestrone, in normal and inflamed human gingiva, *Arch. Oral Biol.,* 19, 425, 1974.
71. **ElAttar, T. M. A. and Lin, H. S.,** Prostaglandin and cyclic AMP levels in human gingiva and effect of estrogens on their synthesis *in vitro, J. Dent. Res.,* 56, B59, 1977.

Acquired Enamel Integuments

PELLICLE FORMATION

Gunnar Rølla

INTRODUCTION

The acquired enamel pellicle is the term usually employed to describe the thin protein film which forms on the surface of teeth in vivo after a thorough cleansing procedure. Bacteria are found after 3 to 5 hr, dependent on the gingival status, oral hygiene, and diet. The exact role of the pellicle is not known, but it is assumed that it protects the inorganic phase of the tooth enamel against exposure to acids present in food and beverages and that it reduces the friction between teeth and also between teeth and the oral mucosa. The pellicle constitutes the surface on which bacteria adsorb during plaque formation, and considerable interest has been concentrated on the possible chemical interaction between the macromolecules of the pellicle layer and the surface of the bacteria, in particular the surface of *Streptococcus mutans* and *S. sanguis*. These microorganisms are frequently present in early plaque.

It is assumed that pellicle formation is based on the same mechanism as adsorption of macromolecules to hydroxyapatite. This mechanism is fairly well known due to the fact that hydroxyapatite powder is frequently used as chromatography material in the purification of proteins.

In the following, the properties of the enamel surface will be discussed and an outline given of the current concept of the adsorption of proteins to hydroxyapatite.

THE ENAMEL SURFACE

The inorganic phase exposed on the surface of teeth consists of hydroxyapatite. The importance of the surface properties of the enamel and the nature of the forces involved in its interactions with macromolecules and bacteria in the oral environment have only been fully realized the last few years, although stated clearly 20 years ago by the Neumans.[1] Traditional dental research has been mainly concentrated on the crystallographic aspects of dental enamel and the effect of fluoride on these. However, recent observations clearly indicate that some aspects of the cariostatic effect of fluoride may be explained in terms of changed surface properties of the enamel.[2-4]

A tooth in the mouth can be compared to a crystal lattice submerged in an electrolyte. It is a well-known physical principle that the electrolyte ions will be adsorbed to oppositely charged sites on a crystal surface. The solubility of the "salts" thus formed gives an indication of the strength of the attractive forces involved; low solubility obviously indicating high affinity. The pattern of the charges on the surface of the lattice is balanced by electrolyte ions in the hydration shell, thus securing the electroneutrality of the crystal.

The dental enamel exhibits much more phosphate than calcium positions on the outer surface (it is calcium deficient) and is therefore negatively charged.[5,6] It has furthermore been stated that whereas the phosphate of the hydroxyapatite surface is charged, the corresponding calcium positions are unpolarized. Calcium in the hydration layer, however, is hydrated and charged.[1] The marked overweight of phosphate in the hydroxyapatite surface causes a corresponding high concentration of positively charged counter-ions in the Stern layer in the hydration shell. Polyvalent cations displace monovalent cations in these positions; there is reason to believe that reactive calcium ions adsorbed as counter-ions to the hydroxyapatite crystal is a major feature of the surface of the teeth in vivo. The low solubility of calcium phosphate in general indicates high affinity of this cation for hydroxyapatite

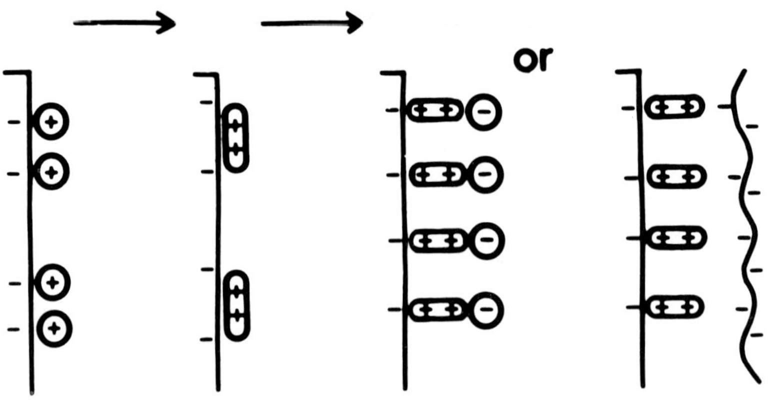

FIGURE 1. An illustration to show the principles of equivalent and equimolar adsorption of calcium on hydroxyapatite, the latter changing the surface charge from negative to positive. A positive charge carried by the counter-ions appears to be the mechanism by which the acidic protein pellicle binds to the surface of teeth. (From Rølla, G. and Bowen, W. H., *Acta Odontol. Scand.*, 36, 219, 1978. With permission.)

phosphate. Divalent cations can be adsorbed to a negatively charged surface on an equivalent or an equimolar basis. The latter principle involves change of net charge and binding of an additional anion to maintain electroneutrality.[7] The principle is illustrated in Figure 1. The equimolar adsorption of divalent cations is presumably caused by an incomplete shielding of the negatively charged surface by equivalently adsorbed ions.

This evidence for fluoride uptake by dental enamel through adsorption to equimolarly adsorbed calcium and some of the fluoride concentration taking place in plaque may be caused in this way as well.[8-11] This model would predict that acidic, calcium binding proteins and negatively charged microorganisms have high affinity for teeth. Experimental evidence for such a mechanism has been reported.[12-16]

Formation of the Acquired Enamel Pellicle

Bernardi and co-authors[17] have presented the current concept of the mechanism of the interaction of proteins with hydroxyapatite. This concept was later modified and extended by Rølla and Bowen.[11] The hydroxyapatite surface appears to be amphoteric; it binds acidic and basic proteins equally well (Figure 2). Adsorbed acidic proteins can be desorbed by phosphate or other anions,[4,15] while adsorbed basic proteins are replaced by calcium ions. Both phenomena are the result of competitive inhibition. The principle of the amphoretic nature of the hydroxyapatite surface can be accounted for by the model illustrated in Figure 2; acidic proteins are most likely adsorbed to calcium bound in the hydration layer of the crystal (Figure 2A, C, D), whereas basic proteins are adsorbed to phosphate groups in the crystal surface proper (Figure 2B). Basic proteins thus have to displace calcium ions in the hydration shell to reach their position[11] (see Figure 3).

The oral environment contains not only enamel surfaces and bacteria, but also present in the system are a large number of different salivary proteins. A number of experiments designed to approach the problem of polymer adsorption to surfaces have shown that the following principles are involved in the interaction between macromolecules and solids:

1. Soluble polymers with high molecular weight have higher affinity for solid surfaces than polymers with low molecular weight, presumably because of the polyfunctional nature of large molecules.[18]
2. Ionic polymers have a higher affinity for solids than nonionic polymers. The reason

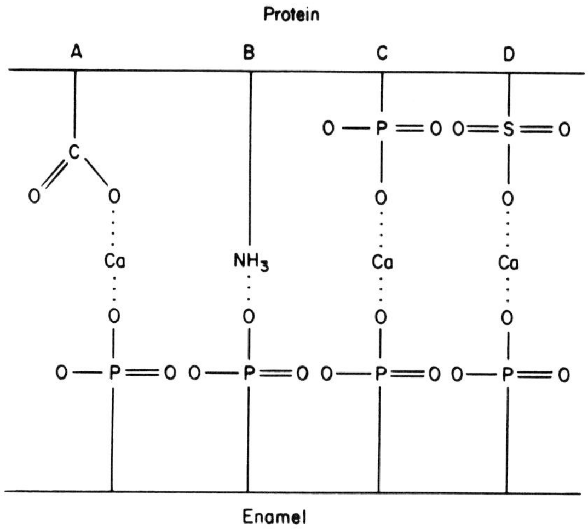

FIGURE 2. The most important mechanisms by which macromolecules are bound to hydroxyapatite surfaces. The surface carries a high density of phosphate groups which determine the selectivity of the reactions with proteins. (From Lehner, T., Ed., *The Borderland Between Caries and Periodontal Disease,* Vol. 2, Academic Press, London, 1980.

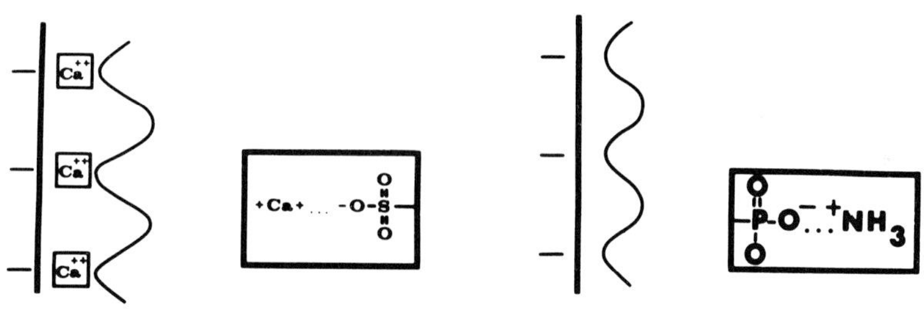

FIGURE 3. A diagrammatic illustration of binding of proteins to dental enamel. An acidic glycoprotein is adsorbed to the counter-ions present in the hydration shell of the mineral. A sulfated glycoprotein is used as an example (left). A basic protein is adsorbed to the enamel surface proper displacing the cations in the hydration shell in the process (right). (From Bennick, A. and Wong, R., *Saliva and Dental Caries*, Kleinberg, I., Ellison, S. A., and Mandel, I. D., Eds., Information Retrieval, New York, 1979, 59. With permission.)

for this may be that the ionic groups cause an extension of the molecules by the repulsion acting between the bound charged groups of the molecule.[19]

3. There is evidence for a chemical interaction of polymers containing carboxyl-, phosphate, and sulfate groups with dental enamel.[19-22]
4. Calcium counter-ions adsorbed to the surface of the enamel presumably react chemically with acidic groups available in the adsorbed polymer. This is presumably the principle by which polycarboxylate cements bind to dental enamel.[20]

There is evidence for a very strong interaction of phosphate- and sulfate-containing macromolecules with hydroxyapatite surfaces.[21,22] Phosphonates and sulfonates have also been shown to function as crystal poisons in many experiments.[23] This effect is most likely based on a strong surface adsorption to hydroxyapatite, presumably combined with a chemical reaction with the calcium ions in the hydration layer of counter-ions, permanently blocking

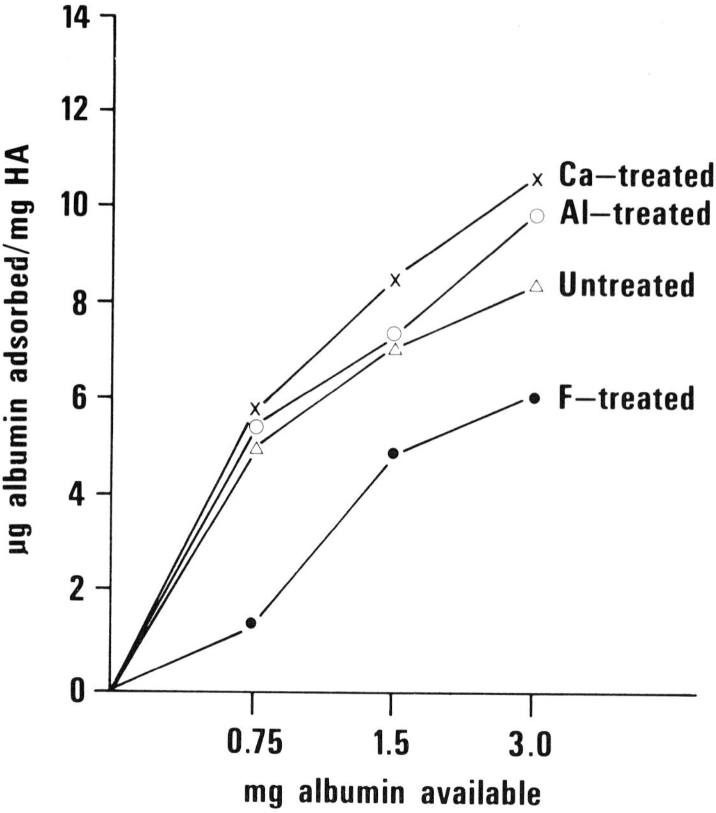

FIGURE 4. Effect of a brief pretreatment of hydroxyapatite (HA) with 1 m*M* of calcium-, aluminum- or fluoride ions on the uptake of an acidic protein (albumin).

further growth or crystallization of the hydroxyapatite crystal lattice.[24] Data concerning selective adsorption of salivary glycoproteins to dental enamel support the concept of selective adsorption of acidic proteins.

The chemical basis for selective adsorption is most likely that tooth surfaces display numerous calcium ions equimolecularly adsorbed. As discussed above, this would give a suitable condition for selective adsorption of salivary polyanions (see Figures 4 and 5).

It has been shown that sulfated glycoproteins with blood group substance activity are found in the acquired pellicle on human teeth in vivo.[15,25] It also appears that phosphoproteins may be present.[26]

The blood group active mucin molecules have high molecular weight, are polyanionic, and thus have the physical and chemical properties for a high affinity to hydroxyapatite, even in the salivary environment of competing inorganic phosphate ions. Sulfated "mucins" have a general protective role on the inner surfaces of the body so their presence on the teeth appears likely also from the teleological point of view.

Phosphoproteins are known to be present in saliva in relatively high amounts. Such proteins have been found in pellicle formed on human teeth, although the exact amounts are uncertain. Phosphoproteins, through their free phosphate groups, have a high affinity for hydroxyapatite. Phosphoproteins are also present in the hard tissues of bone and teeth and are assumed to play an essential structural role in these tissues. It thus seems conceivable that phosphoproteins may be engaged in protecting the surface of teeth in the oral cavity.[27]

A problem is that most salivary phosphoproteins are proline rich and that chemical analysis of pellicle materials contain only moderate amounts of this amino acid.[28] Another problem

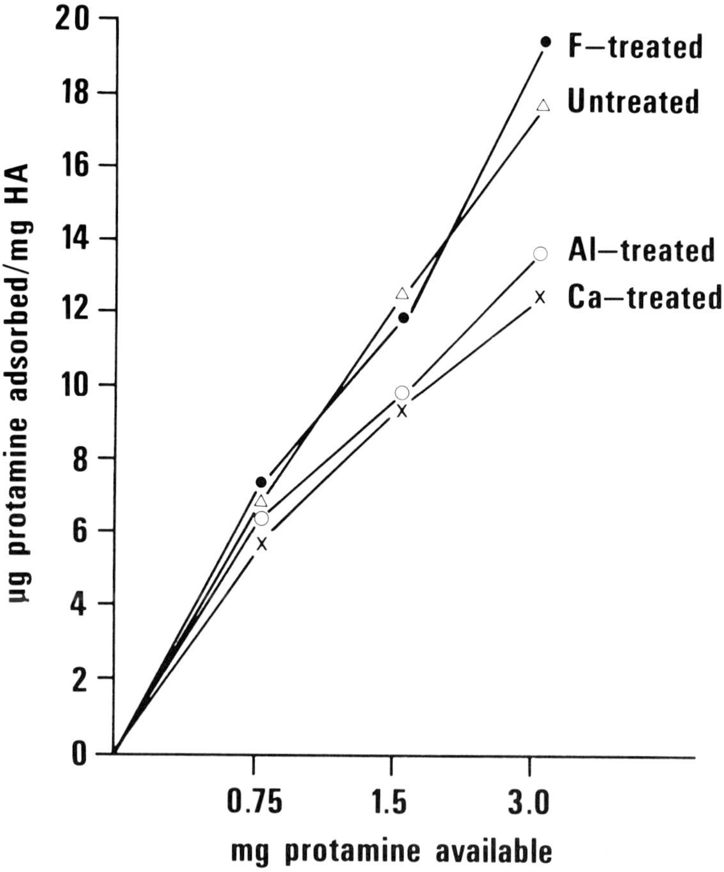

FIGURE 5. Effect of a brief pretreatment of hydroxyapatite (HA) with 1 m*M* of calcium-, aluminum- or fluoride ions on the uptake of a basic protein (protamine).

is that phosphatases are widely distributed in the oral cavity, so it appears likely that the phosphate groups of these proteins may be rather unstable. Phosphoproteins have also been suggested to play a role as inhibitors of precipitation of calcium phosphate in saliva, thus contributing to the high saturation of saliva with this salt.[27]

CONCLUSION

Pellicle formation in vivo is most likely caused by the same mechanism which rules protein adsorption to hydroxyapatite in vitro. Acidic proteins appear to be preferentially adsorbed due to the environment in the mouth (i.e., equimolar adsorption of calcium to the teeth, Figure 1). Phosphoproteins and sulfated glycoproteins are the most anionic polymers in saliva and presumably these play the key role in pellicle formation.

From a teleological viewpoint, it seems conceivable that these proteins are present to protect the teeth from excessive wear and from erosion by dietary acids. The pellicle also protects the oral mucosa from excessive friction by the teeth during mastication and speech.

REFERENCES

1. **Neuman, W. F. and Neuman, M. W.,** *The Chemical Dynamics of Bone Mineral,* University of Chicago Press, Chicago, 1958.
2. **Glantz, P. O.,** On wettability and adhesiveness, *Odontol. Revy,* 20 (Suppl. 17), 1, 1969.
3. **Tinanoff, N., Brady, J. M., and Gross, A.,** The effect of NaF and SnF_2 mouth-rinses on bacterial colonization of tooth enamel: TEM and SEM studies, *Caries Res.,* 10, 415, 1976.
4. **Rølla, G.,** Adsorption of Salivary Glycoproteins and Bacteria to Hydroxyapatite, Coll. Int. CNRS No. 230 Physiochimie et Crystallographie des Apatites d'Interet Biologique, 459, 1973.
5. **Kibby, C. L. and Hall, K. W.,** *The Chemistry of Biosurfaces,* Vol. 2, Hair, M. L., Ed., Marcel Dekker, New York, 1972, 663.
6. **Ash, R., Barrer, R. M., Butler, M. D., and Hudson, J. D. W.,** *Tooth Enamel,* Vol. 2, Fearnhead, R. W. and Stack, M. V., Eds., John Wright, Bristol, 1971, 52.
7. **Weiss, A.,** Equimolecular cation exchange at low-charged ion exchangers, *Kolloidzeitschrift,* 158, 22, 1958.
8. **Rølla, G. and Melsen, B.,** Desorption of protein and bacteria from hydroxyapatite by fluoride and monofluorophosphate, *Caries Res.,* 9, 66, 1975.
9. **Rølla, G., Hsu, D., and Bowen, W. H.,** The influence of fluoride on the uptake of protein by hydroxyapatite, *Caries Res.,* 11, 308, 1977.
10. **Rølla, G.,** Effects of fluoride on initiation of plaque formation, *Caries Res.,* 11 (Suppl. 1), 243, 1977.
11. **Rølla, G. and Bowen, W. H.,** Surface adsorption of fluoride and ionic exchange reactions on hydroxyapatite, *Acta Odontol. Scand.,* 36, 219, 1978.
12. **Mandel, I. D. and Concool, B.,** Salivary protein interactions with cations and anions, *J. Dent. Res.,* 54 (Special Issue A), L94, 1975.
13. **Belcourt, A.,** Adsorption of human salivary proteins on synthetic hydroxyapatite, *Arch. Oral Biol.,* 21, 717, 1976.
14. **Krogstad, S. and Rølla, G.,** Affinity of a sulphated rabbit submandibular glycoprotein for hydroxyapatite *in vitro, J. Biol. Buccale,* 5, 31, 1977.
15. **Rølla, G. and Embery, G.,** Chemistry of the sulfate groups in a sulfated glycoprotein from rabbit submandibular gland, *Scand. J. Dent. Res.,* 85, 237, 1977.
16. **Rølla, G., Robrish, S., and Bowen, W. H.,** Interaction of hydroxyapatite and protein-coated hydroxyapatite with *Streptococcus mutans* and *Streptococcus sanquis, Acta Pathol. Microbiol. Scand. Ser. B.,* 85 (8), 431, 1977.
17. **Bernardi, G., Giro, M. G., and Gaillard, C.,** Chromatography of polypeptides and proteins on hydroxyapatite columns: some new developments, *Biochim. Biophys. Acta,* 278, 409, 1972.
18. **Fontana, B. J.,** *The Chemistry of Biosurfaces,* Vol. 1, Hair, M. L., Ed., Marcel Dekker, New York, 1971, 83.
19. **Fuerstenau, D. W.,** *The Chemistry of Biosurfaces,* Vol. 1, Hair, M. L., Ed., Marcel Dekker, New York, 1971, 143.
20. **Beech, D. R. A.,** A spectroscopic study of the interaction between human tooth enamel and polyacrylic acid (polycarboxylate cement), *Arch. Oral Biol.,* 17, 907, 1972.
21. **Termine, J. D. and Conn, K. M.,** Inhibition of apatite formation by phosphorylated metabolites and macromolecules, *Calcif. Tissue Res.,* 22, 149, 1976.
22. **Embery, G. and Rølla, G.,** Interaction between sulphated macromolecules and hydroxyapatite studied by infrared spectroscopy, *Acta Odontol. Scand.,* 38, 105, 1980.
23. **Bartels, T. and Arends, J.,** *Tooth Surface Interactions and Preventive Dentistry,* Rølla, G., Sønju, T., and Embery, G., Eds., IRL Press, London, 1981, 157.
24. **Voegel, J. C. and Frank, R. M.,** *Tooth Surface Interactions and Preventive Dentistry,* Rølla, G., Sønju, T., and Embery, G., Eds., IRL Press, London, 1981, 147.
25. **Sønju, T., Christensen, T. B., Kornstad, L., and Rølla, G.,** Electron microscopy, carbohydrate analyses and biological activities of the proteins adsorbed in two hours to tooth surfaces *in vivo, Caries Res.,* 8, 113, 1974.
26. **Bennick, A. and Wong, R.,** *Saliva and Dental Caries* (a special supplement to *Microbiology Abstracts),* Kleinberg, I., Ellison, S. A., and Mandel, I. D., Eds., Information Retrieval, New York, 1979, 59.
27. **Hay, D. I. and Morena, E. C.,** in *Saliva and Dental Caries* (a special supplement to *Microbiology Abstracts*), Kleinberg, I., Ellison, S. A., and Mandel, I. D., Eds., Information Retrieval, New York, 1974, 45.
28. **Sønju, T. and Rølla, G.,** Chemical analysis of the acquired pellicle formed in two hours in cleaned human teeth *in vivo, Caries Res.,* 7, 30, 1973.

PLAQUE

G. N. Jenkins

METHODS OF COLLECTING PLAQUE

Although remnants of plaque are sometimes present on demineralized or even ground sections of enamel, their presence is fortuitous and their structure too deranged to be of much value for systematic histological study. Before methods of sectioning enamel were fully developed, plaque was collected on plastic "teeth" mounted on dentures and worn in the mouth for various periods after which it was embedded and histological sections cut in the usual way.[1] Another method was to tie Mylar® strips onto the teeth and after suitable intervals remove them and study chemically or histologically the deposits of plaque or early calculus.[2,3] The method was adopted for fissure plaque by mounting Mylar® film shaped like a fissure on a gold receptacle inserted into preexisting amalgam fillings.[4]

These methods suffer from the disadvantage that the plaque is formed on a foreign surface and not on enamel — a factor known to affect the composition of pellicle and therefore possibly of plaque. Any interactions between enamel and plaque would be prevented and the absence of pellicle from some sections of plaque suggests that such interaction might occur normally.[5] The most undisturbed and natural plaques were obtained by taking extracted teeth upon which plaque had accumulated, fixing them, and sectioning blocks of enamel cut out from the teeth with or without demineralization with HCL or EDTA.[5-7] The undemineralized sections were cut with a diamond knife and studied with the electron microscope (EM) which revealed the intimate contacts between plaque, pellicle, and enamel.

For chemical or bacterial analysis, plaque is collected usually for a minimum of 24 hr, but sometimes for much longer periods, and is scraped off with a suitable instrument preferably made of plastic to avoid any possible contamination with traces of enamel that might be removed by a metal scaler.[8] The morphology of the plaque cannot be preserved by this method so there is little information about chemical differences between the inner, middle, and outer layer. The composition varies widely from place to place in the mouth,[8] so that collections from a whole mouth represent an average — with the risk that over- or under-representation of one specialized site (e.g., lower incisors) might seriously affect the composition of the sample. Ideally, samples from different sites should be kept separate, but, apart from greatly increasing the labor of manipulation and analysis, sample sizes become too small for the accurate analysis of some constituents.

METHODS OF ESTIMATING THE AMOUNTS OF PLAQUE

Precise estimations of the amounts of plaque present in a mouth are difficult because the collection can never be complete and collection from a whole mouth takes several minutes, during which there is risk of loss of weight by evaporation, although the latter difficulty can be avoided by drying the sample to constant weight before weighing. To reduce the time of collection, it is sometimes made from a limited number of teeth, say, one quadrant of the mouth, but naturally the sample size is smaller and may be inadequate for some separations or analysis.

Plaque can be estimated in vivo by estimating the proportion of the enamel surface covered. This is facilitated by use of a disclosing solution (erythrosin has been widely used), but this has the disadvantage that very thin layers weighing very little become stained and give inflated values.[9] The correlation between weight and the visual plaque score is higher without staining.[9]

A disclosing solution consisting of a mixture of methylene blue and triphenyl tetrazonium chloride differentiates between mature plaque 3 or more days old (which stains blue immediately after application of the dye, but becomes red up to 10 min later) and more recently formed plaque which remains blue.[10] The reducing conditions of mature plaque decolorize the methylene blue and make the red dye visible.

Visual estimates of the amount of plaque[11] are expressed crudely as good, fair, or poor oral hygiene, but, more accurately, is scored on 12 areas of the teeth as 0 = no plaque present, 1 = plaque covering not more than one third of the tooth surface, 2 = as 1, but between one and two thirds of the surface, and 3 = plaque covering more than two thirds of the surface (Greene and Vermillion's Oral Hygiene Index).[12] The Plaque Index of Löe[13] is based on estimates of the thickness of plaque on the mesial, distal, buccal, and lingual aspects of each tooth.

Some workers use the N content rather than the weight as a measure of the amount of plaque, but this is also subject to considerable error.[14] The absorbance at 600 nm of plaque dispersed by sonication in 10 volumes of water has also been used.[15]

ATP is naturally present in plaque bacteria and its estimation or the related "adenylate energy charge" has been used as a measure of the viable cell mass.[16] It has the advantage that very small amounts of plaque (<90 μg) can be estimated accurately by the luciferase assay for ATP.

COMPOSITION OF PLAQUE

The composition of plaque is so variable from site to site and from time to time that average values are of limited value and too few data have been published for most constituents to give meaningful ranges. It contains many enzymes, so its composition changes after collection unless precautions are taken. In view of this and of the uncertainty of some of the analytical methods as applied to small samples, almost all the data on plaque composition are crude and capable of detecting only major trends or differences.

The water content, estimated by heating at 105°C to constant weight is less variable than most constituents and ranges from 80 to 85% (mean 82%).[17,18] Plaque consists of a protein matrix with a variable proportion of lipid and polysaccharides in which many species of bacteria, estimated as 400 million per milligram,[19] are embedded, and some epithelial cells, all cells together occupying about 70% of its volume.[20] It is obviously desirable to separate the matrix from the bacteria for some estimations, and although this can never be complete, several methods have been investigated.[18] In the method finally adopted by Krembel et al.,[18] the plaque (10 mg/mℓ) is dispersed mechanically in water at pH 6.5; the bacteria are removed by centrifuging and filtering the supernatant through a 0.45-μm millipore filter. The separation is incomplete because some matrix does not dissolve.

In a second method,[21] the plaque is extracted with alkali (pH 12) which is a good solvent of the matrix and prevents most enzymic breakdown, but also extracts some bacterial constituents.[22,23] The bacteria and other cells are then removed by centrifuging and for some purposes are resuspended in alkali several times. Fox and Dawes[23] studied in detail the effect of alkali and of conditions of temperature, pH, and ionic strength on the extraction of combined samples of pure cultures of predominantly plaque organisms and compared the proportions of N extracted from them and from plaque. About four times as much N was extractable from plaque as from the organisms, suggesting that three quarters of the N of plaque extracts is from the matrix and the rest is from the bacteria, assuming that plaque bacteria are as susceptible to extraction as the probably much healthier organisms grown in culture.[23] The proportion extracted from bacteria increased if the extraction was carried out for longer than 1 hr, but was relatively independent of bacterial concentration, stage of growth of the organisms, or ionic strength, but was increased by rise in temperature, osmotic shock, or by increase in pH (minimum extraction at pH 5.0, maximum of 12.7). Unfortunately at the pH of minimum extraction of the bacteria, the matrix protein is also relatively insoluble, and this suggested the overall optimum conditions for a good separation were a temperature of 0°, extraction time not greater than 30 min and a pH of 12.[23]

Owing to the great variations in techniques of fractionation of plaque used by different workers, no satisfactory average figures can be given nor is it easy to conclude from these figures the composition of the original plaque. All that can be done is to report briefly on the main results obtained by different workers. However, there is agreement that extracts contain a high proportion (about 50% of the dry weight of the original plaque) of high molecular weight protein, probably derived from salivary protein, and a good deal (up to 30% dry weight) of low molecular weight proteins and amino acids, much of it probably extracted from cells.[21]

Krembel et al.[18] reported that the plaque contained (percent of dry weight) 40 to 50% protein (mostly in cells), 13 to 17% carbohydrate, (mostly in the matrix) made up of free sugars and high molecular weight polysaccharides, the amounts depending on the interval between eating and collecting the plaque. Lipid is reported to occupy 10 to 14% of the dry weight of plaque, mostly in the matrix,[18] but the bacteria also contain lipoteicholic[24] acid and lipopolysaccharide (endotoxin);[25] the fatty acids in typical bacteria isolated from plaques have been reported.[24,26,27] Critchley made extracts with water and 50% $CaCl_2$ (which dissolves glycoprotein) and fractions of the extracts were incubated for varying periods to allow the plaque enzymes to break down the constituents.[28] The results showed that plaque contained free amino acids, free sugars, oligo- and polysaccharides, mostly polymers of glucose, glycopeptides, and proteins and that up to 33% of the dry plaque was dialyzable — these small molecules probably formed by enzymic action within the plaque. That they remained in plaque suggested that diffusion through plaque was limited. The amino acid composition resembled that of sublingual saliva,[28] but comparison of the amino acids of complex mixtures of proteins is obviously of limited value.

Hotz et al.,[29] in a paper giving good reference to previous work, reported the results of an extraction with water and 1 *M* KOH of plaque pooled from 3500 school-children. The water-soluble fraction was 29.6% of dry weight and contained 60% carbohydrates, 1.2% N, and more than 4% protein. The water-insoluble fraction (67.1%) contained 11.3% carbohydrates, 7.4% nitrogen and 30.7% was insoluble in alkali. The complex carbohydrates were mostly glucans, but some other hexoses and pentoses were present. The water-insoluble polysaccharides had predominantly 1:3 linkages ("mutans") and represented 1.35% of dry weight: low molecular weight carbohydrates, probably enzymic breakdown products of polymers, were 5.6% of the dry weight.

Sulfated proteins have been isolated in pellicle and plaque from the monkey *Macaca irus* after injecting[35] SO_4 and are probably derived from the submandibular-sublingual or the minor salivary glands.[30] They have been detected in human pellicle and, as they are reported to be capable of aggregating (see later) certain salivary bacteria, it is likely that they occur also in human plaque associated with bacteria.[30]

The detailed structure of many polysaccharides formed in vitro from *Streptococcus mutans*[31,32] or other oral streptococci has been described.[31,33,34] These are all potential constituents of plaque, although they have not necessarily been found in extracts.

Levine et al.[35] made comparisons of UV spectra of immunoelectrophoretic separations and isoelectric focusing of plaque extracts, saliva, and serum and concluded that extracts of plaque from the cervical margins contained a higher proportion of serum protein than salivary proteins. They concluded that the gingival fluid was probably a more important source of protein than saliva. Extracts of plaque showed the presence of free amino acids and putrescine, cadavarine, amino butyric acid, and histamine.[36]

Samples of early plaque contain some epithelial cells and leukocytes, but these autolyze and are not generally present in plaques several days old.[37] Their contents released during autolysis provide another source of plaque protein as well as nutrients for the bacteria.

It is important to stress that food debris is not an important constituent of plaque (except fissure plaque in which food particles, once they enter the fissure, tend to become trapped).[38] Indigestible food may be impacted between the teeth, but plaque grows around these residues and microscopical study shows very little food material within plaque itself.

PLAQUE FLUID

Before discussing the mineral matter in plaque it is convenient to describe plaque fluid and its isolation. The realization that the composition of the fluid phase of plaque, especially of the inner layers, was likely to be critical in controlling caries and gingivitis led to the development of methods isolating and analyzing "plaque fluid".[39,41] Before this it had sometimes been assumed that the fluid phase was identical to saliva[40] bathing the plaque. It is still not possible to collect plaque or plaque fluid separately from the inner, middle, and outer layers of plaque, but representative samples of the fluid from homogenized plaque can be obtained by centrifuging; the conditions used by Tatevossian and Gould were[41,42] 5000 g for 15 min at 2°C in a fixed angle rotor the plaque being stored in ice during its collection (which can take considerable time if collected from a number of subjects); storage in liquid N_2 led to cryodamage and release of ions from the bacteria. A higher yield can be obtained by centrifuging at 18,000 g for 30 min, but at the risk of greater contamination of the fluid by bacterial contents. In the pioneering experiments on plaque fluid,[39] the plaque was "tapped" into a capillary tube so that after centrifuging, the fluid was contained in a readily obtained, fairly long column, but inserting the plaque into the tube took some time during which the plaque was deteriorating and the whole of the sample could never be put into the tube. So later, tubes of 4-mm internal diameter were used and the fluid (50 to 70% of the plaque column) was drawn off into a capillary tube.[41] Its volume can be measured with much greater precision than the weight of plaque so that the analytical results on plaque fluid are much less variable.

The results of analyses clearly show that plaque fluid is totally different from saliva,[41] implying a low level of diffusion, in and out of plaque. When analyzed for amino acids, the chromatograms of monkey plaque indicated high concentrations of an unusual component identified as δ-amino valeric acid,[43] almost certainly of bacterial origin. The concentration of amino acids in plaque fluid was higher in monkeys fed a high sucrose diet, presumably the sugar stimulated bacterial metabolism.[44]

The proteins have been separated by electrophoresis and 8 to 17 bands were detected;[45] IgG, lactoferrin, and IgA were consistently present at concentrations exceeding those of saliva, suggesting that gingival fluid was an important source of plaque proteins. Albumin and amylase were also present.[44]

THE MINERAL MATTER IN PLAQUE

Plaque is fairly high in mineral matter, the ash content being reported to average 10% dry weight and 15% of the matrix.[18] Calcium and phosphate concentrations are especially high[8,46,47] and may play an important part in the role of plaque in caries, as discussed later.

Potassium from cells is the inorganic constituent in highest concentration (2.75% dry weight).[47,48] Approximate concentrations of trace elements show the presence of a wide range: published figures for Cu, Pb, Li, Sr, and Zn have been summarized.[48]

That the concentrations of calcium and inorganic phosphate (P_i) in plaque[49] are greatly in excess of those in saliva[8,46,47] has been known for many years and has since been shown to be related to the caries activity and has been implicated in theories of plaque formation. The question immediately arises as to the form in which these ions are present.

The concentrations of Ca and P_i in plaque fluid show that a considerable proportion is not intracellular or bound to matrix proteins, although some phosphate is present in the former state and some calcium in the latter. The plaque fluid contains concentrations of Ca and P_i exceeding those of the solubility product of calcium phosphate and is therefore, even more than saliva, a supersaturated solution. This state is probably maintained by the presence in plaque fluid of statherin — a salivary protein which prevents the seeded precipitation of calcium phosphate.[50,51]

Two electron microscope studies of plaque have suggested that apatite crystals may exist in plaque matrix, but it seems likely that the use of a fixative containing phosphate and the processing of the specimen may have produced this crystallization.[5,52,53,53a] In Frank and Brendel's paper,[5] the crystals are described as "calculus like" and may therefore be not typical of normal young plaque. Lyophilized plaque collected from students after abstaining from oral hygiene for 1 week was shown by X-ray diffraction to contain hydroxyapatite (HA), the crystals being especially large in high calculus producers.[53] This does not necessarily mean that HA is present in much younger plaque in the undried condition.

The Ca and P_i concentrations of plaque in the lower incisor region are much higher than in upper incisors,[8,49,54] presumably from the proximity of the submandibular saliva with its higher range of pH and concentrations of calcium and phosphate. A tendency for higher concentrations of Ca and P_i in plaque (significant for calcium) between the central lower incisor than between the incisor and canines has also been reported.[49] These differences provide a likely explanation of the lower susceptibility to caries and greater susceptibility to calculus deposition of the lower compared with the upper incisor. These data led to an investigation of the correlation between plaque Ca and P_i in different individuals and their caries activity.[55-57] There is considerable evidence that high levels of these ions in plaque are associated with lower increments of caries[57] and lower total caries experience[56,57] or fewer early and late lesions[58] (but excluding probed lesions and restorations which reduced the sensitivity of the caries score).[58] The question arises whether the converse — a low Ca and P_i concentration in plaque of caries-prone subjects — is a primary factor in a caries-prone mouth or is it, like caries, the result of high acid production releasing these ions? Although weak correlations have been found between the Ca and P_i concentration of saliva and plaque,[59,60] the question has not been thoroughly studied and, in view of the many factors (time of day, rate saliva flow proportion of saliva from the different glands) which affect saliva composition, is not easy to study. A relation between sugar intake and plaque Ca and P_i has also been shown,[59] compatible with the view that the relation between these ions in plaque and caries activity is a result common to high acid production. In a penetrating discussion of this question (marred only by the limited and uncertain data), Schamschula et al.[61] point out that there seem to be homeostatic controls of plaque mineral composition and that the composition is largely independent of the composition of food and drinking water — even rinses with 0.2% NaF solutions affect the plaque F^- for a few minutes only. They incline to the view that unknown factors endogenous to plaque control its mineral composition, and therefore its cariogenicity, although they conclude that saliva composition may have some influence.

The addition of a variety of inorganic phosphates to the diets of rodents reduces experimental caries markedly.[62,63] Clinical trials in man in which calcium phosphate was used were entirely negative,[64] but two trials in which a neutral mixture of sodium phosphates was added to one food only (sugared flakes) for 1 year were reported to produce good reduction in caries in both institutionalized children and adults.[65,66]

Study of the influence of P_i, either incorporated into lumps of sugar or as rinses, on the plaque P_i has suggested a good reason for these essentially negative results.[67] In this study,[67] sucking sugar lumps containing 0.37% P_i caused an average fall of 13% in the P_i concentration of the plaque. When the results were examined closely, it became clear that in plaque initially low in P_i, the P_i concentration rose following the sugar lump, those initially high fell, whereas intermediate concentration did not change. Rinses of 0.37% did raise significantly ($p<0.001$) the P_i of plaque by 59%, but other similar experiments showed that the rise was detectable only for about 10 min. The explanation of these results is probably as follows:[67,68] the concentration of P_i in plaque is often so high that it is difficult to raise the environmental concentration of P_i above the plaque level and thus establish a concentration gradient. This occurs with a 0.37% rinse because rinses are a poor salivary stimulant and are subjected to very little dilution. When sugar lumps are used, however, the rapid flow of saliva greatly dilutes the P_i in the sugar[67] and the P_i of saliva itself falls dramatically when its rate of flow increases.[69] Consequently, it is only possible to increase plaque P_i from sugar supplements when the initial plaque is low in P_i or high concentrations of P_i are used, which are undesirable as additives because they affect the taste of the sugar and may disturb Ca metabolism. Plaques low in P_i are, however, associated with high caries as mentioned above[55,59] so it may be that inorganic P_i supplements would reduce caries in very caries-prone subjects (with a low plaque P_i), but have no benefit on the higher levels present in most plaques. The clinical trials of P_i supplements made no attempt to compare the effect in subjects matched with high and low initial caries scores.

A low pH, e.g., immediately following sugar ingestion, is associated with a reduction in the Ca and P_i of plaque, presumably by reducing the binding of Ca by the plaque and either by washing out of P_i during the metabolism of the sugar or both.[70] Separate experiments comparing rinses with deionized water and with sugar solutions suggest the washing out effect is the larger,[68] but the concentration of P_i returns to its original level within 5 to 10 min by the unknown mechanism endogenous to the plaque — possibly an enzymic release of P_i from organic reservoirs.[68] During these experiments, large and unexplained (possibly spurious) changes in the Na and K concentration of the plaque occurred. The effect of sugar rinses in lowering the Ca and P_i of plaque via the formation of acid, may, in addition to being an explanation for the correlation between Ca and P_i and caries,[59] be one of the ways in which sugar increases caries.

The effect of dietary supplements of P_i in reducing caries in animals is a local action, absent when the P_i is given by stomach tube, and is probably a common ion effect reducing the demineralization produced by acid. The rise in plaque P_i seems too short-lived to have much action on remineralization which presumably occurs between 15 and 30 min after sugar ingestion when the pH is approaching neutrality (see later). Other possible effects of P_i as an additive are desorption of plaque protein or increased buffering effect of the plaque, but these seem too small to be effective.

It is not clear why P_i supplements reduce caries in rodents, but are ineffective in man. Suggestions that rat saliva is lower in P_i than human saliva (and therefore more readily influenced by supplements) were based on samples produced at unphysiological rates of flow by pilocarpine.[71] Analysis of samples of resting saliva obtained by blotting the mouths of rats with filter paper suggested a composition similar to that of man,[72] but the P_i concentration of plaque in rats does not appear to have been studied. A possible explanation is that the P_i was added to the animals' diet so that every mouthful of food was accompanied by a high P_i in the oral environment. In human subjects, the P_i has been included only in certain foods (usually sugar and/or flour) so that cariogenic foods could be present in the plaque when concentration of P_i was not elevated, and experiment shows that a raised P_i concentration is short-lived and would not last from one P_i supplemented food to a meal which did not contain P_i.

THE F^- OF PLAQUE

For some years up to 1960, the anticaries effect of F^- was attributed mostly to an increased acid resistance of enamel. The concentrations of F^- required to inhibit bacteria had been shown to be 2 to 10 ppm and even 100 ppm were required to affect growth,[73,74a] far higher than the 0.1 to 0.4 ppm then thought to be in saliva[74] (recent work shows that the true value is about one tenth of this)[75] and tacitly assumed to be the concentration in plaque. When the F^- concentration of plaque was eventually estimated, it was reported to be astonishingly high (up to 67 ppm wet weight)[76] and to be affected by the F^- of drinking water (averaging 25 ppm in a low F^- area, 49 with 2 ppm in the water).[77] With the introduction of the more accurate ion specific electrode, lower values have been found (up to 10 ppm),[78,79] but even these values show that F was bound, otherwise it would be expected to diffuse into the salivary environment where the concentration is much less than 1% that of plaque. The electrode showed that most of the F was not in the ionic form, and the F was not extracted under conditions which dissolved apatite, disproving that it was bound to apatite — whose existence in fresh plaque is uncertain anyway. This left the bacteria as the probable sites of binding, and growing bacteria in F^--containing media has confirmed this,[81] although there is still no agreement as to which bacterial compartment the F^- is attached[82-84] or the extent to which the bound F^- inhibits the bacteria.[85]

Several studies have shown that the plaque F^- seems to inhibit slightly bacterial acid production[81,86-88] which probably represents one of the several ways in which F^- reduces caries. Plaque from residents of towns with F^- in their water supplies produced a slightly smaller drop in pH when incubated in vitro with sugar than did control plaque.[81] A similar difference was found in plaques collected before and some months after F^- was introduced into the water of one town.[86] Although their difference was small (less than 0.2 pH units), they were consistent and highly significant statistically. Some data suggest that F^- ingestion changes the bacterial composition of plaque, such as reducing the proportion of organisms which synthesize intracellular polysaccharide.[87,89] The marked inhibition of acid production in vitro of bacteria incubated with F^- is thought to involve the transport of sugar by the phosphoenolpyruvate phosphotransferase system rather than a direct effect on glycolysis.[90] The F^- both of plaque and of bacteria is present in several forms: ionic, loosely bound, and tightly bound[80] (although the use of these terms by different authors does not always refer to the same fractions),[82-84] again supporting the view that the F^- of plaque is contained within the bacteria. The concentration of F^- in plaque in high F^- towns have been shown to be inversely correlated with the caries activity of individuals.[74]

However, caution is required in accepting the results, suggesting that plaque F^- reduces acid production. The F^- concentrations of plaque are inversely proportional to plaque weight, suggesting that they decrease as the plaque accumulates,[76,79,91,92] so thick accumulations have lower concentrations than young, thin plaques. Other studies have shown that the thicker the plaque the greater is the pH drop with sugar;[93] presumably acid cannot readily diffuse out of thick plaques and the anaerobic conditions on the inner surface of plaque probably reduce or prevent the oxidation of the acids. The correlation between acid production and F^- concentration might therefore be an artifact: thick plaques show a large pH drop and, independently, thick plaques are low in F^-. However, there is no evidence that F^- (at the levels in water) affects the weight of plaque deposited so the relation between F^- and plaque acid production is probably causal, although the effect is thought by some workers to be too small to be of practical importance.[94]

COMPOSITION OF FISSURE PLAQUE

Owing to the small quantities of fissure plaque formed and difficulties in collecting it, no chemical analysis appears to have been attempted. It has been studied by the EM in small numbers of subjects either on plaques of unknown age formed normally[8,95] or with control of age in fissures of extracted teeth or artificial fissures of Mylar® foil implanted into large occlusal restorations.[95,97,98]

The main difference between fissure and smooth surface plaque 7 days old is the larger amount of food debris, such as plant tissue and muscle fibers, in the former. The bacteria present were mostly cocci and short rods, but at the entrance to the fissure, there were also filamentous bacteria which were absent from the deepest parts. In fissure plaque 200 to 270 days old, the counts were somewhat similar, but *S. mutans*, present even in the absence of caries, was a considerably higher proportion of the streptococci than in 7-day fissure plaque[97] — evidently it colonizes fissures rather slowly. The plaque forming in the natural fissures showed a greater tendency to mineralize than that in Mylar® fissures. Theilade et al.[98] suggest that the presence of the enamel surface in a narrow space from which outward diffusion is limited might be a source of plaque Ca and P_i which would favor mineralization compared with the Mylar® foil.

THE FORMATION OF PLAQUE

Only about 10 to 20 mg of plaque form daily, containing not more than 1 to 2 mg of matrix protein. Clearly the formation of plaque matrix is a very small-scale process, leading to the precipitation of less than 0.1% of the protein in 1 day's average output of saliva.

An early theory of plaque formation suggested that it arose from the evaporation of films of saliva on the teeth.[99] This implied that plaque matrix was identical to the totality of proteins in the saliva surrounding the tooth and that it depended on the alternate drying and wetting of the surface. However, the proteins of plaque are not identical to those of saliva, and plaque forms as readily on the posterior teeth (where drying is unlikely except in mouth-breathers) as on anterior teeth where drying is more likely. It is possible that some deposits on the teeth (not true plaque), e.g., those formed during sleep[99] might arise in this way, but these do not appear to have been investigated or compared with plaque.

Some of the proteins of submandibular and sublingual saliva collected together (the secretions are difficult to collect separately) are unstable and readily precipitate on contact with a foreign surface, on shaking, or even spontaneously. In some subjects, this saliva is cloudy as it emerges from the tube of a collecting device, and this has been found to occur when the pH and phosphate concentrations are in the upper range.[249] Whether these precipitates resemble plaque matrix in composition and whether the persons who form them are also rapid plaque formers has not been investigated. Some workers have supposed that the contact of proteins with hydroxyapatite might lead to conformational changes, insolubility, and deposition.[20] This process might be involved in pellicle formation, but it is less likely that any plaque forms on "naked" apatite. Direct contact between plaque and enamel, which has been shown in some electron micrographs,[5,141] probably occurs following the removal by bacterial activity of a preformed pellicle,[150] although this has been doubted.[141]

Isoelectric Precipitation

Another early theory which has persisted, and with modification may be partly true, is that local bacterial acid production lowers the pH to that of the isoelectric point of some salivary proteins, thus leading to their precipitation.[100] The main objections to this as a main cause of plaque formation are that, at least in places, plaque forms on bacteria-free pellicle and has been reported to form during many hours without food[6] when acid production would be negligible. Nevertheless, a fully established plaque producing much acid might then lead to the precipitation of salivary protein and thus extend the plaque. Acid may favor plaque formation by other mechanisms described later.

A theoretical objection to the isoelectric precipitation theory was raised by Dawes.[101] He reported that when parotid saliva was acidified with 1 *M* lactic or acetic buffer over the pH range 2.0 to 6.0 at intervals of 0.5 units, no precipitation occurred, until the pH reached 3.5 (below the value reached by oral bacteria), but when the unbuffered acids or HCl were used, precipitation did occur as high as pH 6.5. He explained this as the formation of irreversible precipitates by local denaturation in the vicinity of the low pH surrounding the drop of unbuffered acids before it was diluted by mixing. With the buffers, the pH would not fall as low and, Dawes suggested, would not reach the isoelectric point. Apart from the differences in local pH produced by buffered and unbuffered acids, the buffered acids have a much higher ionic strength which might prevent precipitation. That the latter was the correct explanation was shown by the later finding[102] that when KCl was added to the HCl to raise its ionic strength, no precipitate occurred in whole or duct saliva, and when unbuffered organic acids were used, precipitation occurred within the range of oral pH values provided the ionic strength of the acid was less than 0.1 mol/ℓ. Precipitation was variable between subjects and occurred less frequently in parotid than in submandibular saliva. The validity of the acid hypothesis of plaque formation depends therefore on the question of the nature of the bacterial acids present in vivo; they are a mixture of organic acids in a probably low, but uncertain, ionic strength at the rather ill-defined boundary between plaque and saliva. The ionic strength is uncertain, because although diffusion of ions out of plaque is limited, its extent in the immediate surroundings of plaque is unknown. It must be emphasized again that acid precipitation could only occur near an already well-established plaque as the acid production from the small isolated clumps of bacteria in very early plaque would be so diluted by the adjacent saliva that it is most unlikely to lower the pH to values at which significant precipitation can occur.

The isoelectric point of salivary protein covers the range of 3 to 10 with the maximum precipitation at 3.7.[103,104] Alkaline extracts of plaque contain numerous proteins which have a similar range of IEP and a maximum precipitation of pH 3 to 4;[105] however, as these proteins have already been denatured and perhaps modified by bacterial proteolytic enzymes there IEPs may differ from those of the original salivary proteins.

Enzymic Degradation of Glycoproteins

Dawes and Jenkins[106] and Leach[107] reported almost simultaneously that plaque contained no detectable sialic acid (i.e., <0.002%), although some salivary glycoproteins do contain it. This suggested that the removal of sialate from these proteins by the enzyme neuraminidase (EC 32.1.8), widely distributed among oral bacteria, might alter their solubility and lead to their precipitation and this became a widely quoted explanation of plaque matrix formation.[108] The alternative possibility that only proteins not containing sialate might enter plaque matrix was not considered.

This hypothesis was supported by the finding that incubating clear submandibular saliva, collected from the duct, with plaque or salivary sediment led to an increase in turbidity and finally a precipitate.[102] Sugar was not added, and although previous workers had shown that the release of sugars from glycoproteins led to some acid formation, this was on too small a scale to reduce pH to the level at which isoelectric precipitation would occur. Incubation of saliva with crude neuraminidase (cholera fitrate) also causes precipitation, but this occurred almost immediately and, estimation of free sialate showed, before the release of most of the sialate and was not prevented by boiling the enzyme solution.[102] Clearly this precipitate was not caused by enzyme action, and its nature was not investigated further. When pure neuraminidase (Sigma, Grade VI) was used, turbidity did not occur in 11 out of 15 experiments, again casting serious doubt on this hypothesis.[102]

Briscoe et al.[109] incubated with neuraminidase a mixture of parotid and submandibular saliva collected from the duct and reported a turbidity which reached its maximum within 1 min of incubation, whereas only about one half the sialate was released after 7-min incubation. It is therefore doubtful whether the turbidity was related to sialate release. These workers found that the cleavage of sialate from salivary proteins did not increase their tendency for adsorption onto hydroxyapatite (HA) (a test for sources of pellicle rather than plaque) or saliva-treated HA (a test for plaque formation or adherence to pellicle-covered

HA). Other experiments showed that sialate removal tended to reduce, rather than increase, isoelectric precipitation.[102]

In another series of experiments designed to test this hypothesis, the results suggested a different mechanism.[102] Plaque, covered with a thin layer of saliva or water, was incubated and the absorption of the fluid phase at 215 nm was measured, it being predicted that if the enzymes of plaque led to precipitation of salivary proteins, the absorption would fall. The results were complicated because some protein diffused out of plaque (i.e., absorption increased in the plaque and water control tubes), but the sum of the absorptions in a saliva only control and the plaque-water control were greater than the plaque-saliva mixture, indicating that some precipitation had occurred. However, in a modified experiment, in which the plaque was mixed with the saliva instead of being smeared on it, similar results were obtained even with heat inactivated plaque. Evidently some heat-stable constituent of plaque was capable of precipitating salivary proteins (see later). Further study of the carbohydrates of salivary glycoproteins showed that they included fucose, galactose, mannose, *N*-acetyl-glucosamine, and galactosamine and that they were all released and metabolized when saliva was incubated, but survived if the saliva was heated or antibiotics added.[110]

Sialate was removed first and incubation for up to 8 hr was required for complete stripping of the carbohydrate side-chains. The presence of these carbohydrates in long side-chains projecting from the protein core makes these proteins water-soluble and even heat stable. When a tube of submandibular saliva, collected fresh from the duct, is plunged into boiling water, very little turbidity results, but if bacterial enzymes are added first, the salivary proteins become progressively more susceptible to precipitation as the side-chains are removed. These results suggested that the whole side-chain, and not merely the terminal sialate, had to be removed before precipitation occurred. Leach has developed this hypothesis, pointing out that the protein core contains many hydrophobic amino acids which cannot interact in the intact molecules because of the steric hindrance effect of the side chains.[111,112] When the latter are removed, however, there is an entropically motivated release of the structured water surrounding the molecules and the hydrophobic amino acids of adjacent molecules can then adhere and at least some of the resulting aggregates become attached to the tooth surface as plaque matrix. The detailed physical chemistry of this process is described by Tanford.[113,114] Leach considers that this hydrophobic interaction overrides the contribution made by other types of weak bonds, such as van der Waals' forces or hydrogen or ionic bonds.

Among the experimental evidence collected by Leach[111,112] compatible with this theory, although not, of course, proving it, is the following:

1. Chaotropic agents are substances which break down structured water, and since there is less gain in entropy in aggregating hydrophobic residues with poorly structured water, there is less tendency for aggregation to occur. Leach has reported that the typical chaotropic agent ethanediol reduces the aggregation of salivary proteins after incubation (i.e., after removal of their side-chains) as the theory predicts.
2. Chlorhexidine, one of the few compounds known to inhibit plaque growth in concentrations too low to inhibit bacteria or the release of carbohydrates from salivary glycoproteins, consists of alternate hydrophobic and hydrophilic groups. Leach suggests it might inhibit plaque formation because the hydrophobic part of its molecule interacts with the hydrophobic cores of the residues of glycoprotein and this prevents their aggregation. The electrostatic repulsion of the two positive charges in chlorhexidine would also deter aggregation.
3. When the supernatant from saliva was incubated at 37°C until it became tubid and reequilibrated over the temperature range 0 to 40°C, the turbidities increased linearly with temperature, typical of a hydrophobic interaction, and contrary to other weak bond interactions which are inversely related to temperature.

Leach admits the tentative nature of his hypothesis, and it is hoped that other workers competent in this field of physical chemistry will reinvestigate this evidence.

Precipitation by Calcium Ions

Experiments described above[110] suggested that a heat-stable constituent of plaque could cause precipitation from saliva.[102] It is also well established that increasing the concentration of calcium ions in saliva by the addition of calcium lactate or acetate leads to a cloudiness, often assumed to be colloidal calcium phosphate forming as the solubility product is exceeded.[115-117] These facts and the high Ca^{2+} and P_i concentration in plaque suggested that Ca precipitation might be a factor in plaque formation. When drops of saliva evaporate on a glass slide, the deposit is not readily washed off by a stream of water, but if EDTA is added to the saliva before it evaporates, most of the deposits are removed by a stream of water.[118] Evidently the cations (and Ca^{2+} is the cation mostly involved) anchor the protein to the glass and possibly this might occur also on the enamel surface. Detailed study of the precipitates following the addition of Ca^{2+} to saliva[102,119] showed that they formed most readily in submandibular saliva. Precipitation was somewhat variable in parotid saliva, but did occur provided the concentration of Ca^{2+} was increased up to 150 μg/mℓ. The precipitates increased on standing, and the composition of a precipitate pooled from submandibular saliva of 5 subjects was 53% protein, 16% Ca, 4% P, 10% lipid, and 2% carbohydrate. The question arose as to whether this precipitate was initially calcium phosphate gradually adsorbing protein or protein acquiring Ca and P. The evidence on this point is contradictory, as Edgar et al.[102] found that the protein content remained the same, gradually acquiring Ca and P (indicating protein precipitation as the primary effect), whereas Leach[112] reported the opposite. Leach examined the precipitate under the EM and stated that it consisted of needles characteristic of apatite surrounded by an amorphous halo and concluded that the precipitate was not a model for formation of plaque matrix;[112,120] it seems to the present writer, however, that these crystals might be artifacts formed during the processing before electron microscopy. Addition of calcium lactate to parotid saliva after 24-hr dialysis, during which all the P would be removed, produced no precipitate, but with submandibular saliva, the precipitate still occurred,[102] showing that at least under these in vitro conditions, precipitation was independent of calcium phosphate formation. An alternative method by which Ca^{2+} might precipitate protein is by conformational changes affecting its solubility[121,122] by exposing hydrophobic surfaces which can bind together.

The presence of extracellular lipoteichoic acid in plaque and cultures of *S. mutans*, especially if grown with sucrose, confers a negative charge on the bacterial surface. The pellicle is also negatively charged, so that divalent ions such as those of calcium might form bridges between the pellicle and lipoteichoic acid on the bacteria and between the layers of bacteria. This hypothesis has been developed by Winter and co-workers[123] and reviewed critically by Leach.[112,113] Leach gives several reasons for his opinion that calcium bridges cannot be important in the precipitation of protein, although some of them are based on experiments whose results disagree among the different workers. For example, whereas Leach reported that dialysis completely prevented precipitation by calcium, others have found that with submandibular saliva, calcium precipitation still occurred. Also, Leach stated that addition of EDTA to plaque leaves it intact, in spite of removing Ca^{2+}, whereas others have found that it causes disaggregation. Leach believes that many of the effects ascribed to Ca^{2+} are nonspecific effects of increased ionic strength. This was disproved by Edgar,[119] who showed that suspensions of plaque sediment more rapidly after the addition of Ca^{2+} and more slowly in the presence of EDTA, corresponding changes in ionic strength having no effect.

There are two possible sources of Ca^{2+} which might increase salivary concentrations sufficiently to cause protein precipitation. One is gingival fluid, reported to contain two to three times the concentration of saliva,[124,125] so that at the interface where the slow, trickle of resting saliva flows over the gingiva and meets the gingival fluid there would be a localized zone of high C^{2+} concentration, and this is exactly where disclosing solution shows plaque formation begins. Plaque forming at this point is likely to engulf some plasma proteins from the gingival fluid which could explain their presence in plaque. The second source of Ca^{2+} is the plaque itself; when its pH is lowered, some Ca^{2+} is released and would presumably diffuse out and again meet the constant resting flow of saliva;[104] this process would increase preexisting plaque rather than initiate new plaque.

THE INTERACTIONS OF SEVERAL MECHANISMS

There is no agreement yet about the relative importance of these various mechanisms, and it seems very likely that plaque matrix forms by the interaction of several of them. Saliva incubated with neuraminidase is slightly more readily precipitated by Ca^{2+} and slightly less readily precipitated by acid.[102] Removal of sialate and other constituents of the side-chains may expose the hydrophobic amino acids and lead to spontaneous precipitation as well as increase precipitation by Ca^{2+}. The diffusion of acid formed in plaque may influence isoelectric precipitation of the proteins of saliva as well as favor the release of Ca^{2+} which independently also precipitate protein.

THE ENTRY OF THE BACTERIA INTO PLAQUE

The formation of matrix was originally considered as the first and a separate stage of plaque formation, and the entry of the bacteria was thought to be a later more or less random invasion. The species of bacteria in plaque differ from those in saliva, and it is believed that a selective process determines which organisms will or will not enter plaque.[126] This replaces the older view that the bacteria entered randomly and their selective survival and growth depended upon environmental conditions, such as the pH, E_h, and nutrient supply inside the plaque. These factors may influence variations in the flora of plaque in different parts of the mouth, e.g., differences in the flora of upper and lower incisors related to the higher pH of lower incisors owing to their greater access to submandibular saliva.[127]

Two studies based on microscopy of the entry of bacteria into plaque forming after a prophylaxis suggest that they arise from pockets of organisms left, despite the prophylaxis, in narrow furrows or crevices of the enamel.[6,128] This is particularly clear in the scanning electron micrographs (SEMs) of replicas of enamel by Saxton immediately and up to 24 hr after prophylaxis.[128] During this interval, a few organisms in a deep crevice became transformed into a thick deposit in the crevice and overflowed onto the pellicle of the adjacent enamel. McDougall showed a similar result with light micrographs of demineralized sections of a tooth extracted 24 hr after a prophylaxis.[6]

According to one view, a more important source of bacteria is, however, the attachment of clumps of bacteria from saliva, also shown in the SEMs of Saxton.[128] Saliva contains some proteins which, at pH values of about 5.0, lead to the aggregation of certain species of salivary bacteria,[126,129] and those that are aggregated may be the species that enter plaque.

If the concentration of divalent ions (calcium has been tested mainly) is raised from 0.004 *M*, the pH at which aggregation occurs is raised and may be as high as 6.5 under optimum conditions.[130] All these substances would tend to neutralize the negative charge on the bacteria and therefore their mutual repulsion. The SEM shows that some of the aggregates then became attached to the mucosa. It is not often realized that the tongue and gingivae are covered in bacterial plaque not unlike the much more widely studied plaques on enamel. The attachment of the bacteria depends, at least in the tongue, on the degree of keratinization.[131]

The properties and structures of the aggregating factor for *Leptotrichia buccalis* have been studied; it is a high molecular weight protein containing about 80% carbohydrate made up of two sugars and two amino sugars.[132] In secretors it has blood group activity, but its composition is quite similar in the one nonsecretor studied.[132] Different organisms are aggregated by different substances, those for strains of *S. sanguis* and *S. mutans* have been detected visually between the cells by electron microscopy.[133] The aggregating factor for oral streps are absorbed on to HA and the factors influencing the absorption have been studied.[134]

The aggregating activity to *S. mutans* (serotype c) is higher in resting than in stimulated saliva, and in five subjects the activity in resting but not stimulated saliva was inversely related to the rate at which organisms in general (not necessarily *S. mutans*) adhered during 16 hr in vivo to a thoroughly cleaned tooth, as a measure of the rate of plaque formation.[135] At its face value, this result implies that aggregation disfavors plaque formation, but it is puzzling that a factor specific for one organism had a highly significant inverse correlation to the rate of attachment of the unspecified mixed organisms, loosely referred to by Magnusson et al.[135] as plaque formation. This has been interpreted as evidence that whereas some salivary constituents favor plaque formation other constituents, by mimicking the bacterial binding sites on the teeth, may inhibit it.[136] The aggregates formed by *S. mutans* are fragile compared with those of other oral streptococci.[137] Other factors concerned in the specificity of entry of bacteria into plaque are the presence of fimbriae on the surfaces of many oral bacteria[138] sometimes removed by treatment with proteolytic bacteria or lipases which also prevent attachment. Bacteria surfaces may also possess substances with lectin-like properties, i.e., proteins, which combine with carbohydrates.[139] Some carbohydrates, by competing with the binding site in the plaque, can inhibit the attachment of bacteria, e.g., galactose inhibits the binding of *S. mutans* to saliva-treated HA.[140]

Lie[141] studied with the SEM the development of plaque on HA powder embedded in epoxy resin mounted on a splint and worn in the mouth for varying intervals up to 48 hr by volunteers on a low-sucrose diet. The findings differed from those of some previous workers in that he reported that individual bacteria — and not large aggregates — became attached to the pellicle. Their distribution was not usually random, but a few dozen bacteria would appear close together as a monolayer, surrounded by bacteria-free areas. Many of the bacteria were attached to the pellicle or to each other by thread-like processes of uncertain composition; the low-sucrose diet would make it unlikely that they consisted of polysaccharides (see later), although even low concentrations of sucrose might be adequate for polymer formation by small numbers of bacteria. Lie suggests that the role of what have come to be called aggregating factors may be to attach the bacteria to the pellicle rather than to form clumps, and he further points out that aggregation has been observed only in vitro. Although Lie was observing a process occurring in vivo, it was not on a natural enamel surface and this might have affected his findings.

Frank and Brendel[5] also showed that monolayers of bacteria rather than clumps first colonized the pellicle, but they studied only one area of enamel (midway between the gingival margin and the contact point of molars and premolars) which differs from the areas studied by others. Their EMs show plaque bacteria in mature plaque in direct contact with enamel without any intervening pellicle, as reported also in some cases by Lie,[141] but the number of teeth studied and the frequency of plaque without pellicles is not mentioned. Leach and Saxton had previously emphasized that the pellicle differed markedly in different parts of the enamel.[96] They showed a bacterial plaque forming on a pellicle on a labial surface of an incisor and the progressive removal of the pellicle probably by proteolysis by extracellular enzymes from the bacteria as more bacteria enter the plaque.[150]

Low concentrations of saliva increased adherence of *S. mutans* strains on HA, but increasing the concentration lead to a lower uptake and apparently saturation of the bacterial surface, as high concentration had the same effect as medium concentrations.[146] When adherence to increasing amounts of HA was tested, the number of organisms per milligram of HA fell as the HA increased — presumably there were too few organisms to cover the larger amounts of HA. After 5-min preincubation of the saliva and bacteria, however, the numbers adsorbed was constant over a wide range of HA which could be explained by a monolayer on the HA of saliva-saturated organisms.[146]

In confirmation of others,[147,148] Magnusson and Ericson[146] think that agglutination is brought about by glycoprotein, and they excluded a role of 1gA, although others think that 1gA may prevent bacterial adherence. Early suggestions that the aggregating factors varied at different times of the menstrual cycle were not confirmed.[149]

Once attached, the bacteria may then divide, when food is available and other factors are favorable, to determine the eventual total mass of plaque[142] which if undisturbed, may reach a thickness of 0.3 mm on smooth surfaces. Corresponding with the growth of the bacteria, new matrix is also deposited,[141,142] presumably by the mechanisms mentioned above, and bacterial products, especially the polysaccharides mentioned below, may contribute to the matrix. Aggregates of bacteria may continue to become attached even to mature plaque.[143]

The variations in ability of different bacteria species to attach to enamel[143,144] or plaque is suggested as another explanation for the fact that some species (e.g., *S. sanguis*) are more numerous in plaque and others (e.g., *S. salivarius*) more numerous in saliva. Adherence to mucosal surfaces also shows some specificity.[143] With a view to finding methods of preventing plaque formation, substances have been investigated which interfere with the adherence of bacteria to HA, either untreated or coated with saliva or dextran (to simulate enamel covered with pellicle or plaque).[145] Amylase, several fatty acids, and chlorhexidine have this property, and human sera prevent adherence to saliva-coated, but not dextran-coated enamel.

THE BACTERIA OF PLAQUE

Stralfors[151] was probably the first investigator to attempt to classify and enumerate the bacteria of plaque by modern methods and to realize the great technical difficulties involved. The first difficulty is to break up the bacterial colonies in plaque, as otherwise, with plating methods of counting, one colony containing perhaps hundreds of organisms would count as one organism. A second difficulty is to find conditions in which all plaque bacteria will grow equally.[152]

Stralfors estimates that there were about 400 million bacteria per milligram of plaque (wet weight), mostly cocci. In a later more detailed study of the various types, dispersed plaque[153] of unknown age was plated anaerobically on heart infusion blood agar and the colonies were picked out and pure cultures were obtained. They were classified on the basis of their morphology and metabolic reactions. Any organisms not cultivable under the conditions used would not be included thus possibly leading to serious errors. It was reported that the main groups were facultative streptococci, 27%; facultative diphtheriods, 23%; anaerobic diphtheriods, 18%; peptostreptococci, 13%; *Veillonella*, 6%: *Bacteriodes*, 4%; fusobacteria, 4%; *Neisseria*, 3%; and *Vibrios*, 2%. Organisms such as lactobacilli, which comprise less than 1% of the total, were not included in this survey. Hardie[152] reviewed the characteristics of the main groups of plaque bacteria, emphasizing the site-to-site variation. A table comprising the distribution of various bacterial species in gingival crevice, coronal plaque, saliva, and the mucosal surfaces compiled from the results of 14 authors was published by Gibbons and van Houte.[154]

Gram-positive cocci dominate the organisms of plaque during its early growth of up to 1 or 2 days, but if plaque is allowed to accumulate for several more days, the inner layers become anaerobic and the proportion of these organisms diminish to about 50%, so that by 9 days, veillonellae, corynobacteria, and filamentous organisms such as actinomyces become prominent[155,156]. The filamentous bacteria are normally arranged parallel to each other with one end attached to the pellicle or enamel.

BACTERIAL POLYSACCHARIDES IN PLAQUE

When the bacteria are well established in plaque, extracellular polysaccharides may form in the presence of sucrose and become major constituents of the matrix.[157-159] In rodents and human subjects[160,161] (but not in dogs),[162] the volume of plaque is larger after feeding a diet high in sucrose compared with similar diets high in glucose. As a result, the bulk of these polymers increases the amount of matrix, and their stickiness increases the number of organisms becoming attached to plaque. Although in one similar study,[15,163] extending over 4 days, on dental students, a high-sucrose diet did not increase the amount of plaque compared with high-glucose intake, it did increase microbial density, the proportion of *S. mutans* and the activity of plaque dextran- and levan-hydrolases. The absence of an effect on the amount of plaque probably arose because by 4 days the growth had reached the maximal stage at which growth and deposition equalled removal of plaque from friction by cheek, lips, and food (the main factor limiting plaque thickness). Had the progress on the 2nd and 3rd day been followed, it is likely that the amount of plaque would have been found to be larger with the sucrose diet as other workers have found. The negative findings with sucrose on dogs probably arose from differences between the flora of this and the other species studied, corresponding to the fact that dogs rarely suffer from caries.[164]

The dramatic effect of sucrose on plaque led to the speculation that it makes plaque more cariogenic, perhaps by reducing its permeability,[165] which in turn implied that sucrose (the sole precursor of the most important polymers) was more cariogenic than other sugars. It was even mistakenly assumed by some, that these polymers were the only important constituent of plaque matrix and that in their absence plaque would not form. Protein is the essential constituent of the matrix, of course, to which polysaccharides may or may not be added. The microbiological approach to plaque formation, with its emphasis on the agglutination and adhesion of bacteria, has virtually ignored a separate salivary contribution to matrix formation.[166] The salivary agglutinins attached to some of the bacteria entering plaque might be incorporated into plaque but this would seem to be a minor contribution. Another suggestion is that vesicles, superficially resembling plaque matrix when viewed with the EM, attached to and perhaps extruded by certain plaque bacteria may form the matrix. The present reviewer believes that the SEM studies and the chemical evidence strongly suggests that matrix formation, mainly from salivary protein, is a process independent of, and simultaneous with, the deposition of bacteria.

Four main types of polysaccharides have been shown to form in plaque after exposure to sucrose for 5 to 10 min either in vivo or in vitro, and only one of them is formed from other sugars.

1. The first extracellular polysaccharide to be studied is a polymer of glucose with a 1:6 linkage, insoluble in water and 50% alcohol, but soluble in alkali and known as dextran.[33,167] It is formed from sucrose by a dextransucrase, sucrase: α-1,6-glucan:D-fructose 2 glucosyltransferase, (EC 2.4.1.5) bound to the surface of many oral streptococci.[167,168] It can be broken down slowly in plaque by bacterial dextranases present in several species of bacteria (including *S. mutans*), as plaque contains much less dextran after fasting than after feeding.[169,170] *S. mutans* grown in media containing glucose as the sugar does not possess cell-bond glucosyltransferase, but with sucrose the cells show high activity.[171]
2. Guggenheim[170] showed that plaque streptococci synthesized a mixture of 1:6 and 1:3 polymers (the latter he called "mutans") branched in the 3 and 6 positions. The solubility of the mixture depended on the proportion of the two, the 1:3 being the less soluble.[167,170]
3. An aqueous extract of plaque removes polymers of fructose widely known as levans

with a 2:6 linkage branched in the 1 or 6 position.[172] They are quite readily metabolized to acid and may be utilized by plaque bacteria as a food source which may prolong acid production after ingested sugar has been diluted and washed away by saliva. They are relatively soluble in 50% alcohol. Levans synthesized in vitro from oral streptococci, presumably resembling those formed in vivo, have molecular weights ranging from 2.7 to 21.6 $\times$ 10^6 or from *S. salivarius* as high as 31.5 $\times$ 10^6, thus providing a means of separation from the less-soluble dextrans that are present in higher concentration in plaque.[173]

4. Many sugars, both mono- and disaccharides lead to the synthesis within most of the plaque bacteria of iodophilic intracellular polymers (1:4 glucans) that are utilizable as energy reserves since they are much diminished in plaque collected some hours after eating.[174,175]

All the polymers are rapidly formed as EMs of plaque stained for 1:2 glycol groups taken before and 3 to 6 min after a rinse with 10% sucrose show a dramatic increase in both intra- and extracellular polymers.[174] Critchley et al.[165] reported that plaque collected 48 hr after a period during which the ingestion of dietary sugars had been greatly restricted reacted more slowly to sucrose rinses, giving further evidence that sucrose induces the enzymes which synthesize these polymers.[171] While the great effect of the polymers on the composition of plaque cannot be denied, the present writer believes that their importance in caries has not been shown to be more than marginal. Comparison of the caries scores in animals induced by diets high in sucrose or other sugars show great variability, with only a slight and inconsistent tendency for sucrose to be more cariogenic. There is no convincing evidence that the polymers reduce the permeability of natural plaque (see later), although this is difficult to measure.

The belief that these polymers are important in caries[159,176] led to many experiments on the use of dextranase (1,6-α-D-glucan 6-glucanohydrolase) to remove plaque and hopefully to prevent caries, but the results were indecisive.[176-179] After the realization that mutans as well as dextrans were synthesized by plaque, mutanases (1,3-α-D-glucan 3-glucanohydrolase) were also tested.[176,182] The results suggested that if dextranase was incorporated into the diet of gnotobiotic rats infected with glucan-synthesizing streptococci it reduced caries but was ineffective in animals harboring a normal flora. Mouthwashes containing the enzyme has limited[177,178] or no[179] effect in removing plaque in human subjects. The results as a whole suggest that the enzymes, if present during sucrose ingestion, might prevent the accumulation of the polymers, but once formed, their breakdown was difficult to achieve perhaps because the large enzyme molecules could not permeate the plaque. Topical application of mutanases has given more promising results in reducing caries in rats superinfected with *S. mutans* O M Z 176 and *Actinomyces viscosus* Ny 1, although, curiously enough, the amount of plaque was not reduced.[182] It was postulated that the effect was due to changes in the permeability of plaque following the presumed removal of the mutan.

In view of the instability of enzymes and the great cost of these preparations, sufficiently pure for clinical use, and the fact that polymers formed by different organisms vary in their susceptibility to breakdown by dextranases, this approach to caries prevention has been largely abandoned. In any case, in view of the uncertain role of these polymers in caries, its theoretical basis was never soundly established.

Maltose and low molecular weight dextrans act as alternative acceptors for the glucosyl moiety of sucrose and thereby have been reported to reduce the synthesis of the insoluble high molecular weight polymers.[183] Adherence by *S. mutans* to glass has been reduced by addition of maltose to the growth medium,[184] and plaque formation and caries have been reduced in hamsters by including low molecular weight dextrans in their diet.[185,186]

Actinomyces viscosus synthesize slimes from a variety of sugars and some of them aggregate their parent species, but their role in plaque formation is unknown.[187,188] Some organisms attach themselves, not only to plaque matrix, but also to each other — e.g., cocci bind to filamentous bacteria to form "corncobs".[189]

THE PROPERTIES OF PLAQUE

Plaque, as a concentrated mass of protein and bacteria, has a high buffering power, many times greater than that of saliva, especially below pH 6.0.[190] The bicarbonate, which is the main buffer in saliva, is unimportant as a stable constituent of plaque, but is the main salivary constituent entering an acidified plaque and contributing to its rise in pH. In so doing, the bicarbonate is decomposed and does not therefore accumulate in plaque.

The buffering power of plaque undoubtedly reduces the pH drop occurring after sugar ingestion and is also important in reducing the demineralizing effect of acid drinks.[191] In the presence of a thick plaque, acid drinks have a much smaller effect in reducing the pH of the tooth surface,[191] perhaps the only conditions in which the presence of plaque may be beneficial.

THE METABOLISM OF PLAQUE AND ITS RELATION TO DENTAL DISEASE

The most widely studied metabolic change is the acid production which follows sugar ingestion. This had been studied by microelectrodes, usually made of antimony[192,193] inserted into the plaque at intervals after a sugar rinse or the ingestion of foods being investigated. It has the disadvantage that the plaque is disturbed, and it is difficult to place the electrode in exactly the same position for successive readings and quite small movements or displacements can influence the reading. A second method is to collect, at intervals, very small amounts of plaque from most surfaces of the teeth, totalling about 1 mg per sample, and measure their pH outside the mouth in a "one-drop electrode" (a small cup-shaped electrode in which the sample can be suspended in water).[194,195] This method is technically easy and gives information about the average pH of the plaque as a whole, rather than from a limited number of sites. The third method consists of building a glass microelectrode into an extracted tooth and mounting it on a denture; the signal can be transmitted either by a microradio transmitter or by wire to an external pH meter.[196-198] This method gives information about the inner surface of plaque which is most relevant to caries, but has the serious disadvantage that only one site can be studied in a mouth and few subjects are suitable for, or willing to, wear the somewhat complicated appliance.

The results of all three methods are in agreement that resting plaque pH is usually between 6 and 7 and that it falls within 5 to 10 min after the ingestion or a rinse with sugar to about 5 or below and rises slowly (the "Stephan curve"). The main differences between the results of the three methods are that with built-in electrodes, the pH reaches lower levels (even as low as 4.0) and takes a longer time to rise (over 60 min or even longer, whereas with the other methods the rise is usually complete within 20 to 30 min).

The rapid fall is caused by glycolysis in the densely packed bacteria, lactate (both L[+] and D[−] forms) being the main product.[199] Acetic, propionic, and *n*-butyric acids are also present and tend to remain as the pH rises.[199-201] The rise in pH is caused by an outward diffusion of acids, inward diffusion of bicarbonate from saliva (especially if saliva flow has been stimulated), aerobic and anaerobic conversion of lactic acid into the less ionized acids, and also base production (see later).

As described above, synthesis of extracellular polysaccharides occurs rapidly from sucrose and of intracellular polymers from many sugars. The slow breakdown of the levans and intracellular polymers leads to some production of acid and may be an additional factor in delaying the rise in pH. It is frequently found at the end of a "Stephan curve" that the pH stabilizes at a pH slightly lower than the initial value, presumably because slow acid production continues when the factors responsible for neutralization have waned.

As mentioned above, plaque fluid, like saliva, is super-saturated at neutrality with Ca^{2+} and PO_4^{3-} ions, owing to the presence of a salivary protein "statherin". As the pH falls and PO_4^{3-} and HPO_4^{2-} ions gain a hydrogen atom, the concentration of PO_4^{3-}, and thus the degree of saturation, falls until at some value, varying between 5 and 6.5, it ceases to be saturated. At and below this "critical pH" some enamel may dissolve and caries may begin or progress. A trend which opposes the decrease in saturation with the fall in pH is the release of Ca^{2+} ions from bound forms in the plaque tending to compensate for the fall in PO_4^{3-}.[102]

Featherstone et al. have concluded,[204] from physicochemical studies of the caries process, that only unionized lactic acid (and not the lactate ion) can diffuse into enamel and cause demineralization. The critical pH is, in this view, that pH at which a sufficiently high concentration of unionized acid is present to ensure that it diffuses into the enamel in amounts capable of dissolving it. Acetic acid produces deeper lesions in vitro than lactic acid probably because at the pH of plaque, acetic acid is less highly ionized and therefore penetrates enamel more rapidly.[205]

The pH of fasting plaque is often higher than that of the saliva which bathes it, implying that active base production occurs.[206] Saliva and gingival fluid contain urea which is rapidly converted into ammonia, thus tending to raise pH. This is not the result of urease action (which would result in the formation of two parts of NH_3 to one part of CO_2) because more CO_2 is released than NH_3.[207] The urea apparently enters the pathways for synthesizing of amino acids which are then deaminated or decarboxylated to amines.[207]

The bacteria probably derive their nitrogen by proteolytic action of matrix protein and the products of autolysis of dead bacteria. Both saliva and plaque contain proteolytic activity.[208]

Saliva contains a tetra-peptide, gly-gly-lys-arg (pH rise factor or sialin), which reduces the pH fall and accelerates the pH rise of salivary sediment (the centrifuged bacteria from mixed saliva) when incubated with 0.05% (but not 0.5%) sugar.[209] It has a twofold action: it is a substrate for deamination (as are other peptides containing arginine) and it increases the rate of sugar uptake by the bacteria. When the sugar concentration is low, it becomes rapidly exhausted so that acid production stops and base production from deamination is unopposed. With higher levels of sugar, it cannot be used up completely so that acid production continues and its effect exceeds that of ammonia production.[209]

When the pH of plaque is rising, 15 to 20 min after exposure to sugar, the plaque fluid will again become saturated with calcium phosphate and at some point precipitation may occur.[211] This may lead to some remineralization of any carious areas under the plaque, evidence for which is seen in EMs of early caries where large irregularly arranged apatite crystals may be seen in places where the original oriented crystals of the enamel have been dissolved.[212] Fluoride, even in very low concentration, is known to favor remineralization, and this may be a major effect of the F^- in plaque fluid.[212]

PLAQUE AND GINGIVITIS PRECEDING PERIODONTAL DISEASE

For many years the invariable association of severe gingivitis and periodontal disease with calculus led to the assumption that calculus was the irritant, mainly mechanical, which caused these conditions. The study of experimental gingivitis by Löe et al. showed that abstention from tooth-brushing at intervals of 72 hr or greater led to marked gingivitis within 4 to 6 weeks, i.e., plaque 72 hr or more old caused gingivitis,[213] a period far too short for the formation of consolidated calculus (although in rapid calculus formers, some detached areas of mineralization could be expected).[214] Thus, aging plaque, and not calculus, causes the gingivitis which, if allowed to continue unchecked over many years, is believed to lead to periodontal disease and eventual loss of teeth.

It has usually been thought that some bacterial product, or products, enters the gingivae and leads to a complicated immunological response which results in chronic inflammation and eventual loss of supporting tissue.[215,216] Of the plaque constituents tested separately, lipopolysaccharide from *Veillonella* and lipoteichoic acid from *Lactobaccillus fermentis* had the greatest effect.[217-219]

Plaque extracts have a direct toxic effect on the growth of HeLa cells in culture. Fractionation of the extract shows the presence of two active constituents: a thermolabile, non-diffusible substance of molecular weight about 225.[220,221] Incubation of macrophages with freeze-dried sterilized plaque led to three- to fourfold increase in size and release of certain lysosomal enzymes.[222] Although there is no proof that plaque constituents can exert these effects in vivo, they provide a possible cause for chronic gingivitis which has not been considered as fully as the immunological theories.

PERMEABILITY OF PLAQUE

Methods of Study

The permeability of plaque has proved to be a difficult topic to study quantitatively. Apart from the difficulty of studying natural plaque with its morphology undisturbed, there is the further problem that many substances are metabolized by plaque or are bound to it and loss by these means is difficult to differentiate from slow diffusion.

A method which uses the most natural form of plaque, but is applicable only for substances like F^- that combine with enamel, consists of placing pieces of human enamel on a device that can be worn in the mouth to allow plaque to accumulate and then removing the plaque from one specimen (the control) and exposing both to the substance being studied.[223] Comparison of the amount of binding by the two pieces of enamel indicates to what extent the plaque acted as a barrier.

Disks of natural plaque formed on membranes, worn in vivo for 3, 5, and 7 days, have been used to study the inward and outward diffusion of acids formed from sugar.[224] Some experimenters have used pure cultures of some plaque organism grown on enamel as a very approximate substitute for plaque, and others have inserted freshly collected pooled plaque into either capillary tubes, thus forming a long column of plaque,[225,226] or made disks or membranes of plaque in specially constructed diffusion vessels.[227] Penetration of different substances through the columns by analyzing different levels or, with disks, by analysis of the fluid on both sides, then indicates the rate of diffusion.

The presence of cultures of *S. mutans* or of natural plaque on pieces of enamel did not affect the uptake of ^{18}F from $Na^{18}F$ and the saturation uptake of ^{18}F by the plaque was almost instantaneous so that, once saturated, it gave little resistance.[223] This is of some clinical importance because following the practice of the initiators of topical application of fluoride, patients have received a scale and polish before topical application; these results suggest that this is unnecessary and a waste of time. When sucrose was applied to the outer surface of plaque formed in vivo, less acid was released from the inner than from the outer surfaces and the inner surfaces of 7-day plaques did not release any more acid than from 3-day plaques — suggesting, contrary to a widespread belief, that the older plaques (7-day) are no more cariogenic than young (3-day) plaques.[224]

The rapidity of acid production from sugar and of base production from urea clearly shows that these nonpolar substances penetrate plaque rapidly. Charged particles, however, would be in touch with so many substances of the opposite charge that it might be expected that diffusion would be restricted. The column method showed that the penetration of Ca^{2+} was greater than that of PO_4^{3-} and, surprisingly, both penetrated more readily than glucose or glucosamine,[225] showing that many factors influence permeability of individual substances (but the column method does not avoid interaction with plaque constituents). When diffusion through plaque fluid and the centrifuged residue were measured separately, ^{14}C starch moved more rapidly than in water, but the acid metabolites moved more slowly than in water.[226] In the residue, diffusion was related to the degree of packing as influenced by the duration and speed of centrifuging.[226]

The diffusion of xenon, used as a small uncharged and unreactive particle, through disks of plaque was found to be about one half as rapid as in water, indicating a high permeability and without significant subject-to-subject variation,[228] in spite of there being different proportions of dextrans present in the plaque, suggesting that the polysaccharides had little effect on permeability. This is contrary to the conclusion of Hojo et al., who found greater diffusion of acid from colonies of a mutant organism which produced no dextran compared with the normal dextran-producing organism,[229] but colonies of bacteria, lacking a protein matrix, may not be a satisfactory model for plaque. A slightly different technique, in which clearance from, rather than penetration into plaque was measured, showed that diffusion of Na^+ and F^- through plaque was about one quarter that through water and that similar rates were found for sucrose and acetate.[228]

When the diffusion of tritiated water, ^{14}C-labeled sugar, and acids through disks of plaque were compared by Dibdin,[227] he found that tritiated water diffused at about one third to one fourth the rate in aqueous solution and the other substances tested at one fourth and one third of that in water. Diffusion decreased slightly with increased molecular weight and, contrary to previous belief, was not much affected by the charge on the particle. Unionized lactic acid at pH 3.0 diffused only 20% more quickly than in the ionized condition at pH 7.0, in partial confirmation of the conclusions of Featherstone[205] on diffusion through enamel. Growth of the plaque in vitro, with or without sucrose, almost halved diffusion, but did not affect perm-selectivity. Dibden[227] drew attention to the fact that in fissures and contact points, plaque might have a thickness as great as 2 mm[20] through which diffusion would be much slower than in the plaque (0.5 mm or less) used in most experiments. Although the results in general indicate higher permeability and less effect of polysaccharides than had been thought, much more work is required to resolve the effects of thickness and the contradictions on the differences between nonpolarized and charged particles.

THE CONTROL OF PLAQUE

The realization that both the major dental diseases, caries and periodontal disease, are caused by plaque has led to an intensive study of methods of preventing or controlling it. Whereas methods of preventing the pH from falling would be expected to reduce or prevent caries (unless "complexing agents active at neutrality supplement the effects of acids" in causing caries, as some workers suggest),[230] only effective removal of plaque or the prevention of its formation seems likely to prevent periodontal disease. This is confirmed by the low caries and gingivitis scores of subjects given a frequent professional scale and polish (prophylaxis).[231] The following paragraph gives a brief summary of methods of reducing and preventing the total amount of plaque, but does not discuss methods of reducing its pathological effect by modifying its metabolism.

METHODS OF CONTROLLING PLAQUE

Mechanical Methods

The most obvious method of removing plaque and preventing longstanding accumulation is toothbrushing, but, unfortunately, study of the effectiveness of toothbrushing has failed to show that it has any significant effect in preventing caries, although it seems more successful in reducing gingivitis.[232] Most of the studies have been unsatisfactory in that they have not been longitudinal — i.e., have not related toothbrushing habits over a particular period to the caries increase over the same period and most are also dependent on the memory and honesty of the subjects in reporting their frequency of toothbrushing. However, if toothbrushing, as normally carried out, did influence caries, a trend might be expected among the two dozen or so surveys that have been traced in the literature by the present reviewer. In fact, the majority show no effect, and the few that are favorable are cancelled out by the several surveys indicating that those who brushed most frequently had the most caries! Only those surveys carried out since fluoride dentifrices became widely used have given consistently positive correlation, and it seems more likely that the fluoride in the dentifrice was responsible rather than the toothbrushing itself.[233]

The reason for the ineffectiveness of toothbrushing is mostly because it does not remove enough plaque from the sites of caries. Far more time and thoroughness is needed than is usually applied to remove plaque from the gingival margins, fissures, and contact points. This applies also to the use of detergent foods such as apples which, disclosing solutions show, are only effective in removing plaque from smooth surfaces and give an illusory belief that the teeth are clean. As mentioned above, traces of unremoved plaque greatly speed up the development of further plaque: after a thorough scale and polish, several days elapse before significant quantities develop. Nevertheless, it might have been expected that frequent toothbrushing would at least prevent the plaque from becoming sufficiently thick to favor intensive pH drops, but the admittedly unsatisfactory surveys on caries do not suggest this.

There are good reasons for believing that the normal toothbrushing procedures have little effect on the acid produced by individual meals because the pH drop occurs very quickly (within 5 min or less) after a meal containing sugar.[192,206] Unless the teeth are brushed before the pH drops, which means inconveniently close to the end of eating, it could do little to prevent the Stephan curve. If the teeth are brushed when the pH is rising, it might even reduce the remineralization which occurs towards the end of a Stephan curve. A case can be made for recommending toothbrushing before meals because if the plaque is removed or made thinner, then acid production would be diminished when food contacts the tooth and the timing of brushing before meals is not critical. The very thorough professional oral hygiene procedures tested by Lindhe et al.[231] are too expensive in manpower to be practicable on a mass scale, but there is still much scope for investigating new toothbrush designs and the optimum timing and techniques for their use.

Chemical Control of Plaque

Plaque might be controlled chemically by two methods — first, by treating the surface of the tooth so that it does not favor the deposition of plaque matrix or the adherence of bacteria, and second, by attempting to inhibit bacteria or prevent the precipitation of salivary proteins.[234] At the same time, it is probably undesirable to prevent pellicle formation which has a protective effect against acid.

Stannous fluoride, by interacting with enamel probably forming stannous fluorophosphate, which makes the surface less water wettable, has been shown to reduce plaque formation[235] — an effect of the stannous ion often mistakenly attributed to the fluoride.[236] However, there is no evidence that fluoride, at effective anticaries levels in water,[236] reduces plaque formation, although rinses with 0.2% NaF have been reported to reduce the amount of plaque.[237] In vitro experiments which tested the effect of treating enamel with sodium fluoride on its ability to adsorb proteins (as a model for plaque formation) have led to contradictory results and sometimes to confusion between factors involved in plaque and in pellicle formation.[236] Amine fluorides have a more consistent effect,[238] but the relative importance of the fluoride and of the amine is still uncertain.[236] Stannous fluoride suffers from the disadvantage that its prolonged use stains the pellicle (although this is rectified by a scale and polish) and of solution instability so that preparations have a limited shelf life.

Lanthanum ions have a greater affinity for the binding sites of proteins and bacteria that have calcium ions. Winter's group,[239] arguing from their belief that calcium bridges are important in plaque, suggest that lanthanum ions have the effect of displacing calcium from the bridge positions between bacteria, pellicle, or neighboring bacteria and combining with the anionic sites of both surfaces, thus making them positive and mutually repellant. In vitro tests showed that the adherence of bacteria to HA was reduced by concentrations of lanthanum chloride exceeding 5×10^{-5} *M*, but lower concentrations increased adherence presumably by charge neutralization in the same way as do much higher concentrations of calcium. Lanthanum salts have the additional property of promoting the release of phosphate from phosphate esters, and incubation of *S. mutans* with lanthanum salts increased the release of phosphate, some of which is likely to be derived from teichoic acid (believed to be one of the binding sites on the bacterial surface) and thus likely reduce the adherence still further. A preliminary clinical trial on the effect of a mouth rinse containing lanthanum gave promising results.[239]

The effects of dextranases and mutanases on the prevention or reduction of extracellular polysaccharides have already been mentioned. As the enzymes can at best only prevent one aspect of plaque formation and are expensive to purify and difficult to deliver to human subjects, their value is in doubt. A crude preparation of mutanase incorporated in a chewing gum has been used for 1 week in a crossover experiment in 20 human subjects. This resulted in a reduced plaque weight (but not in the area of tooth covered) and gingivitis score, but with some local side effects, probably caused by a protease present as a contaminant.[240] Undoubtedly, the most effective substance known so far for plaque control is chlorhexidine, but it suffers from the disadvantage of an unpleasant taste and a tendency to increase pellicle formation and stain the teeth yellow.[111,241] It acts in concentrations too low to have a direct effect in killing bacteria and is probably effective because it is retained in the mouth for long periods. During this time, it may react with bacteria and salivary proteins, changing their surface potential and then reducing their adherence to enamel and other aggregating properties.[242] There have been very few clinical trials on its effect in reducing caries. In one experiment, the reduction in the DMFS increment during a 2-year trial of daily rinses in caries-prone children with a fluoride-chlorhexidine mixture (from a control of 6.3 to 2.9) was significantly greater than that of fluoride alone (6.3 to 4.3).[243] A 12-month trial in adults with low caries activity gave a suggestive result (increments of new lesions of 0.11 and 0.47 in chlorhexidine and control groups, respectively), but these increments are perhaps too small for accurate measurement.[244]

In view of the importance of retention in the mouth of chlorhexidine, Katz and Park[245] have devised methods by which other antibacterial agents are attached to enamel "conditioned" by dissolving it slightly and then reprecipitating so that the agent becomes incorporated in the enamel. In vitro and animal tests have given promising results using this procedure with fluorophane.

Several antibiotics have been found to reduce plaque (penicillin, vancomycin, and kanamycin) at least in some subjects but the risk of developing resistant strains of organisms is a serious objection to their use.[245] A number of widely distributed plant constituents have been reported (in abstract only) to reduce plaque formation,[247] probably because lectins (substances which bind with and precipitate proteins) bind and inactivate the salivary agglutinating factors.[248] These substances are readily available in almost unlimited quantities and would seem free from disadvantages; it is surprising that their potentialities as plaque-controlling agents have not been more fully investigated.

Immunization to prevent dental caries aims at inhibiting bacteria or some metabolic activity such as polysaccharide synthesis rather than the prevention of plaque formation as a whole and therefore falls outside the scope of this chapter. The search for additional substances which can be readily and cheaply applied, e.g., in dentifrices which either affect the tooth surface (like stannous fluoride) or the salivary bacteria and proteins (like chlorhexidine), but without contraindication is the most challenging field in contemporary dental research.

REFERENCES

1. **Ennever, J., Robinson, H. B. G., and Slade, H. D.,** Studies of the bacterial plaque and dental caries. I. Method for the in vivo production of bacterial plaques suitable for microsectioning, *J. Dent. Res.*, 27, 599, 1948.
2. **Mandel, I. D., Levy, B. M., and Wasserman, B. H.,** Histochemistry of calculus formation, *J. Periodontol.*, 28, 132, 1957.
3. **Mühlemann, H. A. and Schneider, U. K.,** Early calculus formation, *Helv. Odontol. Acta*, 3, 22, 1959.
4. **Löe, H., Karring, Th., and Theilade, E.,** An in vivo method for the study of the microbiology of occlusal fissures, *Caries Res.*, 7, 120, 1973.
5. **Frank, R. M. and Brendel, A.,** Ultrastructure of the approximal dental plaque and the underlying normal and carious enamel, *Arch. Oral Biol.*, 11, 883, 1966.
6. **McDougall, W. A.,** Studies on dental plaque, *Aust. Dent. J.*, 8, 261, 398, 1963.
7. **Meckel, A. M.,** The formation and properties of organic films on teeth, *Arch. Oral Biol.*, 10, 585, 1965.
8. **Dawes, C. and Jenkins, G. N.,** Some inorganic constituents of dental plaque and their relationship to early calculus formation and caries, *Arch. Oral Biol.*, 7, 161, 1962.
9. **Loesche, W. and Green, E.,** Comparison of various plaque parameters in individuals with poor oral hygiene, *J. Periodontal Res.*, 7, 173, 1972.
10. **Katayama, T., Susuki, T., and Okada, S.,** Clinical observation of dental plaque maturation, *J. Periodontol.*, 46, 610, 1975.
11. **Mandel, I. D.,** Indices for measurement of soft accumulations in clinical studies of oral hygiene and periodontal disease, *J. Periodontal Res.*, 9 (Suppl. 14), 7, 1974.
12. **Greene, J. C. and Vermillion, J. R.,** Oral hygiene index: a method for classifying oral hygiene status, *J. Am. Dent. Assoc.*, 61, 172, 1960.
13. **Löe, H.,** The gingival index, the plaque index and the retention index, *J. Periodontol.*, 30, 610, 1967.
14. **Curtis, M. A. and Eastoe, J. E.,** A method for the determination of total nitrogen in small samples of dental plaque, *Arch. Oral Biol.*, 23, 425, 1978.
15. **Folke, L. E., Gawronski, T. H., and Staar, R. H.,** Effect of dietary sucrose on quantity and quality of plaque, *Scand. J. Dent. Res.*, 80, 529, 1972.
16. **Distler, W., Kröncke, A., and Maurer, G.,** Adenosine triphosphate content of human dental plaque as a measure of viable cell mass, *Caries Res.*, 14, 265, 1980.
17. **Jenkins, G. N.,** The influence of environmental fluids on enamel solubility, *J. Dent. Res.*, 45, 662, 1966.
18. **Krembel, J., Frank, R. M., and Deluzarche, A.,** Fractionation of human dental plaque, *Arch. Oral Biol.*, 14, 563, 1969.
19. **Strålfors, A.,** Investigations into the bacterial chemistry of dental plaques, *Odontol. Tidskr.*, 58, 155, 1950.
20. **Winkler, K. L. and Backer Dirks, O.,** Mechanism of the dental plaque, *Int. Dent. J.*, 8, 561, 1958.
21. **Silverman, G. and Kleinberg, I.,** Fractionation of dental plaque and the characterisation of cellular and acellular components, *Arch. Oral Biol.*, 12, 1387, 1967.
22. **Christie, T. M., Lacy, R., and Newbrun, E.,** Staining reaction of dental plaque after various extraction procedures, *J. Dent. Res.*, 54, 487, 1975.
23. **Fox, D. J. and Dawes, C.,** The extraction of protein matrix from human dental plaque, *Arch. Oral Biol.*, 15, 1069, 1970.
24. **Rølla, G., Oppermann, R. V., Bowen, W. H., Ciardi, J. E., and Knox, K. W.,** High amounts of lipoteichoic acid in sucrose-induced plaque in vivo, *Caries Res.*, 14, 235, 1980.
25. **Singer, R. E. and Dutton, W. G.,** A comparison of the effects of endotoxin upon fibroblast proliferation and macromolecular synthesis, *J. Dent. Res.*, 58, 1634, 1979.
26. **Sharma, M. L. and Newbrun, E.,** Chemical composition of the washed cells of *Streptococcus sanguis* (804) and *Streptococcus mutans* (B-14), *J. Dent. Res.*, 54, 482, 1974.
27. **Amdur, B. H., Szabo, E. I., and Socransky, S. S.,** Fatty acids of Gram-positive bacterial rods from human dental plaque, *Arch. Oral Biol.*, 23, 23, 1978.
28. **Critchley, P.,** The breakdown of the carbohydrate and protein matrix of dental plaque, *Caries Res.*, 3, 249, 1969.
29. **Hotz, P., Guggenheim, B., and Schmid, R.,** Carbohydrates in pooled dental plaque, *Caries Res.*, 6, 103, 1972.

30. **Rølla, G., Melsen, B., and Sønju, T.,** Sulphated macromolecules in dental plaque in the monkey *Macaca irus*, *Arch. Oral Biol.*, 20, 341, 1975.
31. **Birkhed, D., Rosell, K.-G., and Granath, K.,** Structure of extracellular water-soluble polysaccharides synthesized from sucrose by oral strains of *Streptococcus mutans*, *Streptococcus salivarius*, *Streptococcus sanguis* and *Actinomyces viscosus*, *Arch. Oral Biol.*, 24, 53, 1979.
32. **Trautner, K. and Felgenhauer, B.,** Fractionation of glucans synthesised by *Streptococcus mutans* in vitro, *Caries Res.*, 13, 1, 1979.
33. **Tosiki Nisizawa, Imal, S., Akada, H., Hinoide, M., and Araya, S.,** Extracellular glucans produced by oral streptococci, *Arch. Oral Biol.*, 21, 207, 1976.
34. **Newbrun, E.,** Extracellular polysaccharides synthesised by glycosyltransferases of streptococci, *Caries Res.*, 6, 132, 1972.
35. **Levine, M., Beeley, J. A., and Beeley, J. G.,** Comparative biochemical studies on the composition of saline extracts of human dental plaque, saliva and serum, *Arch. Oral Biol.*, 21, 741, 1976.
36. **Hyatt, A. T. and Hayes, M. L.,** Free amino acids and amines in human dental plaque, *Arch. Oral Biol.*, 20, 203, 1975.
37. **Tinanoff, N. and Gross, A.,** Epithelial cells associated with the development of dental plaque, *J. Dent. Res.*, 55, 580, 1976.
38. **Theilade, J., Fejerskov, O., and Hørsted, M.,** A transmission electron microscopic study of 7-day old bacterial plaque in human tooth fissures, *Arch. Oral Biol.*, 21, 587, 1976.
39. **Edgar, W. M. and Tatevossian, A.,** The aqueous phase of plaque, in *Tooth Enamel*, Vol. 2, Fearnhead, R. W. and Stack, M. V., Eds., 1969, 229.
40. **Brudevold, F., Amdur, B. H., Vogel, J. J., and Spinelli, M.,** Effect of ingested supplementary phosphates on the tooth surface, *J. Dent. Res.*, 43, 1169, 1964.
41. **Tatevossian, A. and Gould, C. T.,** Methods for sampling and analysis of the aqueous phase of human dental plaque, *Arch. Oral Biol.*, 21, 313, 1976.
42. **Tatevossian, A. and Gould, C. T.,** The composition of the aqueous phase in human dental plaque, *Arch. Oral Biol.*, 21, 319, 1976.
43. **Curtis, M. A. and Eastoe, J. E.,** A preliminary investigation into the identity of a major ninhydrin-positive component of dental plaque fluid from the monkey *Macaca fascicularis*, *Arch. Oral Biol.*, 23, 421, 1978.
44. **Curtis, M. A. and Eastoe, J. E.,** Comparison of free amino acid pools in dental plaque fluid from monkeys (*Macaca fascicularis*) fed on high and low sugar diets, *Arch. Oral Biol.*, 23, 989, 1978.
45. **Tatovossian, A. and Newbrun, E.,** Electrophoretic and immunoelectrophoretic studies of proteins in the aqueous phase of human dental plaque, *Arch. Oral Biol.*, 26, 275, 1981.
46. **Allen, W. and Moore, B. W.,** Calcium Content of Plaque and Saliva, *I.A.D.R. Abstr. 108*, 1957.
47. **Schamschula, R. G., Agus, H., Bunzel, M., Adkins, B. L., and Barmes, D. E.,** The concentrations of selected major and trace minerals in human dental plaque, *Arch. Oral Biol.*, 22, 321, 1977.
48. **Hardwick, J. L. and Martin, C. J.,** A pilot study using mass spectrometry for the estimation of the trace element content of dental tissues, *Helv. Odontol. Acta*, 11, 62, 1967.
49. **Ferguson, D. B. and Thomas, P. D.,** Variability of calcium and phosphorus concentrations in dental plaque collected from human anterior teeth, *Arch. Oral Biol.*, 23, 839, 1978.
50. **Grøn, P. and Hay, D. I.,** Inhibition of calcium phosphate precipitation by human saliva secretions, *Arch. Oral Biol.*, 21, 201, 1976.
51. **Hay, D. I. and Moreno, E. C.,** Macromolecular inhibitors of calcium phosphate precipitation in human saliva, in *Saliva and Dental Caries*, Kleinberg, I., Ellison, S. A., and Mendel, I. D., Eds., IRL, London, 1979.
52. **Kleinberg, I., Chatterjee, R., Kaminsky, F. S., Cross, H. G., Goldenburg, D. J., and Kaufman, H. W.,** Plaque formation and the effect of age, *J. Periodontol.*, 42, 497, 1971.
53. **Kaufman, H. W. and Kleinberg, I.,** X-ray diffraction in dental plaque, *Calcif. Tissue Res.*, 11, 97, 1973.
53a. **Tannenbaum, P. J., Posner, A. S., Mandel, I. D., and Chebik, F.,** Formation of calcium phosphates in saliva and dental plaque, *J. Dent. Res.*, 55, 997, 1976.
54. **Ashley, F. P.,** Calcium and phosphorus levels in human dental plaque — variations according to site of collection, *Arch. Oral Biol.*, 20, 167, 1975.
55. **Schamschula, R. G., Adkins, B. L., Barmes, D. E., Charlton, G., and Davey, B. G.,** Caries experience and the mineral content of plaque in a primitive population in New Guinea, *J. Dent. Res.*, 56c, 62, 1977.
56. **Schamschula, R. G., Bunzel, M., Agus, H. M., Adkins, B. L., Barmes, D. E., and Charlton, G.,** Plaque minerals and caries experience: associations and interrelationships, *J. Dent. Res.*, 57, 427, 1978.
57. **Ashley, F. P.,** Calcium and phosphorus concentrations in dental plaque related to dental caries in 12 to 14-year-old subjects, *Caries Res.*, 9, 351, 1975.
58. **Kleinberg, I., Chatterjee, R., Castaldi, C. R., and Chebik, F.,** Relations between plaque, ash, calcium and phosphate levels and dental caries, *Proc. I.A.D.R. Abstr. 635*, 1969.

59. **Ashley, F. P.,** Relationship of diet, saliva, plaque and caries, *J. Dent. Res.*, 51(Abstr. 3), 1234, 1972.
60. **Ashley, F. P.,** Salivary influences on the concentration of calcium and phosphorus in dental plaque, *J. Dent. Res.*, 55D(Abstr. 146), 142, 1976.
61. **Schamschula, R. G., Adkins, B. L., Barmes, D. E., Charlton, G., and Davey, B. G.,** WHO Study of Dental Caries Etiology in Papua New Guinea, WHO Offset Publ. No. 40, World Health Organization, Geneva 1978.
62. **Nizel, A. E. and Harris, R. S.,** The effect of phosphates on experimental dental caries: a literature review, *J. Dent. Res.*, 43, 1123, 1964.
63. **Lilienthal, B.,** *Phosphates and Dental Caries*, S. Karger, Basel, 1977.
64. **Stralfors, A.,** The effect of calcium phosphate on dental caries in children, *J. Dent. Res.*, 43, 1137, 1144, 1150, 1964.
65. **Stookey, G. K., Carroll, R. A., and Muhler, J. C.,** The clinical effectiveness of phosphate-enriched breakfast cereals on the incidence of dental caries in children: results after 2 years, *J. Am. Dent. Assoc.*, 74, 752, 1967.
66. **Brewer, H. E., Stookey, G. K. and Muhler, J. C.,** A clinical study concerning the anticariogenic effects of NaH_2PO_4-enriched breakfast cereals in institutionalized subjects, *J. Am. Dent. Assoc.*, 80, 121, 1970.
67. **Tatevossian, A., Edgar, W. M., and Jenkins, G. N.,** Changes in the concentrations of phosphates in human plaque after the ingestions of sugar with and without added phosphates, *Arch. Oral Biol.*, 20, 617, 1975.
68. **Tatevossian, A. and Gould, C. T.,** The kinetics of inorganic phosphate in human dental plaque and saliva, *Arch. Oral Biol.*, 24, 461, 1979.
69. **Dawes, C.,** The effects of flow rate and duration of stimulation on the concentrations of protein and the main electrolytes in human parotid saliva, *Arch. Oral Biol.*, 14, 277, 1969.
70. **Luoma, H.,** Uptake of phosphate by caries-active and caries-inactive streptococci, *Arch. Oral Biol.*, 13, 1331, 1968.
71. **Ericsson, Y.,** Some differences between human and rodent saliva, *Arch. Oral Biol.*, 7 (ORCA Suppl.), 327, 1962.
72. **Tatevossian, A. and Wright, W. G.,** The collection and analysis of rat saliva, *Arch. Oral Biol.*, 19, 825, 1974.
73. **Bibby, B. G. and van Kesteren, M.,** The effect of fluoride on mouth bacteria, *J. Dent. Res.*, 19, 391, 1940.
74. **Martin, D. J. and Hill, I. N.,** Evanston caries study: fluorine content of saliva, *J. Dent. Res.*, 29, 291, 1950.
74a. **Wright, D. E. and Jenkins, G. N.,** The effect of fluoride on the acid production of saliva-glucose mixtures, *Br. Dent. J.*, 96, 30, 1954.
75. **Shannon, I. L.,** Biochemistry of fluoride in saliva, *Caries Res.*, 11 (Suppl. 1), 206, 1977.
76. **Hardwick, J. L. and Leach, S. A.,** The fluoride content of dental plaque, *Arch. Oral Biol. Suppl. Proc. 9th ORCA Congress* 151, 1962.
77. **Dawes, C., Jenkins, G. N., Hardwick, J. L., and Leach, S. A.,** The relation between the fluoride concentration in the dental plaque and in drinking water, *Br. Dent. J.*, 119, 164, 1965.
78. **Grøn, P., Yao, K., and Spinelli, M.,** A study of inorganic constituents in dental plaque, *J. Dent. Res.*, 48, 799, 1969.
79. **Agus, H., Schamschula, R. G., Barmes, D. E., and Bunzel, M.,** Association between the total fluoride content of dental plaque and individual caries experience in Australian children, *Community Dent. Oral Epidemiol.*, 4, 210, 1976.
80. **Jenkins, G. N. and Edgar, W. M.,** Distribution and form of F in saliva and plaque, *Caries Res.*, 11 (Suppl. 1), 226, 1977.
81. **Jenkins, G. N., Edgar, W. M., and Ferguson, D. B.,** The distribution and metabolic effects of human plaque fluorine, *Arch. Oral Biol.*, 14, 105, 1969.
82. **Kashket, S. and Rodriguez, V. M.,** Fluoride accumulation by a strain of human oral *Streptococcus sangius*, *Arch. Oral Biol.*, 21, 459, 1976.
83. **Kashket, S. and Bunick, F. J.,** Binding of fluoride in oral streptococci, *Arch. Oral Biol.*, 23, 993, 1978.
84. **Katayama, T., Edgar, W. M., Jenkins, G. N., and Johnson, B.,** Preliminary fractionation of fluoride-binding constituents of human oral *Streptococcus sanguis*, *Arch. Oral Biol.*, 26, 341, 1981.
85. **Edgar, W. M., Cockburn, M. A., and Jenkins, G. N.,** Uptake of fluoride and its inhibitory effects in oral microorganisms in culture, *Arch. Oral Biol.*, 26, 615, 1981.
86. **Edgar, W. M., Jenkins, G. N., and Tatevossian, A.,** The inhibiting effect of fluoride on plaque bacteria, *Br. Dent. J.*, 128, 129, 1970.
87. **Broukal, Z. and Zajicek, O.,** Amount and distribution of extracellular polysaccharides in dental microbial plaque, *Caries Res.*, 8, 97, 1974.
88. **Bowen, W. H.,** The effect of fluoride and molybdenum on caries activity and the composition of plaque in monkeys (*M. irus*), *Caries Res.*, 6, 254, 1972.

89. **de Stoppelar, J. D., van Houte, J., and Backer Dirks, O.,** The relationship between extracellular polysaccharide producing streptococci and smooth surface caries in 13-year-old children, *Caries Res.*, 3, 190, 1969.
90. **Hamilton, I. R.,** Effects of fluoride on enzymatic regulation of bacterial carbohydrate metabolism, *Caries Res.*, 11 (Suppl. 1), 262, 1977.
91. **Birkeland, J. M., Lorkjend, L., and von der Fehr, F. R.,** The influence of fluoride rinses on the fluoride content of dental plaque in children, *Caries Res.*, 5, 169, 1971.
92. **Rugg-Gunn, A. J., Edgar, W. M., Cockburn, M. A., and Jenkins, G. N.,** Correlation between fluoride concentration, sample weight and acid production in dental plaque from children, *Arch. Oral Biol.*, 26, 61, 1981.
93. **Clarke, N. G. and Fanning, E. A.,** Plaque pH and calcium sucrose phosphate: a telemetric study, *Aust. Dent. J.*, 16, 13, 1971.
94. **Silverstone, L. M., Johnson, N. W., Hardie, J. M., and Williams, R. A. D.,** *Dental Caries, Aetiology, Pathology and Prevention,* Macmillan, New York, 1981, 123.
95. **Schroeder, H. E. and de Boever, J.,** The Structure of Microbial Dental Plaque, in *Dental Plaque,* McHugh, W. D., Ed., Churchill Livingstone, London, 1970.
96. **Löe, H., Karring, Th., and Theilade, E.,** An in vivo method for the study of the microbiology of occlusal fissures, *Caries Res.*, 7, 127, 1973.
97. **Theilade, E., Fejerskov, O., Karring, T., and Theilade, J.,** A microbiological study of old plaque in occlusal fissures of human teeth, *Caries Res.*, 12, 313, 1978.
98. **Theilade, J., Fejerskov, O., and Hørsted, M.,** A transmission electron microscopic study of 7-day-old bacterial plaque in human tooth fissure, *Arch. Oral Biol.*, 21, 587, 1976.
99. **Dobbs, E. C.,** Local factors in dental caries, *J. Dent. Res.*, 12, 852, 1932.
100. **Kirk, E. C.,** A consideration of the question of susceptibility and immunity to dental caries, *Dent. Cosmos,* 52, 729, 1910.
101. **Dawes, C.,** Is acid-precipitation of salivary proteins a factor in plaque formation?, *Arch. Oral Biol.*, 9, 375, 1964.
102. **Edgar, W. M., Jenkins, G. N., and Hillam, D.,** Salivary precipitation and the development of plaque matrix, in *Dental Plaque and Surface Interactions in the Oral Cavity,* Leach, S. A., Ed., IRL, London, 1980, 197.
103. **Armstrong, W. G.,** Characterisation studies on the specific human salivary proteins adsorbed in vitro by hydroxyapatite, *Caries Res.*, 5, 215, 1971.
104. **Beeley, J. A.,** Separation of human salivary proteins by isoelectric focusing in polyacrylamide gels, *Arch. Oral Biol.*, 14, 569, 1969.
105. **Silverman, G. and Kleinberg, I.,** Fractionation of dental plaque and the characterisation of its cellular and acellular components, *Arch. Oral Biol.*, 12, 1387, 1967.
106. **Dawes, C. and Jenkins, G. N.,** Studies Related to the Formation of the Dental Plaque, *Proc. 41st Meet. I.A.D.R.*, Abstr. 262, 1963.
107. **Leach, S. A.,** Release and breakdown of sialic acid from human salivary mucin and its role in the formation of dental plaque, *Nature (London),* 189, 486, 1963.
108. **Saxton, C. A. and Kolendo, A. B.,** Degradation of salivary glycoproteins and their contribution to plaque formation, *Arch. Oral Biol.*, 12, 1541, 1967.
109. **Briscoe, J. M., Pruitt, K. M., and Coldwell, R. C.,** The effect of neuraminidase on the properties of salivary proteins, *J. Dent. Res.*, 51, 819, 1972.
110. **Leach, S. A. and Critchley, P.,** Bacterial degradation of glycoprotein sugars in human saliva, *Nature (London),* 209, 506, 1966.
111. **Leach, S. A.,** A biophysical approach to interactions associated with the formation of the matrix of dental plaque, in *Dental Plaque and Surface Interactions in the Oral Cavity,* Leach, S. A., Ed., IRL, London, 1980, 159.
112. **Leach, S. A.,** On the nature of the interactions associated with aggregate phenomena in the mouth, *J. Dent.*, 7, 149, 1979.
113. **Tanford, C.,** *Physical Chemistry of Macromolecules,* Wiley Interscience, New York, 1967.
114. **Tanford, C.,** *The Hydrophobic Effect,* John Wiley & Sons, New York, 1973.
115. **Dawes, C.,** Further studies related to the formation of dental plaque, *J. Dent. Res.*, 43 (Abstr. 216), 835, 1963.
116. **Mukherjee, S.,** Effect of various factors on the rate of phosphate precipitation in an inorganic solution and in saliva, *J. Periodontal Res.*, 3, 146, 1968.
117. **Belcourt, A.,** Role des lipides salivaires dans la précipitation d'une glycoprotéine salivaire sous l'effet du calcium, *J. Biol. Buccale,* 3, 179, 1975.
118. **McGaughey, C. and Stowell, E. C.,** Adsorption of salivary proteins by hydroxyapatite: relations between the effects of calcium ions, hydrogen ions, temperature and exposure time, *J. Dent. Res.*, 50, 542, 1971.
119. **Edgar, W. M.,** Studies on the role of calcium in plaque formation, *J. Dent.*, 7, 174, 1979.

120. **Leach, S. A. and Appleton, C. A.,** Reactions of chlorhexidine in the mouth and its relation to the formation of stained pellicle, *Caries Res.*, 12, 101, 1978.
121. **Bettelheim, F. S.,** On the aggregation of a calcium precipitable glycoprotein from human submaxillary saliva, *Biochem. Biophys. Acta*, 236, 702, 1971.
122. **Boat, T., Weisman, U., Pallacivini, J., and di Sant 'Agnes, P.,** A small human submaxillary saliva glycoprotein containing glucose and phosphorus, *Fed. Proc.*, 29, Abstr., 861, 1970.
123. **Winter, M. R. C., Rølla, G., and White, A. J.,** The effect of inorganic ions on dental plaque, in *Dental Plaque and Surface Interactions in the Oral Cavity*, Leach, S. A., Ed., IRL, London, 1980.
124. **Kaslick, R. S., Chasens, A. I., Mandel, I. D., Weinstein, D., Waldman, R., Pluhar, T., and Lazzara, R.,** Quantitative analysis of sodium, potassium and calcium in gingival fluid from gingiva in varying degrees of inflammation, *J. Periodontal Res.*, 41, 93, 1970.
125. **Biswas, S. D., Holthius, A. F., and Lavelle, C. L. B.,** Crevicular calcium-phosphate levels before and after gingival stimulation, *IRCS Med. Sci.*, 5, 187, 1977.
126. **Gibbons, R. J. and Spinell, D. M.,** Salivary-induced aggregation of dental plaque, in *Dental Plaque*, McHugh, W. D., Ed., Churchill Livingstone, London, 1970, 207.
127. **McNamara, T. F., Friedman, B. K., and Roth, P.,** Salivary access as ecological determinant, in *Saliva and Dental Caries*, Kleinberg, I., Ellison, S. A., and Mandel, I. D., Eds., I. R. L., London, 1979, 211.
128. **Saxton, C. A.,** Scanning electron microscopic study of the formation of dental plaque, *Caries Res.*, 7, 102, 1973.
129. **Hay, D. I., Gibbons, R. J., and Spinell, D. M.,** Characteristics of some high molecular weight constituents with bacterial aggregating activity from whole saliva and dental plaque, *Caries Res.*, 5, 111, 1971.
130. **Silverman, G. and Kleinberg, I.,** Studies on factors affecting the aggregation of the microorganisms in human dental plaque, *Arch. Oral Biol.*, 12, 1407, 1967.
131. **Aufdemarte, T. B. and Cameron, I. L.,** The relation of keratinization to bacterial colonisation on the baboon tongue as demonstrated by scanning electron microscopy, *J. Dent. Res.*, 60, 1008, 1981.
132. **Kondo, W., Sato, M., and Sato, N.,** Properties of the human salivary aggregating factor for *Leptotrichia buccalis* cells, *Arch. Oral Biol.*, 23, 453, 1978.
133. **Kashket, S., Skobe, Z., and Garant, P. R.,** Electron microscopic observations of the saliva-dependent aggregation of oral streptococci, *Arch. Oral Biol.*, 23, 125, 1978.
134. **Ericson, T. and Magnusson, I.,** Affinity for hydroxyapatite of salivary substances inducing aggregation of oral streptococci, *Caries Res.*, 10, 8, 1976.
135. **Magnusson, I., Ericson, T., and Pruitt, K.,** Effect of salivary agglutinins on bacterial colonization of tooth surfaces, *Caries Res.*, 10, 113, 1976.
136. **Van Houte, J.,** Bacterial adhesion in the mouth, in *Dental Plaque and Surface Interaction in the Oral Cavity*, Leach, S. A., Ed., I. R. L., London, 1980, 69.
137. **Kashket, S. and Liberman, E.,** Fragility of salivary aggregates of *Streptococcus mutans*, *Caries Res.*, 13, 307, 1979.
138. **Ellen, R. P., Walker, D. L., and Chan, K. H.,** Association of long surface appendages with adherence-related functions of the Gram-positive species *Actinomyces naeslundii*, *J. Bacteriol.*, 134, 1171, 1978.
139. **Cisar, J. O., Vatter, A. E., and McIntire, F. C.,** Identification of the virulence-associated antigen on the surface fibrils of *Actinomyces viscosus* T14, *Infect. Immun.*, 19, 312, 1978.
140. **Gibbons, R. J. and Qureshi, J. V.,** Inhibition of adsorption of *Streptococcus mutans* strains to saliva-treated hydroxyapatite by galactose and certain amines, *Infect. Immun.*, 26, 1214, 1979.
141. **Lie, T.,** Early dental plaque formation, *J. Periodontal Res.*, 12, 73, 1977.
142. **Rutter, P. R. and Marsh, P. D.,** The transport and adhesion of microorganisms in the oral cavity, in *Dental Plaque and Surface Integerments in the Oral Cavity*, Leach, S. A., Ed., I. R. L., London, 1980, 101.
143. **van Houte, J., Gibbons, R. J., and Pulkinnen, A. J.,** Adherence as an ecological determinant for streptococci in the human mouth, *Arch. Oral Biol.*, 16, 1131, 1971.
144. **Hillman, J. D., van Houte, J., and Gibbons, R. J.,** Sorption of bacteria to human enamel powder, *Arch. Oral Biol.*, 15, 899, 1970.
145. **Liljmark, W. F., Schauer, S. V., and Boomquist, C. G.,** Compounds which affect the adherence of *S. sanguis* and *S. mutans* to hydroxyapatite, *J. Dent. Res.*, 57, 373, 1978.
146. **Magnusson, I. and Ericson, T.,** Effect of salivary agglutinins on reaction between hydroxyapatite and a serotype c strain of *Streptococcus mutans*, *Caries Res.*, 10, 273, 1976.
147. **Ericson, T., Pruitt, K., and Wedel, H.,** The reaction of salivary substances with bacteria, *J. Oral Pathol.*, 4, 307, 1975.
148. **Williams, R. C. and Gibbons, R. J.,** Inhibition of streptococcal attachment to receptors on human buccal epithelial cells by antigenically similar salivary glycoproteins, *Infect. Immun.*, 11, 711, 1979.
149. **Magnusson, I. and Ericson, T.,** Variability in concentration of agglutinins in saliva, *Caries Res.*, 11, 109, 1977.

150. **Leach, S. A. and Saxton, C. A.,** An electron microscopic study of the acquired pellicle and plaque formed on the enamel of human incisors, *Arch. Oral Biol.,* 11, 1081, 1966.
151. **Strålfors, A.,** Investigations with the Bacterial Chemistry of Dental Plaques, *Thule,* 1950; (See also Ref. 19).
152. **Hardie, J. M.,** *Dental Caries, Aetiology, Pathology and Prevention,* Silverstone, L. M., Johnson, N. W., Hardie, J. M., and Williams, R. A. D., Eds., Macmillan, New York, 1981.
153. **Gibbons, R. J., Socransky, S., Aranjo, W. C., and van Houte, J.,** Studies of the predominant cultivable microbiota of dental plaque, *Arch. Oral Biol.,* 9, 365, 1964.
154. **Gibbons, R. J. and van Houte, J.,** Dental caries, *Annu. Rev. Med.,* 26, 121, 1973.
155. **Hardie, J. M. and Bowden, G. H.,** The normal microbiological flora of the mouth, in *The Normal Microbial Flora in Man,* Skinner, K. A. and Carrot, G., Eds., Academic Press, New York.
156. **Katz, H. L.,** Microbiological population shifts in developing human dental plaque, *Arch. Oral Biol.,* 12, 1561, 1967.
157. **Manly, R. S. and Dain, J. A.,** Hydrolysis of polysaccharides by oral micro-organisms, *I.A.D.R. 41st Meet.,* Abstr. 125, 1963.
158. **McDougall, W. A.,** Studies on dental plaque. IV. Levans and the dental plaque, *Aust. Dent. J.,* 9, 1, 1964.
159. **Gibbons, R. J. and Banghart, S. B.,** Synthesis of extracellular dextran by cariogenic bacteria and its presence in human dental plaque, *Arch. Oral Biol.,* 12, 11, 1967.
160. **Carlsson, J. and Egelberg, J.,** Effect of diet on early plaque formation in man, *Odontol., Revy,* 16, 112, 1965.
161. **Carlsson, J. and Sundström, B.,** Variations in composition of early dental plaque following ingestion of sucrose and glucose, *Odontol., Revy,* 19, 161, 1965.
162. **Carlsson, J. and Egelberg, J.,** Local effect of diet on plaque formation and development of gingivitis in dogs, *Odontol. Revy,* 16, 42, 1965.
163. **Dennis, D. A., Gawronski, T. H., Sudo, S. Z., Harris, R. S., and Folke, L. E. A.,** Variations in microbial and biochemical components of four-day plaque during a four-week controlled diet period, *J. Dent. Res.,* 54, 716, 1975.
164. **Lewis, T. M.,** A two-year study on the production of dental caries in dogs, *J. Dent. Res.,* 41, 1278, 1962.
165. **Critchley, P., Wood, J. M., Saxton, C. A., and Leach, S. A.,** The polymerisation of dietary sugars by dental plaque, *Caries Res.,* 1, 112, 1967.
166. **Gibbons, R. J. and Spinell, D. M.,** Salivary-induced aggregation by plaque bacteria, in *Dental Plaque,* McHugh, W. D., Ed., Churchill Livingstone, London, 1970, 207.
167. **Trautner, K. and Felgenhauer, B.,** Fractionation of glucans synthesized by *Streptococcus mutans* in vitro, *Caries Res.,* 13, 1, 1979.
168. **Carlsson, J., Newbrun, E., and Krasse, B.,** Purification and properties of dextran-sucrase from *Streptococcus sanguis, Arch. Oral Biol.,* 14, 469, 1969.
169. **Da Costa, T., Brier, L. C., and Gaida, F.,** Dextran hydrolysis by a fusobacterium strain isolated from human dental plaque, *Caries Res.,* 2, 115, 1968.
170. **Guggenheim, B.,** Enzymatic hydrolysis and structure of water-insoluble glucans produced by glucosyltransferases from a strain of *Streptococcus mutans, Helv. Odontol. Acta,* 14 (Suppl. V), 89, 1970.
171. **Spinell, D. M. and Gibbons, R. J.,** Influence of culture medium on the glucosyl transferase and dextran-binding capacity of *Streptococcus mutans* 6715 cells, *Infect. Immun.,* 10, 1448, 1974.
172. **Higuchi, M., Iwami, Y., Yamada, T., and Araya, S.,** Levan synthesis and accumulation by human dental plaque, *Arch. Oral Biol.,* 15, 563, 1970.
173. **Ehrlich, J., Stilvala, S. S., Bahary, W. S., Garg, S. K., Long, L. W., and Newbrun, E.,** Levans I. Fractionation, solution viscosity and chemical analysis of levan produced by *Streptococcus salivarius, J. Dent. Res.,* 54, 290, 1975.
174. **Saxton, C. A.,** An electron microscope investigation of bacterial polysaccharide synthesis in human dental plaque, *Arch. Oral Biol.,* 14, 1275, 1969.
175. **Gibbons, R. J. and Socransky, S. S.,** Intracellular polysaccharide storage by organisms in dental plaque, *Arch. Oral Biol.,* 7, 73, 1962.
176. **Guggenheim, B., Regolati, B., and Mühlemann, H. R.,** Caries and plaque inhibition by mutanase in rats, *Caries Res.,* 6, 289, 1972.
177. **Caldwell, R. C., Sandham, H. J., Mann, W. V., Jr., Finn, S. B., and Formicola, A. J.,** The effect of dextranase mouthwash on dental plaque in young adults and children, *J. Am. Dent. Assoc.,* 82, 132, 1971.
178. **Lobene, R. R.,** A clinical study of the effect of dextranase on human dental plaque, *J. Am. Dent. Assoc.,* 82, 132, 1971.
179. **Keyes, P. H., Hicks, M. A., Goldman, B. M., McCabe, R. M., and Fitzgerald, R. J.,** Dispersion of dextranous bacterial plaques on human teeth with dextranase, *J. Am. Dent. Assoc.,* 82, 136, 1971.

180. **Bowen, W. H.,** The effect of dextranase on caries activity in monkeys *(Macaca irus)*, *Br. Dent. J.*, 131, 445, 1971.
181. **Fitzgerald, R. J., Spinelli, R. M., and Stoudt, T. H.,** Enzymatic removal of artificial plaques, *Arch. Oral Biol.*, 13, 125, 1968.
182. **Guggenheim, B., Regolati, B., Schmid, R., and Mühlemann, H. R.,** Effects of topical application of mutanase in rat caries, *Caries Res.*, 14, 128, 1980.
183. **Gibbons, R. J. and Nygaard, M.,** Synthesis of insoluble dextrans, *Arch. Oral Biol.*, 13, 1249, 1968.
184. **Newbrun, E., Finzen, F., and Sherma, M.,** Inhibition of adherence of *Streptococcus mutans* to glass surfaces, *Caries Res.*, 11, 153, 1977.
185. **Balekjian, A. Y., Longton, R. W., Cole, J. S., III, and Guidry, M. S.,** The effect of disaccharides on the plaque-forming potential of *Streptococcus mutans*, *J. Dent. Res.*, 56, 1359, 1977.
186. **Gibbons, R. J. and Keyes, P. H.,** Inhibition of insoluble dextran synthesis, *Arch. Oral Biol.*, 14, 721, 1969.
187. **Jordan, H. V., Keyes, P. H., and Lim, S.,** Plaque formation and implantation of *Odontomyces viscosus* in hamster fed different carbohydrates, *J. Dent. Res.*, 48, 824, 1969.
188. **van der Hoeven, J. S.,** A slime-producing microorganism in dental plaque of rats, selected by glucose feeding. Chemical composition of extracellular slime elaborated by *Actinomyces viscosus* strain N_y1, *Caries Res.*, 8, 193, 1974.
189. **Jones, S. J.,** A special relationship between spherical and filamentous micro-organisms in mature human dental plaque, *Arch. Oral Biol.*, 17, 613, 1972.
190. **Strålfors, A.,** The buffer capacity of the dental plaque, *J. Dent. Res.*, 27, 587, 1948.
191. **Marthaler, Th.,** Apples and oral health, *Dent. Sci. Res.*, 8, 101, 1970.
192. **Stephan, R. M.,** Changes in hydrogen-ion concentration on tooth surfaces in carious lesions, *J. Am. Dent. Assoc.*, 27, 718, 1940.
193. **Kleinberg, I.,** The construction and evaluation of modified types of antimony electrodes for intra-oral use, *Br. Dent. J.*, 104, 197, 1958.
194. **Fosdick, L. S., Campaigne, E. E., and Fancher, D.,** Rate of acid formation in carious areas: the etiology of dental caries, *Ill. Dent. J.*, 10, 85, 1941.
195. **Rugg-Gunn, A. J., Edgar, W. M., Geddes, D. A. M., and Jenkins, G. N.,** The effect of different meal patterns upon plaque pH in human subjects, *Br. Dent. J.*, 739, 361, 1975.
196. **Graf, H. and Mühlemann, H. R.,** Telemetry of plaque pH from interdental area, *Helv. Odontol. Acta*, 10, 94, 1966.
197. **Mühlemann, H. R. and de Boever, J.,** Radiotelemetry of the pH of interdental areas exposed to various carbohydrates, in *Dental Plaque*, McHugh, W. D., Ed., Churchill Livingstone, London, 1970, 179.
198. **Igarashi, K., Kariyama, K., and Yamada, T.,** Measurement of pH in human dental plaque in vivo with an ion-sensitive transistor electrode, *Arch. Oral Biol.*, 26, 203, 1981.
199. **Geddes, D. A. M.,** The production of L(+) and D(−) lactic acid and volatile acids by human dental plaque and the effect of plaque buffering and acidic strength on pH, *Arch. Oral Biol.*, 17, 537, 1972.
200. **Distler, W. and Kröncke, A.,** Acid formation by mixed cultures of cariogenic strains of *Strep. mutans* and *Veillonella alcalescens*, *Arch. Oral Biol.*, 25, 655, 1980; 26, 123, 1981.
201. **Hoshino, E., Karino, H., and Yamada, T.,** Lactate metabolism by human dental plaque and *Veillonella* under aerobic and anaerobic conditions, *Arch. Oral Biol.*, 26, 17, 1981.
202. **Fosdick, L. S. and Starke, A. C.,** Solubility of tooth enamel in saliva at various pH levels, *J. Dent. Res.*, 18, 260, 417, 1939.
203. **Ericsson, Y.,** Enamel apatite solubility, *Acta Odontol. Scand.*, 8 (Suppl. 3), 1, 1949.
204. **Featherstone, J. D. B., Duncan, J. F., and Cuttress, T. W.,** A mechanism for dental caries, *Arch. Oral Biol.*, 24, 101, 1979.
205. **Featherstone, J. D. B. and Rogers, B. E.,** Effect of acetic, lactic and other organic acids on the formation of artificial carious lesions, *Caries Res.*, 15, 377, 1981.
206. **Kleinberg, I. and Jenkins, G. N.,** Influence of saliva and meals on plaque pH, *Arch. Oral Biol.*, 9, 493, 1964.
207. **Biswas, S. D. and Kleinberg, I.,** Urea metabolism in salivary sediment, *Arch. Oral Biol.*, 16, 759, 1971.
208. **Mäkinin, K. K.,** Studies on oral enzymes, *Acta Odontol. Scand.*, 24, 579, 605, 619, 733, 1966.
209. **Kleinberg, I., Craw, D., and Komiyamo, K.,** Effect of salivary supernatant on the glycolytic activity of the bacteria in salivary sediment, *Arch. Oral Biol.*, 18, 787, 1973.
210. **Kleinberg, I., Kanapka, J. A., Chatterjee, R., Craw, D., D'Angelo, N., and Sandham, H. J.,** Metabolism of nitrogen by the oral mixed bacteria, in *Saliva and Dental Caries*, Kleinberg, I., Ellison, S. A., and Mandel, I. D., Eds., IRL, London, 1979, 357.
211. **Koulourides, T.,** Dynamics of tooth surface — oral fluid equilibrium, *Adv. Oral Biol.*, 2, 149, 1966.
212. **Silverstone, L. M.,** Remineralization phenomena, *Caries Res.*, 11 (Suppl. 1), 59, 1977.
213. **Löe, H., Theilade, E., and Jensen, S. B.,** Experimental gingivitis in man, *J. Periodontol.*, 36, 177, 1965.

214. **Mühlemann, H. R. and Schneider, U. K.**, Early calculus formation, *Helv. Odontol. Acta*, 3, 22, 1959.
215. **Mergenhagen, S. E., Tempel, T. R., and Snyderman, R.**, Immunologic reactions and periodontal inflammation, *J. Dent. Res.*, 49, 256, 1970.
216. **Lehner, T.**, Immunologic aspects of dental caries and periodontal disease, *Br. Med. Bull.*, 31, 125, 1975.
217. **Wilton, J. M. A. and De Almeida, O. P.**, The comparative effect of dental plaque, lipopolysaccharide, lipoteichoic acid, dextran and levan on leukocytes in the mouse peritoneal cavity, in *The Borderland Between Caries and Periodontal Disease II*, Lehner, T. and Cimasoni, G., Eds., Academic Press, London, 1980, 83.
218. **Schroeder, H. E. and Attstrom, R.**, Pocket formation: an hypothesis, in *The Borderland Between Caries and Periodontal Disease II*, Lehner, T. and Cimasoni, G., Eds., Academic Press, London, 1980, 99.
219. **Bolanos, E., Ghanasekhar, D., Singh, G., Church, H., and Dolby, A. E.**, Cellular autoimmunity in periodontal disease, in *The Borderland Between Caries and Periodontal Disease II*, Lehner, T. and Cimasoni, G., Eds., Academic Press, London, 1980, 135.
220. **Levine, M. and Cowley, G. C.**, Human dental plaque extracts: their crude chemical composition and toxicity to cultured Hela cells, *Arch. Oral Biol.*, 19, 583, 1974.
221. **Duguid, R., Al-Makadsi, F., and Cowley, G. C.**, Cytotoxic material extracted from human dental plaque and oral streptococci, *Arch. Oral Biol.*, 25, 349, 1980.
222. **Page, R. C., Davies, P., and Allison, A. C.**, Affects of plaque on lysosomal hydrolyses, *Arch. Oral Biol.*, 18, 1481, 1973.
223. **Joyston-Bechal, S., Duckworth, R., and Braden, M.**, The effect of artificially produced pellicle and plaque on the uptake of ^{18}F by human enamel in vitro, *Arch. Oral Biol.*, 21, 73, 1976.
224. **Geddes, D. A. M.**, Acid production from inner (enamel) and outer (saliva) surfaces of three-, five- and seven-day-old, human, intact dental plaque, *Caries Res.*, 11, 114, 1977.
225. **Melsen, B., Kaae, O., Rølla, G., Fejerskov, D., and Karring, T.**, Penetration of ions in human dental plaque, *Arch. Oral Biol.*, 24, 75, 1979.
226. **Tatevossian, A.**, Diffusion of radiotracers in human dental plaque, *Caries Res.*, 13, 154, 1979.
227. **Dibdin, G. H.**, Diffusion of sugars and carbohydrate acids through human dental plaque in vitro, *Arch. Oral Biol.*, 26, 515, 1981.
228. **McNee, S. G., Geddes, D. A. M., Main, C., and Gillespie, F. C.**, Measurements of the diffusion coefficient of NaF in human dental plaque in vitro, *Arch. Oral Biol.*, 25, 819, 1980.
229. **Hojo, S., Higuchi, M., and Araya, S.**, Glucan inhibition of diffusion in plaque, *J. Dent. Res.*, 55, 169, 1976.
230. **Plaschaert, A. J. M., Mörch, T., and König, K. G.**, Effect of sodium lactate under conditions of neutral pH on the release of calcium from the enamel surface in vitro, *Caries Res.*, 6, 334, 1972.
231. **Lindhe, J., Axelsson, P., and Tollskog, G.**, The effect of proper oral hygiene on gingivitis and dental caries in school children, *Community Dent. Oral Epidemiol.*, 3, 150, 1975.
232. **Russell, A. L.**, International nutrition surveys: a summary of preliminary dental findings, *J. Dent. Res.*, 42, 233, 1963.
233. **Leske, G. S., Ripa, L. W., and Barenie, J. T.**, Comparisons of caries prevalence of children with different daily toothbrushing frequencies, *Community Dent. Oral Epidemiol.*, 4, 102, 1976.
234. **Parsons, J. C.**, Chemotherapy of dental plaque — a review, *J. Periodontol.*, 45, 177, 1974.
235. **Glantz, P. O.**, On wettability and adhesiveness, *Odontol. Revy*, 17, Suppl. 1, 1969.
236. **Jenkins, G. N.**, Mechanisms of action of fluoride in the control of dental plaque, in *The Borderland Between Caries and Periodontal Disease*, Lehner, T. and Cimasoni, G., Eds., Academic Press, London, 1981, 269.
237. **Birkeland, J. M.**, Effect of fluoride on the amount of dental plaque in children, *Scand. J. Dent. Res.*, 80, 82, 1972.
238. **Marthaler, T. M.**, Caries inhibition after seven years of unsupervised use of amine fluoride dentrifrice, *Br. Dent. J.*, 124, 510, 1968.
239. **Winter, M. R. C., Rølla, G., and White, A. J.**, The effects of inorganic cations on dental plaque, in *Dental Plaque and Surface Interactions in the Oral Cavity*, Leach, S. A., Ed., IRL, London, 1980, 211.
240. **Kelstrup, J., Holm-Pedersen, P., and Poulsen, S.**, Reduction of the formation of dental plaque and gingivitis in humans by crude mutanase, *Scand. J. Dent. Res.*, 86, 93, 1978.
241. **Nordbö, H., Sörensen, R., and Sönju, T.**, Furfurals in chlorhexidine-discolored pellicle, *Scand. J. Dent. Res.*, 85, 606, 1977.
242. **Rølla, G. and Melsen, B.**, On the mechanisms of the plaque inhibition by chlorhexidine, *J. Dent. Res.*, 54B, 57, 1975.
243. **Luoma, H., Murtamaa, J., Nuuja, T., Nyman, A., Nummikoski, P., Ainamo, J., and Lumoma, A. R.**, A simultaneous reduction of caries and gingivitis in a group of school children receiving chlorhexidine-fluoride applications, *Caries Res.*, 12, 117, 1978.
244. **Emilson, C. G. and Fornell, J.**, Effect of toothbrushing with chlorhexidine gel on salivary microflora, oral hygiene and caries, *Scand. J. Dent. Res.*, 84, 308, 1976.

245. **Katz, S. and Park, K. C.,** Enhancement of the antiplaque value of antibacterial agents by enamel-conditioning methods. I. Rationale, mechanism and initial findings, *J. Dent. Res.*, 54, 540, 1975.
246. **Katz, S. and Park, K. C.,** Acquisition of antiplaque properties by treated enamel. II. Animal results. III, *J. Dent. Res.*, 55, 1084, 1088, 1976.
247. **Katz, S., Park, K. C., and Miller, C. H.,** Plaque inhibition by plant cell wall fractions, *J. Dent. Res.*, 55B (Abstr. 827, 828, 829), 268, 1976.
248. **Mirth, D. B., Kingman, C. J., and Bowen, J. H.,** Inhibition of saliva-induced aggregations of *Streptococcus mutans* by wheat germ agglutinin, *Caries Res.*, 13, 121, 1979.
249. **Jenkins, G. N.,** unpublished observations.

Saliva

SALIVARY GLAND METABOLISM

Frank Dowd, Eileen Watson, and Richard P. Suddick

Salivary gland secretory cells are specialized for the functions of secretion of proteins and other macromolecules, and of water and electrolytes. While their metabolic machinery reflects this specialization, it also includes those metabolic pathways which are common to most mammalian cells. These latter metabolic mechanisms serve the universal functions of energy production and utilization, the synthesis of regulatory enzymes and basic cell structures not directly involved in secretory processes, and the management of molecular transfer across membrane boundaries membranes for the purpose of preserving a uniform intracellular environment and volume. In addition, secretory cells have developed specialized biochemical mechanisms for carrying out their specialized secretory functions, and commonly occurring metabolic regulatory molecules serve the specialized functions of these cells. Descriptions of the specialized metabolic activities in the secretory cells of the salivary glands have been developed in the following sections.

RECEPTOR REGULATION OF SECRETION

Cellular mechanisms for regulating secretion of salivary macromolecules, water, and electrolytes are not fully understood, due partly to the cellular heterogeneity of salivary gland cells and functional species variability. Most of our knowledge in this area over the last 10 years comes from research done with rat parotid gland. The parotid gland slice system was used by Batzri and co-workers,[1] who initiated fundamental studies on the biochemical aspects of the regulation of salivary gland function. In recent years, due to technological advances, preparations of dispersed parotid acinar cells[2-4] and parotid acini[5] have been developed and used by many investigators. The advantages of these preparations include homogeneity of cells (acinar), the ability to vary the ionic milieu rapidly and precisely, and the separation of protein secretion from dependence on fluid secretion.

It is now well established that salivary gland function is continuously regulated by the autonomic nervous system (ANS) as well as by circulating hormones. Nerves originating in the CNS innervate salivary glands to secrete water, electrolytes, and macromolecules. Those nerves whose impulses release norepinephrine as a neurotransmitter are termed adrenergic, whereas nerves whose impulses act to release acetylcholine are termed cholinergic; salivary glands are regulated by both branches of the autonomic nervous system. Neurotransmitters released from nerve endings act on particularly sensitive portions of the salivary gland cell, the receptor, to produce responses. They can be activated not only by release of neurotransmitters, but also by agents or drugs that mimic transmitters. Receptors have been classified on the basis of selective antagonists and by the relative potency of representative agonists. The major receptor types to be considered in relation to the salivary gland include cholinergic (more specifically muscarinic), α-adrenergic, and β-adrenergic. The cholinergic and α-adrenergic components of the ANS primarily regulate water and electrolyte secretion, whereas the β-adrenergic component exerts its effects primarily on the macromolecular phase of secretion known as exocytosis. Alpha amylase, a major enzymatic protein secreted by parotid glands, has been extensively used as a marker for macromolecular release, and K^+ release has been widely used as a marker for inorganic or electrolyte release. In addition to the above receptors, there is also a Substance P receptor. Substance P is a cationic undecapeptide which appears to be a transmitter both within the CNS and GI tract.[6] It has been shown to regulate water and electrolyte secretion in salivary glands.[7]

ENERGY PRODUCTION

The abundance of mitochondria in most salivary gland secretory cells indicates a need for and availability of oxidative phosphorylation in cell functions.[8] Acinar cells and especially duct cells in the gland contain numerous mitochondria.[8] Feinstein and Schramm,[9a] using rat parotid slices, showed that energy production occurs 50 times faster by oxidative phosphorylation than by glycolysis. Glycolytic activity is very low.[9a] Furthermore, epinephrine, which induces protein secretion, does not stimulate glucose utilization or glycolysis.[9a] This evidence stresses the need for oxidative phosphorylation in the secretory process and hence for active mitochondria in the secretory cells.

One can divide secretion into two components, proteins and other macromolecules and water and electrolytes. Based on the results of Feinstein and Schramm,[9a] protein synthesis depends primarily on oxidative phosphorylation for its ATP supply. Water and electrolyte secretion, also an energy-requiring process, may derive part of its energy from intracellular ATP. If only because of the preference for oxidative phosphorylation for ATP production in the gland, it can be assumed that cellular energy requirements are supplied primarily by oxidative phosphorylation. Extracellular energy, such as vascular and fluid pressure, may also contribute to water and electrolyte secretion.[9b] The presence of mitochondria in both acinar and duct cells is suggestive once again of the need for oxidative metabolism in all phases of secretion.

ELECTROPHYSIOLOGY — ION MOVEMENTS

Upon stimulation of cholinergic or adrenergic nerves or by addition of agents that mimic the effects of nerve stimulation, released transmitters interact with receptors on the cell membrane to produce a complex alteration in ion movements. These changes consist of increases in membrane permeability and stimulation of the Na-K pump. Douglas and Poisner[10] coined the term "stimulus-secretion coupling" to define the sequence of events linking a stimulus at the cell membrane with the physiological response, secretion. Electrophysiological studies have helped to delineate many of the events associated with stimulus-secretion coupling in the salivary gland. The basic approach to the analysis of the process of secretion was revolutionized with the development of microelectrode techniques. These techniques have allowed investigators to determine some of the earliest events initiated by a stimulus. In addition, use of microiontophoresis allowed confirmation that cholinergic, α-adrenergic, and β-adrenergic receptors are all present on one acinar cell.[11]

One of the best-described electrophysiological responses in salivary gland is the interaction of α-adrenergic, cholinergic, and Substance P with receptors to alter the membrane permeability to K^+ ions, thus producing hyperpolarization and a characteristic loss of K^+ ions from salivary acinar cells.[11-13] In rat parotid acinar cells, as well as in pancreatic cells, these agents also trigger increased membrane permeability to Na^+ ions.[14,15] It has been suggested that stimulation of Na^+ influx by secretogogues is secondary to the influx of Ca^{2+} and that Ca^{2+} activation of ion fluxes may be a general phenomenon in exocrine glands.[16] Activation of β-adrenergic receptors, on the other hand, produces membrane depolarization which is associated with protein release.[13] It is through changes in membrane permeability to ions that intracellular events are "turned on" and the process of exocytosis occurs.

PROCESSES IN MACROMOLECULAR SECRETION

The distinguishing feature of secretory cells is the complex of events leading to exocytosis.[17] At first inspection it may appear that this series of events is limited to a very few cell types. It must be emphasized, however, that many cells are involved in exocytosis and the secretion phenomenon is very commonplace in nature. The types of cellular changes involved in secretion therefore have applicability to many cell types. In the salivary gland, it is believed that exocytosis is the major route of protein and glycoprotein secretion, although a nonvesicular form of secretion has been postulated.[18]

There are two basic types of salivary glands, mucous and serous, classified according to the product secreted and the microscopic structural appearance of the cells. The emphasis in this review will be on the parotid gland, a widely studied, serous-type gland. Although we can assume some differences in packaging and secretion between mucous and serous types of glands, the evidence to date suggests that they are similar in the way they package their secretory product;[8,19] however, the process of exocytosis is somewhat different in the serous and mucous glands (see below). Secretory vesicles from serous-type glands are called granules. Those from mucous glands are usually termed mucin droplets.

The close parallel between parotid glands from most species and the exocrine pancreas has allowed researchers to make some important structure-function analogies between the two glands. This is also true for biochemical events associated with secretion. Experiments designed to follow protein synthesis, packaging, storage, and exocytosis by means of radioactive tracer-labeled amino acids have been performed[17,20-22] in conjunction with microscopic investigations.[23-29] These have been valuable in elucidating the sequence of events leading to secretion. The sequence for these events is similar for both glands, although the time sequence may vary from gland to gland and from species to species.

Exocrine gland secretory cells are characterized by a distinctive polarity derived from their function. They are connected in epithelial sheets arranged in blind-ended tubes, the acinar cells forming the end termini of these tubes. Each cell thus has one pole facing the lumen or secretion side and one facing the interstitial fluid side of the cell, i.e., a luminal border with its membrane and a basal border with its membrane. This sidedness continues within the interior of the secretory cell, with mitochondria concentrated and nuclei displaced toward the basal side, and secretory granules in the luminal end of the cell.[8] The synthetic and packaging machinery lies within and between both regions. It is arranged to prepare protein within secretory granules. Thus, structurally and metabolically, the secretory cell is organized to synthesize macromolecules and to move them in a vectorial fashion to accomplish the process of exocytosis (Figure 1).

PROTEIN SYNTHESIS

Protein synthesis, as in other cells, takes place on ribosomes and is an energy-requiring process.[30] The coding process is analogous to that involved in the synthesis of nonsecretory protein and is susceptible to protein synthesis inhibitors.[30,31] Ribosomes are either arranged on sheets of endoplasmic reticulum (ER) near the nucleus of the cell (rough endoplasmic reticulum, RER) or are free in the cytoplasm. Free and attached ribosomes have separate synthetic functions.[17,32]

Free ribosomes are involved in synthesis of many proteins for intracellular use, whereas attached ribosomes are sites of synthesis for secretory protein, lysosomal protein, and membrane protein of the cell.[17,32,33] We will use the term exportable protein for convenience here to include all protein synthesized in the attached ribosomes. (''Exportable'' is used synonomously with exported; see legends for Figures 2 and 3.) The term ''exportable'', although somewhat inaccurate, is useful because it includes a key class of proteins, namely, the salivary macromolecules, which are synthesized for secretion. The arrangement of the RER is very important because in the cisternal spaces between the sheets of the RER, exportable protein is segregated from nonexportable protein. Transport into the cisternal space is a key process and is controlled by chemical adaptation of the proteins destined for secretion.[33-35]

How does the cell distinguish exportable protein from nonexportable protein? During

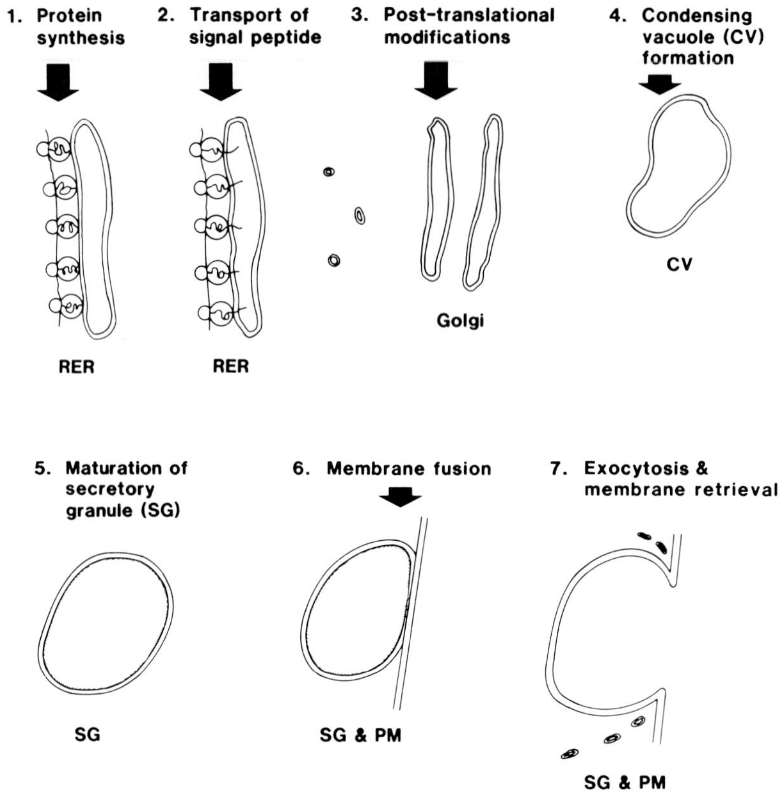

FIGURE 1. Schematic drawing of the secretory process from protein synthesis to membrane retrieval. Arrows indicate metabolic steps where energy is known to be required. Abbreviations not identified in the figure are RER: rough endoplasmic reticulum and PM: plasma membrane. Energy is required for protein synthesis, transport of signal peptides and exportable protein into RER cisternae, posttranslational modifications of proteins, formation of CV, and exocytosis. The exact site of energy required for CV formation is not known although it is somewhere between the RER and CV. GERL as a separate entity is not pictured (see text).

synthesis of proteins a unique process occurs which helps sequester exportable protein by utilizing so-called signal peptides which have been isolated from secretory cells. The function of the signal peptides appears to be the transport of the newly synthesized protein from the cytoplasmic side to the cisternal side of the RER.[33,34] The signal peptide is attached in peptide linkage to the N-terminal end of the growing exportable protein chain. The signal peptides also promote attachment of the free ribosomes to the ER. When sufficient signal peptide has been synthesized in the ribosome along with sufficient length of secretory protein, the protein chain emerges from the large ribosomal subunit to attach to a specific site on the ER.[33-35] The hydrophobic nature of the signal peptides isolated to date is consistent with the task of allowing the growing exportable protein to penetrate and cross the RER membrane.[36-38] Once the signal peptide-exportable protein chain is transferred to the cisternal side of the membrane, a signal peptidase, located in the membrane, cleaves the signal peptide from the exportable peptide chain. The exportable protein does not have to be completely synthesized prior to this hydrolysis and removal of the signal peptide.[33,34] All that is required is that a sufficient length of the polypeptide chain be synthesized to allow the chain to emerge into the cisternal space. This transport system requires metabolic energy and is unidirectional[17,33-35] (Figure 1).

Presecretory molecules, larger than their corresponding proteins in the secretory product,

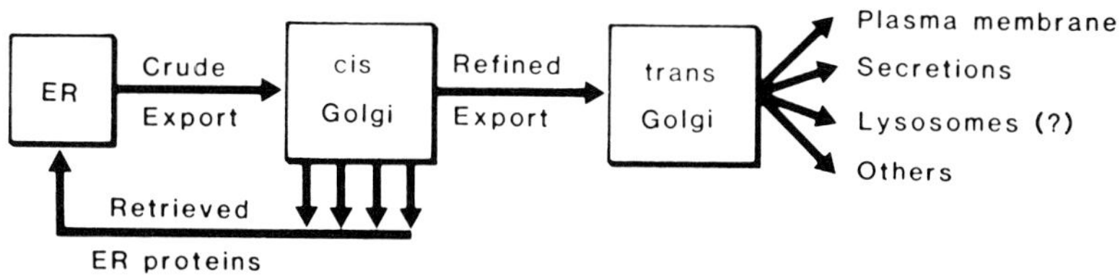

FIGURE 2. The proposed division of labor in the Golgi apparatus. During passage through the *cis* Golgi, characteristic ER membrane proteins would be removed from the export. The now refined mixture of exported proteins would then be passed on to the *trans* Golgi unit for further sorting to yield separate fractions for delivery to multiple cellular compartments. (Reprinted from Rothman, J. E., *Science,* 213, 1212, 1981. Copyright 1981 by the American Association for the Advancement of Science. With permission.)

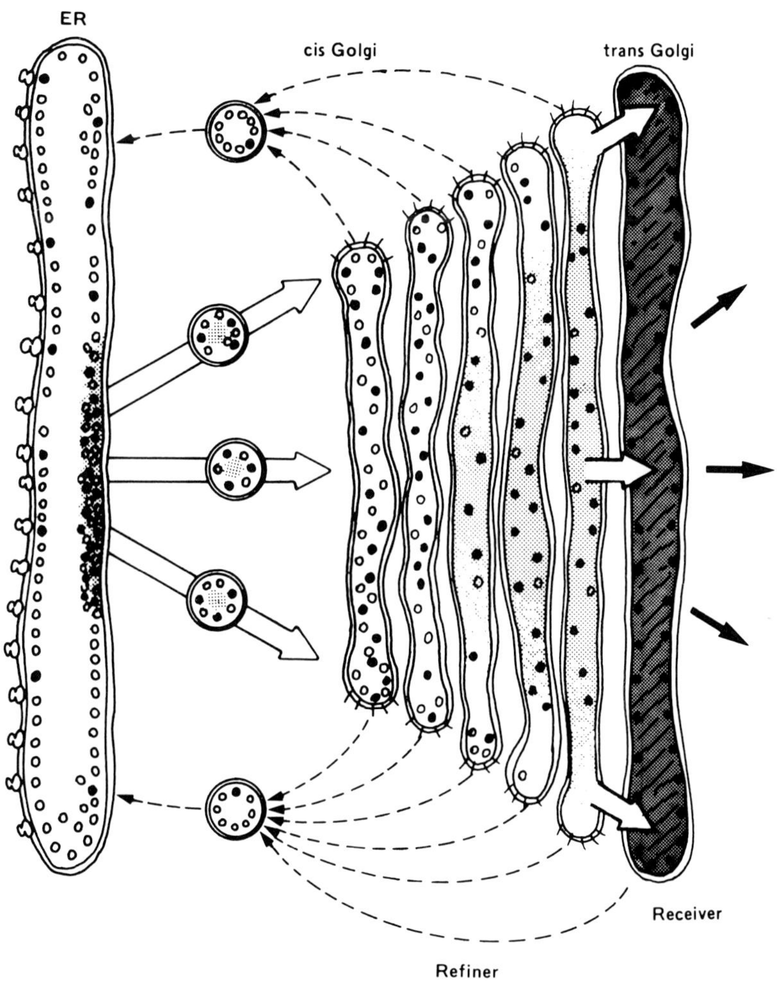

FIGURE 3. An illustration of the process of multistage refinement that may take place in the Golgi stack. The density of shading portrays the progressive transport of exported proteins from ER through the *cis* to the *trans* Golgi. The closed circles represent some of these exported proteins. The open circles (whether drawn as membrane-bound or not) represent ER membrane proteins that would most likely be removed from the rims of Golgi cisternae by budding vesicles that then fuse with ER (dashed arrows). This corresponds to a crosscurrent flow pattern. The thick arrows represent transport steps in which exported proteins are transported vectorially into the *cis* and into the *trans* Golgi compartments. (Redrawn from Rothman, J. E., *Science,* 213, 1212, 1981. Copyright 1981 by the American Association for the Advancement of Science. With permission.)

have been isolated. An example is α-amylase and preamylase from the parotid gland. α-Amylase is the major, starch-splitting digestive enzyme of saliva. The preamylase molecule is approximately 2000 daltons larger than secreted amylase.[37,38] The difference in molecular weight between preamylase and the secreted α-amylase is in good agreement with the deduced molecular weights of signal peptides (10 to 30 amino acids).[34,38]

The foregoing schema is part of the signal hypothesis of Blobel and Dobberstein,[33,34] which encompasses a large number of findings, although there are a few discrepancies between some data and the overall concept.[31] The hypothesis includes the fact that messenger RNA which codes for a particular secretory protein contains a "signal codon" next to the initiation codon. Messenger RNA therefore also codes for the signal peptide. It is possible that some nascent exportable proteins contain their signal peptide somewhere in the middle of the molecule rather than at the N-terminus. Such is the case for chicken ovalbumin secreted in the oviduct.[39,40]

The addition of signal peptides to secretory proteins is a convenient way to control the sequestering of proteins. This process offers two advantages; the newly synthesized protein is concentrated on the cisternal side of the ER, and the process can distinguish exportable protein from nonexportable protein.

PACKAGING OF MACROMOLECULES DESTINED FOR SECRETION

Synthesized exportable macromolecules proceed along a pathway from the cisternal spaces to the terminal portion of the cisternae called transitional elements. The membranes of the transitional elements bud off from the RER and become totally enclosed vesicles. Next, a complex association of secretory proteins takes place in which the products of budding from RER appear to become part of the Golgi complex. The formation of the Golgi is important not only for structural progression to the mature granule, but also because synthetic activity continues here as well. Posttranslational synthesis is important in biosynthesis of many secretory proteins and the Golgi is an important area for this synthesis.

Golgi stacks can be divided into *trans* Golgi and *cis* Golgi based on location within the cell as well as chemical and biochemical characteristics. *Cis* Golgi are Golgi stacks on the ER side, whereas *trans* Golgi are positioned facing the condensing vacuole[17,41-44] (Figures 2 and 3). Furthermore, *trans* Golgi may be distinguished from *cis* Golgi by having a higher density of membrane particles, possessing higher activities of thiamine pyrophosphatase (uridine diphosphatase)[44,45] and galactosyltransferase,[46] and a higher cholesterol content.[17,47]

It has been proposed that the Golgi stacks may act as a multistage refinement organelle, separating out ER membrane protein on the one hand from secretory, lysosomal, and plasma membrane protein on the other.[41] *Cis* Golgi would act to successively remove ER membrane and return the protein to the ER. The *trans* Golgi would receive partially refined protein from the *cis* Golgi and would trap the partially refined mixture for distribution to secretory granules, plasma membranes, or lysosomes.[41] At some point in the process, further recognition and separation would be required to separate out these three elements from each other.

The condensing vacuoles are positioned on the luminal side of the Golgi. They might be considered as developing secretory granules. Proteins from the Golgi are shunted to the condensing vacuoles which are similar in size, but appear less dense than the organelle formed prior to secretion, the secretory granule. Secretory protein is included within the confines of the limiting membranes of the condensing vacuoles. If ATP is removed, condensing vacuole formation does not occur despite the presence of its structural predecessor, RER. The exact site(s) of energy requirement has not been defined[17] (Figure 1).

GERL is a term which refers to smooth membranes situated between the *trans* Golgi and the condensing vacuoles. It is an acronym first used by Novikoff[48] to indicate smooth membranes located in the Golgi region, originating from the ER, and involved in lysosome production. GERL has been shown to have direct communication with the condensing vacuole and therefore may act in its formation.[49] GERL is enzymatically distinguishable from the Golgi in most cases.[49,50] Its major function in the exocrine cells seems to be the development of the condensing vacuoles, although its role in lysosome formation may be the major function in some other cell types. Recent evidence would suggest that GERL is derived from the *trans* Golgi.[51]

If protein synthesis is blocked with drugs, all protein synthesis including the secretory proteins will be blocked. However, the described processes of secretory granule formation and protein secretion, all of which succeed protein synthesis, continue uninhibited until protein is depleted from the cell.[30] This illustrates that the packaging and secretion of proteins are separate processes and independent to a degree from protein synthesis.

Conversion from condensing vacuoles to mature secretory granules involves a noticeable increase in apparent density of the condensing vacuole as seen by electron microscopy. This is due in part to an aggregation of protein. Secretory granules may have higher density for this reason and Golgi may continue to contribute secretory protein to preexisting condensing vacuoles, resulting in an increase in the amount of protein present in the vacuole. Concentration of protein within the vesicle is not dependent on a continuous supply of metabolic energy, since neither secretory granules nor condensing vacuoles swell when ATP production is blocked.[17] This indicates that concentration of protein within the vacuole is not due to an energy-requiring system such as a pump mechanism.

Interaction between cationic and anionic proteins or proteins with calcium could also account for concentration of material in the granules. Interaction of proteins within the granules could reduce osmotic pressure in the granules resulting in expulsion of water and concentration of protein.[17,21,52] Secretory granules would thus appear more dense microscopically. Furthermore secretory granules have been shown to have high levels of calcium, some of it complexed with protein.[53] Glycoproteins, including sulfur containing glycoproteins, could provide anions for this type of an interaction.[17,54,55] Such a mechanism would not require energy input beyond glycoprotein synthesis and condensing vacuole formation. The membrane border of the condensing vacuole appears rough, whereas that of the mature secretory granule appears smooth. This is consistent with the concentration of material in the secretory granule. Tight aggregation of protein may permit compression of the limiting membrane giving the secretory granule a smooth spherical appearance.

Secretory granules isolated intact from salivary gland tissue have been extensively used in various procedures.[56-58] The granules are usually about 1.0 μm in diameter, possessing an intact limiting membrane about 80 to 100° thick. The limiting membranes look very similar to plasma membranes with their trilaminar structure. Their lipid composition is more similar to plasma membranes than to endoplasmic reticulum. Over 95% of the total protein of secretory granules is secretory protein; the remainder is membrane associated.[56] The single secretory protein of highest occurrence in most parotid glands is α-amylase, and it is commonly used as an indicator both of secretory granule content and secretory product.

Radioactive amino acids have been utilized in the exocrine pancreas and parotid glands to follow the temporal sequence from synthesis of protein at the RER to maturation of secretory granules. Under experimental conditions the label reaches a peak in the Golgi at 10 to 30 min after addition of a labeled amino acid and in mature secretory granules by 80 to 180 min, depending on the exocrine gland and the species in question.

CO- AND POSTTRANSLATIONAL MODIFICATIONS OF EXPORTABLE PROTEIN

An important aspect of macromolecule synthesis in the salivary gland is the synthesis of conjugated proteins, *viz.*, glycoproteins and other modifications, including phosphorylation, sulfation, the formation of disulfide bonds, and perhaps limited proteolysis. These processes are probably both co- and posttranslational and therefore to a significant extent follow the synthesis of the peptide chain. This implies that modifications of exportable proteins may continue after the protein has been deposited into the cisternae. In reality, synthesis may continue up to and including the time the protein is located in the Golgi.

The most widely investigated of the nonpeptide synthetic processes is the formation of glycoproteins. They are conveniently divided into two basic categories. One type of glycoprotein is the mucin or glycosaminoglycans. They contain *O*-glycosidic bonds between *N*-acetylgalactosamine (GalNAc) and serine (Ser) or threonine (Thr), and contain a high percentage of carbohydrate. They possess hexuronic acid and hexuronic acid-hexosamine dissacharide repeating sequences.[19,54,60] These glycoproteins account in large measure for the physical properties of mucous secretion. The sequence for the biosynthesis of these compounds is similar to that for *N*-glycosidic compounds (see below); however, some details appear to be unique. Synthesis requires GalNAc which is transferred from UDP GalNAc to Ser or Thr residues. After addition of the sugar, further additions are made either in the form of sialic acid or galactose. Sugars can be added to galactose terminal side chains, but not to sialic acid residues which invariably are terminal. Species differences in mucins are common and may be regulated by the relative amounts of sialyl and galactosyl transferases in the gland.[19]

The second type of salivary glycoproteins contains *N*-glycosidic bonds between *N*-acetylglucosamine (GluNAc) and asparagine (Asn) residues of the protein chain, has a low percentage of carbohydrate, and contains neither hexuronic acid nor a hexuronic acid-hexosamine dissacharide repeating sequence. Proteins such as α-amylase[61] and the basic protein-rich glycoprotein from human saliva[62] are examples of this class.

Glycosylation of the *N*-glycosidic type occurs in two locations. "Core glycosylation" occurs within the endoplasmic reticulum while the final stages of glycosylation termed "terminal glycosylation" occur in the Golgi. Glycosylation is catalyzed by specific glycosyltransferases.[63] "Core glycosylations" of the *N*-glycosidic type are very similar in many different types of tissues, with little variance in the core from one glycoprotein to another.

The sequence of the core oligosaccharide is characteristically two Glu NAcs, nine mannoses, and three glucoses. These are synthesized sequentially on a long-chain lipid carrier, dolicholpyrophosphate.[64,65] From the lipid carrier, the oligosaccharide is transferred as one unit to an Asn residue in an Asn-X-Ser sequence of the peptide chain during its synthesis. Once attached to the polypeptide chain, additions to or subtractions from the chain are often made. Glucose and mannose units are often removed while fucose and sialic acid may be added.

Another important addition to glycoproteins is sulfur which results in the subclass of sulfated glycoproteins in which sulfur is attached covalently either to (GluNAc) or GalNAc.[54,55] The presence of required enzymatic activity suggests that sulfation takes place in the Golgi.[60] Other evidence as to location for addition of sulfate is the use of tracer studies involving ^{35}S-sulfate.[17,54,55,66,67] The fact that these glycoproteins may have protective effects in the GI tract implies a special importance to their biosynthesis.[54,55] Other roles postulated for sulfated glycoproteins include precipitation of salivary bacteria as well as interactions with dental plaque and dental pellicle.[54,55] The reaction sequence leading to sulfate addition is the following.[60]

$$\text{ATP} + \text{SO}_4 \xrightarrow[\text{Sulfurylase}]{\text{ATP}} \text{Adenosine-5}'\text{-phosphosulfate (APS)}$$

$$\text{APS} + \text{ATP} \xrightarrow[\text{Phosphokinase}]{\text{APS}} 3'\text{phospho APS (PAPS)}$$

$$\text{PAPS} + \text{glycoprotein} \xrightarrow[\text{Transferase}]{\text{Sulfo}} \text{Sulfated glycoprotein}$$

Other co- or posttranslational modifications include phosphorylation[68-71] and proteolysis.[72] Phosphorylation has not been widely studied in this system, and little is known about formation of phosphorylated salivary proteins, although it is known they exist.[73] The intense research interest in the function of phosphorylation reactions, however, makes this a likely candidate for future research. Proteolytic changes after synthesis have been noted,[72] but their extent and significance is a matter of conjecture.

EXOCYTOSIS

Secretion of protein is ultimately realized when secretory granule contents are discharged into the lumen of salivary gland ducts. The key event is the fusion of secretory granule membranes and plasma membranes, followed by a conversion of both membranes into one common membrane separating lumen from granule content, and then fission of the common membrane structure, resulting in the expulsion of granule content into the lumen[26,29,74] (Figure 1). This process is classified as a ''merocrine'' event and is characterized as the ''fusion-fission'' process just mentioned. It necessitates the close approximation of the secretory granule membrane with the apical portion of the plasma membrane. A different process occurs in mucous glands, where mucous droplets may discharge their contents through fractures in the luminal membrane with loss of some cytoplasm as is typical of apocrine type glands or by a merocrine process.[8,75,76] There is still uncertainty as to the exact mechanism(s) of mucous gland secretion.

Intracellular structures are located in an ordered arrangement with the basal to luminal progression generally being in the same order as the progression of synthesizing, packaging, and secretory events, i.e., RER → Golgi → condensing vacuole → secretory granule → luminal plasma membrane (Figure 1). The secretion of older secretory granules precedes that of newer synthesized ones except at high secretory rates where secretion is more random.[28] The nature of gland stimulation and the species examined also have some effect on the order of granule release. In the rat parotid gland, for instance, rapid secretion induced by α-adrenergic stimulation results in more random release of granules than slower secretory rates. (Random release being release of both old and newly synthesized granules.)

As secretion takes place, granule membrane becomes one with plasma membrane and the acinar or tubular lumen enlarges at the expense of cell volume. Immediately after secretion of a large amount of protein (e.g., after isoproterenol administration), the lumen of the salivary gland acinar unit is much larger than in the resting state, and often shows several adjacent smaller circular profiles facing the central portion of the lumen due to incorporation of secretory granule membranes into the membranes of the secretory cells. The extent of enlargement of the lumen or invagination into the acinar cells is directly proportional to the strength and duration of the secretory stimulus. Acinar cells appear to be essentially depleted of secretory granules after only a few minutes (e.g., 30) of maximal stimulation.

Microscopically, communication between and fusion of adjoining secretory granules is not often seen in serous glands of most species. However, this seems to occur with greater frequency in mucous droplets of mucous glands.[8] Little is known as to why granules interact with the luminal plasma membrane rather than the lateral plasma membrane of the secretory cell. This question bears heavily on the molecular mechanisms involved in secretion. Although it is known that exocytosis will not proceed without metabolic energy being available, the definitive molecular trigger for the fusion event also is not known. Research suggests that specific sites on either plasma membranes or granule membranes (or both) become available when the gland is stimulated to secrete. Why this happens at selective sites is a matter of conjecture at the present time.

With regard to secretory systems in general, progress has been made in identifying specific proteins which act as possible modifiers of secretion. Experiments utilizing the adrenal gland have been done to evaluate the role of specific proteins which are involved in the secretory process in that gland. An intracellular cytoplasmic receptor of calcium has been reported which acts to promote association of secretory granules with other granules and possibly with plasma membranes. This association microscopically resembles the pentalaminar structure seen immediately prior to exocytosis. Creutz et al.[77,78] have termed this protein ''synexin''. Calcium causes the synexin itself to aggregate which may account for the calcium-induced granule aggregation with synexin acting as the bridge between membranes. Other reports have indicated a role of calcium in fusion of artificial vesicles or in release of amylase by secretory granules.[79-81] Calcium may act as a trigger for exocytosis in vivo. It is interesting to speculate that membrane fusion may also involve a critical concentration of calcium for membrane-membrane interaction between secretory granules and plasma membranes.[82]

A polypeptide has been isolated from adrenal medulla plasma membranes which binds with high affinity to the secretory granules. This polypeptide has a molecular weight of 51,000. It specifically recognizes secretory granules and tends to promote granule fusion. Two other proteins which were isolated by the same procedures also bind with high affinity to the granules, but were minor components in the eluate. The major protein had higher affinity to secretory granules than synexin.[83] These data from the adrenal medulla are important in that they indicate specific proteins which may promote fusion of granules and plasma membrane in vivo. Such proteins have not been identified in salivary glands.

Investigators from other laboratories have shown microscopic changes in the membranes associated with exocytosis. Membrane-membrane interactions in the parotid gland have been examined by Tanaka et al.[26] They showed that particles within the luminal plasma membrane (intramembrane particles, IMP) appear random before stimulation, but after stimulation by isoproterenol the particles arrange asymetrically, becoming cleared from the area of granule-plasma membrane apposition. Thus, a rearrangement of membrane particles, probably proteins, seems to take place with or before secretion.[84] This process could result from protein-protein interaction. It is interesting in light of the synexin work[77,78] and the adrenal secretory granule binding protein work[83] discussed above. It seems probable that specific proteins relay signals for exocytosis. These proteins may themselves aggregate or cause aggregation in membrane proteins, allowing granule membrane and plasma membrane interaction to occur. Protein interaction may contribute by bridging the gap between membranes, drawing them closer by polymerization.[85,86] On the other hand, Tanaka et al.[26] have suggested that these membrane IMP do not directly participate in membrane fusion but may somehow regulate exocytosis by indirect means. This was based on the fact that arrays of IMP were not observed at the exact sites of granule-plasma membrane interaction.

In rat peritoneal mast cells, Lawson et al.,[85] using a freeze-fracture technique, reported a lateral protein displacement of adjacent granule and plasma membrane regions involved in exocytosis. Fusion could conceivably result from such lateral protein displacement resulting in the exposure of lipid-rich areas in the membranes. Such exposure could favor membrane fusion. A lack of intramembrane particles around the plasma membrane fusion site during exocytosis has also been reported in the chromaffin cell.[87] In the study of Aunis et al.,[87] connections between granules and plasma membrane were apparent prior to membrane fusion, suggesting specific recognition sites on either plasma membrane, secretory granules, or both. These types of responses have some similarities to the observations made by Schramm et al.,[88] who reported a process of blebbing or pseudopodia in secretory granules of rat parotid slices after isoproterenol treatment and in isolated secretory granules in the presence of ATP, cyclic AMP, and cyclic AMP-dependent protein kinase.

This description of findings from various secretory systems can be used to formulate a hypothetical secretory signal. Protein-protein interaction could control the secretory response with protein-protein interaction being in turn modified by posttranslational modifications of the proteins involved. It is probable that protein-protein interaction may be influenced by phosphorylation. This scheme is particularly attractive since β-adrenergic agonists increase intracellular cyclic AMP in the parotid gland. This results in activation of cyclic AMP-dependent protein kinase. The reaction catalyzed by this enzyme is protein phosphorylation[89] (see below). The role of phosphorylation in secretion is a subject of active investigation at the present time.

Under certain circumstances, exocrine glands can secrete protein in the absence of sizable numbers of secretory granules.[18] Since protein secretion appears to continue even after degranulation, it has been proposed that protein secretion is possible without the mediation of secretory granules or at least in the absence of exocytosis. Furthermore, it is possible that secretion even in the presence of secretory granules, takes place by extrusion or movement of protein from granule to cytoplasm and then to the lumen of the duct. Nevertheless, the preponderance of evidence favors exocytosis as the major mechanism of protein secretion despite a possible role for alternate secretory pathways.[18]

Secretion is followed by a conservation of at least part of the secretory granule membrane for use in synthesis of new membranes. Secretory granule membranes appear to be constructed from both recently synthesized and older components.[90] A likely mechanism for membrane conservation is the pinching off of vesicular portions of the enlarged acinar lumen[59] (Figure 1). Fragments from these vesicles may then be utilized for granule production at the level of the Golgi complex. Microscopic evidence is consistent with this theory in that small vesicles located near the lateral aspects of the apical plasma membrane are observed during the postsecretory or recovery phase.[59] Furthermore, in the mouse parotid gland it has been shown that isoproterenol stimulation leads to a decrease in adenylate cyclase activity in the plasma membrane, but a corresponding increase of the enzyme activity in lysosomes.[91] One interpretation of these results is that after secretion, plasma membrane protein is being recycled by pinocytosis followed by lysosomal processing. Lysosomes may play a role in digesting membrane material for resynthesis. Conservation of all membrane proteins does not seem to occur to the same extent, however, and therefore reutilization of membrane material appears to be selective.[59,90]

SECOND MESSENGERS

A feature common to secretory systems as well as other tissues involving neurotransmitter or hormonal action is the presence of intracellular control mechanisms which couple the signal received at the cell membrane to the secretory process. These intracellular control mechanisms have been referred to as "second messengers". In salivary glands, cholinergic, α-adrenergic, β-adrenergic, and Substance P stimulation evokes responses in the cell membrane (i.e., electrophysiological changes) which lead to alterations in the concentrations of the second messengers. These changes ultimately lead to enhanced K^+ release or amylase release. Potential sites of regulation by second messengers will be discussed below.

Over the last 10 years, second messengers which have been examined include Ca^{2+}, adenosine 3′5′-cyclic monophosphate (cyclic AMP), and more recently guanosine 3′5′-cyclic monophosphate (cyclic GMP). This is an area of research in which many important contributions to the knowledge of salivary function have been made. Although a role for Ca^{2+} as well as cyclic AMP were recognized as playing a role in secretion as early as the 1960s, more recent work has more firmly established the intracellular role of these agents in macromolecular and inorganic ion release.

ROLE OF CALCIUM

The importance of Ca^{2+} as an intracellular control mechanism has been well recognized in skeletal muscle for many years.[92] Early work by Douglas and Poisner[10] indicated that

Ca^{2+} is required for the secretory process in submandibular glands. Since then the number of secretory processes which have been shown to be dependent on Ca^{2+} has grown rapidly. To establish a role for Ca^{2+} as a second messenger mediating the action of transmitters or secretogogues, several criteria have to be met. These include:

1. Depletion of extracellular or intracellular Ca^{2+} should inhibit the action of the transmitter or secretogogue.
2. The transmitter or secretogogue should increase the influx of Ca^{2+} or the release of intracellularly stored Ca^{2+}.
3. Artificial introduction of Ca^{2+} into cells should mimic the action of secretogogue or transmitter.
4. In the presence of pharmacological agents affecting transmembrane Ca^{2+} movements, there should be parallel changes in Ca^{2+} influx and secretion.

It has not been possible, to date, to directly measure free cytoplasmic Ca^{2+} in salivary glands. However, use of a Ca^{2+} ionophore, A23187, an agent that binds Ca^{2+} ions and transports them across biological membranes, allowed workers to use it as a tool in demonstrating that cytoplasmic Ca^{2+} is indeed important in the secretory process.[93,94] The free intracellular Ca^{2+} concentration has been proposed as a unifying mechanism of stimulus-secretion coupling. Free cytosol Ca^{2+} can be made available by influx of Ca^{2+} into the cell from the extracellular medium[4,94,97] or by release of Ca^{2+} from intracellular storage depots.[98-100] Agonists specific for cholinergic, α-adrenergic, or Substance P receptors trigger a substantial net loss of intracellular K^+ and a small but detectable release of amylase. These effects are dependent on extracellular Ca^{2+} and an increased Ca^{2+} influx is observed; A23187 mimics the effect of these agents on K^+ release and amylase release, and these responses are blocked by agents affecting transmembrane movement of Ca^{2+}. Agonists specific for β-adrenergic receptors trigger macromolecular release and release Ca^{2+} from intracellular stores, possibly from mitochondria[2,101] or ER.[101] Also, depletion of intracellular calcium inhibits the effects of β-adrenergic stimulation on amylase release. In addition β-adrenergic agents increase intracellular levels of cyclic AMP.

Over the past few years it has been noted that Ca^{2+} may not act primarily in its ionic form but may act through binding to a highly specific protein, calmodulin. Calmodulin, a Ca^{2+} regulatory protein, has been shown to regulate several cellular processes in other tissues, including Ca^{2+} transport,[102,103] protein kinase activity and membrane phosphorylation,[104-106,107] phosphodiesterase activity,[107] and microtubule assembly.[108] A recent study has shown that calmodulin as well as a calmodulin-sensitive cyclic nucleotide phosphodiesterase is present in rat parotid gland;[109] other calmodulin regulated activities, however, have not been reported.

A related area of recent speculation is the role of microtubules and microfilaments in secretion. There is indirect evidence that either or both of these structures may function in exocytosis,[110-112] perhaps in conjunction with calmodulin.[108] It is interesting that microfilaments are composed of contractile proteins, principally actomyosin. There is enough evidence to suggest that they have a specialized function in salivary glands.[112-114] Microtubules are somewhat larger in diameter than microfilaments and appear to form a cytoskeleton. Colchicine and cytochalasin B which bind to and disrupt the function of microtubules and microfilaments, respectively, inhibit salivary gland exocytosis.[112-114] Microtubules may act as the framework through which secretory granules travel during protein secretion, and microfilaments may aid in this movement by a contractile process associated with secretion. A contribution of the microtubules and microfilaments to secretion does not rule out other mechanisms. They may for instance act in promoting close proximity of plasma membranes and secretory granule membranes, permitting intermembrane changes to complete the fusion process. A calcium-dependent regulation of microtubule assembly by calmodulin has been demonstrated in rat brain,[108] and thus, it is conceivable that an interaction of Ca^{2+} with calmodulin to form a Ca^{2+}-calmodulin complex may also modulate microtubules and microfilament functions in the salivary gland.

Calcium may also be involved in the regulation of exocytosis at the level of membrane fusion. As discussed above, morphological studies have shown that the release of secretory proteins from rat parotid gland involves fusion of secretory granule membranes with plasma membranes and extrusion of granule contents. From various cell sources, calcium-dependent interactions between plasma membranes and secretory granules and calcium-induced aggregation of secretory granules have been reported.[77,78,81,115]

ROLE OF CYCLIC AMP

It is now well recognized that cyclic AMP plays a significant role in the release of salivary amylase by β-adrenergic agents.[1,116,117] Stimulation of β-adrenergic receptors in parotid glands thus leads to increases in intracellular cyclic AMP levels[1,116,117] and enhanced activity of adenylate cyclase, as noted in broken cell preparations.[118] Studies with highly purified membrane fractions from salivary gland suggest that plasma membranes are the primary site of adenylate cyclase localization.[119]

As with Ca^{2+}, several criteria must be fulfilled in order to designate cyclic AMP as a second messenger. These include:

1. The transmitter or secretogogue raises tissue cyclic AMP levels and produces a physiological response.
2. The elevation in cyclic AMP levels occurs prior to or at least coincides with the physiological response.
3. The stimulus is synergized by drugs that inhibit phosphodiesterase activity.
4. Derivatives of cyclic AMP mimic the action of secretogogues.

Cyclic AMP has been shown to fulfill these criteria in salivary glands.

Phosphodiesterases have been characterized in mouse parotid glands.[120] The enzymes were shown to be a heterogenous group, consisting both of cyclic AMP and cyclic GMP phosphodiesterases and/or relatively low activity compared to pancrease, liver, or brain.[120] The parotid gland also contains a cyclic AMP dependent protein kinase.[120-124] Protein phosphorylation occurs in acinar cell aggregates (dispersed acini), as well as parotid slices,[123,125-129] upon β-adrenergic stimulation. Phosphorylation occurs in plasma membrane-enriched fractions[123,125-129] and is specific to β-adrenergic stimulation,[126] at least in one species. The involvement of protein phosphorylation as a regulatory step in the secretion process has been postulated based on time course and dose-response studies relating cyclic AMP levels and incorporation of ^{32}P into protein and amylase release. Data suggest that perhaps β-adrenergic-stimulated protein phosphorylation may alter the plasma membrane of the secretory cell to trigger the fusion process with secretory granule membranes.[123,125-129]

ROLE OF CYCLIC GMP

A more recently proposed second messenger in salivary gland is cyclic GMP. Recent studies have shown that both α-adrenergic and cholinergic agents which stimulate K^+ and amylase release increase intracellular levels of cyclic GMP.[130-136] This effect is Ca^{2+}-dependent and mimicked by the Ca^{2+} ionophore, A23187.[93,130,131] The physiological role of cyclic GMP in salivary gland as well as in other tissues, however, remains uncertain and very controversial. Although Mangos[133] proposed that cyclic GMP might mediate the effects of cholinergic agents on K^+ release from isolated rat parotid cells, other studies do not support this hypothesis. It was shown that agents that specifically enhance cyclic GMP levels do not increase K^+ release or amylase release.[130-132,137] In mouse parotid, acini agents such as nitroprusside and hydroxylamine, which specifically stimulate guanylate cyclase activity were shown to stimulate amylase release.[5] The reasons for this discrepancy are not known, but may be related to species differences. Levels of cyclic GMP within the cell are dependent on a balance between the enzymes which are involved in the formation and breakdown of the cyclic nucleotides. Little is known about guanylate cyclase[140] in salivary tissue activity. In addition, levels of cyclic GMP dependent protein kinase activity have not been quantitated. Thus, a role for cyclic GMP in regulation of salivary process awaits further documentation.

RADIOLIGAND STUDIES

Since specific antagonists are available for the different receptors,[137,141,142] it has been possible to demonstrate that the receptor types are, in fact, independent of each other. A recent technological advance, radioligand binding, has provided the means for ascertaining the number of cholinergic, adrenergic, and Substance P receptors and thus the density of these receptors. The first successful direct identification of α-adrenergic and β-adrenergic receptors with radioligands occurred in the mid 1970s.[143,144] In the parotid gland, the α-adrenergic receptor has been estimated with [^{3}H]-dihydroergocryptine,[145,146] the β-adrenergic receptor with [^{3}H]-dihydroalprenolol,[147] the cholinergic receptor with [^{3}H]-quinuclidinyl benzylate,[148] and the Substance P receptor with [^{125}I]-physalaemin.[149]

The importance of these studies lies in the fact that although receptors may be independent, their activation may regulate the same effector mechanisms (e.g., calcium influx, calcium release) within the cell. Identification of receptors with the above radioligands has thus allowed determination of the stoichiometric relationships of each receptor to the effector. The techniques of radioligand binding are also of great value in attempting to ascertain the molecular and cellular mechanisms involved in a phenomenon termed "desensitization".[145,146] Briefly stated, desensitization refers to a diminution of a response despite continued presence of agonist. For example, studies in which net K^+ release from isolated rat parotid cells was measured showed that the responsiveness to the α-adrenergic effects of epinephrine declined about 10 min after preincubation with the catecholamine. Using [^{3}H]-dihydroergocryptine, a specific antagonist to α-adrenergic receptors, it was noted that desensitization was accompanied by a parallel decrease in the number of α-receptors.[145,146] In similar studies, Harper and Brooker[132] found the response to the α-adrenergic agonist phenylephrine on α-amylase release declined until complete refractoriness developed by 30 min. The conclusion drawn from these studies was that desensitization occurred at the receptor level.

RELATIONSHIPS TO DISEASES AFFECTING EXOCRINE GLANDS

An area of research is emerging that has taken advantage of the studies described above, i.e., studies dealing with biochemical events associated with stimulus-secretion coupling in salivary gland. This area of research is the development of animal models for the study of diseases affecting exocrine glands. It is well known, for example, that a naturally occurring model for cystic fibrosis does not exist. In cystic fibrosis, patients suffer from altered exocrine secretions. Although the metabolic basis causing this disease is not known, there has been speculation that there are abnormal responses of the exocrine glands to autonomic innervation.[150] Previous studies have utilized skin fibroblasts from cystic fibrotic vs. normal patients. Although research continues with human skin fibroblasts, Martinez and co-workers have developed a pharmacological model in which rats are chronically treated with reserpine.

This treatment produces morphological and biochemical changes in submandibular and parotid glands resembling cystic fibrosis.[151-156] Results show that chronic administration of reserpine impairs the in vitro responses of both α- and β-adrenergic receptors in the rat parotid gland; there is a significantly greater release of K^+ ions[155] and a significantly smaller or reduced release of amylase than that found in untreated animals.[155,156]

Recent studies show that cyclic AMP levels are also elevated in submandibular glands from reserpinized rats, and this effect appears to be related to a decrease in phosphodiesterase activity rather than to an effect on adenylate cyclase activity.[157] Similar results, i.e., decreased amylase release and increased levels of cyclic AMP, have also been noted in mouse parotid gland.[166] In addition, β-receptor characteristics were determined by means of direct binding studies with the β-antagonist [^{3}H]-dihydroalprenolol. Roscher et al.[158] found an increase in the number of β-receptors in submandibular gland after reserpine treatment and a decrease after chronic isoproterenol treatment; chronic isoproterenol treatment also produces changes that resemble cystic fibrosis. Binding affinities in both cases were not changed. The drug treatments are known to modify, by different means, neurotransmitter concentrations at receptor sites. It was thus concluded that changes in the number of β-receptors might be a part of an adaptive process and could be interpreted as a negative feedback system regulating β-adrenergic transmission.

"PHOSPHOLIPID EFFECT"

Enhanced turnover of phosphotidylinositol (PI), or the "phospholipid effect", in response to a host of stimuli was first characterized by Hokin and Hokin in 1953.[159] The "phospholipid effect" is a feature common to many secreting systems, and, therefore, thought to be involved in exocytosis. Although the physiological significance of this effect is not clear, a unifying hypothesis has been proposed suggesting that PI breakdown or turnover is linked to receptor activation and occurs in the cell membrane.[160-163] It has recently been postulated that the PI response is a receptor phenomenon and may be related to activation of surface membrane Ca^{2+} "channels" or "gates".[161] Weiss and Putney[164] suggest that PI turnover may be a reaction more closely associated with Ca^{2+} in Ca channels than a reaction involved in receptor activation or occupation.

SUMMARY

The use of exocrine glands as models for the study of stimulus-secretion coupling has developed considerably in the past 10 years. Recent advances in our knowledge of salivary gland regulation are due in part to the technological advances made. Among those are the development of procedures for preparation of dispersed acinar cells and acini, development of radioligands for the study of the number and affinity of receptors, and development of radioimmunoassay techniques for determination of intracellular cyclic nucleotide levels.[165] Electron microscopic and autoradiographic techniques have allowed a detailed analysis of the secretory cycle.

The steps leading to macromolecule secretion follow a vectorial path beginning near the basal end of the secreting cell and proceeding to the luminal border. Associated with protein synthesis are signal peptides which allow exportable protein to be sequestered within the cisternae of the RER. As the protein is synthesized, shunted through the transitional elements, and deposited in the Golgi, co- and posttranslational modifications of some proteins take place. The most widely studied of these modifications is the addition of carbohydrate moieties to form glycoproteins. The Golgi may also act as multistage refinement organelles. GERL, located in the Golgi region, appears to give rise to condensing vacuoles, which eventually mature to secretory granules. This maturation process may involve complexing of proteins to reduce osmotic pressure within the granule.

Exocytosis follows a fusion-fission process in which the secretory granule membrane joins and becomes one with the plasma membrane, then splits. There may be specific recognition sites on both the secretory granule membranes and plasma membrane to allow membrane attraction and fusion. Evidence indicates dynamic membrane protein interactions are associated with secretion. A cytoplasmic protein, synexin, may, by aggregation, stimulate membrane fusion. Furthermore, calcium enhances synexin aggregation and the proposed role of synexin may help explain the role of calcium in exocytosis.

The role of calcium and cyclic nucleotides as second messengers is clearly an important area of research. In addition, the interaction between Ca^{2+} and calmodulin and the regulation of intracellular events awaits a great deal more research. Current research includes subcellular localization of calmodulin in tissue by indirect immunohistochemical fluorescent techniques. These studies might provide valuable information as to which cellular components might be modulated by calmodulin.

The role of cyclic AMP in macromolecular release is well established. The enzyme involved in cyclic AMP formation, adenylate cyclase, as well as the enzyme involved in cyclic AMP breakdown, phosphodiesterase, require further study. The ability of cyclic AMP-dependent protein kinase to phosphorylate membrane proteins supports the concept that phosphorylation may play a role in exocytosis. It may be that phosphorylation of proteins may promote membrane-membrane interactions between secretory granules and plasma membranes similar to those discussed above and thus have a functional role in exocytosis.

It has also been noted that chronic administration of reserpine or isoproterenol can produce morphological and biochemical changes in salivary glands resembling cystic fibrosis[151-156] and thus the chronically reserpinized or isoproterenol-treated animals have been used as models for cystic fibrosis. Studies with pharmacological models are of great use since they may identify specific processes that may be abnormal in cystic fibrotic patients. In comparison to our knowledge of the normal biochemical processes involved in salivary gland secretion, relatively little is known in diseased states. Finally, the parotid gland as well as the submandibular glands may offer a model which may have application to other exocrine systems.

REFERENCES

1. **Batzri, S., Selinger, A., and Schramm, M.,** Potassium ion release and enzyme secretion: adrenergic regulation by α- and β-receptors, *Science,* 174, 1029, 1971.
2. **Kanagasuntheram, P. and Randle, P. J.,** Calcium metabolism and amylase release in rat parotid acinar cells, *Biochem. J.,* 160, 547, 1976.
3. **Mangos, J. A., McSherry, N. R., Butcher, F., Irwin, K., and Barber, T.,** Dispersed rat acinar cells. I. Morphological and functional characterization, *Am. J. Physiol.,* 229, 553, 1975.
4. **Miller, B. E. and Nelson, D. L.,** Calcium fluxes in isolated acinar cells from rat parotid: the effect of adrenergic and cholinergic stimulation, *J. Biol. Chem.,* 252, 3629, 1977.
5. **Watson, E. L. and Dowd, F.,** Relationship between cyclic GMP levels and amylase release from mouse parotid acini, *Pharmacologist,* 23 (Abstr.), 186, 1981.
6. **Hanley, M. R. and Iversen, L.,** Substance P receptors, in *Neurotransmitter Receptors,* Vol. 9, Yamamura, H. and Enna, S., Eds., Chapman and Hall, London, 1980, 71.
7. **Chang, M. M. and Leeman S. E.,** Isolation of a sialogogic peptide from bovine hypothalamic tissue and its characterization as substance P, *J. Biol. Chem.,* 245, 4784, 1970.
8. **Wilborn, W. H. and Shackleford, J. M.,** Microanatomy of human salivary glands, in *The Biologic Basis of Dental Caries,* Menaker, L., Ed., Harper & Row, Hagerstown, Md., 1980, 3.

9a. **Feinstein, H. and Schramm, M.,** Energy production in rat parotid gland relation to enzyme secretion and effects of calcium, *Eur. J. Biochem.,* 13, 158, 1970.

9b. **Suddick, R. P., Calhoon, T. B., Feller, R. P., and Hyde, R. J.,** The energetics of fluid and electrolyte secretion, in *Saliva and Salivation, Advances in Physiological Sciences,* Zelles, T., Ed., Pergamon Press, New York, 1980, 85.

10. **Douglas, W. W. and Poisner, A. M.,** The influence of calcium on the secretory response of the submaxillary gland to acetylcholine or to noradrenaline, *J. Physiol.*, 165, 528, 1963.
11. **Roberts, M. L. and Petersen, O. H.,** Membrane potential and resistance changes induced in salivary gland acinar cells by microiontophoretic application of acetylcholine and adrenergic agonists, *J. Membr. Biol.*, 39, 297, 1978.
12. **Pedersen, G. L. and Petersen, O. H.,** Membrane potential measurement in parotid acinar cells, *J. Physiol.*, 234, 217, 1973.
13. **Petersen, O. H. and Pedersen, G. L.,** Membrane effects mediated by α- and β-adrenoceptors in mouse parotid acinar cells, *J. Membr. Biol.*, 16, 353, 1974.
14. **Case, R. M., Clausen, T., and Scott-Wilson, C. J.,** Extracellular sodium and stimulus secretion coupling in the rat exocrine pancreas, *J. Physiol.*, 284, 47p, 1978.
15. **Putney, J. W. and Parod, R. J.,** Calcium-mediated effects of carbachol on cation pumping and Na uptake in rat parotid gland slices, *J. Pharmacol. Exp. Ther.* 205, 449, 1980.
16. **Landis, C. A. and Putney, J. W.,** Calcium and receptor regulation of radiosodium uptake by dispersed rat parotid acinar cells, *J. Physiol.*, 297, 369, 1979.
17. **Palade, G. E.,** Intracellular aspects of the process of protein secretion, *Science,* 189, 347, 1975.
18. **Rothman, S. S.,** Enzyme secretion in the absence of zymogen granules, *Am. J. Physiol.*, 228, 1828, 1975.
19. **Schacter, H. and Williams, D.,** Biosynthesis of mucus glycoproteins, in *Proc. Int. Symp. in Mucins in Health and Disease,* Chantler, E., Elder, J., and Elstein, M., Eds., Plenum Press, New York, 1981.
20. **Castle, J. D., Jamieson, J. D., and Palade, G. E.,** Radioautographic analysis of the secretory process in the parotid acinar cell of the rabbit, *J. Cell Biol.*, 53, 290, 1972.
21. **Jamieson, J. D. and Palade, G. E.,** A condensing vacuole conversion and zymogen granule discharge in pancreatic exocrine cells: metabolic studies, *J. Cell Biol.*, 48, 503, 1971a.
22. **Schramm, M. and Bdolah, A.,** The mechanism of enzyme secretion by the cell. III. Intermediate stages in amylase transport as revealed by pulse labeling of slices of parotid gland, *Arch. Biochem. Biophys.*, 104, 67, 1964.
23. **Hand, A.,** The fine structure of von Ebner's gland of the rat, *J. Cell Biol.*, 44, 340, 1970.
24. **Tamarin, A. and Screebny, L. M.,** The rat submaxillary salivary gland. A correlative study of light and electronmicroscopy, *J. Morphol.*, 117, 295, 1965.
25. **Leeson, C.R. and Jacoby, F.,** An electron microscopic gland during its post-natal development and in the adult, *J. Anat.*, 93, 287, 1959.
26. **Tanaka, Y., DeCamilli, P., and Meldolesi, J.,** Membrane interactions between secretion granules and plasmalemma in three exocrine glands, *J. Cell Biol.*, 84, 438, 1980.
27. **Schramm, M. and Danon, D.,** The mechanism of enzyme secretion by the cell. I. Storage of amylase in the zymogen granules of the rat parotid glands, *Biochim. Biophys. Acta,* 50, 102, 1961.
28. **Sharoni, Y., Eimerl, S., and Schramm, M.,** Secretion of old versus new exportable protein in rat parotid slices. Control by neurotransmitters, *J. Cell Biol.*, 71, 107, 1976.
29. **Batzri, S., Selinger, A., Schramm, M., and Robinovitch, M. R.,** Potassium released mediated by the epinephrine α-receptor in rat parotid slices, *J. Cell Biol.*, 248, 361, 1973.
30. **Grand, R. J. and Gross, P. R.,** Independent stimulation of secretion and protein synthesis in rat parotid gland, *J. Biol. Chem.*, 244, 5608, 1969.
31. **Scheele, G., Jacoby, R., and Carne, T.,** Mechanism of compartmentation of secretory proteins: transport of exocrine pancreatic proteins across the microsomal membrane, *J. Cell Biol.*, 87, 611, 1980.
32. **Thibodeau, S. N. and Walsh, K. A.,** Processing of precursor proteins by preparations of oviduct microsomes, *Ann. N.Y. Acad. Sci.*, 343, 180, 1980.
33. **Blobel, G. and Dobberstein, B.,** Transfer of proteins across membranes. I. Presence of proteolytically processed and unprocessed nascent immunoglobulin light chains on membrane-bound-ribosomes of murine myeloma, *J. Cell Biol.*, 67, 835, 1975a.
34. **Blobel, G. and Dobberstein, B.,** Transfer of proteins across membranes. II. Reconstitution of functional rough microsomes from heterologous components, *J. Cell Biol.*, 67, 852, 1975b.
35. **Blobel, G. and Sabatini, D.,** Ribosome-membrane interactions in eukaryotic cells, in *Biomembranes,* Manson, L. A., Ed., Plenum Press, New York, 1971, 193.
36. **Strauss, A. W., Zimmerman, M., Mumford, R. A., and Alberts, A. W.,** Processing of pre-proalbumin and pre-placental lactogen, *Ann. N.Y. Acad. Sci.*, 343, 168, 1980.
37. **Gorecki, M. and Zeelon, E. P.,** Cell-free synthesis of rat parotid preamylase, *J. Biol. Chem.*, 254, 525, 1979.
38. **Karn, R. C., Petersen, T. E., Hjorth, J. P., Nieles, J. T., and Roepstorff, P.,** Characterization of the amino termini of mouse salivary and pancreatic amylases, *FEBS Lett.*, 126, 292, 1981.
39. **Lingappa, V. R., Lingappa, J. R., and Blobel, G.,** Signal sequences for early events in protein secretion and membrane assembly, *Ann. N.Y. Acad. Sci.*, 343, 356, 1980.
40. **Palmiter, J. D., Gagnon, J., and Walsh, K. W.,** Ovalbumin: a secreted protein without a transient hydrophobic leader sequence, *Proc. Natl. Acad. Sci. U.S.A.*, 75, 94, 1978.

41. **Rothman, J. E.,** The golgi apparatus: two organelles in tandem, *Science,* 213, 1212, 1981.
42. **Morre, D. J., Kartenbeck, J., and Franke, W.,** Membrane flow and interconversions among endomembranes, *Biochim. Biophys. Acta,* 559, 71, 1979.
43. **Novikoff, A.,** The endoplasmic reticulum. A cytochemist's view, *Proc. Natl. Acad. Sci. U.S.A.,* 73, 2781, 1976.
44. **Tartakoff, A. M.,** Golgi complex: crossroads for vesicular traffic, *Int. Rev. Exp. Pathol.,* 22, 227, 1980.
45. **Yanazaki, M. and Hayashi, E.,** Allosteric properties of nucleoside diphosphatase and its identify with thiamine pyrophosphatase, *J. Biol. Chem.,* 243, 2934, 1968.
46. **Borgese, N. and Meldolesi, J.,** Localization and biosynthesis of NADH-cytochrome b5 reductase, an integral membrane protein in rat liver cells. I. Distribution of the enzyme activity in microsomes, mitochondria, and Golgi complex, *J. Cell Biol.,* 85, 501, 1980.
47. **Orci, L., Montesano, R., Meda, P., Malaisse-Langae, F., Brown, D., Perrelet, A., and Vassalli, P.,** Heterogeneous distribution of filipin-cholesterol complexes across the cisternae of the Golgi apparatus, *Proc. Natl. Acad. Sci. U.S.A.,* 78, 293, 1981.
48. **Novikoff, A. B.,** GERL its form and function in neurons of rat spinal ganglia, *Biol. Bull.,* 127, 358, 1964.
49. **Hand, A. R. and Oliver, C.,** Cytochemical studies of GERL and its role in secretory granule formation in exocrine cells, *Histochem. J.,* 9, 375, 1977.
50. **Hand, A. R.,** Cytochemical differentiation of the golgi apparatus from GERL, *J. Histochem. Cytochem.,* 28, 82, 1980.
51. **Oliver, C., Auth, R. E., and Hand, A. R.,** Morphological and cytochemical alterations of the Golgi apparatus and GERL in rat parotid acinar cells during ethionine intoxication and recovery, *Am. J. Anat.,* 158, 275, 1980.
52. **Jamieson, J. D. and Palade, G. E.,** Synthesis, intracellular transport, and discharge of secretory proteins in stimulated pancreatic exocrine cells, *J. Cell Biol.,* 50, 135, 1971b.
53. **Wallach, D. and Schramm, M.,** Calcium and the exportable protein in rat parotid gland. Parallel subcellular distribution and concomitant secretion, *Eur. J. Biochem.,* 21, 433, 1971.
54. **Embery, G. and Rølla, G.,** Chemistry of the sulphate groups in a sulphated glycoprotein from rabbit submandibular gland, *Scand. J. Dent. Res.,* 85, 241, 1977.
55. **Embery, G. and Whitehead, E.,** Sulphated glycoproteins in rat saliva, *Arch. Oral Biol.,* 20, 843, 1975.
56. **Castle, J. D., Jamieson, J. D., and Palade, G. E.,** Secretion granules of the rabbit parotid gland. Isolation, subfractionation, and characterization of the membrane and content subfractions, *J. Cell Biol.,* 64, 182, 1975.
57. **Lambert, M., Camus, J., and Christopher, J.,** Phosphorylation of protein components of isolated zymogen granule membranes from the rat pancreas, *FEBS Lett.,* 49, 228, 1974.
58. **Robinovitch, M. R., Lagunoff, D., and Iversen, J. M.,** The presence of mast cell granules in rat parotid secretory granule preparations, *J. Histochem. Cytochem.,* 28, 1351, 1980.
59. **Bogart, B. I.,** Secretory dynamics of the rat submandibular gland. An ultrastructure and cytochemical study of the isoproterenol-induced secretory cycle, *J. Ultrastruct. Res.,* 52, 139, 1975.
60. **Young, R. W.,** The role of the golgi complex in sulfate metabolism, *J. Cell Biol.,* 57, 175, 1973.
61. **Keller, P. J., Kauffman, D. L., Allan, B. J., and Williams, B. L.,** Further studies on the structural differences between the isoenzymes of human parotid α-amylase, *Biochemistry,* 10, 4867, 1971.
62. **Li, H. C. and Levine, M. J.,** Characterization of a glycopeptide from the proline-rich glycoprotein in human parotid saliva, *Arch. Oral Biol.,* 25, 353, 1980.
63. **Schacter, H.,** Glycoprotein biosynthesis and processing, in *Lysosomes and Lysosomal Storage Diseases,* Callahan, J. W. and Lowden, J. A., Eds., Raven Press, New York, 1981, 1.
64. **Montreuil, J.,** Recent data on the structure of the carbohydrate moiety of glycoproteins. Metabolic and biological implications, *Pure Appl. Chem.,* 42, 431, 1975.
65. **Czichi, U. and Lennarz, W. J.,** Localization of the enzyme system for glycosylation of proteins via the lipid-linked pathway in rough endoplasmic reticulum, *J. Biol. Chem.,* 252, 7901, 1977.
66. **Reggio, H. A. and Palade, G.E.,** Sulfated compounds in the zymogen granules of the guinea pig pancreas, *J. Cell Biol.,* 77, 288, 1978.
67. **Berg, N. B. and Austin, B. P.,** Intracellular transport of sulfated macromolecules in parotid acinar cells, *Cell Tissue Res.,* 165, 215, 1976.
68. **Schlesinger, D. H. and Hay, D. I.,** Complete covalent structure of statherin, a tyrosine-rich acidic peptide which inhibits calcium phosphate precipitation from human parotid saliva, *J. Biol. Chem.,* 252, 1689, 1977.
69. **Bennick, A., Wong, R., and Cannon, M.,** Structure and biological activities of salivary acidic proline-rich phosphoproteins, in *Calcium Binding Proteins and Calcium Function,* Wassermann, R. H., Ed., Elsevier, New York, 1977, 391.
70. **Boat, R. F., Wiesman, U. N., and Pallavicini, J. C.,** Purification and properties of the calcium-precipitable protein in submaxillary saliva of normal and cystic fibrosis subjects, *Pediatr. Res.,* 8, 531, 1974.
71. **Mayo, J. W. and Carlson, D. M.,** Protein composition of human submandibular secretions, *Arch. Biochem. Biophys.,* 161, 134, 1974.

72. **Wong, R. S. C. and Bennick, A.,** Purification of parotid protease which cleaves proline-rich protein C to A, *J. Dent. Res.*, 60 (Abstr. 1374), 653, 1981.
73. **Boackle, R. J. and Suddick, R. P.,** Salivary proteins and oral health, in *The Biologic Basis of Dental Caries,* Menaker, L., Ed., Harper & Row, New York, 1980, 113.
74. **Tandler, B. and Erlandson, R. A.,** Ultrastructure of baboon parotid gland, *Anat. Rec.*, 184, 115, 1976.
75. **Tandler, B. and Paulsen, J. H.,** Fusion of the envelope of mucous droplets with the luminal plasma membrane in acinar cells of the cat submandibular gland, *J. Cell Biol.*, 68, 775, 1976.
76. **Tandler, B.,** Salivary glands and the secretory process, in *Textbook of Oral Biology,* Shaw, J. H., Sweeney, E. A., Cappuccino, C. C., and Meller, S. M., Eds., W. B. Saunders, Philadelphia, 1978, 580.
77. **Creutz, C. E., Pazoles, C. J., and Pollard, H. B.,** Identification and purification of an adrenal medullary protein (Synexin) that causes calcium-dependent aggregation of isolated chromaffin granules, *J. Biol. Chem.*, 253, 2858, 1978.
78. **Creutz, C. E., Pazoles, C. J., and Pollard, H. B.,** Self-association of synexin in the presence of calcium correlation with synexin-induced membrane fusion and examination of the structure of synexin aggregates, *J. Biol. Chem.*, 254, 553, 1979.
79. **Ingolia, T. D. and Koshland, D. E., Jr.,** The role of calcium in fusion of artificial vesicles, *J. Biol. Chem.*, 253, 3821, 1978.
80. **Yoshida, H., Miki, N., Ishida, H., and Yamamoto, I.,** Release of amylase from zymogen granules of ATP and a low concentration of Ca^{2+}, *Biochim. Biophys. Acta,* 158, 489, 1968.
81. **Ishida, H., Miki, N., Hata, F., and Yoshida, H.,** Conformational changes in zymogen granules from the parotid glands induced by ATP and a low concentration of calcium, *Jpn. J. Pharmacol.*, 21, 239, 1971.
82. **Ekerdt, R., Dahl, G., and Grantzl, M.,** Membrane fusion of secretory vesicles and liposomes. Two different types of fusion, *Biochim. Biophys. Acta,* 646, 10, 1981.
83. **Meyer, D. I. and Burger, M. M.,** Isolation of a protein from the plasma membrane of adrenal medulla which binds to secretory vesicles, *J. Biol. Chem.*, 254, 9854, 1979.
84. **DeCamilli, P., Peluchetti, D., and Meldolesi, H.,** Dynamic changes of the luminal plasmalemma in stimulated parotid acinar cells. A freeze-fracture study, *J. Cell Biol.*, 70, 59, 1976.
85. **Lawson, C., Raff, M. C., Gomperts, B., Fewtrell, C., and Gilula, N. B.,** Membrane events during membrane fusion. A study of exocytosis in rat peritoneal mast cells, *J. Cell Biol.*, 72, 242, 1977.
86. **Beisson J., Cohen, J., Lefort-Tran, M., Pouphile, M., and Rossignol, M.,** Control of membrane fusion in exocytosis. Physiological studies on a paramecium mutant blocked in the final step of the trichocyst extrusion process, *J. Cell Biol.*, 85, 213, 1980.
87. **Aunis, D., Hesketh, J. E., and Devilliers, G.,** Freeze-fracture study of the chromaffin cell during exocytosis: evidence for connections between the plasma membrane and secretory granules and for movements of plasma membrane-associated particles, *Cell Tissue Res.*, 197, 443, 1979.
88. **Schramm, M., Selinger, A., Salomon, Y., Eytan, E., and Batzri, S.,** Pseudopodia formation by secretory granules, *Nature (London),* 240, 204, 1972.
89. **Glass, D. B. and Krebs, E. G.,** Protein phosphorylation catalyzed by cyclic AMP-dependent and cyclic GMP-dependent protein kinases, *Ann. Rev. Pharmacol. Toxicol.*, 20, 363, 1980.
90. **Wallach, D., Kirschner, N., and Schramm, M.,** Non-parallel transport of membrane proteins and content proteins during assembly of the secretory granule in rat parotid gland, *Biochim. Biophys. Acta,* 375, 87, 1975.
91. **Revis, N. W. and Durham, J. P.,** Adenylate cyclase activity in the parotid gland of the mouse after isoproterenol stimulation, *J. Histochem. Cytochem.*, 27, 1317, 1979.
92. **Martinosi, A.,** Transport of calcium by the sarcoplasmic reticulum, in *Metabolic Pathway,* Vol. 6, *Metabolic Transport,* Hokin, L. E., Ed., Academic Press, New York, 1972, 317.
93. **Butcher, F. R.,** The role of calcium and cyclic nucleotides in α-amylase release from slices of rat parotid: studies with the divalent cation ionophore A23187, *Metabolism,* 24, 409, 1975.
94. **Rossignol, G., Herman, G., Chambaut, A. M., and Keryer, G.,** The calcium ionophore A-23187 as a probe for studying the role of Ca^{2+} ions in the mediation of carbachol effects on rat salivary glands: protein secretion and metabolism of phospholipids and glycogen, *FEBS Lett.*, 43, 241, 1974.
95. **Putney, J. W., Van deWalle, C. M., and Leslie, B. A.,** Receptor control of calcium influx in parotid acinar cells, *Mol. Pharmacol.*, 14, 1046, 1978.
96. **Putney, J. W.,** Stimulation of ^{45}Ca influx in rat parotid gland by carbachol, *J. Pharmacol. Exp. Ther.*, 199, 516, 1976.
97. **Watson, E. L., Williams, J. A., and Siegel, I. A.,** Calcium mediation of cholinergic-stimulated amylase release from mouse parotid gland, *Am. J. Physiol.*, 236, C233, 1979.
98. **Butcher, F. R.,** Regulation of calcium efflux from isolated rat parotid cells, *Biochim. Biophys. Acta,* 630, 254, 1980.
99. **Putney, J. W., Weiss, S. J., Leslie, B. A., and Marier, S. H.,** Is calcium the final mediator of exocytosis in the rat parotid gland? *J. Pharmacol. Exp. Ther.*, 203, 144, 1977.

100. **Watson, E. L., Farnham, C. J. M., Friedman, J., and Farnham, W. F.,** Effects of monensin on amylase released from mouse parotid acini, *Am. J. Physiol.*, 240, C189, 1981.
101. **Dormer, R. L. and Ashcroft, S. J. H.,** Studies on the role of calcium ions in the stimulation by adrenaline of amylase release from rat parotid, *Biochem. J.*, 144, 543, 1974.
102. **Hinds, T. E., Larsen, F. L., and Vincenzi, F. F.,** Plasma membrane Ca^{2+} transport: stimulation by soluble proteins, *Biochem. Biophys. Res. Commun.*, 81, 455, 1978.
103. **Jarrett, H. W. and Penniston, J. T.,** Partial purification of the Ca^{2+}- Mg^{2+} ATPase activator from human erythrocytes: its similarity to the activator of 3′, 5′ - cyclic nucleotide phosphodiesterone, *Biochem. Biophys. Res. Commun.*, 77, 1210, 1977.
104. **DeLorenzo, R. J.,** Calcium-dependent phosphorylation of specific synaptosomal fraction proteins: possible role of phosphoproteins in mediating neurotransmitter release, *Biochem. Biophys Res. Commun.*, 71, 590, 1976.
105. **DeLorenzo, R. J. and Freedman, S. D.,** Calcium dependent neurotransmitter release and protein phosphorylation in synaptic vesicles, *Biochem. Biophys. Res. Commun.*, 80, 183, 1978.
106. **DeLorenzo, R. J., Freedman, S. D., Yohe, W. B., and Maurer, S. C.,** Stimulation of Ca^{2+}-dependent neurotransmitter release and presynaptic nerve terminal protein phosphorylation by isolated synaptic vesicles, *Proc. Natl. Acad. Sci. U.S.A.*, 76, 1838, 1979.
107. **Cheung, W. Y.,** Cyclic 3′,5′ - nucleotide phosphodiesterase, *Biochem. Biophys. Res. Commun.*, 38, 533, 1970.
108. **Marcum, J. M., Dedman, J. R., Brinkley, B. R., and Means, A. R.,** Control of microtubule assembly-disassembly by calcium-dependent regulator protein, *Proc. Natl. Acad. Sci. U.S.A.*, 82, 287, 1978.
109. **Ku, K. Y. and Butcher, F. R.,** Detection of a calmodulin-sensitive cyclic nucleotide phosphodiesterase in rat parotid gland, *Biochim. Biophys. Acta*, 631, 70, 1980.
110. **Chambaut-Guerin, A. M., Muller, P., and Rossignol, G.,** Microtubules and protein secretion in rat lacrimal glands. Relationship between colchicine binding and its inhibitory effect on the intracellular transport of proteins, *J. Biol. Chem.*, 253, 3870, 1978.
111. **Chambaut-Guerin, A. M., Muller, P., and Rossignol, B.,** Microtubules and protein secretion in rat lacrimal glands. Inhibitory effect of the tubulin colchicine complex isolated from lacrimal glands upon brain tubulin polymerization. Identification of the complex by gel electrophoresis, *J. Biol. Chem.*, 254, 10,734, 1979.
112. **Butcher, F. R. and Goldman, R. H.,** Effect of cytocholasin B and colchicine on the stimulation of α-amylase release from rat parotid tissue slices, *Biochem. Biophys. Res. Commun.*, 48, 23, 1972.
113. **Butcher, F. R. and Goldman, R. H.,** Effect of cytocholasin B and colchicine on α-amylase release from rat parotid slices. Dependence of the effect on N^6, O^2 - dibutyryl adenosine 3′,5′-cyclic monophosphate concentration, *J. Cell Biol.*, 60, 519, 1974.
114. **Patzelt, C., Brown, O., and Jeanrenaud, B.,** Inhibitory effect of colchicine on amylase secretion by rat parotid glands. Possible localization in the golgi area, *J. Cell Biol.*, 73, 578, 1977.
115. **Milutinovic, S., Argent, B. E., Schulz, I., and Sacks, G.,** Studies on isolated subcellular components of cat pancreas, *J. Membr. Biol.*, 36, 281, 1977.
116. **Butcher, F. R., Goldman, J. A., and Nemerovski, M.,** Effect of adrenergic agents on α-amylase release and adenosine 3′,5′-monophosphate accumulation in rat parotid tissue slices, *Biochim. Biophys. Acta*, 392, 82, 1975.
117. **Malamud, D.,** Amylase secretion from mouse parotid and pancreas: role of cyclic AMP and isoproterenol, *Biochim. Biophys. Acta*, 279, 373, 1972.
118. **Schramm, M. and Naim, E.,** Adenyl cyclase of rat parotid: activation by fluoride and norepinephrine, *J. Biol. Chem.*, 245, 3225, 1970.
119. **Durham, J. P., Galanti, J., and Revis, N. W.,** The purification and characterization of plasma membranes and the subcellular distribution of adenylate cyclase in mouse parotid gland, *Biochim. Biophys. Acta*, 394, 388, 1975.
120. **Chiu, H.-I., Franks, D. J., Rowe, R., and Malamud, D.,** Cyclic AMP metabolism in mouse parotid glands. Properties of adenylate cyclase, protein kinase and phosphodiesterase, *Biochim. Biophys. Acta*, 451, 29, 1976.
121. **Butcher, F. R., Thayer, M., and Goldman, J. A.,** Effect of adenosine 3′,5′-cyclic monophosphate derivatives on amylase release, protein kinase, and cyclic nucleotide phosphodiesterase activity from rat parotid tissue, *Biochim. Biophys. Acta*, 421, 289, 1976.
122. **Dowd, F. J. and Shannon, I. L.,** Kinetic properties of a cyclic AMP-dependent protein kinase from pig parotid gland, *Arch. Oral Biol.*, 21, 423, 1976.
123. **Kanamori, T. and Hayakawa, T.,** Cyclic AMP-dependent ^{32}P incorporation in a protein in rat parotid slices, *Biochem. Int.*, 1, 395, 1980.
124. **Kanamori, T., Hayakawa, T., and Nagatsu, T.,** Adenosine 3′,5′-monophosphate-dependent protein kinase and amylase secretion from rat parotid gland, *Biochem. Biophys. Res. Commun.*, 57, 394, 1974.

125. **Baum, B. S., Frieberg, J. M., Ito, H., Roth, G. S., and Filburn, C. R.,** β-Adrenergic regulation of protein phosphorylation and its relationship to exocrine secretion in dispersed rat parotid gland acinar cells, *J. Biol. Chem.*, 256, 9731, 1981.
126. **Dowd, F. J., Watson, E. L., Horio, B., Lau, Y-S., and Park, K.,** Phosphorylation of rabbit parotid microsomal protein occurs only with β-adrenergic stimulation, *Biochem. Biophys. Res. Commun.*, 101, 281, 1981.
127. **Jahn, R., Ungar, C., and Soling, H. D.,** Specific protein phosphorylation during stimulation of amylase secretion by β-agonists or dibutyryl adenosine 3′,5′-monophosphate in the rat parotid gland, *Eur. J. Biochem.*, 112, 345, 1980.
128. **Mednieks, M. I. and Hand, A. R.,** Signal processing in salivary glands: morphological and biochemical expression of cyclic nucleotide-mediated events, *J. Dent. Res.*, 60(Abstr.), 343, 1981.
129. **Teoh, T. S. and Spearman, T. N.,** Protein phosphorylation in rat parotid gland minces following stimulation by isoproterenol, *Fed. Proc.*, 40(Abstr.), 447, 1981.
130. **Butcher, F. R., McBride, P. A., and Rudich, L.,** Cholinergic regulation of cyclic nucleotide levels, amylase release and K^+ efflux from rat parotid glands, *Mol. Cell Endocrinol.*, 5, 243, 1976.
131. **Butcher, F. R., Rudich, L., Emler, C., and Nemerovski, M.,** Adrenergic regulation of cyclic nucleotide levels, amylase release and potassium efflux in rat parotid gland, *Mol. Pharmacol.*, 12, 862, 1976.
132. **Harper, J. F. and Brooker, G.,** Amylase secretion from the rat parotid: refractoriness to muscarinic and adrenergic agonists, *Mol. Pharmacol.*, 14, 1031, 1978.
133. **Mangos, J. A.,** Role of cGMP in isolated rat parotid acinar cells, *J. Dent. Res.*, 57, 989, 1978.
134. **Templeton, D., Butcher, F. R., Turner, C. J., Muir, T. C., and Durham, J. P.,** Effect of electrical stimulation of the autonomic nerves on the levels and synthesis of cyclic nucleotides in the rat salivary glands: relationship to enhanced growth, *J. Cyclic Nucl. Res.*, 3, 107, 1977.
135a. **Watson, E. L., Friedman, J., and Siegel, I. A.,** Mediation of β-adrenergic-stimulated amylase release from mouse parotid gland, *Life Sci.*, 26, 1919, 1980.
135b. **Watson, E. L., Jacobson, K. L., and Dowd, F.,** Cyclic nucleotides and amylase release from mouse and rabbit parotid acini, *Proc. West. Pharmacol. Soc.*, 24, 19, 1981.
136. **Wojcik, J. D., Grand, R. J., and Kimberg, D. V.,** Amylase secretion by rabbit parotid gland. Role of cyclic AMP and cyclic GMP, *Biochim. Biophys. Acta*, 411, 250, 1975.
137. **Rudich, L. and Butcher, F. R.,** Effect of substance P and eledoisin on K^+ efflux amylase release and cyclic nucleotide levels in slices of rat parotid gland, *Biochim. Biophys. Acta*, 444, 704, 1976.
138. **Deguchi, T.,** Activation of guanylate cyclase in cerebral cortex of rat by hydroxylamine, *J. Biol. Chem.*, 252, 596, 1977.
139. **Schultz, K. B., Schultz, K., and Schultz, G.,** Sodium nitroprusside and other smooth muscle-relaxants increase cyclic GMP levels in rat ductus deferens, *Nature (London)*, 265, 750, 1977.
140. **Durham, J. P.,** Guanylate cyclase: assay and properties of the particulate and supernatant enzymes in mouse parotid, *Eur. J. Biochem.*, 61, 535, 1976.
141. **Leslie, B. A., Putney, J. W., and Sherman, J. A.,** α-Adrenergic, β-adrenergic and cholinergic mechanisms for amylase secretion by rat parotid gland *in vitro*, *J. Physiol.*, 260, 351, 1976.
142. **Schramm, M. and Selinger, Z.,** The function of α- and β-adrenergic receptors and a cholinergic receptor in the secretory cell of rat parotid gland, *Adv. Cytopharmacol.*, 2, 29, 1974.
143. **Greenberg, D. A., U'Prichard, D. C., and Snyder, S. H.,** α-Adrenergic receptor binding in mammaliam brain: differential labeling of agonist and antagonist states, *Life Sci.*, 19, 69, 1976.
144. **Williams, L. T. and Lefkowitz, R. J.,** Alpha-adrenergic receptor identification by [^{3}H] dihydroergocryptine binding, *Science*, 192, 791, 1976.
145. **Strittmatter, W. J., Davis, J. N., and Lefkowitz, R. J.,** α-Adrenergic receptors in rat parotid cells. I. Correlation of [^{3}H]-dihydroergocryptine binding and catecholamine-stimulated potassium efflux, *J. Biol. Chem.*, 252, 5472, 1977a.
146. **Strittmatter, W. J., Davis, J. N., and Lefkowitz, R. J.,** α-Adrenergic receptors in rat parotid cells. II. Desensitization of receptor binding sites and potassium release, *J. Biol. Chem.*, 252, 5478, 1977b.
147. **Au, D. K., Malbon, C. C., and Butcher, F. R.,** Identification and characterization of β-adrenergic receptors in rat parotid membranes, *Biochim. Biophys. Acta*, 500, 361, 1977.
148. **Putney, J. W. and Van deWalle, C. M.,** The relationship between muscarinic receptor binding and ion movements in the rat parotid gland, *J. Physiol.*, 299, 521, 1980.
149. **Putney, J. W., Van deWalle, C. M., and Wheeler, C. S.,** Binding of ^{125}I-physalaemin to rat parotid acinar cells, *J. Physiol.*, 301, 205, 1980.
150. **Barbero, G. J. and Chernick, W. S.,** The role of the autonomic nervous system in cystic fibrosis, in *Research on Pathogenesis of Cystic Fibrosis of the Pancreas (Mucoviscidosis)*, di Sant' Agnese, P. A., Ed., International Conference at National Institutes of Health, Bethesda, Md., 1964, 208.
151. **Mangos, J. A., Boyd, R. L., Bouchlas, D., and Farnham, W.,** Isolated parotid acinar cells from d,l-isoproterenol-treated rats. A cellular model for cystic fibrosis, *J. Dent. Res.*, 60, 80, 1981.

152. **Mangos, J. A., McSherry, N. R., Benke, P. J., and Spock, A.,** Studies of the pathogenesis of cystic fibrosis. The isoproterenol-treated rat as an experimental model, in *Cystic Fibrosis,* Proc. 5th Int. Conf., Lawson, D., Ed., Cystic Fibrosis Research Trust, London, 1969, 25.
153. **Martinez, J. R., Adelstein, F., Quissell, D., and Barbero, G. J.,** The chronically reserpinized rat as a possible model for cystic fibrosis. I. Submaxillary gland morphology and ultrastructure, *Pediatr. Res.,* 9, 463, 1974.
154. **Martinez, J. R., Adshead, P. C., Quissel, D. O., and Barbero, G. J.,** The chronically reserpinized rat as a possible model for cystic fibrosis. II. Composition and cilioinhibitory effects of submaxillary saliva, *Pediatr. Res.,* 9, 470, 1975.
155. **Martinez, J. R., Braddock, M. E., Martinez, A. M., and Cooper, C.,** Effect of chronic reserpine administration of K^+ and amylase release from the rat parotid gland, *Pediatr. Res.,* 13, 1124, 1979.
156. **Martinez, J. R. and Petersen, O. H.,** Impaired amylase release from the parotid gland of rats treated with reserpine, *Experientia,* 35, 1343, 1979.
157. **Morgan C. W., Bylund, D. B., Forte, L. R., and Martinez, J. R.,** Neurotransmitter regulation of the c-AMP enzymatic system in salivary glands, *Fed. Proc.,* 20(Abstr.), 1727, 1981.
158. **Roscher, A. A., Wiesmann, U. N., and Honegger, U. E.,** Changes in β-adrenergic receptors in submaxillary glands of chronically reserpine or isoproterenol-treated rats, *J. Pharmacol. Exp. Ther.,* 216, 419, 1981.
159. **Hokin, M. R. and Hokin, L. E.,** Enzyme secretion and the incorporation of P^{32} into phospholipids of pancreas slices, *J. Biol. Chem.,* 203, 967, 1953.
160. **Michell, R. H.,** Inositol phospholipids and cell surface receptor function, *Biochim. Biophys. Acta,* 415, 81, 1975.
161. **Michell, R. H.,** Inositol phospholipids in membrane function, *Trends Biochem. Sci.,* 4, 128, 1979.
162. **Michell, R. H., Jafferji, S. S., and Jones, L. M.,** Receptor occupancy dose-response curve suggests that phosphatidylinositol breakdown may be intrinsic to the mechanism of the muscarinic cholinergic receptor, *FEBS Lett.,* 69, 1, 1976.
163. **Michell, R. H., Jafferji, S. S., and Jones, L. M.,** The possible involvement of phosphatidylinositol breakdown in the mechanism of stimulus-response coupling at receptors which control cell-surface calcium gates, *Adv. Exp. Biol. Med.,* 83, 447, 1977.
164. **Weiss, S. J. and Putney, J. W.,** The relationship of phosphatidylinositol turnover to receptors and calcium ion channels in rat parotid acinar cells, *Biochem. J.,* 194, 463, 1981.
165. **Steiner, A. L., Parker, C. W., and Kipnis, D. M.,** Radioimmunoassay for cyclic nucleotides. I. Preparation of antibodies and iodinated cyclic nucleotides, *J. Biol. Chem.,* 25, 1106, 1972.
166. **Watson, E. L.,** Unpublished observation.

INORGANIC COMPONENTS OF HUMAN SALIVA

Ira L. Shannon

INTRODUCTION

When the biochemical constituents of saliva are to be discussed, it is first necessary to identify very specifically the particular component of the total oral fluid that is to be considered. Whole mixed saliva is very simple to collect by having the subject expectorate into a collection vessel or by removal of the fluid as it accumulates in the floor of the mouth. Quantitative biochemical measurements on this whole saliva are, however, open to criticism; the fluid contains not only the fluid emanating from the salivery glands, but also large numbers of bacterial cells, exfoliating cells from the host, products that may be entering the mouth from the gingival crevices, trace amounts of foods or liquids that have been ingested, and, in short, portions of anything that happens to be in the mouth at the time. While certain of these contaminating factors can be removed by centrifugation, there are unquestionably other contaminants that dissolve in the saliva and cannot be differentiated from the true salivary components. (It is therefore not possible to consider whole saliva as representing only the conglomerate of fluids emanating from the salivary glands.) For this reason, whole saliva is employed less and less in quantitative biochemical studies of saliva.

It is much more desirable to collect oral fluid from the individual salivary glands for chemical analysis. Devices have been designed for the collection of submandibular saliva from the human,[1-3] but there is the ever-present problem of to what degree the fluid from the sublingual gland has been included. It is also true that this submandibular-sublingual fluid is rather difficult to manage in the laboratory due to its physical characteristics.

Parotid saliva is by far the easiest to collect of the isolated salivary gland fluids. It is an excellent fluid for laboratory manipulations, and the collector has absolute assurance that he is obtaining his fluid from only a single gland source. Devices for the collection of parotid saliva are based upon the original designs of Carlson and Crittenden[4] in 1910. Several modifications have been made of the original collector[5-7] such that individualized collectors are now available that make it possible for a subject to collect his own parotid fluid without assistance during prolonged periods of isolation.[8] All of these devices are basically two-chambered caps, with the overall diameter varying from 2.0 to 2.5 cm. The innermost chamber is positioned over the orifice of Stensen's duct opposite the upper first molar, and the device is held in position by vacuum produced by a rubber bulb connected to the outer ring of the collector. As parotid saliva flows into the inner chamber of the device, it is directed to a collection vessel through plastic tubing. The collector may be left in position for several hours without difficulty, and saliva may be allowed to flow under resting (unstimulated) conditions or under any of many different exogenous stimulation schemes that have been employed. It is a great advantage that exogenous stimulants or drugs can be placed in the mouth during collection without fear of contamination of the parotid saliva samples.

The following discussion is limited to the major electrolytes of human parotid saliva. Emphasis will be placed upon quantitative measurements of each variable, the effect of flow rate on variable concentration, and a brief discussion of possible secretory mechanisms that are suggested by these quantitative observations.

EXOGENOUSLY STIMULATED PAROTID SALIVA

Eliciting agents of many types are employed to collect saliva. Gustatory stimulants are particularly popular if the secretion from an isolated gland is being collected. The first data to be presented on stimulated parotid fluid involved collection by the following procedure. Since it has been previously determined that peppermint-, cherry-, and grape-flavored candy drops, applied in that order, would produce an ascending rate of gland function, this was the pattern employed. A 5-min waste (accommodation) sample, which was discarded, under a particular stimulus was first collected, followed by a 10-min sample for analysis. The candy drop was replaced with a fresh one at 5-min intervals. The first experiment was designed to ascertain the effect of changes in stimulated rate of flow in 314 subjects on the concentration of various saliva constituent variables.

For these 314 subjects and with the peppermint candy drop stimulus, the mean rate of parotid flow was approximately 0.30 mℓ/min. There were highly significant increases in rate of flow when the peppermint, cherry (0.55 mℓ/min), and grape candy (0.80 mℓ/min) stimulants were applied in that order. Along with this increased rate of flow, there were significant ($p<0.01$) parallel increases in sodium, chloride, calcium, and bicarbonate concentrations. That is, as the volume of fluid exiting from the gland increased, the concentration of these constituents per unit of volume, also increased, the concentration of these constituents per unit of volume, also increased. Since the bicarbonate buffer system is dominant in controlling acid-base balance, both buffer capacity and pH increased with the bicarbonate concentration.

Significant ($p<0.01$) negative correlations were found between flow rate and phosphorus level and flow rate and magnesium concentration. To complete the circle of variation in flow rate effects, the level of potassium was found to be completely independent of flow rate variations.

It is thus clear that the biochemical constituents of blood and saliva behave quite differently. The inorganic constituents of blood plasma vary within only very narrow physiologic limits, and, even in disease, the mammalian system is not prepared to tolerate wide variations in these plasma factors; the saliva components, on the other hand, may have very wide physiologic ranges. In many instances these variations depend fundamentally upon the intensity of stimulation with which the fluid is collected. It is also intriguing that the concentrations of different variables in saliva may be increased, decreased, or unaffected by changes in rate of gland function. Certainly this is not a static system of simple conduits.

UNSTIMULATED PAROTID SALIVA

In the light of the foregoing examples of alterations in levels of inorganic components in stimulated parotid saliva as a function of flow rate, it is not surprising to find that there are rather remarkable constituent responses when this fluid is collected under unstimulated conditions. A dual chambered metal collection device[7] is placed over the orifice of Stensen's duct with an absolute minimum of manipulation. Subjects are required to sit quietly over the collection period. Every effort is made to collect that fluid which is, as nearly as possible, the true resting secretion of the gland. It was previously reported[9] that when collections were made from 4589 subjects, the flow rate mean was 0.040 mℓ/min (SD = 0.031). This is approximately 1/20th of a medium rate of stimulated flow from the same gland.

Table 1 provides data resulting from unstimulated parotid saliva sampling of relatively large groups of subjects. All participants were healthy young adult males who were screened medically prior to acceptance. Collections were undertaken at 7:00 a.m. with all subjects fasting. When the mean values in Table 1 are compared to those at the highest rate of stimulated flow in the previous experiment, distinct changes are evident. Along with this decreased rate of gland function, there were significant decreases in concentrations of sodium, chloride, and bicarbonate along with significant increases in levels of potassium, calcium, phosphorus, and magnesium. This is another demonstration of the ability of flow rate changes to affect the constituent variable concentrations in human saliva.

Table 1
PHYSIOLOGIC CONCENTRATIONS FOR INORGANIC COMPONENTS OF UNSTIMULATED PAROTID FLUID

Variable	Number of subjects	Mean	SD
Sodium (meq/ℓ)	513	2.61	2.00
Potassium (meq/ℓ)	513	36.7	12.51
Chloride (meq/ℓ)	527	24.8	7.64
Calcium (mg/100 mℓ)	512	3.02	1.00
Phosphorus (mg/100 mℓ)	1026	31.6	14.61
Magnesium (mg/100 mℓ)	312	0.26	0.11
Bicarbonate (meq/ℓ)	502	1.04	0.96

The effect of rate of flow of unstimulated parotid saliva on constituent concentration was then studied by appropriate grouping of the subjects. Resting parotid saliva was collected over a 90-min sampling interval and the number of subjects in each group ranged from 312 to 513. For this particular analysis the subjects in each group were divided into six groups, based entirely upon rate of flow for the single unstimulated parotid saliva sample given. The groups included subjects that produced resting saliva at rates of less than 0.021, 0.021 to 0.030, 0.031 to 0.050, 0.051 to 0.080, and more than 0.080 mℓ/min. At the lowest rate of flow, the potassium concentration was 42.0 meq/ℓ, and as flow rate increased, the potassium concentration decreased in stepwise fashion to a low of 25.5 meq/ℓ. Results for chloride were very similar to those for potassium: as flow rate increased the level of chloride decreased significantly from 29.7 to 17.3 meq/ℓ. Calcium exhibited a virtually identical pattern, with its level decreasing from 3.71 to 2.39 mg/100 mℓ; phosphorus likewise decreased from 49.1 to 19.5 mg/100 mℓ; magnesium fell from 0.41 to 0.12 mg/100 mℓ; and bicarbonate increased from 0.54 to 2.17 meq/ℓ as the rate of flow increased. This decrease in bicarbonate levels at the lower rates of flow is a straightforward expression of decreased glandular energy expenditure. The one constituent variable upon which the rate of unstimulated flow did not appear to exert a significant effect was sodium. Along with the virtual quadrupling of rate of flow, there was no significant change (2.61 meq/ℓ) in either direction in the sodium level. This freedom from flow rate effect was confirmed in two additional experiments and utilized as a principal basis for substantiating an alternate mechanistic concept of salivary glandular mechanics. In summary, when the rate of unstimulated parotid flow was used to divide patients into six groups, sodium was the only variable that did not change in concentration as flow rate shifted. The change in bicarbonate is explained as energy expenditure dependent and is thereby quite different from the other variables. The changes in potassium, chloride, calcium, phosphorus, and magnesium were unidirectional — as flow rate decreased the concentration of each was significantly elevated in step with the flow rate change. The first simple explanation for such a change might well be the reabsorption of water in excess of constituent reabsorption as the saliva remained in the duct system over these longer and longer periods of time (at lower and lower rates of flow).

Table 2
EFFECT OF ATROPINE ON PAROTID FLOW RATE AND ELECTROLYTES

Parotid flow (mℓ/min)					Parotid fluid constituent					
Before atropine		After atropine				Before atropine		After atropine		
Mean	SD	Mean	SD	P	Identification and unit	Mean	SD	Mean	SD	P
0.052	0.028	0.012	0.006	0.01	Sodium (meq/ℓ)	2.05	1.38	2.05	1.30	N.S.
0.052	0.028	0.012	0.006	0.01	Potassium (meq/ℓ)	27.7	6.24	43.0	11.00	0.01
0.052	0.028	0.012	0.006	0.01	Chloride (meq/ℓ)	18.5	3.78	30.0	7.83	0.01
0.068	0.032	0.020	0.010	0.01	Calcium (meq/ℓ)	1.71	0.56	2.46	0.88	0.01
0.068	0.032	0.020	0.010	0.01	Magnesium (meq/ℓ)	0.18	0.08	0.38	0.17	0.01
0.034	0.026	0.010	0.006	0.01	Phosphorus (mg/100 mℓ)	31.9	7.09	48.4	13.88	0.01
0.044	0.035	0.011	0.006	0.01	Bicarbonate (meq/ℓ)	0.77	0.92	0.24	0.31	0.01

ATROPINE EFFECTS

For the study of constituent changes as a result of flow rate alterations in exogenously stimulated saliva, it is quite simple to collect multiple samples at varying rates of flow from the same individual. Such an undertaking is not possible when unstimulated saliva is to be studied. Thus, extremely little work has been done on multiple unstimulated saliva samples collected from the same individual.

Since it was shown that sodium concentration behaved in a fashion all its own in between-subject studies of resting parotid saliva, it was of interest to check the response of this variable in consecutive unstimulated saliva collected at differing rates of flow from the same subject. After an appropriate accommodation period, a 45-min control sample was collected at 7:30 a.m. from each healthy young adult male subject. Every effort was made to collect, as nearly as possible, the true resting secretion of the gland. Each subject was then administered 1/50 grain of atropine sulfate by mouth and, after a 45-min waiting period, a second 45-min unstimulated sample was collected in the same fashion. Twenty subjects provided pairs of samples that were analyzed for sodium, potassium, and chloride; 20 more for calcium and magnesium; a like number for phosphorus; and a final 20 for bicarbonate concentration. Thus, 80 subjects participated in this phase of the experiment.[10]

The results are presented in terms of alterations in flow rates and concentration of the particular constituents as a result of atropine administration (Table 2). Probability values are also presented. In all instances there was a significant ($p<0.01$) decrease in parotid flow associated with the atropine dosage. Along with these decreases in gland function, potassium concentration increased from 27.7 to 43.0 meq/ℓ, chloride level was elevated from 18.5 to 30.0 meq/ℓ, calcium increased from 1.71 to 2.46 meq/ℓ, and magnesium rose from 0.18 to 0.38 meq/ℓ. These changes are, of course, in exact accord with the findings of the previous experiment. Further, bicarbonate concentration decreased significantly from 0.77 to 0.24 meq/ℓ, a result attributable to a decrease in glycolysis resulting from the drug. Osmolality, specific gravity, and total solids concentration all increased significantly as flow rate decreased following atropine dosage. In this experiment, as in the former work with unstimulated saliva, a significant drug-induced depression in parotid flow rate (from 0.052 to 0.012 mℓ/min) induced no change in either direction in the concentration of sodium in the exiting fluid. The level was approximately 2.0 meq/ℓ both before and after atropine dosage. This work was extended to include many additional saliva components, and when the work was complete, a tentative hypothesis was developed that these concentration increases seen with flow rate decreases must result from water reabsorption primarily. The fact that sodium concentration was unaffected by these same flow rate changes suggested that it was reabsorbed in proportion to water reabsorption; the premise could also be that water reabsorption was dependent upon and proportional to the active reabsorption of sodium.

While the previous work with atropine sulfate-induced flow rate depression reinforced the points supporting proportionate reabsorption of water and sodium, it was true that the basic physiologic state had certainly been altered by the administration of this potent parasympatholytic agent. There was a need for a system to reduce unstimulated parotid flow significantly without such alteration. A previous study[11] indicated a suppressant effect on parotid flow rate by light deprivation which permitted experiments designed to explore the development of a physiologic system for modifying unstimulated flow.

Parotid saliva samples (20-min resting) were collected under routine laboratory conditions from trained donors. A blindfold was positioned and a second sample was taken under other identical conditions. This sampling was repeated in this order for 5 days and then it was reversed (blindfolded period first) over a second 5 days. Overall mean flow rate without the blindfold was 0.057 mℓ/min (SD = 0.027); with the blindfold in position the flow rate mean decreased significantly to 0.014 mℓ/min (SD = 0.010). For all subjects, no matter if blindfolded during the first or second collection period, there was a highly significant flow rate depression associated with light deprivation.[12] This observation was confirmed in additional work[13,14] and the model was utilized to study possible changes in electrolyte levels as a result of darkness-induced flow rate depression. In this case room darkening rather than blindfolding was employed. The control flow rate mean for the same five subjects under conditons as outlined above was 0.056 mℓ/min (SD = 0.028) and it decreased significantly ($p<0.01$) to 0.028 mℓ/min (SD = 0.016) under darkened room conditions. With this decrease in gland function, the potassium concentration increased from 19.9 meq/ℓ (SD = 0.01) to 38.3 meq/ℓ (SD = 3.71), and chloride increased from 19.3 meq/ℓ (SD = 0.96) to 24.9 meq/ℓ (SD = 2.02). Each of these changes was highly significant. The mean for sodium concentration under routine lighting was 5.8 meq/ℓ (SD = 0.18); with light deprivation the mean was 6.0 meq/ℓ (SD = 3.57). The difference between these sodium means was not significant. It was thus reaffirmed without the use of a pharmacologic agent that alterations in rate of flow of resting parotid saliva did not alter sodium levels in this fluid in either direction.

These light deprivation studies provided the following information: blindfolding consistently produced a significant depression in flow rate which was also observed when room darkening was used as a light deprivation method; when blindfolding was introduced during gustatory stimulation of secretion, a significant decrease in flow rate was observed but the relative magnitude of the effect was far less than that seen under resting conditions. The question arises as to the neuroanatomical pathways involved in this effect of light deprivation on parotid flow rate.

It has been accepted over the years that parotid gland secretory function is under purely parasympathetic control.[15] The flow-controlling parasympathetic reflex is well authenticated, the anatomy of this arc extending from the cell body of the parasympathetic neuron in the inferior salivary nucleus to the postsynaptic neuron located on or very near the gland through the auriculotemporal nerve. The sympathetic innervation of the parotid originates at the T1-T2 level, follows the sympathetic chain to the superior cervical ganglion, and, from this point fibers pass along the blood vessels to the parotid gland.

It is probable that sympathetic innervation is important in the control of the resting function of the human parotid gland. Photic stimulation of retinal receptors is also responsible for a component of sympathetic neuronal activity that plays a part in this control. There are two major reasons for this suggestion. First, specific sympathetic pathways are known to exist from the retina to the parotid gland by way of the superior cervical ganglion. Secondly, there are several striking similarities in the innervation and light-induced cyclic changes between the parotid and the pineal gland. The pineal is a neuroendocrine organ controlled by light through nerve fibers which leave the main visual pathway at the optic chiasma to travel via the inferior optic tract to the nucleus of Bochenek, and thence down the brain stem and spinal cord to sympathetic cell bodies in the intermediolateral nucleus of T1 and T2. From there, impulses may be relayed to the superior cervical ganglion, and to the pineal gland which is richly supplied with sympathetic nerves passing directly from the superior cervical ganglion. The superior cervical ganglion is also the source of sympathetic postganglionic neurons which innervate the salivary glands.[16] Thus, both the pineal and the parotid receive sympathetic postganglionic neurons from the superior cervical ganglion.

Wurtman et al.[16] have demonstrated an impressive diurnal parallel in catecholamine content of the pineal and salivary glands. Their results have been confirmed and extended by Moore and Smith.[17] In rats, maintained in diurnal lighting, norepinephrine content of both the pineal and the salivary glands was highest at the end of a dark period and lowest at the end of a light period. When the animals were placed in constant darkness, the norepinephrine levels remained high, whereas in constant light they remained low.

It is thus suggested that the darkness-induced flow rate depression is mediated through the sympathetic nervous system via pathways similar to those established for the pineal. Additional findings have suggested the possibility that such sympathetic efferents may be, in part, cholinergic.

When the process of fluid and electrolyte secretion is examined it becomes pertinent to determine its durability. Are there significant changes in parotid flow rate or in parotid saliva electrolyte levels as stimulation is prolonged? In this experiment nine healthy young adult males served as subjects. Parotid saliva samples were taken under the stimulus of sour grape candy drops and, after the first 10-min sample was discarded, each subject provided a series of nine 20-min collections. The candy drop was replaced with a fresh one at 5-min intervals. Mean parotid flow for all samples for all nine subjects was 0.591 mℓ/min (SD = 0.294). There was no significant change in rate of flow over the prolonged sampling period. Neither was there a significant alteration in concentration for either sodium, potassium, chloride, calcium, phosphorus, magnesium, or bicarbonate. Clearly, the fluid generation system for the human parotid is sufficient to support steady stimulated secretion over an extended time period; likewise the secretion mechanism exhibits no difficulty in maintaining a constant level of output over this same 3-hr sampling interval.

MECHANISMS FOR ELECTROLYTE SECRETION

It was pointed out by Heidenhain[18] over 100 years ago that the concentration of salts in saliva of the dog increased as the rate of function was increased. In 1954, Thaysen et al.[19] developed a hypothesis to explain these concentration changes. They collected human parotid saliva under parasympathomimetic stimulation and presented their classic observations on sodium, potassium, chloride, and bicarbonate concentrations. They found that levels of sodium, chloride, and bicarbonate increased with increases in flow and that potassium was independent of flow rate changes. It was proposed that the original saliva was formed in the acinar portion of the gland and that its composition was unchanged by intensity of stimulation. Sodium appeared to be reabsorbed in the duct system and since this reabsorption mechanism was of limited magnitude, increasing gland stimulation overwhelmed its potential and sodium concentration increased.

Micropuncture and microcannulation studies on animals have, in general, supported such a mechanism. Such studies have been conducted[20-22] and have indeed provided presumptive evidence for ion transport secretory systems at specific sites in the gland. These studies indicate that there is a fluid present in the acinar-intercalated duct region of the gland that is isosmotic and contains plasma-like levels of sodium and potassium. It is indicated further that in the ducts sodium is reabsorbed in excess of water to produce a hypotonic, low sodium fluid.[23]

The exact degree to which animal study results will extrapolate to the human is subject to question. It is known that tremendous interspecies differences exist or, indeed, are characteristic in salivary glands. A few examples will suffice to make the point: Stimulation of the sympathetic nerves induces a brisk and copious flow from the submandibular gland of the cat[24] but no flow at all from the parotid of the dog.[25] Continuous spontaneous secretion is characteristic of the bovine parotid but not the submandibular gland in this same species. Spontaneous secretion takes place from the parotid of ruminants but not from other types of animals. In cats, dogs, and rats the sublingual gland secretes continuously but in the rabbit it is the submandibular that gives this continuous flow. Such spontaneous secretion persists even under general anesthesia.[26] Differences in salivary composition between species are legend; some of the major salivary components in man are either extremely low or missing completely from the saliva of animals that are used very frequently in glandular experiments. Certainly, care must be exercised in moving without experimental verification from species to species in predicting glandular mechanics.

In contrast to extrapolating data from animals to the human, there is the process of developing models based upon inferences from actual human data. The latter is the procedure that was employed. The premise was not accepted that, in the human, the acinar fluid is isotonic and contains plasma-like concentrations of electrolytes. Rather, it was believed that there are fluid-secreting cells in the human salivary gland systems that elaborate a hypotonic solution initially. Neither was it accepted that reabsorption of sodium in the human parotid is necessarily similar to that of the submandibular gland of the rat — that hypotonicity results in that system from reabsorption of sodium in excess of water. The model proposed is that in the human the reabsorption of sodium is in proportion to water removal from the fluid passing through the gland.

Fundamental is the disagreement on how the salivary glands are able to secrete a fluid that is hypotonic to plasma. The assumption over the years has been that the acinus is the primary secretory site for both fluid and electrolytes and that, as previously stated, this is an isotonic fluid that is acted upon as it passes through the gland toward the orifice. Even if it is granted that the samples collected in micropuncture and microcannulation studies on salivary gland of experimental animals[20,22,23,27-39] do represent pure acinar or intercalated duct fluid, unchanged in any way by the collection procedure itself, all of the problems are not solved. The two chief problems are the assumption the acinus is the primary fluid-secreting site in the gland, and that the actual parenchymal site from which the sample was collected was behaving during the collection process exactly as were other similar but undisturbed parenchymal sites with the gland during secretion.

Acceptance of this mechanism was widespread long before the micropuncture experiments mentioned above. If the concept of an isosmotic primary fluid of secretion is accepted, then sodium reabsorption in excess of potassium secretion is a satisfactory explanation for the ever-present hypotonicity of saliva. A point in favor is that most glands and organs that are capable of transcellular fluid transfer do produce or transfer an isotonic fluid. This is, however, not an invariable rule; the hypotonicity of the secretion of gastric glands and the hypertonicity of fluid secretion by salt glands in birds and reptiles, are well established.

Primarily as a working hypothesis to stimulate further research, the possibility was suggested that in the human parotid system the primary fluid of secretion is hypotonic when secreted, and that this site of secretion is probably in the striated duct system of the gland. There is, of course, no completely authenticating evidence that this is true, but by inferential reasoning based upon the human data, a rather strong case can be made.

Osmolality of saliva is very important in the hypothesis and this variable deserves primary attention. When stimulated parotid saliva is collected at three distinct rates of flow, in ascending order, the degree of hypotonicity of this fluid can be appreciated. For instance, it was seen that the mean for osmolality reached a high value of 127 mOsm/kg. The accepted normal for osmolality of plasma is about 290 mOsm/kg with a range of from about 281 to 297 mOsm/kg.

Since a positive correlation between stimulated flow and osmolality was observed, studies were undertaken to increase rate of flow even higher to see how osmolality responded. Eleven individuals were selected from a larger group because of their propensity for rapid saliva generation. A total of 1100 samples (grape candy drop stimulated) was collected from these individuals during a 4-week period. Mean flow rate was 2.31 mℓ/min (SD = 0.88) and the mean for osmolality was 135.8 mOsm/kg (SD = 32.7).[37] The subject with the highest flow rate mean for his 100 samples also had the highest osmolality mean, but this figure was only 180.4 mOsm/kg. Thus, even at these extremely high rates of flow the osmolality of parotid saliva remains in the area of only 60% of that found for plasma.

At these relatively high rates of gland function, the "contact time" available for the luminal column of fluid to undergo constituent alteration across duct cell membranes becomes much shorter; therefore, the fluid collected at the duct orifice under these conditions should resemble actual cellular secretion more than a lower flow sample. This preliminary piece of evidence is not offered as proof that saliva, when initially secreted, is not isosmotic with plasma. Rather, it is offered to point out that saliva that remains in the duct system for an infinitely short time is still remarkably hypotonic. This, at least, gives a basis for questioning the concept that the saliva was necessarily isosmotic when initially secreted and suggests the possibility of the secretion of a hypotonic primary fluid.

It has been shown that highly significant positive correlations exist between flow rate and total solids, specific gravity, and osmolality in stimulated parotid saliva. With this point in mind, and with the restriction that only one sample can be collected from each individual, a similar approach can be taken to examination of data on saliva collected completely without exogenous stimulants. In experiments designed to obtain data for osmolality in unstimulated saliva, the individuals were separated into five distinct flow rate groups. This makes possible only the estimation of between-subject correlation coefficients since each subject provided only one sample, but there are significant implications in these observations. Similar results are obtained when specific gravity or total solids are plotted against flow rate. The important observation is that the positive correlations with flow noted in stimulated saliva have been completely reversed at these very low rates of gland function and that highly significant ($p<0.01$) negative correlations are present. Since these measurements are nonspecific and subject to influences of the individual variables in saliva, an experiment was designed to study the responses of the individual constituents at resting and lower rates of flow.

A total of 300 healthy young adults participated in an experiment directed toward ascertaining the effect of an atropine-induced flow rate depression on the constituents of human resting parotid saliva.[38]

The oral administration of 1/50 grain of atropine sulfate brought about a highly significant ($p<0.01$) decrease in flow and associated significant ($p<0.01$) increases in total solids, specific gravity, and osmolality. As flow rate was decreased from 0.053 mℓ/min (SD = 0.031) to 0.023 mℓ (SD = 0.014), the osmolality mean increased from 48.6 mOsm/kg (SD = 15.8) to 83.5 mOsm/kg (SD = 35.8). Constituent variables that were measured also increased significantly ($p<0.01$) in concentration along with the atropine-induced decrease in flow. The only chemical variable that remained constant despite the flow rate change was sodium. As flow rate fell from 0.052 mℓ/min (SD = 0.028) to 0.012 mℓ/min (SD = 0.006), the concentration of sodium remained constant at 2.05 meq/ℓ.

The experiments with atropine-induced flow rate depression and with a similar state produced by light deprivation, have made it possible for us to develop a picture of the very rapidly changing nature of human parotid secretion over the minute but critically important low range of physiologic flow rates. The single most important characteristic about the changing nature of the secretion is that virtually all of the constituents become diluted as flow rate increases in this low range of flow. This fact is reflected and verified by the associated decreases in osmolality, total solids, and specific gravity. The other important observation is that the sodium concentration remains constant.

These facts together with similar information on human parotid saliva constituents during reflex stimulated secretion and also with the results of the animal studies conducted by Suddick et al. suggesting that intraglandular vascular pressure provides an energy source for fluid secretion[40,41] has led to the proposal of ideas and models on fluid and electrolyte secretion that diverge sharply from widely held and long cherished traditional views.[42] The essence of these concepts is that the fluid-secreting tubules are able to elaborate a hypotonic electrolyte solution directly (as opposed to the concept that a primary isotonic acinar secretion is modified in transit through the ducts to hypotonicity as a result of sodium reabsorption). From the viewpoint of membrane transfer and energetics, this concept is probable only if the glandular vascular pressure provides energy for fluid transfer in the human parotid gland in the same manner which studies by the author suggest is true for the submandibular gland of the rat.[40-42] However, the foregoing observations, (1) that a dilution of constituents proceeds as secretion rate increases within the resting flow range; (2) that the fluid becomes even more hypotonic as flow rates increase; and (3) that the sodium concentration does not change over this same flow rate range, all support the concept of the primary secretion of a hypotonic fluid.

We realize that these observations are not beyond question and that different interpretations may indeed later prove appropriate. We do not believe that a concept is beyond question simply because it becomes a ''classic'' in the minds of physiologists. At the risk of being branded heretics, we have gone so far as to question the interpretation of Ludwig's classic work with the dog submaxillary gland. In 1851, he found that the pressure generated when the gland was stimulated to secrete may be higher than the arterial pressure as measured by direct cannulation of the femoral artery. The classic interpretation has been that this demonstrated that secretion is an active glandular process and that secretion does not derive its fundamental energy from the glandular vascular pressure. Although several questions might be directed toward the experiments themselves, we believe that long ago someone should have asked: Why does secretion progressively diminish and then cease just at or very near the measured femoral artery pressure? Why doesn't secretory pressure far exceed or fall far below femoral artery pressure? Is it merely a coincidence that the secretion of saliva in the dog produces pressures that always seem to be very near to the femoral artery pressures?

Does the asking of these questions affirm that Ludwig's conclusions are necessarily inaccurate and that we have been laboring under an erroneous concept for more than a century? Certainly not, but we do think that they are pertinent questions and further, that conclusions able to stand the test of time can be built only on frameworks that can meet the same criteria. It is in this light that we offer this concept of primary secretion of a hypotonic saliva in the human parotid gland.

REFERENCES

1. **Truelove, E. L., Bixler, D., and Merritt, A. D.,** A simplified method for collecting submandibular saliva, *J. Dent. Res.*, 46, 1400, 1967.
2. **Schneyer, L. H.,** Method for collection of separate submaxillary and sublingual salivas in man, *J. Dent. Res.*, 34, 257, 1955.
3. **Block, P., and Brothman, S.,** A method of submaxillary saliva collection without cannulation, *N.Y. State Dent. J.*, 28, 116, 1962.
4. **Carlson, A. J. and Crittenden, A. L.,** The relation of ptyalin concentration in the diet and the rate of secretion of the saliva, *Am. J. Physiol.*, 26, 169, 1910.

5. **Lashley, K. S.,** Reflex secretion of the parotid gland, *J. Exp. Psychol.,* 1, 461, 1916.
6. **Shannon, I. L., Prigmore, J. R., and Chauncey, H. H.,** Modified Carlson-Crittenden device for the collection of parotid fluid, *J. Dent. Res.,* 41, 778, 1962.
7. **Shannon, I. L., and Chauncey, H. H.,** A parotid fluid collection device with improved stability characteristics, *J. Oral Ther. Pharmacol.,* 4, 93, 1967.
8. **Shannon, I. L. and Terry, J. M.,** A self-positioning device for the collection of parotid fluid from isolated human subjects, *J. Dent. Res.,* 44, 126, 1965.
9. **Shannon, I. L.,** A formula for human parotid fluid collected without exogenous stimulation, *J. Dent. Res.,* 46, 309, 1967.
10. **Shannon, I. L., Suddick, R. P., and Chauncey, H. H.,** Effect of atropine-induced flow rate depression on the composition of unstimulated human parotid fluid, *Arch. Oral Biol.,* 14, 761, 1969.
11. **Barylko-Pikielna, N., Pangborn, R. M., and Shannon, I. L.,** Effect of cigarette smoking on parotid secretion, *Arch. Environ. Health,* 17, 731, 1968.
12. **Shannon, I. L., Feller, R. P., and Suddick, R. P.,** Light deprivation and parotid flow in the human, *J. Dent. Res.,* 51, 1642, 1972.
13. **Shannon, I. L., and Suddick, R. P.,** Effects of light and darkness on human parotid saliva flow rate and chemical composition, *Arch. Oral Biol.,* 18, 601, 1973.
14. **Shannon, I. L. and Feller, R. P.,** Effects of light of specific spectral characteristics on human resting parotid gland function, *Arch. Oral Biol.,* 19, 1077, 1974.
15. **Brown, C. C.,** The parotid puzzle: a review of the literature on human salivation and its applications to psychophysiology, *Psychophysiology,* 7, 65, 1970.
16. **Wurtman, R. J., Axelrod, J., and Kelley, D. E.,** *The Pineal,* Academic Press, New York, 1968, 108 and 139.
17. **Moore, R. Y. and Smith, R. A.,** Postnatal development of a norepinephrine response to light in the rat pineal and salivary glands, *Neuropharmacology,* 10, 315, 1971.
18. **Heidenhain, R.,** Ueber die Wirkung einiger Gifte auf die Nerven der glandula submaxillaris, *Arch. Ges. Physiol.,* 5, 309, 1872.
19. **Thaysen, J. N., Thorn, N. A., and Schwartz, J. L.,** Excretion of sodium, potassium, chloride and carbon dioxide in human parotid saliva, *Am. J. Physiol.,* 178, 155, 1954.
20. **Holtzgreve, H., Martinez, J. R., and Vogel, A.,** Micropuncture and histologic study of submaxillary glands of young rats, *Pfluegers Arch.,* 290, 134, 1966.
21. **Mangos, J. A., Braun, G., and Hamann, K. F.,** Micropuncture study of sodium and potassium excretion in the rat parotid saliva, *Pfluegers Arch.,* 291, 99, 1966.
22. **Martinez, J. R., Holtzgreve, H., and Frick, A.,** Micropuncture study of submaxillary glands of adult rats, *Pfluegers Arch.,* 290, 124, 1966.
23. **Young, J. A., Fromter, E., Schogel, E., and Hamann, K. F.,** Micropuncture and perfusion studies of fluid and electrolyte transport in the rat submaxillary gland, in *Secretory Mechanisms of Salivary Glands,* Schneyer, L. H., and Schneyer, C. A., Eds., Academic Press, New York, 1967, 11.
24. **Langley, J. N.,** The salivary glands, in *Textbook of Physiology,* Schaefer, Pentland, Edinburgh, 1898, 475.
25. **Babkin, B. P.,** *Secretory Mechanism of the Digestive Glands,* 2nd ed., Hoeber, New York, 1950.
26. **Emmelin, N. and Zotterman, Y.,** *Oral Physiology,* Pergamon Press, New York, 1972.
27. **Mangos, J. A. and Braun, G.,** Micropuncture study of the rat parotid gland, in *Moderne Probleme der Paediatrie,* Vol. 10, Rossi, E. and Stroll, E., Eds., S. Karger, Basel, 1967, 107.
28. **Mangos, J. A. and Braun, G.,** Excretion of total solute, sodium and potassium in the saliva of the rat parotid gland, *Pfluegers Arch.,* 290, 184, 1966.
29. **Martin, C. J. and Young, J. A.,** A micropuncture investigation of the effects of sympathomimetic and a parasympathomimetic drug on water and electrolyte fluxes in the main excretory duct of the submaxillary gland, *Pfluegers Arch.,* 327, 303, 1971.
30. **Martinez, J. R.,** Water and electrolyte secretion by submaxillary gland, in *Exocrine Glands,* Botelho, S. Y., Brooks, E. P., and Shelley, W. B., Eds., University of Pennsylvania Press, Philadelphia, 1969, 20.
31. **Martinez, J. R.,** Action of ouabain on submaxillary secretion in the dog, *J. Pharmacol. Exp. Ther.,* 178, 616, 1971.
32. **Martinez, J. R. and Ullrich, K. J.,** Micropuncture studies of rat submaxillary gland, in DiSant'Agnese Research on Pathogenesis of Cystic Fibrosis, National Institute of Arthritis and Metabolic Diseases, Bethesda, Md., 1964, 129.
33. **Schogel, E. and Young, J. A.,** Micropuncture and perfusion investigation of sodium and potassium transport in the rat submaxillary gland, *J. Physiol. (London),* 183, 73, 1966.
34. **Young, J. A.,** Microperfusion investigation of chloride fluxes across the epithelium of the main excretory duct of the rat submaxillary gland, *Pfluegers Arch.,* 303, 366, 1968.
35. **Young, J. A., Fromter, E., and Schogel, E.,** Electrolyte fluxes and potential differences in the main excretory duct of the rat submaxillary gland, *Pfluegers Arch.,* 289, 81, 1966.

36. **Young, J. A., Fromter, E., Schogel, E., and Hamman, K. F.,** A microperfusion investigation of sodium reabsorption and potassium secretion by the main excretory duct of the rat submaxillary gland, *Pfluegers Arch.,* 295, 157, 1967.
37. **Young, J. A., Fromter, E., Schogel, E., and Hamman, K. F.,** Micropuncture and perfusion investigation of active sodium and potassium transport in the rat submaxillary gland, *Bull. of the Post-Graduate Committee in Med., University of Sydney,* 5, 23, 181, 1967.
38. **Young, J. A. and Martin, J. C.,** Effects of sympatho. and parasympathomimetic drugs on the electrolyte concentration of primary and final saliva in the rat submandibular gland, *Pfluegers Arch.,* 327, 285, 1971.
39. **Young, J. A. and Schogel, E.,** Micropuncture investigation of sodium and potassium excretion in rat submaxillary saliva, *Pfluegers Arch.,* 291, 85, 1966.
40. **Han, S. S., Sreebny, L., and Suddick, R. P., Eds.,** Secretion?, in *Symposium on the Mechanism of Exocrine Secretion,* University of Michigan Press, Ann Arbor, 1973.
41. **Suddick, R. P., Dowd, F. J., and Shannon, I. L.,** Effects of arterial and venous occlusion on submaxillary gland secretion in rats, *Arch. Oral Biol.,* 16, 509, 1971.
42. **Shannon, I. L., Suddick, R. P., and Dowd, F. J.,** *Saliva: Composition and Secretion (Monograph in Oral Science),* S. Karger, Basel, 1974, 1.

HUMAN GLANDULAR SALIVARY PROTEINS

Donald I. Hay

INTRODUCTION

Electrophoretic studies have shown that human salivary secretions contain as many as 30 to 40 individual macromolecular constituents.[1,2] A few of these are multiple forms of the same protein,[3] but most are unique macromolecules. Some are well characterized, their functions have been established, and they have analogues in other body secretions (e.g., amylase, immunoglobulin A, lysozyme, lactoperoxidase, and lactoferrin). Functions for others have been proposed and they appear to be unique to saliva (e.g., the anionic proline-rich proteins and statherin). Also, biological functions have been attributed to other salivary macromolecules, but these have yet to be completely identified and characterized (e.g., proteins responsible for bacterial aggregating activity or those mediating attachment to the tooth surface). In addition, there remain many salivary macromolecules for which no function is known (e.g., the "basic" proline-rich glycoprotein, many enzymes, and numerous small basic and acidic proteins and peptides). Since the mouth represents a unique interface with the body's external environment, it seems likely that specialized protective functions remain to be discovered which will prove to be associated with those macromolecules for which no function has yet been proposed.

This review will focus on those human salivary proteins, present in the glandular secretions, for which functions are known or appear to be established. An exception to this is secretory immunoglobulin A, which has been the subject of a recent comprehensive review.[4] Enzymes and other proteins with no known function, or which are poorly characterized, will be discussed in less detail. Recent reviews[5,6] of salivary constituents cover additional material. Entities such as growth factors, enzymes absent from human saliva, such as ribonuclease and protease present in mouse submandibular glands, and polypeptides of general physiological significance, such as the kallikrein-kinin system, have also been the subjects of recent reviews[7-12] and will not be discussed here.

α-AMYLASE

α-Amylase hydrolyzes α-1,4 glucan bonds in polysaccharides containing three or more α-1,4 linked D-glucose residues. The products are glucose and maltose, providing the enzyme is present at sufficiently high concentrations (e.g., at salivary levels), otherwise maltose and maltotriose are the major end-products.[13] Amylase has a digestive function and has also been shown to have antibacterial activity against a few specific organisms. Presumably, it also plays a role in removing carbohydrate debris from the teeth. Amylases are widely distributed in animals, plants, and lower organisms and have been the subject of previous reviews.[14,15]

The amylolytic activity of saliva, recognized over 150 years ago,[16] was until recently considered to be the most important attribute of this secretion. The human salivary enzyme has been isolated and crystallized.[17] Although the crystalline enzyme is homogenous in the ultracentrifuge, it was separated into at least two fractions by moving boundary electrophoresis.[18] Further studies of this heterogeneity revealed other isozymes.[19] The basis of this molecular heterogeneity was found to be due to the occurrence of glycosylated and nonglycosylated forms of the enzyme and to variations in the degree of amidation of the two kinds of isozymes.[20,21] Although isozymes of differing electrophoretic mobilities appeared to be normal salivary constituents, transformation of isozymes to more acidic forms could be experimentally induced. It was shown that deamidation reactions were responsible,[21] a proposition which was supported by results of studies of pancreatic[22] and salivary isozymes[23] and by the specific deamidation of amylase by peptido-glutaminases.[24]

The carbohydrate side chains of the glycoamylases are asparagine-linked,[21] and their structures have been investigated.[25,26] Enzymatic properties are apparently unaffected by glycosylation or by the naturally occurring variations in amidation. The significance of these variations is unknown. It has also been found that neonate salivary amylase consisted of more anionic forms that the adult enzyme.[27]

Genetic studies indicate that salivary amylase is coded for by a single gene locus and that the isozymes are posttranslationally derived by deamidation and glycosylation.[28] Electrophoretically detected genetic variants are rare, but three phenotypes have been detected by differential inhibition of the enzyme.[29] The pancreatic enzyme occurs as a glycoprotein and is coded for at a single locus distinct from that of salivary amylase. Although the amino acid composition of the two enzymes is similar, a significant difference in the primary structure has been inferred from peptide maps.[30] Radioimmunoassay indicates that in the human the two enzymes cross-react only slightly,[31] confirming an earlier observation.[32] The molecular weights of the glycosylated and nonglycosylated human salivary enzymes have been reported to range from 62,000 to 63,000 and 56,000 to 59,000 daltons, respectively,[21,26] and the isoelectric points of the isozymes have been shown to range from 5.7 to 6.88.[33]

α-Amylases investigated so far contain at least one tightly bound calcium ion[34] which is considered to play a key role in maintaining the molecular configuration of the enzyme and enzymatic activity.[35] Removal of the ion leads to a reversible loss of activity, loss of stability, and an increased vulnerability to proteolysis. Historically, pancreatic amylase was the first enzyme found to be modulated by an effector ligand,[36] chloride in this case. It is also of interest that amylase consists of a single polypeptide chain, so the allosteric binding site is present on the same chain as the active site, unlike oligomeric enzymes. The binding of a chloride ion to a single binding site on the amylase molecule strongly activates the enzyme, simultaneously causes a 240-fold increase in the affinity of amylase for its bound calcium ion, and reduces proton exchange between the enzyme and solvent.[37] These results indicate a conformational change, but this was too subtle to be detected by fluorescence or circular dichroism measurements. Structural studies of human salivary amylase have not been done, though the three-dimensional crystal structure of the porcine pancreatic enzyme[38] and its partial amino acid sequence[39] have been reported recently.

α-Amylase concentrations in the saliva of neonates is low, but rises sharply at 0.9 to 1.9 years, and reaches adult levels at 5 to 6 years.[40] Amylase is considered to be a digestive enzyme, continuing to act in the stomach until inactivated by gastric secretions. It is difficult to understand, therefore, why amylase should be present in tears,[41] serum, urine and milk,[42] and genital secretions.[43] The observation that the growth and survival of 13 strains of *Neisseria gonorrhoeae* are inhibited by salivary amylase,[44] its widespread distribution in the body, and its synthesis alongside other macromolecules with antibacterial activity suggests the possibility of other functions for this enzyme.

LACTOFERRIN

Lactoferrin is a red, iron-binding protein, synthesized by acinar epithelial cells and neutrophils. It binds ferric ions, thus depriving microorganisms of this essential nutrient. It may also inhibit microbial growth or kill microorganisms by other mechanisms which are not yet understood. Some organisms are able to compete with lactoferrin for the available Fe^{3+}.

Lactoferrin was first identified in milk.[45] It was isolated from bovine[46] and human milk,[47-51] identified in saliva,[52] and its presence in salivary glands and other exocrine secretions was demonstrated by immunological methods.[53,54] Lactoferrin resembles human transferrin in its ability to bind two atoms of iron per mole, with the simultaneous binding of 2 mol of bicarbonate. Both proteins consist of a single polypeptide chain, with a molecular weight of 76,000, and are clearly related to one another. They differ, however, in that they show no immunological cross-reactivity unless denatured, and lactoferrin retains its iron-binding properties to below pH 4, under which conditions transferrin releases its bound iron.[55-57] The isoelectric points of lactoferrin and transferrin are 9 and 5.9, respectively.[58]

Most studies involving purified lactoferrin have used material prepared from milk, and several sophisticated preparative methods have been developed.[59,60] There appears to be only one report of the isolation of the salivary analogue.[61] Extensive studies of the role of lactoferrin in mammary secretions, and its role in protection and nutrition of the neonate have been conducted and recently reviewed.[62] Lactoferrin is secreted to a large extent as the iron-free apo-protein, and its secretion is greatly enhanced in some disease states.[63,64] As a consequence of its iron-binding properties, it inhibits growth of microorganisms by depriving them of iron. This activity has been demonstrated with many organisms,[65-67] including *Streptococcus mutans*, and is readily reversed by saturating the molecule with iron.[68,69] This ability of the host to withhold nutrients has been termed ''nutritional immunity''.[70,71] It also appears that the activity of lactoferrin may go beyond effects attributable to simple chelation of iron since it can exert a bactericidal effect at levels less than those required to bind the available iron.[72] This effect was only observed with the iron-free molecule, the iron-saturated molecule was inactive. The mechanisms involved in this kind of bactericidal activity are not yet understood, but apo-lactoferrin became bound to the bacterial cell surfaces in these experiments, while iron-saturated lactoferrin did not. Apparently, access to the cell surface is important in this kind of killing, since sucrose-grown *S. mutans* was less sensitive than the glucose-grown organism. Resistance to lactoferrin activity can be induced in some organisms,[73] and conditions of lactoferrin treatment strongly affect the extent to which organisms are killed.[74] Organisms which resist lactoferrin activity may be capable of synthesizing iron-binding molecules (enterochelins) which successfully compete with lactoferrin for the available iron,[75,76] a property which appears to be associated with virulence.

An enzyme-linked immunosorbent assay (ELISA) technique for determination of lactoferrin has been developed[77] and compared with other methods. While the latter methods have been satisfactory for some investigations,[63,64] it was pointed out that salivary lactoferrin concentrations tend to be less than 0.01 mg/mℓ and the more sensitive ELISA technique is to be preferred.

LACTOPEROXIDASE

Salivary lactoperoxidase (LPO) and thiocyanate, acting together with microbially produced hydrogen peroxide, yield oxidized thiocyanate ion derivatives which are toxic to many microorganisms. The toxic products affect both the original hydrogen peroxide-producing and adjacent organisms, resulting in diminished bacterial activity.

The original observations of this enzyme came from studies of milk,[78-80] and the inhibitory effects of saliva on growth and metabolic activity of lactobacilli.[81,82] Subsequent studies[83-85] showed the inhibitory system to consist of a heat-stable, dialyzable factor and a heat-labile macromolecule. The cofactor was shown to be thiocyanate,[86,87] and the macromolecule proved to be lactoperoxidase,[88] identical to the enzyme previously isolated from bovine milk,[89] identified in bovine salivary gland extract,[90] and similar to the myeloperoxidase of phagocytic cells. The enzyme is present in the acinar cells of the parotid and submaxillary glands and has been isolated from salivary gland tissue,[91] saliva,[92] and the Harderian and lacrimal glands.[93]

Although added peroxide had been shown to be required for in vitro LPO inhibition of microbial activity, it was found to be unnecessary in many of the foregoing studies. Since the hydrogen peroxide degrading enzyme, catalase, neutralized the growth-inhibiting activity of milk and saliva,[88] it was proposed that microbially produced hydrogen peroxide was an important factor in inhibition. It was shown that addition of hydrogen peroxide greatly increased the spectrum of organisms affected by the LPO-thiocyanate system.[94] Since *S. sanguis*[95] and *S. mitis*[96] produce hydrogen peroxide and are among the predominant organisms in the oral cavity and on the tooth surface, the LPO-thiocyanate system may be expected to be active in the oral cavity and to affect a wide spectrum of organisms because of the availability of hydrogen peroxide.

Further studies showed that the LPO system could inhibit growth of many microbial species,[97-99] including *S. mutans*.[100] Attempts were made to potentiate the system with additional hydrogen peroxide, and some reduction of plaque accumulation, gingivitis, and dental caries were reported.[101] LPO is readily adsorbed on to the surfaces of bacterial cells, enamel powder, and hydroxyapatite[102-105] and retains its activity in the adsorbed state. It has been detected in acquired pellicle[106] and dental plaque.[107]

Studies of the mechanisms responsible for the activity of the LPO system showed that bacterial hexokinase was strongly inhibited.[108,109] It was also proposed that oxidized derivatives of thiocyanate were the active agent,[110] which was later shown to be hypothiocyanite.[111,112] It was also found that enzymes other than hexokinase were affected, including enzymes responsible for carbohydrate transport across the cell membrane.[113] The proposed action of the LPO system involves LPO-catalyzed oxidation of thiocyanate, either directly to hypothiocyanite or to thiocyanogen $(SCN)_2$, which hydrolyzes spontaneously to hypothiocyanous acid (HOSCN) or hypothiocyanite.[114,115] The hypothiocyanite ion is highly reactive towards free sulfhydryl groups in low molecular weight compounds[111] and proteins,[114] including hexokinase.[116] The initial reaction between protein sulfhydryl groups and hypothiocyanite was found to be a reversible formation of a sulfenyl thiocyanate followed by irreversible oxidation of the sulfhydryl to sulfenic acid and release of the thiocyanate, which would then be available for further reaction.[115] Depending on conditions, other groups in proteins may be affected.[114] At lower pH values (approximately 5 to 6), hypothiocyanous acid (HOSCN, pH 5.3) will exist and may pass into bacterial cells more readily than its charged anion.[117] Hypothiocyanite is stable in the absence of reactive sulfhydryl and can accumulate in human saliva,[118,119] Concentrations as high as 60 μM were detected in freshly collected human oral fluid. Most values, however, fell between 0 to 20 μM, with a median value of 10 μM. Incubation of these samples substantially increased the hypothiocyanite concentration (to 5 to 90 μM, median 36 μM), an increase which was inhibited by dietary sugars, particularly by sucrose.

Human salivary LPO is heterogeneous and was separated into three active subfractions by DEAE cellulose chromatography, isoelectric focusing, and electrophoresis. Isoelectric points of 6.4, 7.1, and 9.7 were reported.[120,121] Multiple sources of peroxidases in the oral cavity have been identified.[122] The enzyme and thiocyanate have been detected in human tear fluid, but the complete system is not generally present in tears of other species.[123]

LYSOZYME (MURAMIDASE)

Lysozyme preferentially hydrolyzes the β-1,4 glucosidic linkages between *N*-acetylmuramic acid and *N*-acetylglucosamine which occur in the peptidoglycan structure of the cell walls of certain microorganisms. This leads to lysis in sensitive species. A less well-understood action of lysozyme leads to partial loss of bacterial cell contents and reduced viability in other species. Lysozyme may also aggregate some organisms and so play a role in their preferential removal from the oral cavity. The presence of extracellular structures, such as the capsular material common in many oral species, presumably provides protection against the action of this enzyme.

Lysozyme activity was first observed in nasal and other secretions,[124] but most early information came from studies of the related avian enzyme obtained from egg white. Considerable information is now available on its structure, physicochemical properties, and mechanisms leading to substrate hydrolysis. It was the first enzyme to be well characterized in the latter respect and has been recently reviewed.[125] It is widespread in external secretions and tissues of mammals. A plant enzyme has been found to have analogous activity, but is chemically distinct.[126]

Many bacterial species which have adapted to conditions in the oral cavity can resist the action of lysozyme at concentrations many times greater than those occurring in saliva,[127] and no differences were found in the lysozyme concentrations of saliva from caries-free and caries-active subjects.[128] More subtle effects of lysozyme, however, have recently been reported.[129] Treatment of *S. mutans* with avian lysozyme led to loss of cell contents, as indicated by loss of a labeled low molecular weight marker (^{14}C- nicotinamide). These results were interpreted to mean that lysozyme had penetrated the cell wall and was interacting with and increasing the permeability of the cell membrane. As pointed out by these authors, such an interaction between the membrane and the highly cationic lysozyme may be anticipated by analogy with other studies,[130] although there is a clear possibility that membrane damage could also occur due to interactions between hydrophobic residues of the protein and membrane lipid.[131] It was also noted[132] that sodium chloride treatment of lysozyme-treated *S. mutans* cells led to lysis. These results could not be completely explained in terms of cell wall lysis by the enzyme, and it was suggested that lysis possibly involved chaotropic effects. Simply put, the presence of chaotropes results in water structure with less order, and, for thermodynamic reasons, this enhances the transfer of hydrophobic or apolar groups, such as cell membrane constituents, into the aqueous phase. Chaotropic effects of ions increase with decreasing charge density, so thiocyanate or iodide should be more effective than similarly charged smaller ions. In fact, thiocyanate was found to be the most effective of the ions tested. In addition, it was pointed out[132] that hypothiocyanite, originating from lactoperoxidase activity, might be even more effective because of its greater size and correspondingly lower charge density.

ANIONIC PROLINE-RICH PROTEINS (PRP)

The PRP are a group of saliva-specific phosphoproteins which possess the unusual ability of inhibiting crystal growth of calcium phosphate salts. Since saliva is supersaturated with respect to most calcium phosphate salts, it has been proposed that a possible biological function of the PRP is to prevent unwanted deposition of these salts onto tooth surfaces, to which these proteins are known to bind preferentially. Thus, together with statherin, they act to stabilize the supersaturated saliva and so help maintain a protective environment for the teeth.

Early studies of human salivary proteins showed them to be unusually rich in proline.[133] Bands which later proved to be due to the PRP were detected by disk gel electrophoresis by several investigators.[134-136] The PRP were separated on an analytical scale by fractionation on G50 Sephadex® followed by disk electrophoresis of the fractions. They were found to be present in parotid and submandibular salivas, and oral fluid (whole saliva), but their concentrations in the latter secretion were well below those found in the glandular secretions.[137] This unexpected result, later confirmed by immunological methods,[138] is evidently due to the relatively rapid postsecretory degradation of the PRP by enzymes associated with the oral microflora.

The four major PRP have been isolated from parotid saliva and named PRP 1 to 4.[139] Simultaneously, two proteins named A and C were isolated from the same source and proved to be identical to PRP 3 and 1, respectively.[140] Minor PRP (A-H) have been identified, and two of these were isolated and partially characterized.[141] Isolation of the PRP was achieved using a combination of salt precipitation, anion exchange and gel filtration chromatography, and preparative electrophoresis.

Studies of proteins A and C identified them as phosphoproteins, each containing two phosphate residues per mole of protein. A small amount of carbohydrate was reported to be present, but this was less than the equivalent of one or two glucose residues per mole. Molecular weights were reported as 11,000 and 16,300 for A and C, respectively, correcting lower figures previously reported. Immunochemical studies showed that the two proteins cross-reacted, with protein A showing a reaction of partial identity with C. It was proposed that part of the difference between A and C was due to the presence of an additional molecular segment at the carboxy-terminus of protein C. Both proteins bound calcium, an activity attributed in part to the presence of bound phosphate. The removal of phosphate with alkaline phosphatase decreased the amount of calcium binding. Under conditions approximating those in saliva, 2 mol of calcium were bound per mole of protein. Digestion of both A and C with trypsin and collagenase failed to affect their calcium binding properties. The sedimentation coefficient of C, but not of A, was higher in the presence of calcium, suggesting the possibility of a calcium-induced aggregation or conformational change, but no such changes were detected using circular dichroism or nuclear magnetic resonance. In spite of the high proline content, neither poly-proline nor triple-helical collagen-like structure could be detected using the foregoing techniques.[142-145]

Structural studies of the PRP were initially focused on the determination of the sequences of their amino-terminal segments because of their special properties[146] and the determination of the complete primary structure of protein A (PRP 3).[147,148] Later, the sequence of PRP 4[149] and Protein C (PRP 1) were reported.[150] Sequences of peptides, obtained from oral fluid, have been described,[151,152] and some of these may represent fragments derived from the PRP by microbial degradation. The sequence of PRP 2 has been completed, but not yet published. Results from sequence studies of monkey PRP have also been reported.[153,154]

The four human PRP are very closely related. PRP 1 and 2, each 150 residues long, are virtually identical except that residue 4, asparagine in PRP 1, is aspartate in PRP 2. The sequences of the smaller proteins, PRP 3 and 4, correspond exactly to the first 106 residues of PRP 1 and 2, respectively. The four molecules exhibit a high degree of structural and charge asymmetry. With the exception of the carboxy-terminal negative charge, the negatively charged residues, including the two phosphoserines present, are confined to the amino-terminal portion of the molecule, with most of them (13 of 16) being located in the first 30 residues. Similarly, with the exception of arginine at residue 30, the positively charged residues are located in the carboxy-terminal part of the molecule. Most of the proline, glutamine, and glycine, which form 106 of the 150 residues of PRP 1, occur in the carboxy-terminal 120 residues. Consequently, the amino-terminal 30-residue segment differs considerably in composition and properties from the remainder of the molecule.

Since the PRP form a major fraction (20 to 30%) of the total salivary protein, they may be expected to play a significant role in the mouth. It has been proposed that the PRP, together with statherin, are important in controlling both the behavior of calcium and phosphate in saliva and the interaction of these ions with the teeth. This proposal centers on the fact that saliva is supersaturated with respect to the basic calcium phosphate salts which form the dental enamel.[155] Consequently, a protective environment is provided for the teeth, in that driving forces exist in the saliva which:

1. Suppress dissolution of the dental enamel into its fluid environment
2. Will oppose dissolution of enamel during attack by the acid end-products of bacterial metabolism, as in dental caries
3. Will recalcify and repair at least some early carious lesions if conditions are appropriate[156,157]

An anomaly inherent in this proposal is that the same driving forces should cause spontaneous precipitation of calcium phosphate salts in the saliva, and, even if the driving forces are insufficient to initiate spontaneous precipitation, seeded crystal growth of calcium phosphate salts on enamel surfaces would be inevitable.

Since this undesirable precipitation does not occur, saliva was examined for the presence of inhibitors of calcium phosphate precipitation. These studies[158,159] led to the identification of statherin as a potent inhibitor of spontaneous precipitation and the identification of both statherin and the PRP as inhibitors of seeded crystal growth of calcium phosphate salts. The presence and activities of these two molecules explain how the desirable features of salivary supersaturation can express themselves in the oral cavity, while the undesirable features are suppressed. The selective recalcification of early carious lesions was explained by proposing that a lesion which retains a surface layer of sound enamel will recalcify, while a lesion which has been penetrated by the inhibitors will not. Similarly, calcification of plaque was explained by proposing that inhibitors, if present in plaque, would be destroyed by microbial enzymatic activity. Since calcium and phosphate ions have relatively unrestricted access to plaque, precipitation to form calculus is inevitable, unless, as seems possible, the plaque microflora or its products affect the behavior of these ions.

The function proposed for the PRP, that they prevent undesirable precipitation of calcium phosphate salts on the tooth surface, requires:

1. That they adsorb to calcium phosphate crystal surfaces
2. That in their adsorbed state they inhibit calcium phosphate crystal growth
3. That these processes occur on the tooth surface under normal oral conditions

Unusual adsorption behavior of the PRP has been noted.[160] Both statherin and the PRP are selectively adsorbed from saliva by hydroxyapatite, indicating that of the 40 or so proteins present in saliva, these 2 appeared to possess an exceptionally high affinity for apatitic minerals. Adsorption of statherin and PRP 3 onto hydroxy- and fluorapatites has been studied.[161] The experimental data fitted an adsorption model based on the Langmuir adsorption isotherm, and values were obtained for adsorption affinities and the number of adsorption sites for each protein. Interestingly, higher affinities were found for fluorapatite and fluorhydroxyapatites than for hydroxyapatite, a result which was explained in terms of competitive adsorption isotherms. Thus, adsorption involves displacement of adsorbed water, followed by adsorption of the macromolecule. Since fluoride-containing apatites possess a lower surface free energy, their interaction with water is decreased, thus enhancing adsorption of the macromolecules. Conceivably, this difference in behavior could lead to differences in the formation of the acquired enamel pellicle on fluoridated and nonfluoridated tooth surfaces, and so play a part in the cariostatic properties of fluoride. Other aspects of the adsorption behavior of these molecules has been reviewed elsewhere.[162,163]

Quantitative in vitro studies of adsorption of PRP onto hydroxyapatite from saliva, using specific antisera, confirmed the selective adsorption of these proteins.[164,165] These studies also showed that other salivary proteins interfere with adsorption, presumably by competition, and that adsorption was influenced by phosphate, but that other salivary parameters had little effect. Presence of the PRP on the natural tooth surface has been reported[147] and investigated in detail.[138] These studies showed the presence of the PRP in long-term in vivo pellicle, and the PRP were located histologically, using an indirect immunoperoxidase technique, in the acinar cells of the parotid and submandibular glands in both man and the monkey, *Macaca fascicularis*.

The PRP inhibit calcium phosphate crystal growth.[159,166] This inhibition was measured in well-defined systems, using as seed material hydroxyapatite crystals pretreated with PRP and statherin at different concentrations. Inhibition of crystal growth was closely related to the extent of seed crystal surface coverage by the inhibitor, as determined by adsorption measurements. The inhibitory activity of the PRP was sufficient to explain the lack of crystal growth on the tooth surface in vivo, under normal salivary conditions. Thus, present experimental data are consistent with the proposed function of the PRP.

The relationships between the molecular structures of the PRP and several of their properties have been extensively studied. The molecular segment responsible for inhibition of calcium phosphate crystal growth was shown to lie in the first amino-terminal 30 residues.[146,167] This property is consistent with the high content of acidic amino acids and the presence of two phosphoserine residues in this segment. The same segment was identified as containing the active site by which the PRP bound to hydroxyapatite,[168] and it was found that the adsorbed PRP could be treated with degradative enzymes, resulting in loss of the proline-rich part, but that the adsorbed segment was protected from degradation. This presumably has important consequences for the survival of the biological activity of these proteins. It was also shown that the calcium binding activity of the PRP was located in the same segment of the molecule.[169]

The major PRP exhibit genetic polymorphism,[170] with PRP 1 and 3 and PRP 2 and 4 representing the phenotypes of two homozygotes, with all four proteins being present in the salivas of heterozygotes. Significantly, this pairing corresponds to the substitution noted at residue 4, i.e., the difference between the pairs consists of a single substitution; PRP 1 and 3 possess an asparagine and PRP 2 and 4 an aspartic acid residue at position 4. It has been proposed[171] that the PRP 3 and 4 are possibly posttranslational derivatives of PRP 1 and 2, respectively. Little attention has so far been directed to the minor PRP, except for genetic studies, which have been reviewed elsewhere.[3] Possible links between heritable salivary proteins, including the PRP, and dental disease have been reported.[172]

STATHERIN

Statherin is a 43-residue, tyrosine-rich phosphoprotein which possesses the ability to inhibit both spontaneous and secondary precipitation (crystal growth) of calcium phosphate salts. Because of this unusual property, it has been proposed that its function is to prevent precipitation from the salivary secretions, which are normally supersaturated with respect to most calcium phosphate salts, and so help provide a protective and stabilized environment for the teeth.

The first observations of human salivary statherin came from studies of its selective adsorption behavior onto apatitic minerals and dental enamel, prior to its actual identification.[160,173,174] It was also observed in calcium-induced precipitates from normal saliva and in spontaneous precipitates in saliva from cystic fibrosis patients.[175] Statherin was isolated from human parotid saliva by a combination of ammonium sulfate precipitation, gel filtration chromatography, and preparative electrophoresis.[176] It proved to be a small phosphoprotein, with a molecular weight of 5380, an isoelectric point of 4.22, and an unusual amino acid composition; 25% of the molecule is formed from aromatic amino acids, mainly tyrosine, a further 25% is formed from acidic residues, mainly present as glutamine, and it contains 16% of proline.

It has been proposed that the function of statherin is to inhibit precipitation of calcium phosphates from saliva, which is supersaturated with respect to these salts.[158,159] The implications of salivary supersaturation and inhibitory activity have already been discussed. Statherin at physiological concentrations not only inhibits crystal growth in seeded systems, but also inhibits spontaneous (unseeded) precipitation from solutions supersaturated with respect to calcium phosphate salts, unlike the PRP which lack this ability. This property has led to the proposal that statherin may function in the secretory granules and gland ducts to stabilize the supersaturated saliva prior to secretion, and this may be as important as its possible postsecretory activity in the oral cavity.[167]

The complete primary structure of statherin has been determined.[177] It consists of 43 residues, and the molecule possesses the same asymmetry of structure and charge noted with the PRP. The negatively charged, phosphoserine containing amino-terminal segment, however, is only 5 residues long, compared to 30 residues or more for the PRP. The essentially uncharged, 30-residue, carboxy-terminal segment contains all the tyrosine, proline, and glutamine present in the molecule, with these three amino acids forming 70% of this segment.

The detailed mechanisms by which it inhibits precipitation are not understood, but it has been proposed that precipitation inhibitors in general bind to the first-formed calcium phosphate nuclei or embryo crystals and block their growth or transformation, so delaying precipitation. In cases where the formation of embryo crystals is rate-limiting, the delay may be substantial, as is the case with calcium phosphate precipitation. The ability of statherin to strongly adsorb to apatitic minerals, and by inference to embryo crystals, is obviously an important property with respect to its ability to inhibit spontaneous precipitation of calcium phosphate salts. The adsorption of statherin onto apatitic minerals has been investigated,[161] and its effect on seeded crystal growth has been studied.[166] Some information has been obtained with respect to the active segment of the molecule. Tryptic cleavage releases the acidic, phosphoserine-containing amino-terminal segment. The inhibitory activity of this fragment in spontaneously precipitating systems is far less than that of the parent molecule. This suggests that the uncharged, tyrosine-rich, carboxy-terminal segment plays an important role, although, by itself, it is without inhibitory activity.[167]

PROTEINS INVOLVED IN BACTERIAL ADHERENCE AND AGGREGATION

The concepts of bacterial adherence to oral surfaces, and bacterial aggregation and deletion from the mouth as factors in virulence and oral defense were first proposed by Gibbons.[178] They have had far-reaching effects on concepts of microbial ecology as discussed in recent comprehensive reviews.[179,180] With respect to adhesion and attachment, the key points proposed were that bacteria interact in a selective fashion with oral surfaces, with specificity being determined by reactive groups on both the colonized surface and on the bacterial cell surface. Since oral surfaces are normally coated with a film of adsorbed salivary proteins, these proteins must carry the determinants which control attachment. Clearly, an excess of analogues of the proteins concerned, present in the oral fluid, could block the reactive groups on the bacterial cell surfaces, and so prevent colonization. These same proteins, or others or both, may also cause aggregation of bacterial cells, so preventing them from colonizing oral surfaces, resulting in their deletion from the mouth.

Although considerable progress has been made, the specific salivary proteins responsible for the foregoing phenomena have yet to be identified and fully characterized. Early studies[181,182] identified a relatively high molecular weight glycoprotein fraction as playing an important role. Further studies[183-186] implicated a sulfated glycoprotein fraction from whole saliva. This fraction possessed blood group activity, a significant observation considering that sulfated glycoprotein with blood group activity has been identified on the surfaces of human[187] and monkey teeth and in plaque.[188] Other salivary constituents, however, have been shown to be active in the assay systems used,[189,190] and the carbohydrate moieties responsible for blood group activity do not appear to be the same as those responsible for bacterial interactions.[191] It seems probable that several salivary macromolecules, together with postsecretory modification, will prove to play significant roles in adhesion and aggregation phenomena.[192]

BASIC PAROTID GLYCOPROTEIN

This protein was one of the earliest to be identified in and isolated from human parotid saliva.[193-196] It has a molecular weight of 36,500 and is notable for its high content of proline and acidic amino acids, most of the latter being amidated. Partial structural studies[197-198] have been done. Although it is a major constituent of parotid saliva, no biological function has been established for this protein. It has been proposed that it protects or enhances the activity of salivary amylase,[199] but this has not been confirmed. Another study showed that the protein binds to the cell surfaces of *S. sanguis* and *S. mitis*,[200] but does not cause aggregation of these and other organisms.[191]

OTHER PROTEINS

In addition to the foregoing, saliva contains a considerable number of other macromolecules. Most of these have no known function, or functions which have been suggested for them have yet to be confirmed. They include a remarkably diverse group of basic proteins and peptides,[201-203] several acidic proteins and a neutral protein from submandibular saliva,[204,205] a calcium-precipitable protein from the salivas of normal subjects and cystic fibrosis patients,[206-208] a cysteine-containing glycoprotein from submandibular-sublingual saliva,[209] sulfate-containing glycoproteins,[186,188] a zinc-binding protein, named gustin, considered to play a role in taste sensation,[210,211] a calcium-binding glycoprotein from sublingual-submaxillary saliva,[212] and histidine-rich polypeptides,[213] some of which have been isolated.[214,215] Studies and reviews of high molecular weight human salivary glycoproteins have also appeared.[6,216-219] The large number of enzymes detected in or isolated from human saliva have been summarized elsewhere.[5]

REFERENCES

1. **Steiner, J. C. and Keller, P. J.,** An electrophoretic analysis of the protein components of human parotid saliva, *Arch. Oral Biol.,* 13, 1213, 1968.
2. **Meyer, T. S. and Lamberts, B. L.,** The use of wool fast blue BL for the electrophoresis of human parotid saliva proteins in acrylamide gel, *Arch. Oral Biol.,* 13, 839, 1968.
3. **Azen, E. A.,** Genetic protein polymorphisms in human saliva: an interpretive review, *Biochem. Genet.,* 16, 79, 1978.
4. **McGhee, J. R. and Michalek, S. M.,** Immunobiology of dental caries: microbial aspects and local immunity, *Annu. Rev. Microbiol.,* 35, 595, 1981.
5. **Arglebe, C.,** Biochemistry of human saliva, *Adv. Oto-Rhino-Laryngol.,* 26, 97, 1981.
6. **Mandel, I. D.,** Human submaxillary, sublingual, and parotid glycoproteins and enamel pellicle, in *The Glycoconjugates*, Vol. 1, Mammalian Glycoproteins and Glycolipids, Horowitz, M. and Pigman, W., Eds., Academic Press, New York, 1977, 154.
7. **Barka, T.,** Biologically active polypeptides in submandibular glands, *J. Histochem. Cytochem.,* 28, 836, 1980.
8. **Polak, J. M. and Bloom, S. R.,** Neuropeptides of salivary glands, *J. Histochem. Cytochem.,* 28, 871, 1980.
9. **Bing, J., Poulsen, K., Hackenthal, E., Rix, E., and Taugner, R.,** Renin in the submandibular gland: a review, *J. Histochem. Cytochem.,* 28, 874, 1980.
10. **Orstavik, T. B.,** The kallikrein-kinin system in exocrine organs, *J. Histochem. Cytochem.,* 28, 881, 1980.
11. **Murphy, R. A., Watson, A. Y., Metz, J. and Forssmann, G.,** The mouse submandibular gland: an exocrine organ for growth factors, *J. Histochem. Cytochem.,* 28, 890, 1980.
12. **Schacter, M.,** Kallikreins (kininogenases) — a group of serine proteases with bioregulatory actions, *Pharmacol. Rev.,* 31, 1, 1980.
13. **Roberts, P. J. P. and Whelan, W. J.,** The mechanism of carbohydrase action, *Biochem. J.,* 76, 246, 1960.
14. **Takagi, T., Toda, H., and Isemura, T.,** in *The Enzymes,* Vol. 3, Boyer, P. D., Ed., Academic Press, New York, 1971, 235.
15. **Thoma, J. A., Spradlin, J. E., and Dygert, S.,** in *The Enzymes*, Vol. 3, Boyer, P. D., Ed., Academic Press, New York, 1971, 115.

16. **Leuchs, E. F.,** *Poggendorff's Ann. d. Phys. Chem.*, 22, 623, 1831.
17. **Meyer, K. H., Fischer, E. H., Staub, A., and Bernfeld, P.,** Sur les enzymes amylolytiques. X. Isolement et crystallisation del'α-amylase de salive humaine, *Helv. Chem. Acta,* 31, 2158, 1948.
18. **Muus, J.,** Studies on the salivary amylase with special reference to the interaction with chloride ions, *C. R. Trav. Lab. Carlsberg,* 28, 317, 1953.
19. **Muus, J. and Vnenchak, J. M.,** Isozymes of salivary amylase, *Nature (London),* 204, 283, 1964.
20. **Kauffman, D. L., Zager, N. I., Cohen, E., and Keller, P. J.,** The isoenzymes of human parotid amylase, *Arch. Biochem. Biophys.*, 137, 325, 1970.
21. **Keller, P. J., Kauffman, D. L., Allan, B. J., and Williams, B. L.,** Further studies on the structural differences between the isoenzymes of human parotid α-amylase, *Biochemistry,* 10, 4867, 1971.
22. **Lorentz, K.,** Salivary isoamylases: deamidation products of amylase, *Clin. Chim. Acta,* 93, 161, 1979.
23. **Lorentz, K. and Flatter, B.,** Studies on isoamylase formation in biological fluids, *Enzyme,* 24, 163, 1979.
24. **Ogawa, M., Kosaki, G., Matsuura, K., Fujimoto, K. I., Minamiura, N., Yamamoto, T., and Kikuchi, M.,** Modification of human pancreatic amylase isozymes by peptido-glutaminases I and II, *Clin. Chim. Acta,* 87, 17, 1978.
25. **Watanabe, S. and Keller, P. J.,** Isolation and partial characterization of glycopeptides from human parotid amylase A, *Biochim. Biophys. Acta,* 336, 62, 1974.
26. **Yamashita, K., et al.,** Structural studies of the sugar chains of human parotid α-amylase, *J. Biol. Chem.*, 255, 5635, 1980.
27. **Bellavia, S. L., Moreno, J., Sanz, E., Picas, E. I., and Blanco, A.,** α-Amylase activity of human neonate and adult saliva, *Arch. Oral Biol.*, 24, 117, 1979.
28. **Karn, R. C., Rosenblum, B. B., Ward, J. C., Merritt, A. D., and Shulkin, J. D.,** Immunological relationships and post-translational modifications of human salivary amylase and pancreatic amylase isozymes, *Biochem. Genet.*, 12, 485, 1974.
29. **Gertzman, G. B., Rivas, M. L., and Strumeyer, D. H.,** Population distribution of human salivary amylases: differential inhibition by wheat protein, *J. Dent. Res.*, 59, 1010, 1980.
30. **Stiefel, D. J. and Keller, P. J.,** Preparation and some properties of human pancreatic amylase including a comparison with human parotid amylase, *Biochim. Biophys. Acta,* 302, 345, 1973.
31. **Boehm-Truitt, M., Harrison, E., Wolf, R. O., and Notkins, A. L.,** Radioimmunoassay for human salivary amylase, *Anal. Biochem.*, 85, 476, 1978.
32. **Sanders, G. T. and Rutter, W. J.,** Molecular properties of rat pancreatic and parotid α-amylase, *Biochemistry,* 11, 130, 1972.
33. **Rosenmund, H. and Kaczmarek, M. J.,** Isolation and characterization of isoenzymes of human salivary and pancreatic α-amylase, *Clin. Chim. Acta,* 71, 185, 1976.
34. **Stein, E. A., Hsiu, J., and Fischer, E. H.,** Alpha-amylases as calcium metallo-enzymes. I. Preparation of calcium-free apoamylases by chelation and electrodialysis, *Biochemistry,* 3, 56, 1964.
35. **Hsiu, J., Fischer, E. H., and Stein, E. A.,** Alpha-amylases as calcium-metalloenzymes. II. Calcium and catalytic activity, *Biochemistry,* 3, 61, 1964.
36. **Fischer, E. H. and Stein, E. A.,** *The Enzymes,* 2nd ed., Boyer, P. D., Lardy, H., and Myrback, K., Academic Press, New York, 1960, 313.
37. **Levitzki, A. and Steer, M. L.,** The allosteric activation of mammalian α-amylase by chloride, *Eur. J. Biochem.*, 41, 171, 1974.
38. **Payan, F., Haser, R., Pierrot, M., Frey, M., Astier, J. P., Abadie, B., Duee, E., and Buisson, G.,** The three dimensional structure of α-amylase from porcine pancreas at 5Å resolution — the active-site location, *Acta Crystallogr.*, B36, 416, 1980.
39. **Pasero, L., Abadie, B., Mazzei, Y., Moinier, D., Bizzozero, J. P., Foubereau, M., and Marchis-Mouren, G.,** *Ann. Biol. Anim. Biochim. Biophys.*, 19, 1033, 1979.
40. **O'Donnell, M. D. and Miller, N. J.,** Plasma pancreatic and salivary-type amylase and immunoreactive trypsin concentrations: variations with age and reference ranges for children, *Clin. Chim. Acta,* 104, 265, 1980.
41. **Van Haeringen, N. J., Ensink, F., and Glasius, E.,** Amylase in human tear fluid: origin and characteristics, compared with salivary and urinary amylases, *Exp. Eye Res.*, 21, 395, 1975.
42. **Fridhandler, L., Berk, J. E., and Ueda, M.,** Isolation and measurement of pancreatic amylase in human serum and urine, *Clin. Chem.*, 18, 1493, 1972.
43. **Hanafy, H. M.,** Possible role of amylase enzyme in prostatic and seminal fluid, *Urol. Int.*, 34, 11, 1979.
44. **Mellersh, A., Clark, A., and Hafiz, S.,** Inhibition of *Neisseria gonorrheae* by normal human saliva, *Br. J. Vener. Dis.*, 55, 20, 1979.
45. **Sorensen, M. and Sorenson, S. P. L.,** The proteins in whey, *C.R. Lab. (Carlsberg) (Ser. Chim.),* 23, 55, 1939.
46. **Groves, M. L.,** The isolation of a red protein from milk, *J. Am. Chem. Soc.*, 82, 3345, 1960.
47. **Gruettner, R., Schaefer, K. H., and Schroeter, W.,** On the production of iron-binding proteins in human milk, *Klin. Wochenschr.*, 38, 1162, 1960.

48. **Johansson, B.,** Isolation of an iron-containing red protein from human milk, *Acta Chem. Scand.*, 14, 510, 1960.
49. **Montreuil, J. and Mullet, S.,** Isolation of lactosiderophilin from human milk, *C. R.*, 250, 1736, 1960.
50. **Montreuil, J., Tonnelat, J., and Mullet, S.,** Preparation and properties of lactosiderophilin (lactotransferrin) of human milk, *Biochim. Biophys. Acta*, 45, 413, 1960.
51. **Blanc, B. and Isliker, H.,** Isolation and characterisation of the red siderophilic protein from maternal milk: lactotransferrin, *Bull. Soc. Chem. Biol.*, 43, 929, 1961.
52. **Masson, P. L., Carbonara, A. O., and Heremans, J.F.,** Studies on the proteins of human saliva, *Biochim. Biophys. Acta*, 107, 485, 1965.
53. **Masson, P., Heremans, J. F., and Dive, C.,** An iron binding protein common to many external secretions, *Clin. Chim. Acta*, 14, 735, 1966.
54. **Tourville, D. R., Alder, R. H., Bienenstock, J., and Tomasi, T. B.,** The human secretory immunoglobulin system. Immunohistochemical localization of gamma A secretory piece and lactoferrin in normal tissues, *J. Exp. Med.*, 129, 411, 1969.
55. **Masson, P. L. and Heremans, J. F.,** Metal combining properties of human lactoferrin (Red milk protein). I. The involvement of bicarbonate in the reaction, *Eur. J. Biochem.*, 6, 579, 1968.
56. **Querinjean, P., Masson, P. L., and Heremans, J. F.,** Molecular weight, single chain structure and amino-acid composition of human lactoferrin, *Eur. J. Biochem.*, 20, 420, 1971.
57. **Bluard-Deconinck, J. M., Masson, P. L., Osinski, P. A., and Heremans, J. F.,** Amino-acid sequence of cysteic peptides of lactoferrin and demonstration of similarities between lactoferrin and transferrin, *Biochim. Biophys. Acta*, 365, 311, 1974.
58. **Malamud, D. and Drysdale, J. W.,** Isoelectric points of protein: a table, *Anal. Biochem.*, 86, 620, 1978.
59. **Lonnerdal, B., Carlsson, J., and Porath, J.,** Isolation of lactoferrin from human milk by metal-chelate affinity chromatography, *FEBS Lett.*, 75, 89, 1977.
60. **Blackberg, L. and Hernell, O.,** Isolation of lactoferrin from human whey by a single chromatographic step, *FEBS Lett.*, 109, 180, 1980.
61. **Brandtzaeg, P.,** Human secretory immunoglobulins. VII. Concentration of parotid IgA and other secretory proteins in relation to the rate of flow and duration of secretory stimulus, *Arch. Oral Biol.*, 16, 1295, 1971.
62. **Bezkorovainy, A.,** Anti-microbial properties of iron-binding proteins, *Adv. Exp. Med. Biol.*, 135, 139, 1981.
63. **Tabak, L., Mandel, I. D., Karlan, D., and Baurmash, H.,** Alterations in lactoferrin in salivary gland disease, *J. Dent. Res.*, 57, 43, 1978.
64. **Tabak, L., Mandel, I. D., Herrera, M., and Baurmash, H.,** Changes in lactoferrin and other proteins in a case of chronic recurrent parotitis, *J. Oral Pathol.*, 7, 91, 1978.
65. **Oram, J. D. and Reiter, B.,** Inhibition of bacteria by lactoferrin and other iron chelating agents, *Biochim. Biophys. Acta*, 170, 351, 1968.
66. **Kirkpatrick, C. H., Green, I., and Rich, R.,** Inhibition of growth of *Candida albicans* by iron-unsaturated lactoferrin, *J. Infect. Dis.*, 124, 539, 1971.
67. **Reiter, B., Brock, J. H., and Steel, E. D.,** Inhibition of *E. coli* by bovine colostrum and post-colostral milk. II. The bacteriostatic effect of lactoferrin on a serum susceptible and serum resistant strain of *E. coli*, *Immunology*, 28, 83, 1975.
68. **Cole, M. F., Arnold, R. R., Mestecky, J., Prince, S., Kulhavy, R., and McGhee, J. R.,** Studies with human lactoferrin and *Streptococcus mutans*, *Proc. Microbial Aspects of Dental Caries*, Stiles, H. M., Loesche, W. J., and O'Brien, T. C., Eds., (Spec. Suppl.), *Microbiol. Abstr.*, 1, 359, 1976.
69. **Gladstone, G. P. and Walton, E.,** Effect of iron on the bactericidal proteins from rabbit PMN leucocytes, *Nature (London)*, 277, 849, 1970.
70. **Weinberg, E. D.,** Iron and susceptibility to infectious diseases, *Science*, 184, 952, 1974.
71. **Weinberg, E. D.,** Iron and infection, *Microbiol. Rev.*, 42, 45, 1978.
72. **Arnold, R. R., Cole, M. F., and McGhee, J. R.,** A bactericidal effect for human lactoferrin, *Science*, 197, 263, 1977.
73. **Arnold, R. R., Brewer, M., and Gauthier, J. J.,** Bactericidal activity of human lactoferrin: sensitivity of a variety of microorganisms, *Infect. Immun.*, 28, 893, 1980.
74. **Arnold, R. R., Russell, J., Champion, W. J., Brewer, M., and Gauthier, J.,** Bacterial activity of human lactoferrin: differentiation from the stasis due to iron deprivation, *Infect. Immun.*, 35, 792, 1982.
75. **Bullen, J. J., Rogers, H. J., and Leigh, L.,** Iron binding proteins in milk and resistance to *Escherichia coli* in infants, *Br. Med. J.*, 1, 69, 1972.
76. **Bishop, J. G., Schanbacher, F. L., Ferguson, L. C., and Smith, K. L.,** In vitro growth inhibition of mastitis causing coliform bacteria by bovine apo-lactoferrin, *Infect. Immun.*, 14, 911, 1976.
77. **Dipaola, C. and Mandel, I. D.,** Lactoferrin concentration in human parotid saliva as measured by an enzyme-linked immunosorbent assay (ELISA), *J. Dent. Res.*, 59, 1463, 1980.
78. **Hesse, W.,** *Z. Hyg. u. Infectionskrankh.*, 17, 238, 1894.
79. **Jones, F. S. and Little, R. B.,** The bactericidal properties of cows milk, *J. Exp. Med.*, 45, 319, 1927.

80. **Reiter, B., Pickering, A., Oram, J. D., and Pope, G. S.,** Peroxidase-thiocyanate inhibition of streptococci in raw milk. Symposium: "Antibiotics: structure and function", *J. Gen. Microbiol.*, 33, xii, 1963.

81. **Clough, O. W., Bibby, B. G., and Berry, G. P.,** The effect of saliva on *Lactobacillus acidophilus, J. Dent. Res.*, 17, 493, 1938.

82. **Bibby, B. G., Hine, M. K., and Clough, O. W.,** The antibacterial action of human saliva, *J. Am. Dent. Assoc.*, 25, 1290, 1938.

83. **Wright, R. C. and Tramer, J.,** Factors influencing the activity of cheese starters. The role of milk peroxidase, *J. Dairy Res.*, 25, 104, 1958.

84. **Jago, G. R. and Morrison, M.,** Lactoperoxidase, *J. Biol. Chem.*, 238, 2843, 1963.

85. **Zeldow, B. J.,** Studies on the antibacterial activity of human saliva. I. A bacteriocidin for lactobacilli, *J. Dent. Res.*, 38, 789, 1959.

86. **Dogon, I. L., Kerr, A. C., and Amdur, B. H.,** Characterization of an antibacterial factor in human parotid secretions active against *Lactobacillus casei, Arch. Oral Biol.*, 7, 81, 1962.

87. **Zeldow, B. J.,** Studies on the antibacterial action of human saliva. III. Cofactor requirements of a *Lactobacillus bactericidin, J. Immunol.*, 90, 12, 1963.

88. **Klebanoff, S. J. and Leubke, R. G.,** The antilactobacillus system of human saliva. Role of salivary peroxidase, *Proc. Soc. Exp. Biol. Med.*, 118, 483, 1965.

89. **Morrison, M. and Hultquist, D. E.,** Lactoperoxidase. II. Isolation, *J. Biol. Chem.*, 238, 2847, 1963.

90. **Morrison, M. and Allen, P. Z.,** The identification and isolation of lacteroperoxidase from salivary gland, *Biochem. Biophys. Res. Commun.*, 13, 490, 1963.

91. **Morrison, M., Allen, P. Z., Bright, J., and Jayasinge, W.,** Identification and isolation of lactoperoxidase from salivary gland, *Arch. Biochem. Biophys.*, 111, 126, 1965.

92. **Slowey, R. R., Eidelman, S., and Klebanoff, S. J.,** Antibacterial activity of the purified peroxidase from human saliva, *Bacteriology*, 96, 575, 1968.

93. **Morrison, M. and Allen, P. Z.,** Lactoperoxidase: identification and isolaton from Harderian and lacrimal glands, *Science*, 152, 1626, 1966.

94. **Klebanoff, S. J., Clem, W. H., and Luebke, R. G.,** The peroxidase-thiocyanate-hydrogen peroxide antimicrobial system, *Biochem. Biophys. Acta*, 117, 63, 1966.

95. **Holmberg, K. and Hallander, H. O.,** Production of bactericidal concentrations of hydrogen peroxide by *Streptococcus mutans, Arch. Oral Biol.*, 18, 423, 1973.

96. **Hamon, C. B. and Klebanoff, S. J.,** A peroxidase-mediated, *Streptococcus mitis* dependent antimicrobial system in saliva, *J. Exp. Med.*, 137, 438, 1973.

97. **Mandel, I. D.,** Nonimmunologic aspects of caries resistance, *J. Dent. Res.*, 55, C22, 1976.

98. **Arnold, R. R., Pruitt, K. M., Cole, M. F., Adamson, J. M., and McGhee, J. R.,** Salivary antibacterial mechanisms in immunodeficiency, *Proc. Saliva and Dental Caries*, Kleinberg, I., Ellison, S. A., and Mandel, I. D., Eds., (Spec. Supp.), *Microbiol. Abstr.*, 449, 1979.

99. **Germaine, G. R. and Tellefson, L. M.,** Effect of human saliva on glucose uptake by *Streptococcus mutans* and other oral organisms, *Infect. Immun.*, 31, 598, 1981.

100. **Morrison, M. and Steele, W. F.,** Lactoperoxidase, the peroxidase in the salivary gland, in *Biology of the Mouth*, Person, P., Ed., American Association for the Advancement of Science, Washington, D.C., 1968, 89.

101. **Hugoson, A., Koch, G., Thilander, H., and Hoogendoorn, H.,** Lactoperoxidase in the prevention of plaque accumulation, gingivitis and dental caries (III), *Odont. Revy*, 25, 69, 1974.

102. **Pruitt, K. M. and Adamson, M.,** Enzyme activity of salivary lactoperoxidase adsorbed to human enamel, *Infect. Immun.*, 17, 112, 1977.

103. **Tenovuo, J., Valtakoski, J., and Knuuttila, M. L. E.,** Antibacterial activity of lactoperoxidase adsorbed by human salivary sediment and hydroxyapatite, *Caries Res.*, 11, 257, 1977.

104. **Tenovuo, J. and Kurkijarvi, K.,** Immobilised lactoperoxidase as a biologically active and stable form of an antimicrobial anzyme, *Arch. Oral Biol.*, 26, 309, 1981.

105. **Tenovuo, J.,** Formation of the bacterial inhibitor, hypothiocyanite ion, by cell-bound lactoperoxidase, *Caries Res.*, 13, 137, 1979.

106. **Kraus, F. W. and Mestecky, J.,** Salivary proteins and the development of dental plaque, *J. Dent. Res.*, 55, 149, 1976.

107. **Cole, M. F., Hsu, S. D., Baum, B. J., Bowen, W. H., Sierra, L. I., Aquirre, M., and Gillespie, G.,** Specific and non-specific immune factors in dental plaque from young and old populations, *Infect. Immun.*, 31, 998, 1981.

108. **Oram, J. D. and Reiter, B.,** The inhibition of streptococci by lacteroperoxidase, thiocyanate and hydrogen peroxide. The effect of the inhibitory system on susceptible and resistant strains of group N streptococci, *Biochem. J.*, 100, 373, 1966.

109. **Oram, J. D. and Reiter, B.,** The inhibition of streptococci by lactoperoxidase, thiocyanate and hydrogen peroxide. The oxidation of thiocyanate and the nature of the inhibitory compound, *Biochem. J.*, 100, 382, 1966.

110. **Hogg, D. M. and Jago, G. R.,** The antibacterial action of lactoperoxidase, *Biochem. J.*, 117, 779, 1970.
111. **Hoogendoorn, H., Piessens, J. P., Scholtes, W., and Stoddard, L. A.,** Hypothiocyanite ion; the inhibitor formed by the system lactoperoxidase-thiocyanate-hydrogen peroxide, *Caries Res.*, 11, 77, 1977.
112. **Aune, T. M. and Thomas, E. L.,** Accumulation of hypothiocyanite ion during peroxidase-catalyzed oxidation of thiocyanate ion, *Eur. J. Biochem.*, 80, 209, 1977.
113. **Mickelson, M. N.,** Antibacterial action of lactoperoxidase-thiocyanate-hydrogen peroxide on *Streptococcus agalactiae, Appl. Environ. Microbiol.*, 38, 821, 1979.
114. **Aune, T. M., Thomas, E. L., and Morrison, M.,** Lactoperoxidase-catalyzed incorporation of thiocyanate ion into a protein substrate, *Biochemistry*, 16, 4611, 1977.
115. **Aune, T. M. and Thomas, E. L.,** Oxidation of protein sulfhydryls by products of peroxidase-catalyzed oxidation of thiocyanate ion, *Biochemistry*, 17, 1005, 1978.
116. **Adamson, M. and Pruitt, K. M.,** Lactoperoxidase-catalyzed inactivation of hexokinase, *Biochim. Biophys. Acta*, 658, 238, 1981.
117. **Thomas, E. L.,** Lactoperoxidase-catalyzed oxidation of thiocyanate: equilibria between oxidized forms of thiocyanate, *Biochemistry*, 20, 3273, 1981.
118. **Thomas, E. L., Bates, K. P., and Jefferson, M. M.,** Hypothiocyanite ion: detection of the antimicrobial agent in human saliva, *J. Dent. Res.*, 59, 1466, 1980.
119. **Thomas, E. L., Bates, K. P., and Jefferson, M. M.,** Peroxidase antimicrobial system of human saliva: requirements for accumulation of hypothiocyanite, *J. Dent. Res.*, 60, 785, 1981.
120. **Iwamoto, Y., Nakamura, R., Watanabe, T., and Tsunemitsu, A.,** Heterogeneity of peroxidase related to antibacterial activity in human parotid saliva, *J. Dent. Res.*, 51, 503, 1972.
121. **Azen, E. A.,** Salivary peroxidase activity and thiocyanate concentrations in human subjects with genetic variants of salivary peroxidase, *Arch. Oral Biol.*, 23, 801, 1978.
122. **Makinen, K. K. and Tenovuo, J.,** Chromatographic separation of human salivary peroxidases, *Acta Odontol. Scand.*, 34, 141, 1976.
123. **Van Haeringen, N. J., Ensink, F. T. E., and Glasius, E.,** The peroxide-thiocyanate-hydrogen peroxide system in tear fluid and saliva of different species, *Exp. Eye Res.*, 28, 343, 1979.
124. **Fleming, A.,** On a remarkable bacteriolytic element found in tissues and secretions, *Proc. R. Soc. London Ser. B*, 93, 306, 1922.
125. **Osserman, E. F., Canfield, R. E., and Beychok, S.,** *Lysozyme*, Academic Press, New York, 1974.
126. **Meyer, K., Hahnel, E., and Steinberg, A.,** Lysozyme of plant origin, *J. Biol. Chem.*, 163, 733, 1946.
127. **Gibbons, R., de Stoppelar, J. D., and Harder, L.,** Lysozyme insensitivity of bacteria indigenous to the oral cavity of man, *J. Dent. Res.*, 45, 877, 1966.
128. **Mandel, I. D. and Zengo, A. N.,** Genetic and chemical aspects in caries resistance, in Comparative Immunology of the Oral Cavity, U.S. Government Printing Office, Washington, D.C., 1973.
129. **Pollock, J. J., Goodman Bicker, G., Katona, L. I., Cho, M. I., and Iacono, V. J.,** Lysozome bacteriolysis, *Proc. Saliva and Dental Caries*, Kleinberg, I., Ellison, S. A. and Mandel, I. D., Eds., (Spec. Supp.), *Microbiol. Abstr.* 429, 1979.
130. **Harold, F. M.,** Stabilization of *Streptococcus faecalis* protoplasts by spermine, *J. Bacteriol.*, 88, 1416, 1964.
131. **Spitznagel, J. K.,** Bactericidal mechanisms of the granulocyte, in *Progress in Clinical and Biological Research*, Greenwalt, T. J. and Jamieson, G. A., Eds., A. R. Liss, New York, 1977, 103.
132. **Pollock, J. J., Iacono, V. J., Goodman Bicker, H., MacKay, B. J., Katona, L. I., Taichman, L. B., and Thomas, E.,** The binding, aggregation and lytic properties of lysozyme, *Proc. Microbial Aspects of Dental Caries*, Vol. 2, Stiles, H. M., Loesche, W. J., and O'Brien, T. C., Eds., (Spec. Supp.), *Microbiol. Abstr.*, 325, 1976.
133. **Mandel, I. D., Thompson, R., and Ellison, S. A.,** The carbohydrate components of human submaxillary saliva, *Arch. Oral Biol.*, 9, 601, 1963.
134. **Caldwell, R. C. and Pigman, W.,** Disc electrophoresis of human saliva in polyacrylamide gel, *Arch. Biochem. Biophys.*, 110, 91, 1965.
135. **Gifford, G. T. and Yuknis, L.,** Protein patterns in human saliva, *J. Chromatogr.*, 20, 150, 1965.
136. **Steiner, J. C. and Keller, P. J.,** An electrophoretic analysis of the protein components of human parotid saliva, *Arch. Oral Biol.*, 13, 1213, 1968.
137. **Hay, D. I.,** Some observations on human saliva proteins and their role in the formation of the acquired enamel pellicle, *J. Dent. Res.*, 48, 806, 1969.
138. **Kousvelari, E. E., Baratz, R. S., Burke, B., and Oppenheim, F. G.,** Immunochemical identification and determination of proline-rich proteins in salivary secretions, enamel pellicle, and glandular tissue specimens, *J. Dent. Res.*, 59, 1430, 1980.
139. **Oppenheim, F. G., Hay, D. I., and Franzblau, C.,** Proline-rich proteins from human saliva. I. Isolation and partial characterisation, *Biochemistry*, 10, 4233, 1971.
140. **Bennick, A. and Connell, G. E.,** Purification and partial characterisation of four proteins from human parotid saliva, *Biochem. J.*, 123, 455, 1971.

141. **Hay, D. I. and Oppenheim, F. G.,** The isolation from human parotid saliva of a further group of proline-rich proteins, *Arch. Oral Biol.*, 19, 627, 1974.
142. **Bennick, A.,** Chemical and physical characteristics of a phosphoprotein from human parotid saliva, *Biochem. J.*, 145, 557, 1975.
143. **Bennick, A.,** The binding of calcium to a salivary phosphoprotein, protein A, common to human parotid and submandibular secretions, *Biochem. J.*, 155, 163, 1976.
144. **Bennick, A.,** Chemical and physical characterization of a phosphoprotein, protein C, from human saliva and comparison with a related protein A, *Biochem. J.*, 163, 229, 1977.
145. **Bennick, A.,** The binding of calcium to a salivary phosphoprotein, protein C, and comparison with calcium binding to protein A, a related salivary phosphoprotein, *Biochem. J.*, 163, 241, 1977.
146. **Schlesinger, D. H., Jacobs, R., and Hay, D. I.,** Primary structure of protein inhibitors of calcium phosphate precipitation from human salivary secretions, in *Peptides,* Proc. 5th American Peptide Symp., Goodman, M. and Meienhofer, J., Eds., John Wiley & Sons, New York, 1977, 56.
147. **Bennick, A., Wong, R., and Cannon, M.,** Structure and biological activities of salivary acidic proline-rich phosphoproteins, in *Proc. Int. Symp. Calcium Binding Proteins and Calcium Function in Health and Disease,* Wasserman, R. H., Corradino, R. A., Carafoli, E., Kretsinger, R. H., MacLennan, D. H., and Siegal, F. L., Eds., Elsevier, Amsterdam, 1977, 391.
148. **Wong, R. S., Hofmann, T., and Bennick, A.,** The complete primary structure of a proline-rich phosphoprotein from human saliva, *J. Biol. Chem.*, 254, 4800, 1979.
149. **Schlesinger, D. H. and Hay, D. I.,** Complete primary structure of a proline-rich phosphoprotein (PRP-4), a potent inhibitor of calcium phosphate precipitation in human parotid saliva, in *Peptides: Structure and Biological Function,* Proc. 6th American Peptide Symp., Gross, E. and Meienhofer, J., Eds., Pierce Chemical Company, Rockford, Ill., 1979, 133.
150. **Wong, R. S. and Bennick, A.,** The primary structure of a salivary calcium-binding proline-rich phosphoprotein (protein C), a possible precursor of a related salivary protein A, *J. Biol. Chem.*, 255, 5943, 1980.
151. **Isemura, S., Saitoh, E., and Sanada, K.,** Isolation and amino-acid sequences of proline-rich peptides of human whole saliva, *J. Biochem.*, 86, 79, 1979.
152. **Isemura, S., Saitoh, E., and Sanada, K.,** The amino-acid sequence of a salivary proline-rich peptide, P-C, and its relationship to a salivary proline-rich phosphoprotein, protein C, *J. Biochem. (Tokyo),* 87, 1071, 1980.
153. **Oppenheim, F. G., Kousvelari, E., and Troxler, R. F.,** Immunological cross-reactivity and sequence homology between salivary proline-rich proteins in humans and macaque monkey *Macaca fascicularis* parotid saliva, *Arch. Oral Biol.*, 24, 595, 1979.
154. **Schlesinger, D. H. and Hay, D. I.,** Primary structure of the active tryptic fragments of human and monkey salivary anionic proline-rich proteins, *Int. J. Peptide Protein Res.*, 17, 34, 1981. k0

Gron, P., Saturation of human saliva with calcium phosphates, *Arch. Oral Biol.*, 18, 1385, 1973. 155.

156. **Backer-Dirks, O.,** Post-eruptive changes in enamel, *J. Dent. Res.*, 45, 503, 1966.
157. **Von Der Fehr, F. R., Loe, H., and Theilade, F.,** Experimental caries in man, *Caries Res.*, 4, 131, 1970.
158. **Gron, P. and Hay, D. I.,** Inhibition of calcium phosphate precipitation by human salivary secretions, *Arch. Oral Biol.*, 21, 201, 1976.
159. **Hay, D. I. and Gron, P.,** Inhibitors of calcium phosphate precipitation in human whole saliva, in *Microbial Aspects of Dental Caries,* Vol. 1, Stiles, H. M., Loesche, W. J., and O'Brien, T. C., Eds., Information Retrieval, Washington, D.C., 1976, 143.
160. **Hay, D. I.,** The interaction of human parotid salivary proteins with hydroxyapatite, *Arch. Oral Biol.*, 18, 1517, 1973.
161. **Moreno, E. C., Kresak, M., and Hay, D. I.,** Adsorption of two human parotid salivary macromolecules on hydroxy-, fluorhydroxy- and fluorapatites, *Arch. Oral Biol.*, 23, 525, 1978.
162. **Hay, D. I. and Moreno, E. C.,** Differential adsorption and chemical affinities of proteins for apatitic surfaces, *J. Dent. Res.*, 58b, 930, 1979.
163. **Hay, D. I. and Moreno, E. C.,** Macromolecular inhibitors of calcium phosphate precipitation in human saliva. Their roles in providing a protective environment for the teeth, in *Proc. Saliva and Dental Caries,* Kleinberg, I., Ellison, S. A., and Mandel, I. D., Eds., (Spec. Supp.), *Microbiol. Abstr.*, 45, 1979.
164. **Bennick, A. and Cannon, M.,** Quantitative study of the interaction of salivary acidic proline-rich proteins with hydroxyapatite, *Caries Res.*, 12, 159, 1978.
165. **Bennick, A., Cannon, M., and Madapallimattam, G.,** Factors affecting the adsorption of salivary acidic proline-rich proteins to hydroxyapatite, *Caries Res.*, 15, 9, 1981.
166. **Moreno, E. C., Varughese, K., and Hay, D. I.,** Effect of human salivary proteins on the precipitation kinetics of calcium phosphate, *Calcif. Tissue Int.*, 28, 7, 1979.
167. **Hay, D. I., Moreno, E. C., and Schlesinger, D. H.,** Phosphoprotein-inhibitors of calcium phosphate precipitation from salivary secretions, *Inorg. Perspect. Biol. Med.*, 2, 271, 1979.

168. **Bennick, A., Cannon, M., and Madapallimattam, G.,** The nature of the hydroxyapatite-binding site in salivary acidic proline-rich proteins, *Biochem. J.*, 183, 115, 1979.
169. **Bennick, A., McLaughlin, A. C., Grey, A. A., and Madapallimattam, G.,** The location and nature of calcium-binding sites in salivary acidic proline-rich phosphoproteins, *J. Biol. Chem.*, 256, 4741, 1981.
170. **Azen, E. A. and Oppenheim, F. G.,** Genetic polymorphism of proline-rich human salivary proteins, *Science*, 180, 1067, 1973.
171. **Karn, R. C., Friedman, R. D., and Merritt, A. D.,** Human salivary proline-rich (Pr) proteins: a post-translational derivation of the phenotypes, *Biochem. Genet.*, 17, 1061, 1979.
172. **Friedman, R. D., Azen, E. A., Yu, P. L., Green, P. A., Karn, R. C., and Merritt, A. D.,** Heritable salivary proteins and dental disease, *Hum. Hered.*, 30, 372, 1980.
173. **Hay, D. I.,** The adsorption of salivary proteins by hydroxyapatite and enamel, *Arch. Oral Biol.*, 12, 937, 1967.
174. **Pruitt, K. M., Caldwell, R. C., Jamieson, A. D., and Taylor, R. E.,** The interaction of the salivary proteins with the tooth surface, *J. Dent. Res.*, 48, 818, 1969.
175. **Gugler, E., Pallavicini, C. J., Swerdlow, H., and Di Sant' Agnese, P. A.,** The role of calcium in the submaxillary saliva of patients with cystic fibrosis, *J. Pediatr.*, 71, 585, 1967.
176. **Hay, D. I.,** The isolation from human parotid saliva of a tyrosine-rich acidic peptide which exhibits high affinity for hydroxyapatite surfaces, *Arch. Oral Biol.*, 18, 1531, 1973b.
177. **Schlesinger, D. H. and Hay, D. I.,** Human salivary statherin: a peptide inhibitor of calcium phosphate precipitation, in *Calcium Binding Proteins and Calcium Function,* Wasserman, R. H., Corradino, R. A., Carafoli, E., Kretsinger, R. H., MacLennan, D. H., and Siegal, F. L., Eds., Elsevier/North-Holland, 1977, 401.
178. **Gibbons, R. J. and Spinell, D. M.,** Salivary-induced aggregation of plaque bacteria, in *Dental Plaque,* McHugh, W. D., Ed., Churchill Livingstone, Edinburgh, 1970, 207.
179. **Gibbons, R. J. and vanHoute, J.,** Bacterial adherence in microbial ecology, *Annu. Rev. Microbiol.*, 29, 19, 1975.
180. **Beachey, E. H.,** Bacterial adherence, in *Receptors and Recognition,* Vol. 6, Series B, Chapman and Hall, London, 1980.
181. **Hay, D. I., Gibbons, R. J., and Spinell, D. M.,** Characteristics of some high molecular weight constituents with bacterial aggregating activity from whole saliva and dental plaque, *Caries Res.*, 5, 111, 1971.
182. **Kashket, S. and Donaldson, C. G.,** Saliva-induced aggregation of oral streptococci, *J. Bacteriol.*, 112, 1127, 1972.
183. **Williams, R. C. and Gibbons, R. J.,** Inhibition of streptococcal attachment to receptors on human buccal epithelial cells by antigenically similar salivary glycoproteins, *Infect. Immun.*, 11, 711, 1975.
184. **Gibbons, R. J. and Qureshi, J. V.,** Interactions of *Streptococcus mutans* and other oral bacteria with blood group reactive substances, in *Microbial Aspects of Dental Caries,* Stiles, H. M., Loesche, W. J., and O'Brien, T. C., Eds., (Spec. Suppl.)., *Microbiol. Abstr.*, 1, 163, 1976.
185. **Kondo, W., Sato, M., and Sato, N.,** Properties of the human salivary aggregating factor for *Leptotrichia buccalis* cells, *Arch. Oral Biol.*, 23, 453, 1978.
186. **Hogg, S. D. and Embery, G.,** The isolation and partial characterization of a sulphated glycoprotein from human whole saliva which aggregates strains of *Streptococcus sanguis* but not *Streptococcus mutans, Arch. Oral Biol.*, 24, 791, 1979.
187. **Sonju, T., Christensen, T. B., Kornstad, L., and Rolla, G.,** Electron microscopy, carbohydrate analysis and biological activities of the proteins adsorbed in 2 hr to tooth surfaces *in vivo, Caries Res.*, 8, 113, 1974.
188. **Rolla, G. and Embery, G.,** Sulfated glycoproteins in the acquired pellicle and plaque from *Macaca fascicularis* demonstrated with labelled sulfate, *Scand. J. Dent. Res.*, 85, 237, 1977.
189. **Ericson, T., Carlen, A., and Dagerskog, E.,** Salivary aggregating factors, in *Microbial Aspects of Dental Caries,* Stiles, H. M., Loesche, W. J., and O'Brien, T. C., Eds., (Spec. Supp.), *Microbiol. Abstr.*, 1, 151, 1976.
190. **Eggert, F. M.,** The nature of secretory agglutinins and aggregating factors. I-IV, *Int. Arch. Allergy Appl. Immun.*, 61, 192, 1980; 62, 34, 1980.
191. **Levine, M. J., Herzberg, M. C., Levine, M. S., Ellison, S. A., Stinson, M. W., Li, H. C., and VanDyke, T.,** Specificity of salivary-bacterial interactions: role of terminal sialic acid residues in the interactions of salivary glycoproteins with *Streptococcus sanguis* and *Streptococcus mutans, Infect. Immun.*, 19, 107, 1978.
192. **Mandel, I. D.,** Salivary factors in caries prediction, *Proc. Methods of Caries Prediction,* Bibby, B. G. and Shern, R. J., Eds., (Spec. Supp.), *Microbiol. Abstr.*, 147, 1978.
193. **Mandel, I. D. and Ellison, S. A.,** The proteins of human parotid and submandibular saliva, *Ann. N.Y. Acad. Sci.*, 106, 271, 1963.
194. **Mandel, I. D. and Ellison, S. A.,** Organic components of human parotid and submandibular saliva, *Ann. N.Y. Acad. Sci.*, 131, 802, 1965.

195. **Levine, M. J., Weill, J. C., and Ellison, S. A.,** The isolation and analysis of a glycoprotein from parotid saliva, *Biochim. Biophys. Acta,* 188, 165, 1969.
196. **Andjic, M. J., Bonte, M. and Havez, M. R.,** Purification et étude d'une glycoprotéin basique de la saliva parotidienne humaine, *C.R. Acad. Sci. Paris,* 270, 564, 1970.
197. **Degand, P., Boersma, A., Roussel, P., Richet, C., and Biserte, G.,** The peptide moiety of the region carrying the carbohydrate chains of the human parotid glycoprotein, *FEBS Lett.* 54, 189, 1975.
198. **Li, H. C. and Levine, M. J.,** Characterization of a glycopeptide from the proline-rich glycoprotein of human parotid saliva, *Arch. Oral Biol.,* 25, 353, 1980.
199. **Degand, P., Aubert, J. P., Boersma, A., Richet, C., Loucheux-Lefebvre, M. H., and Biserte, G.,** Parotid amylase activity: a possible role for proline-rich proteins, *FEBS Lett.,* 63, 137, 1976.
200. **Shibata, S., Nagata, K., Nakamura, R., Tsunemitsu, A., and Misaki, A.,** Interaction of parotid saliva basic glycoprotein with *Streptococcus sanguis* ATCC 10557, *J. Periodontol.,* 51, 499, 1980.
201. **Levine, M. and Keller, P. J.,** The isolation of some basic proline-rich proteins from human parotid saliva, *Arch. Oral Biol.,* 22, 27, 1977.
202. **Azen, E. A. and Peters, E. H.,** Isolation and partial characterization of human parotid basic proteins, *Biochem. Genet.,* 15, 925, 1977.
203. **Kauffman, D. L. and Keller, P. J.,** The basic proline-rich proteins in human saliva from a single subject, *Arch. Oral Biol.,* 24, 249, 1979.
204. **Juriaanse, A. E. and Booij, M.,** Isolation and characterization of the main neutral proteins from human submandibular saliva, *Arch. Oral Biol.,* 24, 825, 1979.
205. **Juriaanse, A. E. and Booij, M.,** Isolation and characterization of three acidic proteins from human submandibular saliva, *Arch. Oral Biol.,* 24, 621, 1979.
206. **Bettelheim, F. A.,** On the aggregation of a calcium-precipitable glycoprotein from human submandibular saliva, *Biochim. Biophys. Acta,* 236, 702, 1970.
207. **Boat, T. F., Wiesman, V. N., and Pallavicini, J. C.,** Purification and properties of the calcium-precipitable protein in submaxillary saliva of normal and cystic fibrosic subjects, *Pediatr. Res.,* 8, 531, 1974.
208. **Belcourt, A.,** Etude d'une glycoproteine salivaire humaine precipitable par le calcium, *Eur. J. Biochem.,* 53, 185, 1975.
209. **Shomers, J. P., Tabak, L. A., Levine, M. J., Mandel, I. D., and Hay, D. I.,** Properties of cysteine-containing phosphoproteins from human submandibular-sublingual saliva, *J. Dent. Res.,* 61, 397, 1982.
210. **Henkin, R. I., Lippoldt, R. E., Bilstad, J., and Edelhoch, H.,** A zinc protein isolated from human parotid saliva, *Proc. Natl. Acad. Sci., U.S.A.,* 72, 488, 1975.
211. **Shatzman, A. R. and Henkin, R. J.,** Metal-binding characteristics of the parotid salivary protein gustin, *Biochem. Biophys. Acta,* 623, 107, 1980.
212. **Rolla, G. and Jonsen, J.,** A glycoprotein component from human sublingual-submaxillary saliva, *Caries Res.,* 2, 306, 1968.
213. **Balekjian, A. Y. and Longton, R. W.,** Histones isolated from human parotid fluid, *Biochem. Biophys. Res. Commun.,* 50, 676, 1973.
214. **Hay, D. I.,** Fractionation of human parotid salivary proteins and the isolation of an histidine-rich acidic peptide which shows high affinity for hydroxyapatite surfaces, *Arch. Oral Biol.,* 20, 553, 1975.
215. **Baum, B. J., Bird, J. L., Millar, D. B., and Longton, R. W.,** Studies on histidine-rich polypeptides from human parotid saliva, *Arch. Biochem. Biophys.,* 177, 427, 1976.
216. **Baig, M. M., Winzler, R. J., and Rennert, W. M.,** Isolation of mucin from human submaxillary secretions, *J. Immunol.,* 111, 1826, 1973.
217. **Mayo, J. W. and Carlson, D. M.,** Isolation and properties of four α-amylase isozymes from human submandibular saliva, *Arch. Biochem. Biophys.,* 163, 498, 1974.
218. **Oemrawsingh, I. and Roukema, P. A.,** Composition and biological properties of mucins isolated from human submandibular glands, *Arch. Oral Biol.,* 19, 753, 1974.
219. **Roukema, P. A. and Nieuw Amerongen, A. V.,** Sulphated glycoproteins in human saliva, *Proc. Saliva and Dental Caries,* Kleinberg, I., Ellison, S.A., and Mandel, I.D., Eds., (Spec. Supp.), *Microbiol. Abstr.,* 67, 1979.

Temporomandibular Joint

THE TEMPOROMANDIBULAR JOINT

Ronald C. Auvenshine

INTRODUCTION

The temporomandibular joint (TMJ) can be anatomically classified as a ginglymo-arthrodial-type of diarthrotic joint. It functionally is a ball-and-socket joint in which the socket has been modified to a fossa so as to provide for a wide range of motion. The TMJ can be divided into four main areas of research attention. These areas specifically deal with the development of the joint and the three different types of tissue within the joint. These four areas are as follows:

1. Embryogenesis of the TMJ
2. Articular surfaces of the condylar head and glenoid fossa
3. The interarticular meniscus
4. The synovial membrane

The small size of this joint as well as its inaccessibility to sample taking (biopsy) and dissection have made study of this structure difficult. Relatively little information exists in the literature concerning the TMJ. Much of the literature prior to 1965 dealt mainly with the embryogenesis and developmental aspects of the TMJ. Since 1965, with the introduction of electron microscopy and autoradiography, more studies have been performed to investigate more detailed features of the joint. Human TMJ research is relatively void except for some cadavoric work explaining anatomical relationship rather than biochemical techniques.

This chapter will familiarize the reader with some current trends in the application of biochemical techniques in TMJ research. The four basic areas: (1) embryogenesis of TMJ, (2) articular surfaces, (3) interarticular meniscus, and (4) synovial membrane will be dealt with individually.

EMBRYOGENESIS OF TMJ

The mandibular joint develops from both intramembranous and endochondrial ossification. Intramembranous ossification takes place within the body and ramus of the mandible. Specific details of the events surrounding the formation and development of this condyle, its cartilage, and surrounding tissues have been reviewed by Macalister,[1] Symons,[2] Baume,[3] and Zorb and Carlsson.[4]

The body and ramus of the mandible develop by intramembranous bone formation lateral to Meckel's cartilage. The condylar head is derived from a secondary blastema. The blastema consists of a grouping of "skelatoblasts" which are biochemically programmed to differentiate into chondroblasts and osteoblasts.[5] The blastema develops by the process of endochondrial bone formation. The union between the condylar blastema and the developing ramus occurs at the condylar neck. The condylar neck remains as the primary growth center of the mandible until the growth process is completed. Therefore, the condylar neck acts as the epiphyseal plate for interstitial growth in length of the mandible.

The exact time of cellular differentiation within the developing TMJ varies from species to species. The exact mechanism of initiation is of great interest,[7,8] although the role of Meckel's cartilage has yet to be fully explained in mandibular development.[9]

One of the first reports of succinic dehydrogenase activity of the squamosomandibular articulation was reported by Gartner et al.[10] Succinate dehydrogenase, a mitrochondrial enzyme, is one of the principle enzymes concerned with oxidative phosphorylation and is the chief indicator of cellular metabolic activity. Prior to the report of Gartner, succinic dehydrogenase activity had not been studied in the mandible; however, the developing system provided an interesting media of study.

The application of electron microscopic histochemical methods have been applied to the TMJ. Meikle[11] reported on the presence and distribution of the enzymes acid phosphatase, aryl sulfatase, and alkaline phosphatase in the condylar cartilage of the rat mandible.

Other histochemical techniques are being applied to the TMJ. Autoradiography with ^{3}H-proline has been used to study collagen formation and growth in the mandibular joint of the guinea pig.[12] Radioactive sulfate ($^{35}SO_4$) incorporation studies have been used to investigate matrix formation of the mandibular condyle of the rat.[13] Størksen et al.[13] applied autoradiography as a means of measuring cellular proliferation and matrix formation. Katz,[14] on the other hand, utilized autotransplantation techniques in which the developing condylar cartilage was transplanted to the abdominal wall and the preparations were subsequently analyzed for DNA.

Plutonium transfer from bone surfaces of the mandibular condyle using autoradiography has been reported by Priest and Giannola.[15] The results showed that plutonium was unlikely to be accumulated by macrophages as a result of their participation in bone resorption. Morris and Appleton[16] utilized the potassium pyroantimonate technique for selective subcellular localization of calcium in the postnatal rat mandibular condyle. Electron dense calcium pyroantimonate precipitates were observed utilizing electron microscopy. The purpose of this study was to selectively localize calcium in the chondrocytes of the developing mandibular condylar growth cartilage.

ARTICULAR SURFACES

The articular cartilage of the TMJ is covered by three layers of dense fibrous connective tissue. The three layers are arranged in an overlaping web-like pattern.[17] Beneath the layer of dense connective tissue lies the articular cartilage composed of chondrocytes embedded in a ground substance. The chondrocytes manufacture the ground substances which is rich in chondroitin sulfate. The chondrocytes of articular cartilage are rich in lipid. The purpose of the lipid is related to the economy of the cell and enables the chondrocytes to maintain a stable metabolism for long periods of time on a limited supply of oxygen. The integrity of the joint depends on the metabolism of the chondrocyte. With age, changes in the cellular metabolism of the chondrocyte results in changes in the ground substance.[18] As the ground substance breaks down, underlying layers of osteocytes are exposed, resulting in degenerative bone disease.

A study was reported which used organ cultures of human articular cartilage to investigate the synthesis and presence of glycosaminoglycan synthesis.[19] However, very little information exists as to the synthesis of the glycosaminoglycans in the TMJ. For reasons, probably based on size and availability of human TMJ tissue, this tissue has not been used for organ culture. More research should be proposed to investigate the effects of force and remodeling on the cellular chemistry of the articular cartilage of this joint. Clinically significant is the question of cellular plasticity of the growing as well as the adult TMJ. Understanding the synthesis of glycosaminoglycan and the various proportions of these compounds with relation to sulfated vs. nonsulfated forms may lead to better clinical management of growth and degenerative changes of this joint.

THE INTERARTICULAR MENISCUS

The interarticular disk or meniscus of the TMJ is a sheet of connective tissue. The meniscus of the TMJ has a definite anatomy which allows it to conform to the condylar head and the roof of the fossa. It is thickest posteriorly and attached to the anterior tympanic plate by means of the bilaminar bone. The disk is thinnest centrally where it serves as a cushion for joint movement. Anteriorly it agains thickens as it blends with the tendinous insertion of the superior head of the lateral pterygoid muscle. The majority of this dense fibrous connective tissue of the adult TMJ meniscus is avascular and aneural. This is apparently a response to a functional load. The soft peripheral part of this disk consists of loose fibrous connective tissue rich in blood vessels and nerves.

Few studies other than those dealing with the anatomy of this disk have been reported. Kopp[20] reported on the topographical distribution of sulfated glycosaminoglycans with respect to differences in age, sex, and osteoarthritis of the autopsy material. One conclusion reached by Kopp in his study was that the main localization of sulfated GAGs seemed to be in the central load-bearing part of the disk where chondrial cells were responsible for its synthesis. This author also stated that macroscopic thinning of this disk in the central part might be related to a decrease in highly sulfated GAGs.

More research should be conducted to investigate the bilaminar zone. Little information has been published regarding this important and delicate tissue. The superior strata of this bilaminar zone consists of elastic fibers.[21] The presence of elastin provides this tissue with a modulus of elasticity. The presence or the mechanism of production of this elastin has yet to be investigated thoroughly.

THE SYNOVIAL FLUID

The synovial fluid is produced in the synovial tissue by the synovial cells. It is improper to classify this fluid as an ultrafiltrate of blood. The synovial fluid consists mainly of a plasma dialysate enriched with a glycosaminoglycan, hyaluronic acid, synthesized by the synovial cell. There are few cellular elements within the synovial fluid.

The synovial fluid acts as a lubricant for the joint surfaces. It also actively contributes to the nutrition of the avascular superficial parts of the condyle, fossa, and disks. Several studies have been reported concerning the synovial fluid of the TMJ. Griffin[22] reported on the blood supply to the synovial membrane. He stated that the joint and muscles which move it have a similar blood supply and that the flow of synovial fluid within this joint was a result of motion. He described the need for circulation and exchange of nutrients within the joint space and the requirement for a turnover of the fluid.

Schmid's paper[23] represents one of the best publications on the composition of synovial fluid of the TMJ. Ultrastructural techniques have also been applied to the TMJ. Kakudo[24] investigated ultrastructural observation of horseradish peroxidase in synovial living cells of the rat TMJ. Even though his results were inconclusive, his work did institute a broad step in the right direction utilizing a sophisticated technique to study cellular plasticity and biochemical mechanisms.

REFERENCES

1. **Macalister, A.,** The development of the TMJ, *Aust. Dent. J.*, 59, 21, 1955.
2. **Symons, N.,** The development of the human TMJ, *J. Anat. London*, 86, 326, 1952.
3. **Baume, L.,** Ontogenesis of the Human TMJ. I. Development of the condyles, *J. Dent. Res.*, 41, 1327, 1962.
4. **Zorb, G. and Carlsson, G. E.,** *Temporomandibular Joint Function and Dysfunction*, 1st ed., C.V. Mosby, St. Louis, 1979, 43.
5. **Petrovic, A.,** Mechanisms and regulation of mandibular condylar growth, *Acta Morphol. Neerl. Scand.*, 10, 25, 1972.

6. **Bhaskar, S., Weinmann, J., and Schour, I.,** Role of Meckel's cartilage in the development and growth of the rat mandible, *J. Dent. Res.*, 32, 398, 1953.
7. **Frommer, J. and Margolies, M.,** Contribution of Meckel's cartilage to ossification of the mandible in mice, *J. Dent. Res.*, 50, 1260, 1971.
8. **Hall, B.,** Cellular differentiation in skeletal tissues, *Biol. Rev. Cambridge Philos. Soc.*, 45, 455, 1970.
9. **Meikle, M.,** The mineralization of condylar cartilage in the rat mandible: an electron microscopic enzyme histochemical study, *Arch. Oral Biol.*, 21, 33, 1976.
10. **Gartner, L., Hiatt, J., Khan, M., and Provenza, D. V.,** Development of the squamosomandibular articulation in the Mongolian gerbil (*Meriones unguiculartus*). II. Succinate dehydrogenase activity, *Histochem. J.*, 11, 553, 1979.
11. **Meikle, M.,** In vivo transplantation of the mandibular joint of the rat; an autoradiographic investigation into cellular changes at the condyle, *Arch. Oral Biol.*, 18, 1011, 1973.
12. **Oberg, T., Fojers, C., Friberg, U., and Lohmander, S.,** Collagen formation and growth in the mandibular joint of the guinea pig as revealed by autoradiography with ^{3}H-proline, *Acta Odontol. Scand.*, 27, 425, 1969.
13. **Størksen, K., Aukland, S., and Kvinnsland, S.,** Matrix formation in craniofacial cartilages of the rat ^{35}S — sulfate incorporation studies, *Acta Odontol. Scand.*, 37, 29, 1979.
14. **Katz, M. and Kvinnsland, S.,** Matrix formation in the mandibular condyle of the rat (^{35}S) — sulfate incorporation studies, *Acta Odontol. Scand.*, 37, 137, 1979.
15. **Priest, N. and Giannola, S.,** Uptake and loss of plutonium from osteoclasts and macrophages in the mandibular condyle of the rat, *Calcif. Tissue Int.*, 30, 15, 1980.
16. **Morris, D. and Appleton, J.,** Ultrastructural localization of calcium in the mandibular condylar growth cartilage of the rat, *Calcif. Tissue Int.*, 30, 27, 1980.
17. **Hansson, T. and Nordstrom, B.,** Thickness of soft tissue layers and the articular disk in the temporomandibular joint with deviations in form, *Acta Odontol. Scand.*, 35, 281, 1977.
18. **Sharp, C., Gee, E., and Griffin, C.,** The osteogenic potential of the human condyle, *Aust. Dent. J.*, 10, 287, 1965.
19. **McKenzie, L., Horsburgh, B., Ghosh, P., and Taylor, T.,** Organ culture of human articular cartilage: studies on sulphated glycosaminoglycan synthesis, *In Vitro*, 13, 423, 1977.
20. **Kopp, S.,** Topographical distribution of sulphated glycosaminoglycans in human temporomandibular joint disks, *J. Oral Pathol.*, 5, 265, 1976.
21. **Rees, L.,** The structure and function of the temporomandibular joint, *Br. Dent. J.*, 96, 125, 1954.
22. **Griffin, C.,** The mechanism of the blood supply to synovial membrane, *Aust. Dent. J.*, 4, 379, 1959.
23. **Schmid, F. and Ogata, R.,** The composition and examination of synovial fluid, *J. Prosthet. Dent.*, 18, 449, 1967.
24. **Kakudo, K., Takasu, J., and Yamamoto, S.,** Ultrastructural observations of horseradish peroxidase in synovial lining cells of the rat's temporomandibular joint, *J. Dent. Res.*, 56(11), 1376, 1977.

Index

INDEX

A

C

D

E

F

G

H

M

N

O

P

T

U

V

W

X

Y

Z